MACHINES AND MECHANISMS

APPLIED KINEMATIC ANALYSIS

Fourth Edition

David H. Myszka

University of Dayton

Prentice Hall

Boston   Columbus   Indianapolis   New York   San Francisco   Upper Saddle River
Amsterdam   Cape Town   Dubai   London   Madrid   Milan   Munich   Paris   Montreal   Toronto
Delhi   Mexico City   Sao Paulo   Sydney   Hong Kong   Seoul   Singapore   Taipei   Tokyo

Vice President & Editorial Director:
Vernon R. Anthony
Acquisitions Editor: David Ploskonka
Editorial Assistant: Nancy Kesterson
Director of Marketing: David Gesell
Marketing Manager: Kara Clark
Senior Marketing Coordinator: Alicia
Wozniak
Marketing Assistant: Les Roberts
Senior Managing Editor: JoEllen Gohr
Associate Managing Editor: Alexandrina
Benedicto Wolf
Production Editor: Maren L. Miller

Project Manager: Susan Hannahs
Art Director: Jayne Conte
Cover Designer: Suzanne Behnke
Cover Image: Fotolia
Full-Service Project Management:
Hema Latha, Integra Software
Services, Pvt Ltd
Composition: Integra Software
Services, Pvt Ltd
Text Printer/Bindery: Edwards Brothers
Cover Printer: Lehigh-Phoenix Color
Text Font: 10/12, Minion

Credits and acknowledgments borrowed from other sources and reproduced, with permission, in this textbook appear on the appropriate page within the text. Unless otherwise stated, all artwork has been provided by the author.

Many of the designations by manufacturers and seller to distinguish their products are claimed as trademarks. Where those designations appear in this book, and the publisher was aware of a trademark claim, the designations have been printed in initial caps or all caps.

Library of Congress Cataloging-in-Publication Data
Myszka, David H.
Machines and mechanisms : applied kinematic analysis / David H. Myszka.—4th ed.
p. cm.
Includes bibliographical references and index.
ISBN-13: 978-0-13-215780-3
ISBN-10: 0-13-215780-2
1. Machinery, Kinematics of.   2. Mechanical movements.   I. Title.
TJ175.M97 2012
621.8'11—dc22

2010032839

10 9 8 7 6 5 4 3 2

Prentice Hall
is an imprint of

www.pearsonhighered.com

ISBN 10:   0-13-215780-2
ISBN 13: 978-0-13-215780-3

PREFACE

The objective of this book is to provide the techniques necessary to study the motion of machines. A focus is placed on the application of kinematic theories to real-world machinery. It is intended to bridge the gap between a theoretical study of kinematics and the application to practical mechanisms. Students completing a course of study using this book should be able to determine the motion characteristics of a machine. The topics presented in this book are critical in machine design process as such analyses should be performed on design concepts to optimize the motion of a machine arrangement.

This fourth edition incorporates much of the feedback received from instructors and students who used the first three editions. Some enhancements include a section introducing special-purpose mechanisms; expanding the descriptions of kinematic properties to more precisely define the property; clearly identifying vector quantities through standard boldface notation; including timing charts; presenting analytical synthesis methods; clarifying the tables describing cam follower motion; and adding a standard table used for selection of chain pitch. The end-of-chapter problems have been reviewed. In addition, many new problems have been included.

It is expected that students using this book will have a good background in technical drawing, college algebra, and trigonometry. Concepts from elementary calculus are mentioned, but a background in calculus is not required. Also, knowledge of vectors, mechanics, and computer application software, such as spreadsheets, will be useful. However, these concepts are also introduced in the book.

The approach of applying theoretical developments to practical problems is consistent with the philosophy of engineering technology programs. This book is primarily oriented toward mechanical- and manufacturing-related engineering technology programs. It can be used in either associate or baccalaureate degree programs.

Following are some distinctive features of this book:

1. Pictures and sketches of machinery that contain mechanisms are incorporated throughout the text.

2. The focus is on the application of kinematic theories to common and practical mechanisms.

3. Both graphical techniques and analytical methods are used in the analysis of mechanisms.

4. An examination copy of Working Model®, a commercially available dynamic software package (see Section 2.3 on page 32 for ordering information), is extensively used in this book. Tutorials and problems that utilize this software are integrated into the book.

5. Suggestions for implementing the graphical techniques on computer-aided design (CAD) systems are included and illustrated throughout the book.

6. Every chapter concludes with at least one case study. Each case illustrates a mechanism that is used on industrial equipment and challenges the student to discuss the rationale behind the design and suggest improvements.

7. Both static and dynamic mechanism force analysis methods are introduced.

8. Every major concept is followed by an example problem to illustrate the application of the concept.

9. Every Example Problem begins with an introduction of a real machine that relies on the mechanism being analyzed.

10. Numerous end-of-chapter problems are consistent with the application approach of the text. Every concept introduced in the chapter has at least one associated problem. Most of these problems include the machine that relies on the mechanism being analyzed.

11. Where applicable, end-of-chapter problems are provided that utilize the analytical methods and are best suited for programmable devices (calculators, spreadsheets, math software, etc.).

Initially, I developed this textbook after teaching mechanisms for several semesters and noticing that students did not always see the practical applications of the material. To this end, I have grown quite fond of the case study problems and begin each class with one. The students refer to this as the "mechanism of the day." I find this to be an excellent opportunity to focus attention on operating machinery. Additionally, it promotes dialogue and creates a learning community in the classroom.

Finally, the purpose of any textbook is to guide the students through a learning experience in an effective manner. I sincerely hope that this book will fulfill this intention. I welcome all suggestions and comments and can be reached at dmyszka@udayton.edu.

ACKNOWLEDGMENTS

I thank the reviewers of this text for their comments and suggestions: Dave Brock, Kalamazoo Valley Community College; Laura Calswell, University of Cincinnati; Charles Drake, Ferris State University; Lubambala Kabengela, University of North Carolina at Charlotte; Sung Kim, Piedmont Technical College; Michael J. Rider, Ohio Northern University; and Gerald Weisman, University of Vermont.

Dave Myszka

CONTENTS

INTRODUCTION TO MECHANISMS AND KINEMATICS

OBJECTIVES

Upon completion of this chapter, the student will be able to:

1. Explain the need for kinematic analysis of mechanisms.

2. Define the basic components that comprise a mechanism.

3. Draw a kinematic diagram from a view of a complex machine.

4. Compute the number of degrees of freedom of a mechanism.

5. Identify a four-bar mechanism and classify it according to its possible motion.

6. Identify a slider-crank mechanism.

1.1 INTRODUCTION

Imagine being on a design and development team. The team is responsible for the design of an automotive windshield wiper system. The proposed vehicle is a sports model with an aerodynamic look and a sloped windshield. Of course, the purpose of this wiper system is to clean water and debris from the windshield, giving clear vision to the driver. Typically, this is accomplished by sweeping a pair of wipers across the glass.

One of the first design tasks is determining appropriate movements of the wipers. The movements must be sufficient to ensure that critical portions of the windshield are cleared. Exhaustive statistical studies reveal the view ranges

of different drivers. This information sets guidelines for the required movement of the wipers. Fundamental decisions must be made on whether a tandem or opposed wipe pattern better fits the vehicle. Other decisions include the amount of driver- and passenger-side wipe angles and the location of pivots. Figure 1.1 illustrates a design concept, incorporating an opposed wiper movement pattern.

Once the desired movement has been established, an assembly of components must be configured to move the wipers along that pattern. Subsequent tasks include analyzing other motion issues such as timing of the wipers and whipping tendencies. For this wiper system, like most machines, understanding and analyzing the motion is necessary for proper operation. These types of movement and motion analyses are the focus of this textbook.

Another major task in designing machinery is determining the effect of the forces acting in the machine. These forces dictate the type of power source that is required to operate the machine. The forces also dictate the required strength of the components. For instance, the wiper system must withstand the friction created when the windshield is coated with sap after the car has been parked under a tree. This type of force analysis is a major topic in the latter portion of this text.

1.2 MACHINES AND MECHANISMS

Machines are devices used to alter, transmit, and direct forces to accomplish a specific objective. A chain saw is a familiar machine that directs forces to the chain with the objective of cutting wood. A *mechanism* is the mechanical portion of a

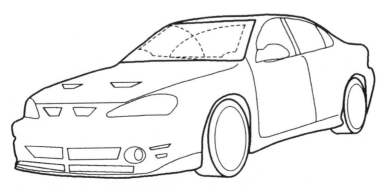

FIGURE 1.1 Proposed windshield wiper movements.

FIGURE 1.2 Adjustable height platform (Courtesy Advance Lifts).

machine that has the function of transferring motion and forces from a power source to an output. It is the heart of a machine. For the chain saw, the mechanism takes power from a small engine and delivers it to the cutting edge of the chain.

Figure 1.2 illustrates an adjustable height platform that is driven by hydraulic cylinders. Although the entire device could be called a machine, the parts that take the power from the cylinders and drive the raising and lowering of the platform comprise the mechanism.

A mechanism can be considered rigid parts that are arranged and connected so that they produce the desired motion of the machine. The purpose of the mechanism in Figure 1.2 is to lift the platform and any objects that are placed upon it. *Synthesis* is the process of developing a mechanism to satisfy a set of performance requirements for the machine. *Analysis* ensures that the mechanism will exhibit motion that will accomplish the set of requirements.

1.3 KINEMATICS

Kinematics deals with the way things move. It is the study of the geometry of motion. Kinematic analysis involves determination of position, displacement, rotation, speed, velocity, and acceleration of a mechanism.

To illustrate the importance of such analysis, refer to the lift platform in Figure 1.2. Kinematic analysis provides insight into significant design questions, such as:

- What is the significance of the length of the legs that support the platform?
- Is it necessary for the support legs to cross and be connected at their midspan, or is it better to arrange the so that they cross closer to the platform?
- How far must the cylinder extend to raise the platform 8 in.?

As a second step, dynamic force analysis of the platform could provide insight into another set of important design questions:

- What capacity (maximum force) is required of the hydraulic cylinder?

- Is the platform free of any tendency to tip over?
- What cross-sectional size and material are required of the support legs so they don't fail?

A majority of mechanisms exhibit motion such that the parts move in parallel planes. For the device in Figure 1.2, two identical mechanisms are used on opposite sides of the platform for stability. However, the motion of these mechanisms is strictly in the vertical plane. Therefore, these mechanisms are called *planar mechanisms* because their motion is limited to two-dimensional space. Most commercially produced mechanisms are planar and are the focus of this book.

1.4 MECHANISM TERMINOLOGY

As stated, mechanisms consist of connected parts with the objective of transferring motion and force from a power source to an output. A *linkage* is a mechanism where rigid parts are connected together to form a chain. One part is designated the *frame* because it serves as the frame of reference for the motion of all other parts. The frame is typically a part that exhibits no motion. A popular elliptical trainer exercise machine is shown in Figure 1.3. In this machine, two planar linkages are configured to operate out-of-phase to simulate walking motion, including the movement of arms. Since the base sits on the ground and remains stationary during operation, the base is considered the frame.

Links are the individual parts of the mechanism. They are considered rigid bodies and are connected with other links to transmit motion and forces. Theoretically, a true rigid body does not change shape during motion. Although a true rigid body does not exist, mechanism links are designed to minimally deform and are considered rigid. The footrests and arm handles on the exercise machine comprise different links and, along with connecting links, are interconnected to produce constrained motion.

Elastic parts, such as springs, are not rigid and, therefore, are not considered links. They have no effect on the kinematics of a mechanism and are usually ignored during

FIGURE 1.3 Elliptical trainer exercise machine (photo from www.precor.com).

kinematic analysis. They do supply forces and must be included during the dynamic force portion of analysis.

A *joint* is a movable connection between links and allows relative motion between the links. The two *primary joints,* also called *full joints,* are the revolute and sliding joints. The *revolute joint* is also called a *pin* or *hinge joint.* It allows pure rotation between the two links that it connects. The *sliding joint* is also called a *piston* or *prismatic joint.* It allows linear sliding between the links that it connects. Figure 1.4 illustrates these two primary joints.

A cam joint is shown in Figure 1.5a. It allows for both rotation and sliding between the two links that it connects. Because of the complex motion permitted, the cam connection is called a *higher-order joint,* also called *half joint.* A gear connection also allows rotation and sliding between two gears as their teeth mesh. This arrangement is shown in Figure 1.5b. The gear connection is also a higher-order joint.

A *simple link* is a rigid body that contains only two joints, which connect it to other links. Figure 1.6a illustrates a simple link. A *crank* is a simple link that is able to complete

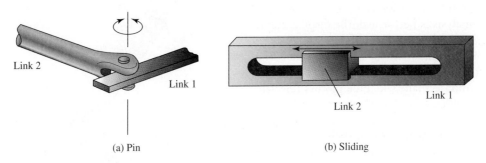

(a) Pin   (b) Sliding

FIGURE 1.4 Primary joints: (a) Pin and (b) Sliding.

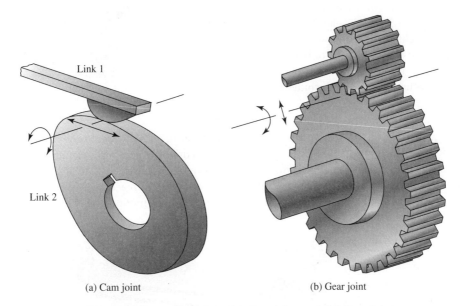

(a) Cam joint   (b) Gear joint

FIGURE 1.5 Higher-order joints: (a) Cam joint and (b) Gear joint.

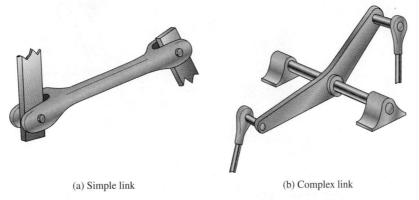

(a) Simple link   (b) Complex link

FIGURE 1.6 Links: (a) Simple link and (b) Complex link.

a full rotation about a fixed center. A *rocker* is a simple link that oscillates through an angle, reversing its direction at certain intervals.

A *complex link* is a rigid body that contains more than two joints. Figure 1.6b illustrates a complex link. A *rocker arm* is a complex link, containing three joints, that is pivoted near its center. A *bellcrank* is similar to a rocker arm, but is bent in the center. The complex link shown in Figure 1.6b is a bellcrank.

A *point of interest* is a point on a link where the motion is of special interest. The end of the windshield wiper, previously discussed, would be considered a point of interest. Once kinematic analysis is performed, the displacement, velocity, and accelerations of that point are determined.

The last general component of a mechanism is the actuator. An *actuator* is the component that drives the mechanism. Common actuators include motors (electric and hydraulic), engines, cylinders (hydraulic and pneumatic), ball-screw motors, and solenoids. Manually operated machines utilize human motion, such as turning a crank, as the actuator. Actuators will be discussed further in Section 1.7.

Linkages can be either *open* or *closed chains*. Each link in a closed-loop kinematic chain is connected to two or more other links. The lift in Figure 1.2 and the elliptical trainer of Figure 1.3 are closed-loop chains. An open-loop chain will have at least one link that is connected to only one other link. Common open-loop linkages are robotic arms as shown in Figure 1.7 and other "reaching" machines such as backhoes and cranes.

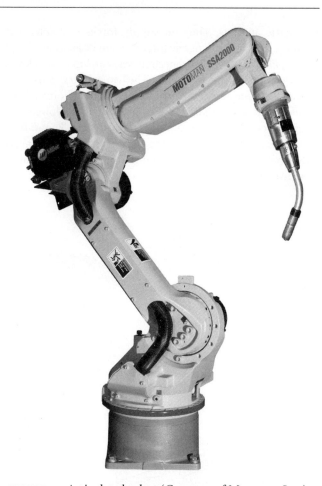

FIGURE 1.7 Articulated robot (Courtesy of Motoman Inc.).

1.5 KINEMATIC DIAGRAMS

In analyzing the motion of a machine, it is often difficult to visualize the movement of the components in a full assembly drawing. Figure 1.8 shows a machine that is used to handle parts on an assembly line. A motor produces rotational power, which drives a mechanism that moves the arms back and forth in a synchronous fashion. As can be seen in Figure 1.8, a pictorial of the entire machine becomes complex, and it is difficult to focus on the motion of the mechanism under consideration.

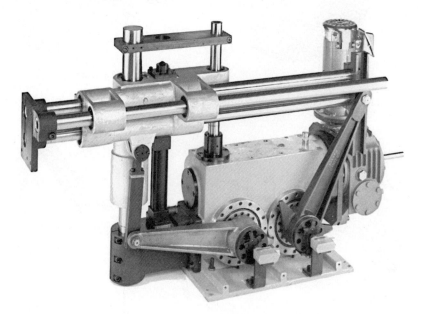

FIGURE 1.8 Two-armed synchro loader (Courtesy PickOmatic Systems, Ferguson Machine Co.).

It is easier to represent the parts in skeleton form so that only the dimensions that influence the motion of the mechanism are shown. These "stripped-down" sketches of mechanisms are often referred to as *kinematic diagrams*. The purpose of these diagrams is similar to electrical circuit schematic or piping diagrams in that they represent variables that affect the primary function of the mechanism.

Table 1.1 shows typical conventions used in creating kinematic diagrams.

A kinematic diagram should be drawn to a scale proportional to the actual mechanism. For convenient reference, the links are numbered, starting with the frame as link number 1. To avoid confusion, the joints should be lettered.

TABLE 1.1	Symbols Used in Kinematic Diagrams	
Component	**Typical Form**	**Kinematic Representation**

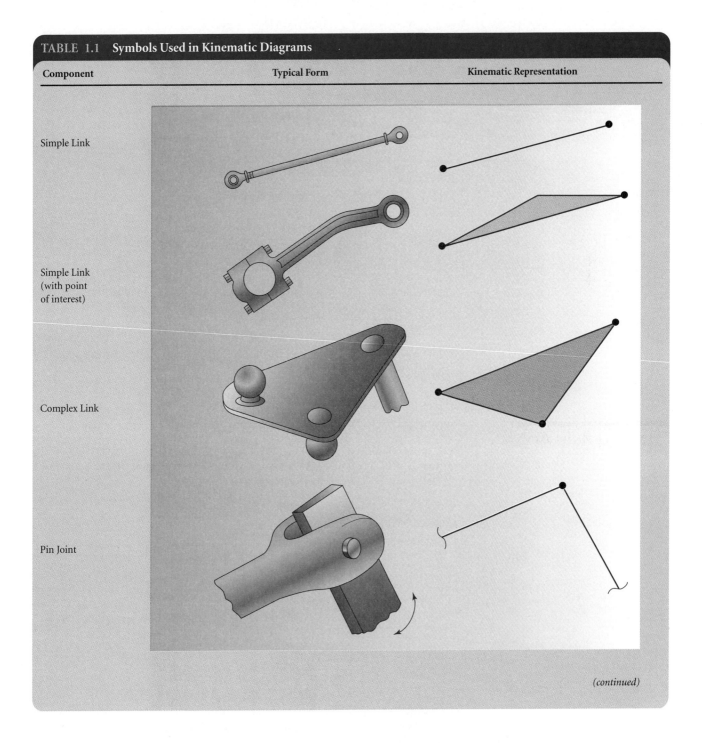

Simple Link		
Simple Link (with point of interest)		
Complex Link		
Pin Joint		

(continued)

TABLE 1.1 (Continued)

Component	Typical Form	Kinematic Representation
Slider Joint		
Cam Joint		
Gear Joint		

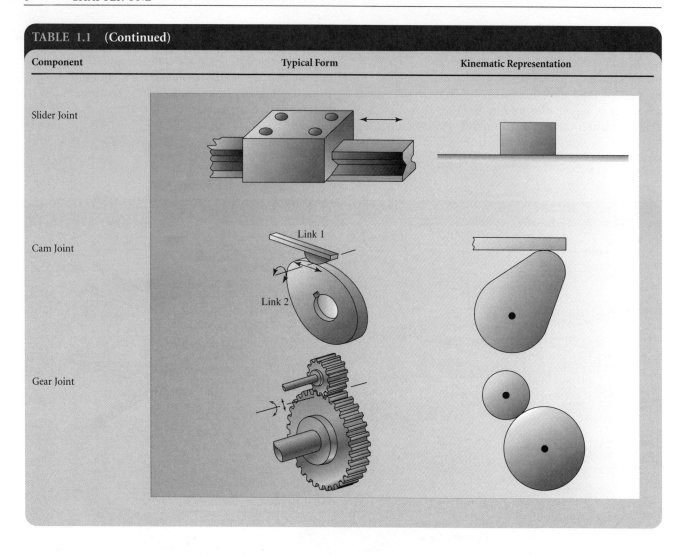

EXAMPLE PROBLEM 1.1

Figure 1.9 shows a shear that is used to cut and trim electronic circuit board laminates. Draw a kinematic diagram.

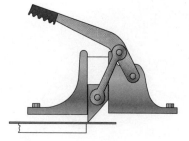

FIGURE 1.9 Shear press for Example Problem 1.1.

SOLUTION:　　1.　*Identify the Frame*

The first step in constructing a kinematic diagram is to decide the part that will be designated as the frame. The motion of all other links will be determined relative to the frame. In some cases, its selection is obvious as the frame is firmly attached to the ground.

In this problem, the large base that is bolted to the table is designated as the frame. The motion of all other links is determined relative to the base. The base is numbered as link 1.

2. ***Identify All Other Links***

   Careful observation reveals three other moving parts:

   > Link 2: Handle
   > Link 3: Cutting blade
   > Link 4: Bar that connects the cutter with the handle

3. ***Identify the Joints***

   Pin joints are used to connect link 1 to 2, link 2 to 3, and link 3 to 4. These joints are lettered *A* through *C*. In addition, the cutter slides up and down, along the base. This sliding joint connects link 4 to 1, and is lettered *D*.

4. ***Identify Any Points of Interest***

   Finally, the motion of the end of the handle is desired. This is designated as *point of interest X.*

5. ***Draw the Kinematic Diagram***

   The kinematic diagram is given in Figure 1.10.

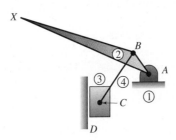

FIGURE 1.10 Kinematic diagram for Example Problem 1.1.

EXAMPLE PROBLEM 1.2

Figure 1.11 shows a pair of vise grips. Draw a kinematic diagram.

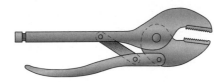

FIGURE 1.11 Vise grips for Example Problem 1.2.

SOLUTION:

1. ***Identify the Frame***

   The first step is to decide the part that will be designated as the frame. In this problem, no parts are attached to the ground. Therefore, the selection of the frame is rather arbitrary.

   The top handle is designated as the frame. The motion of all other links is determined relative to the top handle. The top handle is numbered as link 1.

2. ***Identify All Other Links***

   Careful observation reveals three other moving parts:

   > Link 2: Bottom handle
   > Link 3: Bottom jaw
   > Link 4: Bar that connects the top and bottom handle

3. ***Identify the Joints***

   Four pin joints are used to connect these different links (link 1 to 2, 2 to 3, 3 to 4, and 4 to 1). These joints are lettered *A* through *D*.

4. ***Identify Any Points of Interest***

   The motion of the end of the bottom jaw is desired. This is designated as point of interest *X*. Finally, the motion of the end of the lower handle is also desired. This is designated as point of interest *Y*.

5. ***Draw the Kinematic Diagram***

The kinematic diagram is given in Figure 1.12.

FIGURE 1.12 Kinematic diagram for Example Problem 1.2.

1.6 KINEMATIC INVERSION

Absolute motion is measured with respect to a stationary frame. *Relative motion* is measured for one point or link with respect to another link. As seen in the previous examples, the first step in drawing a kinematic diagram is selecting a member to serve as the frame. In some cases, the selection of a frame is arbitrary, as in the vise grips from Example Problem 1.2. As different links are chosen as a frame, the relative motion of the links is not altered, but the absolute motion can be drastically different. For machines without a stationary link, relative motion is often the desired result of kinematic analysis.

In Example Problem 1.2, an important result of kinematic analysis is the distance that the handle must be moved in order to open the jaw. This is a question of relative position of the links: the handle and jaw. Because the relative motion of the links does not change with the selection of a frame, the choice of a frame link is often not important. Utilizing alternate links to serve as the fixed link is termed *kinematic inversion.*

1.7 MOBILITY

An important property in mechanism analysis is the number of degrees of freedom of the linkage. The degree of freedom is the number of independent inputs required to precisely position all links of the mechanism with respect to the ground. It can also be defined as the number of actuators needed to operate the mechanism. A mechanism actuator could be manually moving one link to another position, connecting a motor to the shaft of one link, or pushing a piston of a hydraulic cylinder.

The number of degrees of freedom of a mechanism is also called the *mobility,* and it is given the symbol *M.* When the configuration of a mechanism is completely defined by positioning one link, that system has one degree of freedom. Most commercially produced mechanisms have one degree of freedom. In constrast, robotic arms can have three, or more, degrees of freedom.

1.7.1 Gruebler's Equation

Degrees of freedom for planar linkages joined with common joints can be calculated through *Gruebler's equation:*

$$M = \text{degrees of freedom} = 3(n - 1) - 2j_p - j_h$$

where:

n = total number of links in the mechanism

j_p = total number of primary joints (pins or sliding joints)

j_h = total number of higher-order joints (cam or gear joints)

As mentioned, most linkages used in machines have one degree of freedom. A single degree-of-freedom linkage is shown in Figure 1.13a.

Linkages with zero, or negative, degrees of freedom are termed *locked mechanisms.* These mechanisms are unable to move and form a structure. A *truss* is a structure composed of simple links and connected with pin joints and zero degrees of freedom. A locked mechanism is shown in Figure 1.13b.

Linkages with multiple degrees of freedom need more than one driver to precisely operate them. Common multi-degree-of-freedom mechanisms are open-loop kinematic chains used for reaching and positioning, such as robotic arms and backhoes. In general, multi-degree-of-freedom linkages offer greater ability to precisely position a link. A multi-degree-of-freedom mechanism is shown in Figure 1.13c.

(a) Single degree-of-freedom ($M = 1$)   (b) Locked mechanism ($M = 0$)   (c) Multi-degree-of-freedom ($M = 2$)

FIGURE 1.13 Mechanisms and structures with varying mobility.

EXAMPLE PROBLEM 1.3

Figure 1.14 shows a toggle clamp. Draw a kinematic diagram, using the clamping jaw and the handle as points of interest. Also compute the degrees of freedom for the clamp.

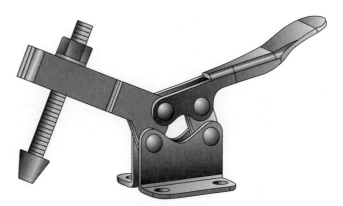

FIGURE 1.14 Toggle clamp for Example Problem 1.3.

SOLUTION: 1. ***Identify the Frame***

The component that is bolted to the table is designated as the frame. The motion of all other links is determined relative to this frame. The frame is numbered as link 1.

2. ***Identify All Other Links***

Careful observation reveals three other moving parts:
   Link 2: Handle
   Link 3: Arm that serves as the clamping jaw
   Link 4: Bar that connects the clamping arm and handle

3. ***Identify the Joints***

Four pin joints are used to connect these different links (link 1 to 2, 2 to 3, 3 to 4, and 4 to 1). These joints are lettered *A* through *D*.

4. ***Identify Any Points of Interest***

The motion of the clamping jaw is desired. This is designated as point of interest *X*. Finally, the motion of the end of the handle is also desired. This is designated as point of interest *Y*.

5. ***Draw the Kinematic Diagram***

The kinematic diagram is detailed in Figure 1.15.

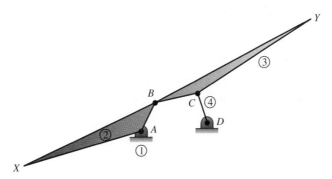

FIGURE 1.15 Kinematic diagram for Example Problem 1.3.

6. ***Calculate Mobility***

Having four links and four pin joints,

$$n = 4, j_p = 4 \text{ pins}, j_h = 0$$

and

$$M = 3(n - 1) - 2j_p - j_h = 3(4 - 1) - 2(4) - 0 = 1$$

With one degree of freedom, the clamp mechanism is constrained. Moving only one link, the handle, precisely positions all other links in the clamp.

EXAMPLE PROBLEM 1.4

Figure 1.16 shows a beverage can crusher used to reduce the size of cans for easier storage prior to recycling. Draw a kinematic diagram, using the end of the handle as a point of interest. Also compute the degrees of freedom for the device.

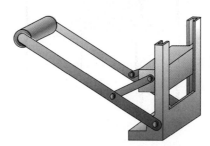

FIGURE 1.16 Can crusher for Example Problem 1.4.

SOLUTION:

1. ***Identify the Frame***

   The back portion of the device serves as a base and can be attached to a wall. This component is designated as the frame. The motion of all other links is determined relative to this frame. The frame is numbered as link 1.

2. ***Identify All Other Links***

   Careful observation shows a planar mechanism with three other moving parts:
   Link 2: Handle
   Link 3: Block that serves as the crushing surface
   Link 4: Bar that connects the crushing block and handle

3. ***Identify the Joints***

   Three pin joints are used to connect these different parts. One pin connects the handle to the base. This joint is labeled as *A*. A second pin is used to connect link 4 to the handle. This joint is labeled *B*. The third pin connects the crushing block and link 4. This joint is labeled *C*.

   The crushing block slides vertically during operation; therefore, a sliding joint connects the crushing block to the base. This joint is labeled *D*.

4. ***Identify Any Points of Interest***

   The motion of the handle end is desired. This is designated as point of interest *X*.

5. ***Draw the Kinematic Diagram***

   The kinematic diagram is given in Figure 1.17.

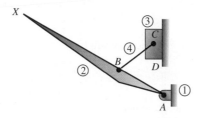

FIGURE 1.17 Kinematic diagram for Example Problem 1.4.

6. *Calculate Mobility*

It was determined that there are four links in this mechanism. There are also three pin joints and one slider joint. Therefore,

$$n = 4, j_p = (3 \text{ pins } + 1 \text{ slider}) = 4, j_h = 0$$

and

$$M = 3(n - 1) - 2j_p - j_h = 3(4 - 1) - 2(4) - 0 = 1$$

With one degree of freedom, the can crusher mechanism is constrained. Moving only one link, the handle, precisely positions all other links and crushes a beverage can placed under the crushing block.

EXAMPLE PROBLEM 1.5

Figure 1.18 shows another device that can be used to shear material. Draw a kinematic diagram, using the end of the handle and the cutting edge as points of interest. Also, compute the degrees of freedom for the shear press.

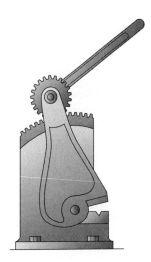

FIGURE 1.18 Shear press for Example Problem 1.5.

SOLUTION:

1. *Identify the Frame*

The base is bolted to a working surface and can be designated as the frame. The motion of all other links is determined relative to this frame. The frame is numbered as link 1.

2. *Identify All Other Links*

Careful observation reveals two other moving parts:
   Link 2: Gear/handle
   Link 3: Cutting lever

3. *Identify the Joints*

Two pin joints are used to connect these different parts. One pin connects the cutting lever to the frame. This joint is labeled as *A*. A second pin is used to connect the gear/handle to the cutting lever. This joint is labeled *B*.

The gear/handle is also connected to the frame with a gear joint. This higher-order joint is labeled *C*.

4. *Identify Any Points of Interest*

The motion of the handle end is desired and is designated as point of interest *X*. The motion of the cutting surface is also desired and is designated as point of interest *Y*.

5. *Draw the Kinematic Diagram*

The kinematic diagram is given in Figure 1.19.

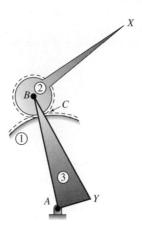

FIGURE 1.19 Kinematic diagram for Example Problem 1.5.

6. *Calculate Mobility*

To calculate the mobility, it was determined that there are three links in this mechanism. There are also two pin joints and one gear joint. Therefore,

$$n = 3 \quad j_p = (2 \text{ pins}) = 2 \quad j_h = (1 \text{ gear connection}) = 1$$

and

$$M = 3(n - 1) - 2j_p - j_h = 3(3 - 1) - 2(2) - 1 = 1$$

With one degree of freedom, the shear press mechanism is constrained. Moving only one link, the handle, precisely positions all other links and brings the cutting edge onto the work piece.

1.7.2 Actuators and Drivers

In order to operate a mechanism, an actuator, or driver device, is required to provide the input motion and energy. To precisely operate a mechanism, one driver is required for each degree of freedom exhibited. Many different actuators are used in industrial and commercial machines and mechanisms. Some of the more common ones are given below:

Electric motors (AC) provide the least expensive way to generate continuous rotary motion. However, they are limited to a few standard speeds that are a function of the electric line frequency. In North America the line frequency is 60 Hz, which corresponds to achievable speeds of 3600, 1800, 900, 720, and 600 rpm. Single-phase motors are used in residential applications and are available from 1/50 to 2 hp. Three-phase motors are more efficient, but mostly limited to industrial applications because they require three-phase power service. They are available from 1/4 to 500 hp.

Electric motors (DC) also produce continuous rotary motion. The speed and direction of the motion can be readily altered, but they require power from a generator or a battery. DC motors can achieve extremely high speeds—up to 30,000 rpm. These motors are most often used in vehicles, cordless devices, or in applications where multiple speeds and directional control are required, such as a sewing machine.

Engines also generate continuous rotary motion. The speed of an engine can be throttled within a range of approximately 1000 to 8000 rpm. They are a popular and highly portable driver for high-power applications. Because they rely on the combustion of fuel, engines are used to drive machines that operate outdoors.

Servomotors are motors that are coupled with a controller to produce a programmed motion or hold a fixed position. The controller requires sensors on the link being moved to provide feedback information on its position, velocity, and acceleration. These motors have lower power capacity than nonservomotors and are significantly more expensive, but they can be used for machines demanding precisely guided motion, such as robots.

Air or hydraulic motors also produce continuous rotary motion and are similar to electric motors, but have more limited applications. This is due to the need for compressed air or a hydraulic source. These drive devices are mostly used within machines, such as construction equipment and aircraft, where high-pressure hydraulic fluid is available.

Hydraulic or pneumatic cylinders are common components used to drive a mechanism with a limited linear stroke. Figure 1.20a illustrates a hydraulic cylinder. Figure 1.20b shows the common kinematic representation for the cylinder unit.

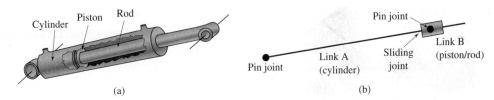

FIGURE 1.20 Hydraulic cylinder.

The cylinder unit contains a rod and piston assembly that slides relative to a cylinder. For kinematic purposes, these are two links (piston/rod and cylinder), connected with a sliding joint. In addition, the cylinder and rod end usually have provisions for pin joints.

Screw actuators also produce a limited linear stroke. These actuators consist of a motor, rotating a screw. A mating nut provides the linear motion. Screw actuators can be accurately controlled and can directly replace cylinders. However, they are considerably more expensive than cylinders if air or hydraulic sources are available. Similar to cylinders, screw actuators also have provisions for pin joints at the two ends. Therefore, the kinematic diagram is identical to Figure 1.20b.

Manual, or hand-operated, mechanisms comprise a large number of machines, or hand tools. The motions expected from human "actuators" can be quite complex. However, if the expected motions are repetitive, caution should be taken against possible fatigue and stain injuries.

EXAMPLE PROBLEM 1.6

Figure 1.21 shows an outrigger foot to stabilize a utility truck. Draw a kinematic diagram, using the bottom of the stabilizing foot as a point of interest. Also compute the degrees of freedom.

FIGURE 1.21 Outrigger for Example Problem 1.6.

SOLUTION:

1. **Identify the Frame**

   During operation of the outriggers, the utility truck is stationary. Therefore, the truck is designated as the frame. The motion of all other links is determined relative to the truck. The frame is numbered as link 1.

2. **Identify All Other Links**

   Careful observation reveals three other moving parts:

   > Link 2: Outrigger leg
   > Link 3: Cylinder
   > Link 4: Piston/rod

3. **Identify the Joints**

   Three pin joints are used to connect these different parts. One connects the outrigger leg with the truck frame. This is labeled as joint A. Another connects the outrigger leg with the cylinder rod and is labeled as joint B. The last pin joint connects the cylinder to the truck frame and is labeled as joint C.

   One sliding joint is present in the cylinder unit. This connects the piston/rod with the cylinder. It is labeled as joint D.

4. ***Identify Any Points of Interest***

The stabilizer foot is part of link 2, and a point of interest located on the bottom of the foot is labeled as point of interest *X*.

5. ***Draw the Kinematic Diagram***

The resulting kinematic diagram is given in Figure 1.22.

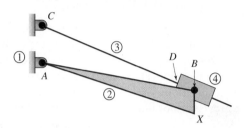

FIGURE 1.22 Kinematic diagram for Example Problem 1.6.

6. ***Calculate Mobility***

To calculate the mobility, it was determined that there are four links in this mechanism, as well as three pin joints and one slider joint. Therefore,

$$n = 4, j_p = (3 \text{ pins} + 1 \text{ slider}) = 4, j_h = 0$$

and

$$M = 3(n - 1) - 2j_p - j_h = 3(4 - 1) - 2(4) - 0 = 1$$

With one degree of freedom, the outrigger mechanism is constrained. Moving only one link, the piston, precisely positions all other links in the outrigger, placing the stabilizing foot on the ground.

1.8 COMMONLY USED LINKS AND JOINTS

1.8.1 Eccentric Crank

On many mechanisms, the required length of a crank is so short that it is not feasible to fit suitably sized bearings at the two pin joints. A common solution is to design the link as an eccentric crankshaft, as shown in Figure 1.23a. This is the design used in most engines and compressors.

The pin, on the moving end of the link, is enlarged such that it contains the entire link. The outside circumference of the circular lobe on the crankshaft becomes the moving pin joint, as shown in Figure 1.23b. The location of the fixed bearing, or bearings, is offset from the eccentric lobe. This eccentricity of the crankshaft, *e*, is the effective length of the crank. Figure 1.23c illustrates a kinematic

model of the eccentric crank. The advantage of the eccentric crank is the large surface area of the moving pin, which reduces wear.

1.8.2 Pin-in-a-Slot Joint

A common connection between links is a pin-in-a-slot joint, as shown in Figure 1.24a. This is a higher-order joint because it permits the two links to rotate and slide relative to each other. To simplify the kinematic analysis, primary joints can be used to model this higher-order joint. The pin-in-a-slot joint becomes a combination of a pin joint and a sliding joint, as in Figure 1.24b. Note that this involves adding an extra link to the mechanism. In both cases, the relative motion between the links is the same. However, using a kinematic model with primary joints facilitates the analysis.

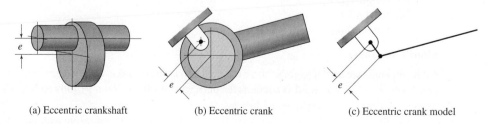

(a) Eccentric crankshaft   (b) Eccentric crank   (c) Eccentric crank model

FIGURE 1.23 Eccentric crank.

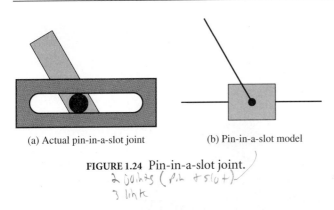

(a) Actual pin-in-a-slot joint   (b) Pin-in-a-slot model

FIGURE 1.24 Pin-in-a-slot joint.

[handwritten: 2 joints (pin + slot) 3 link]

1.8.3 Screw Joint

A screw joint, as shown in Figure 1.25a, is another common connection between links. Screw mechanisms are discussed in detail in Chapter 12. To start with, a screw joint permits two relative, but dependent, motions between the links being joined. A specific rotation of one link will cause an associated relative translation between the two links. For example, turning the screw one revolution may move the nut along the screw threads a distance of 0.1 in. Thus, only one independent motion is introduced.

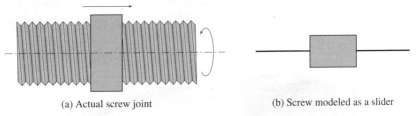

(a) Actual screw joint   (b) Screw modeled as a slider

FIGURE 1.25 Screw joint.

A screw joint is typically modeled with a sliding joint, as shown in Figure 1.25b. It must be understood that out-of-plane rotation occurs. However, only the relative translation between the screw and nut is considered in planar kinematic analysis.

An actuator, such as a hand crank, typically produces the out-of-plane rotation. A certain amount of rotation will cause a corresponding relative translation between the links being joined by the screw joint. This relative translation is used as the "driver" in subsequent kinematic analyses.

EXAMPLE PROBLEM 1.7

Figure 1.26 presents a lift table used to adjust the working height of different objects. Draw a kinematic diagram and compute the degrees of freedom.

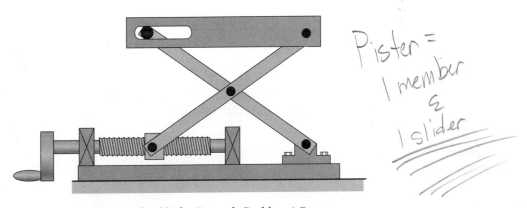

[handwritten: Piston = 1 member & 1 slider]

FIGURE 1.26 Lift table for Example Problem 1.7.

SOLUTION: 1.   *Identify the Frame*

The bottom base plate rests on a fixed surface. Thus, the base plate will be designated as the frame. The bearing at the bottom right of Figure 1.26 is bolted to the base plate. Likewise, the two bearings that support the screw on the left are bolted to the base plate.

From the discussion in the previous section, the out-of-plane rotation of the screw will not be considered. Only the relative translation of the nut will be included in the kinematic model. Therefore, the screw will also be considered as part of the frame. The motion of all other links will be determined relative to this bottom base plate, which will be numbered as link 1.

2. ***Identify All Other Links***

Careful observation reveals five other moving parts:

Link 2: Nut
Link 3: Support arm that ties the nut to the table
Link 4: Support arm that ties the fixed bearing to the slot in the table
Link 5: Table
Link 6: Extra link used to model the pin in slot joint with separate pin and slider joints

3. ***Identify the Joints***

A sliding joint is used to model the motion between the screw and the nut. A pin joint, designated as point *A*, connects the nut to the support arm identified as link 3. A pin joint, designated as point *B*, connects the two support arms—link 3 and link 4. Another pin joint, designated as point *C*, connects link 3 to link 6. A sliding joint joins link 6 to the table, link 5. A pin, designated as point *D*, connects the table to the support arm, link 3. Lastly, a pin joint, designated as point *E*, is used to connect the base to the support arm, link 4.

4. ***Draw the Kinematic Diagram***

The kinematic diagram is given in Figure 1.27.

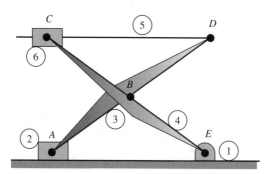

FIGURE 1.27   Kinematic diagram for Example Problem 1.7.

5. ***Calculate Mobility***

To calculate the mobility, it was determined that there are six links in this mechanism. There are also five pin joints and two slider joints. Therefore

$$n = 6 \quad j_p = (5 \text{ pins} + 2 \text{ sliders}) = 7 \quad j_h = 0$$

and

$$M = 3(n - 1) - 2j_p - j_h = 3(6 - 1) - 2(7) - 0 = 15 - 14 = 1$$

With one degree of freedom, the lift table has constrained motion. Moving one link, the handle that rotates the screw, will precisely position all other links in the device, raising or lowering the table.

1.9 SPECIAL CASES OF THE MOBILITY EQUATION

Mobility is an extremely important property of a mechanism. Among other facets, it gives insight into the number of actuators required to operate a mechanism. However, to obtain correct results, special care must be taken in using the Gruebler's equation. Some special conditions are presented next.

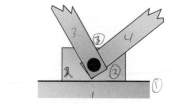

(a) Three rotating links   (b) Two rotating and one sliding link

FIGURE 1.28   Three links connected at a common pin joint.

1.9.1 Coincident Joints

Some mechanisms have three links that are all connected at a common pin joint, as shown in Figure 1.28. This situation brings some confusion to kinematic modeling. Physically, one pin may be used to connect all three links. However, by definition, a pin joint connects two links.

For kinematic analysis, this configuration must be mathematically modeled as two separate joints. One joint will connect the first and second links. The second joint will then connect the second and third links. Therefore, when three links come together at a common pin, the joint must be modeled as two pins. This scenario is illustrated in Example Problem 1.8.

EXAMPLE PROBLEM 1.8

Figure 1.29 shows a mechanical press used to exert large forces to insert a small part into a larger one. Draw a kinematic diagram, using the end of the handle as a point of interest. Also compute the degrees of freedom.

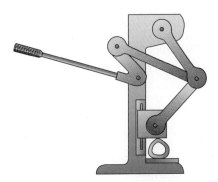

FIGURE 1.29 Mechanical press for Example Problem 1.8.

SOLUTION:

1. *Identify the Frame*

   The bottom base for the mechanical press sits on a workbench and remains stationary during operation. Therefore, this bottom base is designated as the frame. The motion of all other links is determined relative to the bottom base. The frame is numbered as link 1.

2. *Identify All Other Links*

   Careful observation reveals five other moving parts:

   Link 2: Handle

   Link 3: Arm that connects the handle to the other arms

   Link 4: Arm that connects the base to the other arms

   Link 5: Press head

   Link 6: Arm that connects the head to the other arms

3. *Identify the Joints*

   Pin joints are used to connect the several different parts. One connects the handle to the base and is labeled as joint *A*. Another connects link 3 to the handle and is labeled as joint *B*. Another connects link 4 to the base and is labeled as *C*. Another connects link 6 to the press head and is labeled as *D*.

   It appears that a pin is used to connect the three arms—links 3, 4, and 6—together. Because three separate links are joined at a common point, this must be modeled as two separate joints. They are labeled as *E* and *F*.

   A sliding joint connects the press head with the base. This joint is labeled as *G*.

4. *Identify Any Points of Interest*

   Motion of the end of the handle is desired and is labeled as point of interest *X*.

5. *Draw the Kinematic Diagram*

   The kinematic diagram is given in Figure 1.30.

6. *Calculate Mobility*

   To calculate the mobility, it was determined that there are six links in this mechanism, as well as six pin joints and one slider joint. Therefore,

$$n = 6, j_p = (6 \text{ pins} + 1 \text{ slider}) = 7, j_h = 0$$

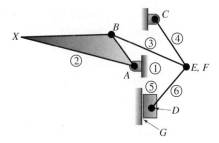

FIGURE 1.30 Kinematic diagram for Example Problem 1.8.

and

$$M = 3(n - 1) - 2j_p - j_h = 3(6 - 1) - 2(7) - 0 = 15 - 14 = 1$$

With one degree of freedom, the mechanical press mechanism is constrained. Moving only one link, the handle, precisely positions all other links in the press, sliding the press head onto the work piece.

1.9.2 Exceptions to the Gruebler's Equation

Another special mobility situation must be mentioned. Because the Gruebler's equation does not account for link geometry, in rare instances it can lead to misleading results. One such instance is shown in Figure 1.31.

Notice that this linkage contains five links and six pin joints. Using Gruebler's equation, this linkage has zero degrees of freedom. Of course, this suggests that the mechanism is locked. However, if all pivoted links were the same size, and the distance between the joints on the frame and coupler were identical, this mechanism would be capable of motion, with one degree of freedom. The center link is

FIGURE 1.31 Mechanism that violates the Gruebler's equation.

redundant, and because it is identical in length to the other two links attached to the frame, it does not alter the action of the linkage.

There are other examples of mechanisms that violate the Gruebler's equation because of unique geometry. A designer must be aware that the mobility equation can, at times, lead to inconsistencies.

1.9.3 Idle Degrees of Freedom

In some mechanisms, links exhibit motion which does not influence the input and output relationship of the mechanism. These *idle degrees of freedom* present another situation where Gruebler's equation gives misleading results. An example is a cam with a roller follower as shown in Figure 1.32. Gruebler's equation specifies two degrees of freedom (4 links, 3 pins, 1 higher-order joint). With an actuated cam rotation, the pivoted link oscillates while the roller follower rotates about its center. Yet, only the motion of the pivoted link serves as the output of the mechanism.

The roller rotation is an idle degree of freedom and not intended to affect the output motion of the mechanism. It is a design feature which reduces friction and wear on the surface of the cam. While Gruebler's equation specifies that a cam mechanism with a roller follower has a mobility of two, the designer is typically only interested in a single degree of freedom. Several other mechanisms contain similar idle degrees of freedom.

FIGURE 1.32 A cam with a roller follower.

1.10 THE FOUR-BAR MECHANISM

The simplest and most common linkage is the four-bar linkage. It is a combination of four links, one being designated as the frame and connected by four pin joints.

Because it is encountered so often, further exploration is in order.

The mechanism for an automotive rear-window wiper system is shown in Figure 1.33a. The kinematic diagram is shown in Figure 1.33b. Notice that this is a four-bar mechanism

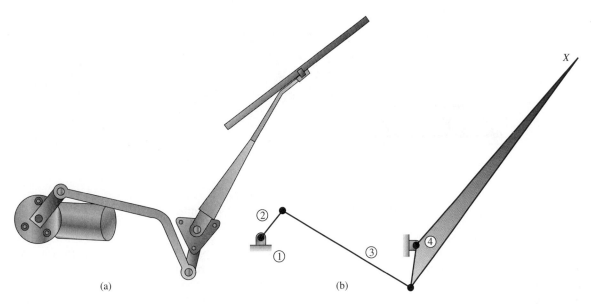

(a)   (b)

FIGURE 1.33 Rear-window wiper mechanism.

because it is comprised of four links connected by four pin joints and one link is unable to move.

The mobility of a four-bar mechanism consists of the following:

$$n = 4, j_p = 4 \text{ pins}, j_h = 0$$

and

$$M = 3(n - 1) - 2j_p - j_h = 3(4 - 1) - 2(4) - 0 = 1$$

Because the four-bar mechanism has one degree of freedom, it is constrained or fully operated with one driver. The wiper system in Figure 1.33 is activated by a single DC electric motor.

Of course, the link that is unable to move is referred to as the frame. Typically, the pivoted link that is connected to the driver or power source is called the *input link*. The other pivoted link that is attached to the frame is designated the *output link* or *follower*. The *coupler* or *connecting arm* "couples" the motion of the input link to the output link.

1.10.1 Grashof's Criterion

The following nomenclature is used to describe the length of the four links.

s = length of the shortest link

l = length of the longest link

p = length of one of the intermediate length links

q = length of the other intermediate length links

Grashof's theorem states that a four-bar mechanism has at least one revolving link if:

$$s + l \leq p + q$$

Conversely, the three nonfixed links will merely rock if:

$$s + l > p + q$$

All four-bar mechanisms fall into one of the five categories listed in Table 1.2.

TABLE 1.2 Categories of Four-Bar Mechanisms

Case	Criteria	Shortest Link	Category
1	$s + l < p + q$	Frame	Double crank
2	$s + l < p + q$	Side	Crank-rocker
3	$s + l < p + q$	Coupler	Double rocker
4	$s + l = p + q$	Any	Change point
5	$s + l > p + q$	Any	Triple rocker

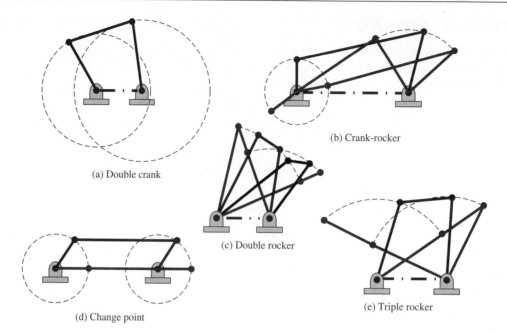

(a) Double crank

(b) Crank-rocker

(c) Double rocker

(d) Change point

(e) Triple rocker

FIGURE 1.34 Categories of four-bar mechanisms.

The different categories are illustrated in Figure 1.34 and described in the following sections.

1.10.2 Double Crank

A double crank, or crank-crank, is shown in Figure 1.34a. As specified in the criteria of Case 1 of Table 1.2, it has the shortest link of the four-bar mechanism configured as the frame. If one of the pivoted links is rotated continuously, the other pivoted link will also rotate continuously. Thus, the two pivoted links, 2 and 4, are both able to rotate through a full revolution. The double crank mechanism is also called a drag link mechanism.

1.10.3 Crank-Rocker

A crank-rocker is shown in Figure 1.34b. As specified in the criteria of Case 2 of Table 1.2, it has the shortest link of the four-bar mechanism configured adjacent to the frame. If this shortest link is continuously rotated, the output link will oscillate between limits. Thus, the shortest link is called the *crank*, and the output link is called the *rocker*. The wiper system in Figure 1.33 is designed to be a crank-rocker. As the motor continuously rotates the input link, the output link oscillates, or "rocks." The wiper arm and blade are firmly attached to the output link, oscillating the wiper across a windshield.

1.10.4 Double Rocker

The double rocker, or rocker-rocker, is shown in Figure 1.34c. As specified in the criteria of Case 3 of Table 1.2, it

has the link opposite the shortest link of the four-bar mechanism configured as the frame. In this configuration, neither link connected to the frame will be able to complete a full revolution. Thus, both input and output links are constrained to oscillate between limits, and are called rockers. However, the coupler is able to complete a full revolution.

1.10.5 Change Point Mechanism

A change point mechanism is shown in Figure 1.34d. As specified in the criteria of Case 4 of Table 1.2, the sum of two sides is the same as the sum of the other two. Having this equality, the change point mechanism can be positioned such that all the links become collinear. The most familiar type of change point mechanism is a parallelogram linkage. The frame and coupler are the same length, and so are the two pivoting links. Thus, the four links will overlap each other. In that collinear configuration, the motion becomes indeterminate. The motion may remain in a parallelogram arrangement, or cross into an antiparallelogram, or butterfly, arrangement. For this reason, the change point is called a singularity configuration.

1.10.6 Triple Rocker

A triple rocker linkage is shown in Figure 1.34e. Exhibiting the criteria in Case 5 of Table 1.2, the triple rocker has no links that are able to complete a full revolution. Thus, all three moving links rock.

EXAMPLE PROBLEM 1.9

A nosewheel assembly for a small aircraft is shown in Figure 1.35. Classify the motion of this four-bar mechanism based on the configuration of the links.

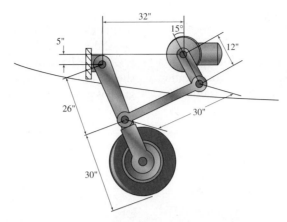

FIGURE 1.35 Nosewheel assembly for Example Problem 1.9.

SOLUTION: 1. ***Distinguish the Links Based on Length***

In an analysis that focuses on the landing gear, the motion of the wheel assembly would be determined relative to the body of the aircraft. Therefore, the aircraft body will be designated as the frame. Figure 1.36 shows the kinematic diagram for the wheel assembly, numbering and labeling the links. The tip of the wheel was designated as point of interest *X*.

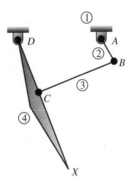

FIGURE 1.36 Kinematic diagram for Example Problem 1.9.

The lengths of the links are:

$$s = 12 \text{ in.}; \quad l = 32 \text{ in.}; \quad p = 30 \text{ in.}; \quad q = 26 \text{ in.}$$

2. ***Compare to Criteria***

The shortest link is a side, or adjacent to the frame. According to the criteria in Table 1.2, this mechanism can be either a crank-rocker, change point, or a triple rocker. The criteria for the different categories of four-bar mechanisms should be reviewed.

3. ***Check the Crank-Rocker (Case 2) Criteria***

Is:

$$s + l < p + q$$
$$(12 + 32) < (30 + 26)$$
$$44 < 56 \quad \rightarrow \quad \{\text{yes}\}$$

Because the criteria for a crank-rocker are valid, the nosewheel assembly is a crank-rocker mechanism.

1.11 SLIDER-CRANK MECHANISM

Another mechanism that is commonly encountered is a slider-crank. This mechanism also consists of a combination of four links, with one being designated as the frame. This

mechanism, however, is connected by three pin joints and one sliding joint.

A mechanism that drives a manual water pump is shown in Figure 1.37a. The corresponding kinematic diagram is given in Figure 1.37b.

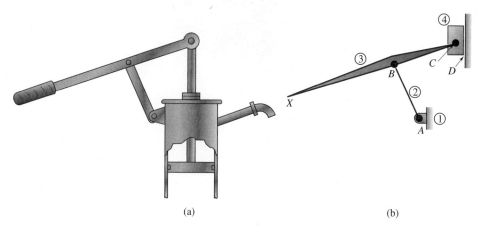

(a)	(b)

FIGURE 1.37 Pump mechanism for a manual water pump: (a) Mechanism and (b) Kinematic diagram.

The mobility of a slider-crank mechanism is represented by the following:

$$n = 4, j_p = (3 \text{ pins} + 1 \text{ sliding}) = 4, j_h = 0$$

and

$$M = 3(n - 1) - 2j_p - j_h = 3(4 - 1) - 2(4) - 0 = 1.$$

Because the slider-crank mechanism has one degree of freedom, it is constrained or fully operated with one driver. The pump in Figure 1.37 is activated manually by pushing on the handle (link 3).

In general, the pivoted link connected to the frame is called the *crank*. This link is not always capable of completing a full revolution. The link that translates is called the *slider*. This link is the piston/rod of the pump in Figure 1.37.

The coupler or connecting rod "couples" the motion of the crank to the slider.

1.12 SPECIAL PURPOSE MECHANISMS

1.12.1 Straight-Line Mechanisms

Straight-line mechanisms cause a point to travel in a straight line without being guided by a flat surface. Historically, quality prismatic joints that permit straight, smooth motion without backlash have been difficult to manufacture. Several mechanisms have been conceived that create straight-line (or nearly straight-line) motion with revolute joints and rotational actuation. Figure 1.38a shows a Watt linkage and Figure. 1.38b shows a Peaucellier-Lipkin linkage.

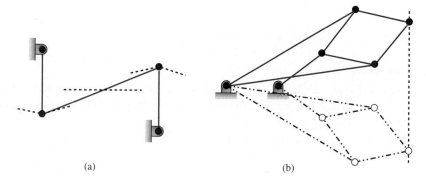

(a)	(b)

FIGURE 1.38 Straight-line mechanisms

1.12.2 Parallelogram Mechanisms

Mechanisms are often comprised of links that form parallelograms to move an object without altering its pitch. These mechanisms create parallel motion for applications such as

balance scales, glider swings, and jalousie windows. Two types of parallelogram linkages are given in Figure 1.39a which shows a scissor linkage and Figure1.39b which shows a drafting machine linkage.

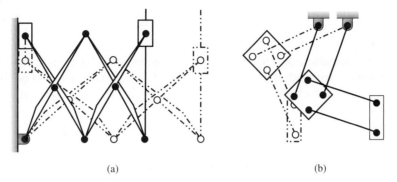

FIGURE 1.39 Parallelogram mechanisms.

1.12.3 Quick-Return Mechanisms

Quick-return mechanisms exhibit a faster stroke in one direction than the other when driven at constant speed with a rotational actuator. They are commonly used on machine tools that require a slow cutting stroke and a fast return stroke. The kinematic diagrams of two different quick-return mechanisms are given in Figure 1.40a which shows an offset slider-crank linkage and Figure 1.40b which shows a crank-shaper linkage.

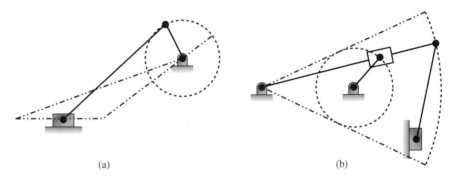

FIGURE 1.40 Quick-return mechanisms.

1.12.4 Scotch Yoke Mechanism

A scotch yoke mechanism is a common mechanism that converts rotational motion to linear sliding motion, or vice versa. As shown in Figure 1.41, a pin on a rotating link is engaged in the slot of a sliding yoke. With regards to the input and output motion, the scotch yoke is similar to a slider-crank, but the linear sliding motion is pure sinusoidal. In comparison to the slider-crank, the scotch yoke has the advantage of smaller size and fewer moving parts, but can experience rapid wear in the slot.

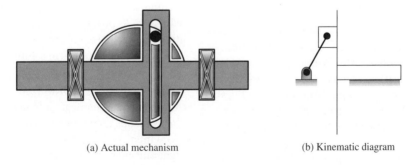

(a) Actual mechanism   (b) Kinematic diagram

FIGURE 1.41 Scotch yoke mechanism.

1.13 TECHNIQUES OF MECHANISM ANALYSIS

Most of the analysis of mechanisms involves geometry. Often, graphical methods are employed so that the motion of the mechanism can be clearly visualized. Graphical solutions involve drawing "scaled" lines at specific angles. One example is the drawing of a kinematic diagram. A graphical solution involves preparing a drawing where all links are shown at a proportional scale to the actual mechanism. The orientation of the links must also be shown at the same angles as on the actual mechanism.

This graphical approach has the merits of ease and solution visualization. However, accuracy must be a serious concern to achieve results that are consistent with analytical techniques. For several decades, mechanism analysis was primarily completed using graphical approaches. Despite its popularity, many scorned graphical techniques as being imprecise. However, the development of computer-aided design (CAD) systems has allowed the graphical approach to be applied with precision. This text attempts to illustrate the most common methods used in the practical analysis of mechanisms. Each of these methods is briefly introduced in the following sections.

1.13.1 Traditional Drafting Techniques

Over the past decades, all graphical analysis was performed using traditional drawing techniques. Drafting equipment was used to draw the needed scaled lines at specific angles. The equipment used to perform these analyses included triangles, parallel straight edges, compasses, protractors, and engineering scales. As mentioned, this method was often criticized as being imprecise. However, with proper attention to detail, accurate solutions can be obtained.

It was the rapid adoption of CAD software over the past several years that limited the use of traditional graphical techniques. Even with the lack of industrial application, many believe that traditional drafting techniques can still be used by students to illustrate the concepts behind graphical mechanism analysis. Of course, these concepts are identical to those used in graphical analysis using a CAD system. But by using traditional drawing techniques, the student can concentrate on the kinematic theories and will not be "bogged down" with learning CAD commands.

1.13.2 CAD Systems

As mentioned, graphical analysis may be performed using traditional drawing procedures or a CAD system, as is commonly practiced in industry. Any one of the numerous commercially available CAD systems can be used for mechanism analysis. The most common two-dimensional CAD system is AutoCAD. Although the commands differ between CAD systems, all have the capability to draw highly accurate lines at designated lengths and angles. This is exactly the capability required for graphical mechanism analysis. Besides increased accuracy, another benefit of CAD is that the lines do not need to be scaled to fit on a piece of drawing paper. On the computer, lines are drawn on "virtual" paper that is of infinite size.

Additionally, the constraint-based sketching mode in solid modeling systems, such as Inventor, SolidWorks, and ProEngineer, can be extremely useful for planar kinematic analysis. Geometric constraints, such as length, perpendicularity, and parallelism, need to be enforced when performing kinematic analysis. These constraints are automatically executed in the solid modeler's sketching mode.

This text does not attempt to thoroughly discuss the system-specific commands used to draw the lines, but several of the example problems are solved using a CAD system. The main goal of this text is to instill an understanding of the concepts of mechanism analysis. This goal can be realized regardless of the specific CAD system incorporated. Therefore, the student should not be overly concerned with the CAD system used for accomplishing graphical analysis. For that matter, the student should not be concerned whether manual or computer graphics are used to learn mechanism analysis.

1.13.3 Analytical Techniques

Analytical methods can also be used to achieve precise results. Advanced analytical techniques often involve intense mathematical functions, which are beyond the scope of this text and of routine mechanism analysis. In addition, the significance of the calculations is often difficult to visualize.

The analytical techniques incorporated in this text couple the theories of geometry, trigonometry, and graphical mechanism analysis. This approach will achieve accurate solutions, yet the graphical theories allow the solutions to be visualized. This approach does have the pitfall of laborious calculations for more complex mechanisms. Still, a significant portion of this text is dedicated to these analytical techniques.

1.13.4 Computer Methods

As more accurate analytical solutions are desired for several positions of a mechanism, the number of calculations can become unwieldy. In these situations, the use of computer solutions is appropriate. Computer solutions are also valuable when several design iterations must be analyzed.

A computer approach to mechanism analysis can take several forms:

- *Spreadsheets* are very popular for routine mechanism problems. An important feature of the spreadsheet is that as a cell containing input data is changed, all other results are updated. This allows design iterations to be completed with ease. As problems become more complex, they can be difficult to manage on a spreadsheet. Nonetheless, spreadsheets are used in problem solution throughout the text.

- *Commercially available dynamic analysis programs*, such as Working Model, ADAMS (Automatic Dynamic Analysis of Mechanical Systems), or Dynamic Designer, are available. Dynamic models of systems can be created from menus of general components. Limited versions of dynamic analysis programs are solid modeling systems. Full software packages are available and best suited when kinematic and dynamic analysis is a large component of the job. Chapter 2 is dedicated to dynamic analysis programs.

- *User-written computer programs* in a high-level language, such as Matlab, Mathematica, VisualBasic, or C++, can be created. The programming language selected must have direct availability to trigonometric and inverse trigonometric functions. Due to the time and effort required to write special programs, they are most effective when a complex, yet not commonly encountered, problem needs to be solved. Some simple algorithms are provided for elementary kinematic analysis in Chapter 8.

PROBLEMS

Problems in Sketching Kinematic Diagrams

1–1. A mechanism is used to open the door of a heat-treating furnace and is shown in Figure P1.1. Draw a kinematic diagram of the mechanism. The end of the handle should be identified as a point of interest.

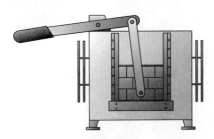

FIGURE P1.1 Problems 1 and 26.

1–2. A pair of bolt cutters is shown in Figure P1.2. Draw a kinematic diagram of the mechanism, selecting the lower handle as the frame. The end of the upper handle and the cutting surface of the jaws should be identified as points of interest.

FIGURE P1.2 Problems 2 and 27.

1–3. A folding chair that is commonly used in stadiums is shown in Figure P1.3. Draw a kinematic diagram of the folding mechanism.

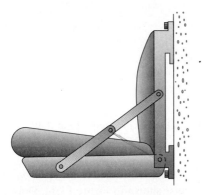

FIGURE P1.3 Problems 3 and 28.

1–4. A foot pump that can be used to fill bike tires, toys, and so forth is shown in Figure P1.4. Draw a kinematic diagram of the pump mechanism. The foot pad should be identified as a point of interest.

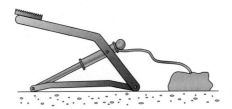

FIGURE P1.4 Problems 4 and 29.

1–5. A pair of pliers is shown in Figure P1.5. Draw a kinematic diagram of the mechanism.

FIGURE P1.5 Problems 5 and 30.

1–6. Another configuration for a pair of pliers is shown in Figure P1.6. Draw a kinematic diagram of the mechanism.

FIGURE P1.6 Problems 6 and 31.

1–7. A mechanism for a window is shown in Figure P1.7. Draw a kinematic diagram of the mechanism.

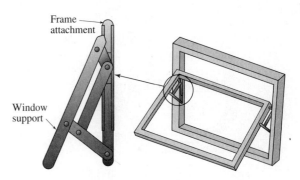

Frame attachment

Window support

FIGURE P1.7 Problems 7 and 32.

1–8. Another mechanism for a window is shown in Figure P1.8. Draw a kinematic diagram of the mechanism.

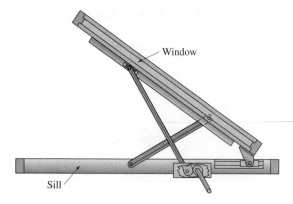

FIGURE P1.8 Problems 8 and 33.

1–9. A toggle clamp used for holding a work piece while it is being machined is shown in Figure P1.9. Draw a kinematic diagram of the mechanism.

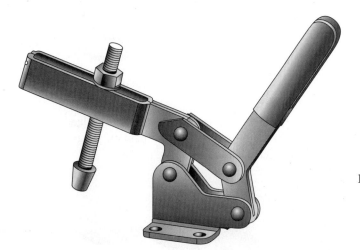

FIGURE P1.9 Problems 9 and 34.

1–10. A child's digging toy that is common at many municipal sandboxes is shown in Figure P1.10. Draw a kinematic diagram of the mechanism.

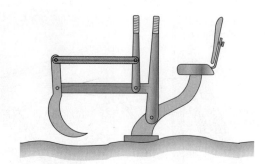

FIGURE P1.10 Problems 10 and 35.

1–11. A reciprocating saw, or saws all, is shown in Figure P1.11. Draw a kinematic diagram of the mechanism that produces the reciprocating motion.

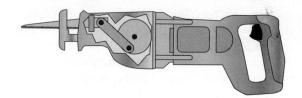

FIGURE P1.11 Problems 11 and 36.

1–12. A small front loader is shown in Figure P1.12. Draw a kinematic diagram of the mechanism.

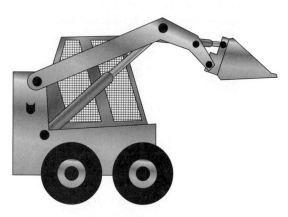

FIGURE P1.12 Problems 12 and 37.

1–13. A sketch of a microwave oven carrier used to assist people in wheelchairs is shown in Figure P1.13. Draw a kinematic diagram of the mechanism.

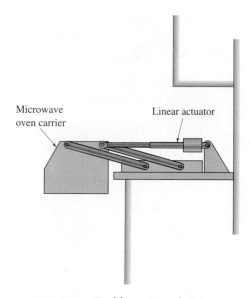

FIGURE P1.13 Problems 13 and 38.

1–14. A sketch of a truck used to deliver supplies to passenger jets is shown in Figure P1.14. Draw a kinematic diagram of the mechanism.

FIGURE P1.14 Problems 14 and 39.

1–15. A sketch of a device to move packages from an assembly bench to a conveyor is shown in Figure P1.15. Draw a kinematic diagram of the mechanism.

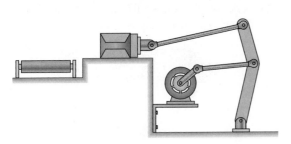

FIGURE P1.15 Problems 15 and 40.

1–16. A sketch of a lift platform is shown in Figure P1.16. Draw a kinematic diagram of the mechanism.

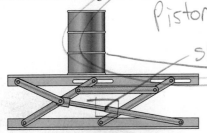

FIGURE P1.16 Problems 16 and 41.

1–17. A sketch of a lift platform is shown in Figure P1.17. Draw a kinematic diagram of the mechanism.

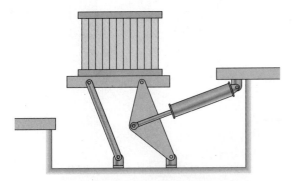

FIGURE P1.17 Problems 17 and 42.

1–18. A sketch of a backhoe is shown in Figure P1.18. Draw a kinematic diagram of the mechanism.

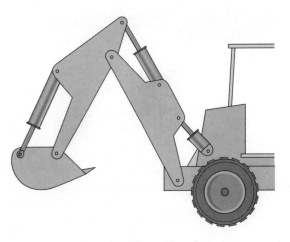

FIGURE P1.18 Problems 18 and 43.

1–19. A sketch of a front loader is shown in Figure P1.19. Draw a kinematic diagram of the mechanism.

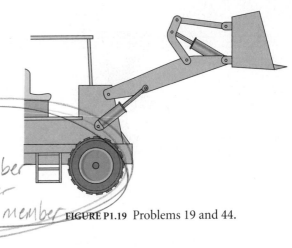

FIGURE P1.19 Problems 19 and 44.

1–20. A sketch of an adjustable-height platform used to load and unload freight trucks is shown in Figure P1.20. Draw a kinematic diagram of the mechanism.

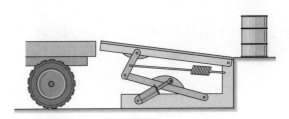

FIGURE P1.20 Problems 20 and 45.

1–21. A sketch of a kitchen appliance carrier, used for undercounter storage, is shown in Figure P1.21. Draw a kinematic diagram of the mechanism.

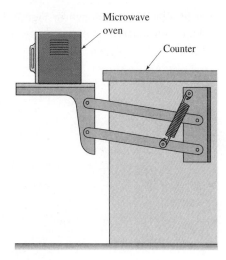

FIGURE P1.21 Problems 21 and 46.

1–22. An automotive power window mechanism is shown in Figure P1.22. Draw a kinematic diagram of the mechanism.

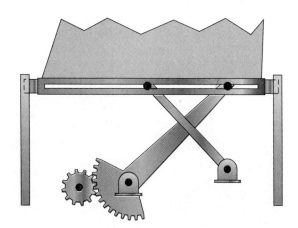

FIGURE P1.22 Problems 22 and 47.

1–23. A sketch of a device to close the top flaps of boxes is shown in Figure P1.23. Draw a kinematic diagram of the mechanism.

1–24. A sketch of a sewing machine is shown in Figure P1.24. Draw a kinematic diagram of the mechanism.

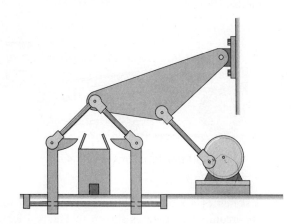

FIGURE P1.23 Problems 23 and 48.

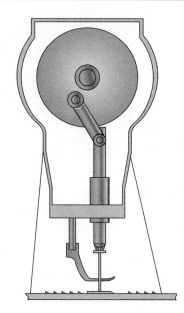

FIGURE P1.24 Problems 24 and 49.

1–25. A sketch of a wear test fixture is shown in Figure P1.25. Draw a kinematic diagram of the mechanism.

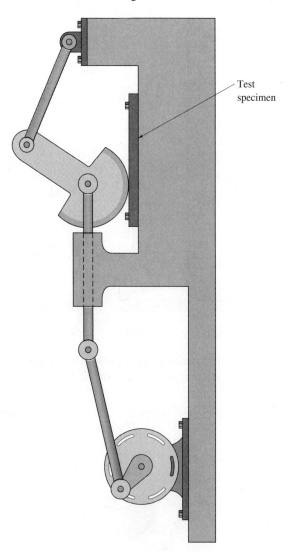

Test specimen

FIGURE P1.25 Problems 25 and 50.

Problems in Calculating Mobility

Specify the number of links and the number of joints and calculate the mobility for the mechanism shown in the figure.

1–26. Use Figure P1.1
1–27. Use Figure P1.2
1–28. Use Figure P1.3
1–29. Use Figure P1.4
1–30. Use Figure P1.5
1–31. Use Figure P1.6
1–32. Use Figure P1.7
1–33. Use Figure P1.8
1–34. Use Figure P1.9
1–35. Use Figure P1.10
1–36. Use Figure P1.11
1–37. Use Figure P1.12
1–38. Use Figure P1.13
1–39. Use Figure P1.14
1–40. Use Figure P1.15
1–41. Use Figure P1.16
1–42. Use Figure P1.17
1–43. Use Figure P1.18
1–44. Use Figure P1.19
1–45. Use Figure P1.20
1–46. Use Figure P1.21
1–47. Use Figure P1.22
1–48. Use Figure P1.23
1–49. Use Figure P1.24
1–50. Use Figure P1.25

Problems in Classifying Four-Bar Mechanisms

1–51. A mechanism to spray water onto vehicles at an automated car wash is shown in Figure P1.51.

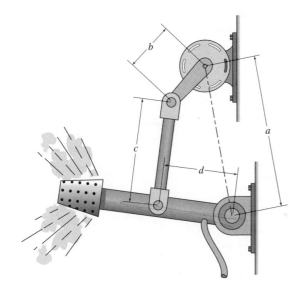

FIGURE P1.51 Problems 51 to 54.

Classify the four-bar mechanism, based on its possible motion, when the lengths of the links are $a = 12$ in., $b = 1.5$ in., $c = 14$ in., and $d = 4$ in.

1–52. For the water spray mechanism in Figure P1.51, classify the four-bar mechanism, based on its possible motion, when the lengths of the links are $a = 12$ in., $b = 5$ in., $c = 12$ in., and $d = 4$ in.

1–53. For the water spray mechanism in Figure P1.51, classify the four-bar mechanism, based on its possible motion, when the lengths of the links are $a = 12$ in., $b = 3$ in., $c = 8$ in., and $d = 4$ in.

1–54. For the water spray mechanism in Figure P1.51, classify the four-bar mechanism, based on its possible motion, when the lengths of the links are $a = 12$ in., $b = 3$ in., $c = 12$ in., and $d = 5$ in.

CASE STUDIES

1–1. The mechanism shown in Figure C1.1 has been taken from a feed device for an automated ball bearing assembly machine. An electric motor is attached to link A as shown. Carefully examine the configuration of the components in the mechanism. Then answer the following leading questions to gain insight into the operation of the mechanism.

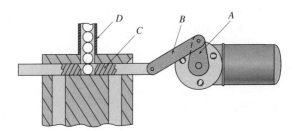

FIGURE C1.1 (Courtesy Industrial Press, Inc.).

1. As link A rotates clockwise 90°, what will happen to slide C?
2. What happens to the ball trapped in slide C when it is at this position?
3. As link A continues another 90° clockwise, what action occurs?
4. What is the purpose of this device?
5. Why are there chamfers at the entry of slide C?
6. Why do you suppose there is a need for such a device?

1–2. Figure C1.2 shows a mechanism that is typical in the tank of a water closet. Note that flapper C is hollow and filled with trapped air. Carefully examine the configuration of the components in the mechanism. Then answer the following leading questions to gain insight into the operation of the mechanism.

1. As the handle A is rotated counterclockwise, what is the motion of flapper C?
2. When flapper C is raised, what effect is seen?
3. When flapper C is lifted, it tends to remain in an upward position for a period of time. What causes this tendency to keep the flapper lifted?
4. When will this tendency (to keep flapper C lifted) cease?

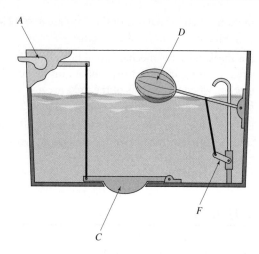

FIGURE C1.2 (Courtesy Industrial Press, Inc.).

5. What effect will cause item D to move?
6. As item D is moved in a counterclockwise direction, what happens to item F?
7. What does item F control?
8. What is the overall operation of these mechanisms?
9. Why is there a need for this mechanism and a need to store water in this tank?

1–3. Figure C1.3 shows a mechanism that guides newly formed steel rods to a device that rolls them into reels. The rods are hot when formed, and water is used to assist in the cooling process. The rods can be up to several thousand feet long and slide at rates up to 25 miles per hour through channel S.

Once the reel is full, the reel with the coiled rod is then removed. In order to obtain high efficiency, the rods follow one another very closely. It is impossible to remove the reel

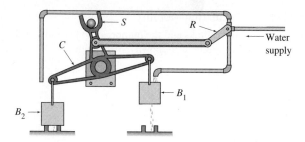

FIGURE C1.3 (Courtesy Industrial Press, Inc.).

in a short time interval; therefore, it is desirable to use two reels in alternation. This mechanism has been designed to feed the rods to the reels.

Buckets B_1 and B_2 have holes in the bottom. The water flow from the supply is greater than the amount that can escape from the holes. Carefully examine the configuration of the components in the mechanism, then answer the following leading questions to gain insight into the operation of the mechanism.

1. In the shown configuration, what is happening to the level of water in bucket B_1?
2. In the shown configuration, what is happening to the level of water in bucket B_2?
3. What would happen to rocker arm C if bucket B_2 were forced upward?
4. What would happen to rocker arm R if bucket B_2 were forced upward?
5. What does rocker arm R control?
6. What is the continual motion of this device?
7. How does this device allow two separate reels to be used for the situation described?
8. Why do you suppose that water is used as the power source for the operation of this mechanism?

BUILDING COMPUTER MODELS OF MECHANISMS USING WORKING MODEL® SOFTWARE

OBJECTIVES

Upon completion of this chapter, the student will be able to:

1. Understand the use of commercially available software for mechanism analysis.
2. Use Working Model® to build kinematic models of mechanisms.
3. Use Working Model® to animate the motion of mechanisms.
4. Use Working Model® to determine the kinematic values of a mechanism.

2.1 INTRODUCTION

The rapid development of computers and software has altered the manner in which many engineering tasks are completed. In the study of mechanisms, software packages have been developed that allow a designer to construct virtual models of a mechanism. These virtual models allow the designer to fully simulate a machine. Simulation enables engineers to create and test product prototypes on their own desktop computers. Design flaws can be quickly isolated and eliminated, reducing prototyping expenses and speeding the cycle of product development.

Software packages can solve kinematic and dynamic equations, determine the motion, and force values of the mechanism during operation. In addition to numerical analysis, the software can animate the computer model of the mechanism, allowing visualization of the design in action.

This chapter primarily serves as a tutorial for simulating machines and mechanisms using Working Model® simulation software. Although the kinematic values generated during the analysis may not be fully understood, the visualization of the mechanism can be extremely insightful. The material presented in the next several chapters will allow the student to understand the numerical solutions of the dynamic software. Proficiency in this type of mechanism-analysis software, coupled with a solid understanding of kinematic and dynamic analysis, will provide a strong basis for machine design.

2.2 COMPUTER SIMULATION OF MECHANISMS

Along with Working Model®, other dynamic analysis programs are available. These include ADAMS® (Automatic Dynamic Analysis of Mechanical Systems), Dynamic Designer®, LMS Virtual.Lab, and Analytix®. All these computer programs allow creation of a mechanism from menus, or icons, of general components. The general components include those presented in Chapter 1, such as simple links, complex links, pin joints, sliding joints, and gear joints. The mechanism is operated by selecting actuator components, such as motors or cylinders, from menus.

In machine design, one of the reasons for the widespread adoption of solid modeling is that it sets the stage for many ancillary uses: Working drawings can be nearly automatically created, renderings that closely resemble the real machine are generated, and prototypes can be readily fabricated. Many products that work with the solid modeling software are available to analyze the structural integrity of machine components. Similarly, studying the motion and forces of moving mechanisms and assemblies is becoming almost an automatic side effect of solid modeling. Figure 2.1 illustrates a solid model design being analyzed with Dynamic Designer within the Autodesk Inventor® Environment.

Regardless of software, the general strategy for performing the dynamic analysis can be summarized as follows:

1. Define a set of rigid bodies (sizes, weights, and inertial properties). These could be constructed in the solid modeling design package.
2. Place constraints on the rigid bodies (connecting the rigid bodies with joints).
3. Specify the input motion (define the properties of the driving motor, cylinder, etc.) or input forces.
4. Run the analysis.
5. Review the motion of the links and forces through the mechanism.

Of course, the specific commands will vary among the different packages. The following sections of this chapter will focus on the details of mechanism analysis using Working Model 2D®. As with any software, knowledge is gained by experimenting and performing other analyses beyond the tutorials. Thus, the student is encouraged to explore the software by "inventing" assorted virtual machines.

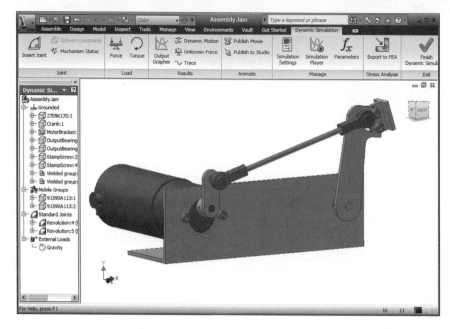

FIGURE 2.1 Dynamic analysis of a solid model.

2.3 OBTAINING WORKING MODEL SOFTWARE

Working Model 2D is created and distributed by Design Simulation Technologies. Copies of the software can be purchased, at substantial educational discounts, online at http://www.workingmodel.com or http://www.design-simulation.com. A free demonstration version of Working Model 2D is also available for download. This demo version enables students to create fully functioning "virtual prototypes" of complex mechanical designs. However, some features are disabled, most notably the Save and Print functions. Regardless, this version can provide an excellent introduction to building computer models of mechanisms. Design Simulation Technologies, Inc. can be contacted at 43311 Joy Road, #237, Canton, MI 48187, (714) 446–6935.

As Working Model 2D is updated, the menus and icons may appear slightly different from the tutorials in this text. However, using some intuition, the student will be able to adapt and successfully complete mechanism simulations.

2.4 USING WORKING MODEL TO MODEL A FOUR-BAR MECHANISM

As mentioned, Working Model is a popular, commercially available motion simulation package. It rapidly creates a model on a desktop computer that represents a mechanical system and performs dynamic analysis. This section uses Working Model to build a model of a four-bar linkage and run a simulation [Ref. 16]. It is intended to be a tutorial; that is, it should be followed while actually using Working Model. The student is then encouraged to experiment with the software to perform other analyses.

Step 1: Open Working Model

1. Click on the Working Model program icon to start the program.

2. Create a new Working Model document by selecting "New" from the "File" menu.
   Working Model displays the user interface. Toolbars used to create links, joints, and mechanism actuators appear along the sides of the screen. Tape controls, which are used to run and view simulations, appear at the bottom of the screen.

3. Specify the units to be used in the simulation. Select "Numbers and Units" in the "View" menu. Change the "Unit System" to English (pounds).
   The units for linear measurements will be inches, angles will be measured in degrees, and forces will be specified in pounds.

Step 2: Create the Links

This step creates the three moving links in a four-bar mechanism. The background serves as the fixed, fourth link.

1. Construct the linkage by creating the three nonfixed links. Double-click on the rectangle tool on the toolbar.
   The tool is highlighted, indicating that it can be used multiple times.

2. Using the rectangle tool, sketch out three bodies as shown in Figure 2.2.
   Rectangles are drawn by positioning the mouse at the first corner, clicking once, then moving the mouse to the location of the opposite corner and clicking again. Rectangles are parametrically defined and their precise sizes will be specified later.

3. Open the "Properties" box and "Geometry" box in the "Window" menu.

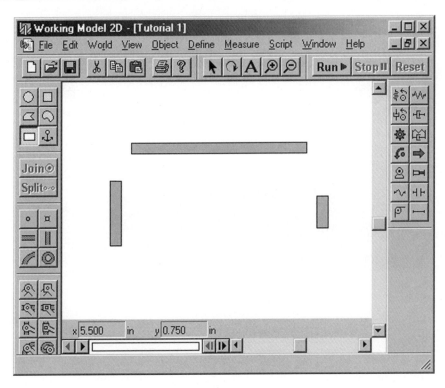

FIGURE 2.2 Three links sketched using the rectangle tool.

This displays information about the links and allows editing of this information.

4. Use the "Properties" box to change the center of the horizontal link to $x = 0$, $y = 0$, $\phi = 0$.
   The location of the rectangle should change upon data entry.

5. Use the "Geometry" box to change the width of the horizontal rectangle to 8.5 and the height to 0.5 in.
   The shape of the rectangle will change.

6. Likewise, use the "Properties" box and "Geometry" box to change the long vertical link to be centered at $x = -5$, $y = -3$ and have a width of 0.5 and a height of 3. Also change the short vertical link to be centered at $x = 5$, $y = -3$ and have a width of 0.5 and a height of 1.5.
   Again, the shape and location of the rectangle should change upon data entry.

7. Close both the "Properties" box and "Geometry" box windows.

8. The zoom icon (magnifying glass) can be used to properly view the links.

Step 3: Place Points of Interest on the Links

This step teaches the usage of the "Object Snap" tool to place points precisely. The "Object Snap" tool highlights commonly used positions, like the center of a side, with an "X." When a point is placed using "Object Snap," the point's position is automatically defined with parametric equations. These equations ensure that the point maintains its relative location even after resizing or other adjustments.

1. Double-click on the point tool. The icon is a small circle.
   The point tool is highlighted, indicating that it can be used multiple times without needing to be reselected before each new point is sketched.

2. Move the cursor over one of the links.
   Notice that an "X" appears around the pointer when it is centered on a side, over a corner, or over the center of a rectangle. This feature is called "Object Snap" and highlights the commonly used parts of a link.

3. Place the cursor over the upper portion of one of the vertical links. When an "X" appears around the pointer (Figure 2.3), click the mouse button.

4. Place additional points as shown in Figure 2.3.
   Make sure that each of these points is placed at a "snap point" as evidenced by the "X" appearing at the pointer.

5. Select the pointer tool. The icon is an arrow pointed up and to the left.

6. Double-click on one of the points that were sketched in steps 3 or 4 to open the "Properties" window.

7. Notice that the points "snapped" to a distance of half the body width from the three edges. This will result in effective link length of 8.0, 2.5, and 1.0 in.

Step 4: Connect the Points to Form Pin Joints

This step joins the points to create pin joints. A pin joint acts as a hinge between two bodies. The SmartEditor prevents joints from breaking during a drag operation.

1. Select the anchor tool.

2. Click on the horizontal link to anchor the link down.

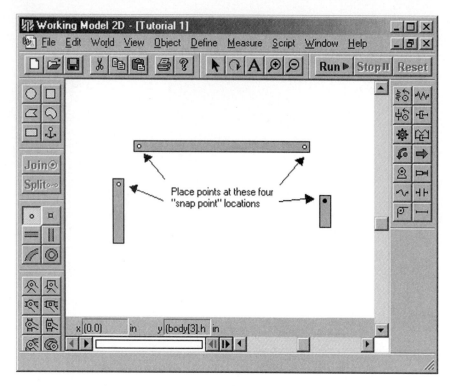

FIGURE 2.3 Point locations.

An anchor is used to tell the SmartEditor not to move this body during construction. After the pin joints have been created, the anchor will be deleted.

3. Select the pointer tool.

4. With the pointer tool selected, click and drag on the background to make a selection box that surrounds the two left points as shown in Figure 2.4. Release the

mouse button, and the two points should now be highlighted (darkened).

This method of selecting objects is called "box select." Any object that is contained completely within the box when the mouse is released is highlighted.

5. Click on the "Join" button in the toolbar, merging the two points into a pin joint.

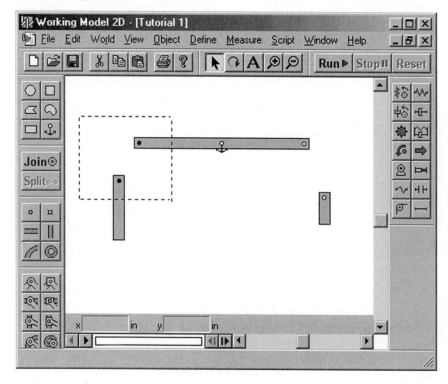

FIGURE 2.4 Select two points to join as a pin joint.

The SmartEditor creates a pin joint between the two points selected, moving the unanchored link into place. The moved link may no longer be vertical. This is fixed in a moment.

6. Perform steps 4 and 5 for the two right points to create another pin joint.
   Once again, the horizontal link remains in this original position, and the SmartEditor moves the vertical link to create the pin joint.

7. Select the left vertical link by clicking on it with the point tool.
   Four black boxes appear around the link, indicating that it has been selected. The boxes are called handles and can be dragged to resize an object.

8. Using the coordinates bar at the bottom of the screen, enter a "0" in the ϕ (rotation) field.
   The coordinates fields at the bottom of the screen are useful to obtain information on Working Model objects. These fields can also be used to edit object information. Changing the rotation to 0° adjusts the bar back to its original, vertical position.

9. If needed, complete steps 7 and 8 on the right vertical link.

10. Select the anchor used to keep the horizontal link in position during building, and press the delete key to remove it.
    The anchor is no longer needed and should be removed.

11. Select the "Pin Joint" tool and place a pin joint, using the snap point, at the lower end of the left, vertical link as indicated in Figure 2.5. The "Pin Joint" tool appears as two links joined by a circle.
    The "Pin Joint" tool is similar to the point tool used to create the last two pin joints. The pin tool automatically creates two points, attaches them to the bodies beneath the cursor (or the body and the background, as in this

case), and creates a joint in one seamless step. This pin joint joins the rectangle to the background.

12. Double-click on the pin joint to open the "Properties" window. Verify that the pin was placed half the body width from the lower edge. This gives an effective link length of 2.5 in.

Step 5: Add a Motor to the Linkage

This step adds the motor to one of the links, actuating the linkage.

1. Click on the motor tool in the toolbox. This tool appears as a circle, sitting on a base and with a point in its center.
   The motor tool becomes shaded, indicating that it has been selected. The cursor should now look like a small motor.

2. Place the cursor over the "snap point" on the lower end of the right, vertical link. Click the mouse.
   A motor appears on the four-bar linkage, as shown in Figure 2.5. Similar to a pin joint, a motor has two attachment points. A motor automatically connects the top two bodies. If only one body were to lay beneath the motor, it would join the body to the background. A motor would then apply a torque between the two bodies to which it is pinned.

3. Double-click on the motor to open the "Properties" box. Verify that the pin was placed half the body width from the lower edge. This gives an effective link length of 1.0 in.

4. Specify the motor velocity to be 360 deg/s. This equates to 60 rpm.

5. Click on "Run" in the toolbar.
   The four-bar linkage begins slowly cranking through its range of motion.

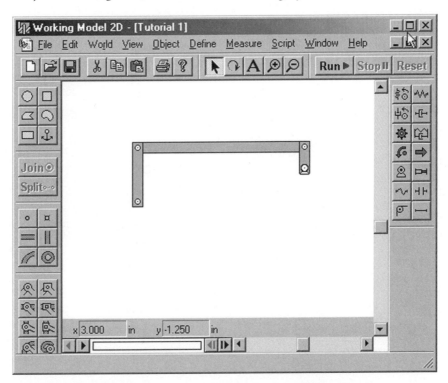

FIGURE 2.5 Adding the final pin joint and motor to the linkage.

6. Click on "Reset" in the toolbar.
   The simulation will reset to frame 0.

7. Double-click on the motor to open the "Properties" box.
   This can also be accomplished by selecting the motor and choosing "Properties" from the "Window" menu to open the "Properties" box.

8. Increase the velocity of the motor to 600 deg/s by typing this value in the "Properties" box.
   Users can define a motor to apply a certain torque, to move to a given rotational position, or to turn at a given velocity or acceleration. Rotation, velocity, and acceleration motors have built-in control systems that automatically calculate the torque needed.

9. Click on "Run" in the toolbar.
   The four-bar linkage once again begins to crank, this time at a much higher velocity.

Step 6: Resize the Links

This step uses the Coordinates Bar on the bottom of the screen to adjust the size and angle of the links. This section highlights Working Model's parametric features. Notice that when a link is resized, all points stay in their proper positions and all joints stay intact. Because they were located with the "Object Snap," these points are positioned with equations and automatically adjust during design changes.

1. If not already selected, click on the pointer tool.

2. Click once on the vertical left-hand link to select it.

3. Enter a slightly larger number in the "h" (height) box of the selected link in the Coordinates Bar at the bottom of the screen.

The link resizes on the screen. Notice how the SmartEditor automatically resizes, repositions, and rebuilds the model based on the parametric equations entered for each joint location.

4. Similarly, resize the other links and move the position of the joints. Watch the SmartEditor rebuild the model.
   Different configurations of a model can be investigated using Working Model's parametric features.

Step 7: Measure a Point's Position

1. Click on "Reset" in the toolbar.
   The simulation stops and resets to frame 0.

2. Select the point tool from the toolbar. It appears as a small, hollow circle.

3. Place the cursor over the horizontal link of the four-bar linkage and press the mouse button.
   A point is attached to the bar. This is a single point and does not attach the bar to the background. It is simply a "point of interest."

4. When a point is not already selected (darkened), select it by clicking on it.

5. Create a meter to measure the position of this point by choosing "Position" from the "Measure" menu.
   A new meter appears. Position meters default to display digital (numeric) information. A digital meter can be changed to a graph by clicking once on the arrow in the upper left-hand corner.

6. Click on "Run" in the toolbar.
   The simulation immediately begins running and measurement information appears in the meter, as shown in Figure 2.6. Meter data can be "exported" as an ASCII

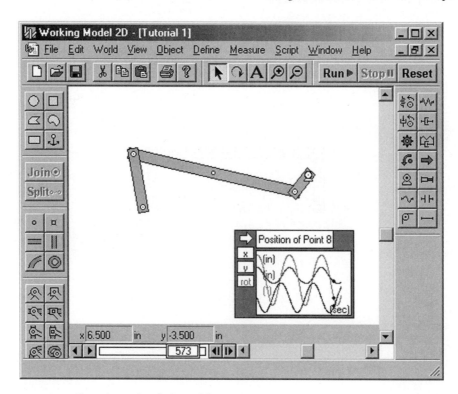

FIGURE 2.6 Running a simulation with a meter.

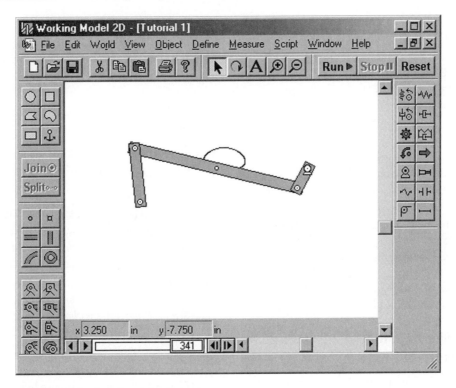

FIGURE 2.7 Tracing the path of a point.

file, copied onto the clipboard, and pasted into a spreadsheet program for further analysis. In this case, the spreadsheet would receive four columns of information: Time, X, Y, and Rotation. One row would appear for each integration time step calculated.

7. Modify the simulation and rerun it.
   Working Model's seamless integration between the editing and running of the dynamics engine allows the user to quickly investigate many different simulation configurations. As an example, modify the mass of the horizontal bar using the "Properties" box, and rerun the simulation. The pin locations can be modified and links resized; then the velocities and forces can be measured. This four-bar linkage can even be investigated in zero gravity by turning off gravity under the "World" menu.

Step 8: Trace the Path of a Point of Interest

This step creates a trace of the movement of a selected point.

1. Select all objects using the box select method described earlier.
   All elements appear black.
2. Select the "Appearance" option in the "Window" menu.
3. In the "Appearance" window, turn off "Track Center of Mass," "Track Connect," and "Track Outline."
   These features can be turned off by clicking over the appropriate check mark.
4. Click on the background to deselect all objects.
5. Select only the point of interest created in step 7.
   Only this point should appear black.

6. Select the "Appearance" option in the "Window" menu.
7. In the "Appearance" window, turn on "Track Connect." Make sure only the one point is selected.
   This feature can be turned on by clicking over the appropriate check mark.

Run the simulation. The screen should look like Figure 2.7.

Step 9: Apply What has been Learned

This demonstration illustrates how to create and run simple simulations in Working Model. The student is encouraged to experiment with this simulation or to create an original mechanism. Working Model has an incredible array of features that allows the creation of models to analyze the most complex mechanical devices.

2.5 USING WORKING MODEL TO MODEL A SLIDER-CRANK MECHANISM

This section serves as a tutorial to create a slider-crank mechanism. It should be followed while actually using Working Model. Again, the student is encouraged to experiment with the software to perform other analyses.

Step 1: Open Working Model as in Step 1 of the Previous Section

Step 2: Create the Links

This step creates the three moving links in the slider-crank mechanism. Again, the background serves as the fixed, fourth link.

1. Create a new Working Model document by selecting "New" from the "File" menu.

2. Specify the units to be used in the simulation. Select "Numbers and Units" in the "View" menu. Change the "Unit System" to SI (degrees).
   The units for linear measurement will be meters, angles will be measured in degrees, and forces will be measured in Newtons.

3. Construct the linkage by creating the three nonfixed links. Double-click on the rectangle tool in the toolbar.
   The tool is highlighted, indicating that it can be used multiple times.

4. Using the rectangle tool, sketch out three bodies as shown in Figure 2.8.
   Position the mouse at the first corner, click once, then move the mouse to the location of the opposite corner and click again. Rectangles are parametrically defined and their precise sizes are specified later.

Step 3: Use the Slot Joint to Join the Sliding Link to the Background

1. Select the "keyed slot" joint icon. The icon appears as a rectangle riding over a horizontal slot.

2. Move the cursor over the snap point at the center of the rectangular sliding link. Click the mouse button. The screen should look like Figure 2.9.

3. Select the pointer tool.

4. Double-click on the slot.
   This opens the "Properties" window for the slot.

5. Change the angle to –45°.
   The incline of the slot changes.

Drag the other links until the screen appears similar to Figure 2.10.

Step 4: Connect the Other Links to Form Pin Joints

This step creates points and joins them to create pin joints. A pin joint acts as a hinge between two bodies.

1. Select the anchor tool.

2. Click on the vertical link to anchor the link down.
   An anchor tells the SmartEditor not to move this body during construction. After the pin joints have been created, the anchor will be deleted.

3. Double-click on the point tool. The icon is a small circle.
   The point tool is highlighted, indicating that it can be used multiple times without needing to be reselected before each new point is sketched.

4. Place the cursor over the upper portion of one of the vertical links. When an "X" appears around the pointer (Figure 2.11), click the mouse button.

5. Place additional points at the ends of the horizontal link, as shown in Figure 2.11.
   Make sure that each of these points is placed at a "snap point" as evidenced by the "X" appearing at the pointer.

6. Place another point at the center of the sliding rectangle.
   This point is used to create a pin joint to the coupler.

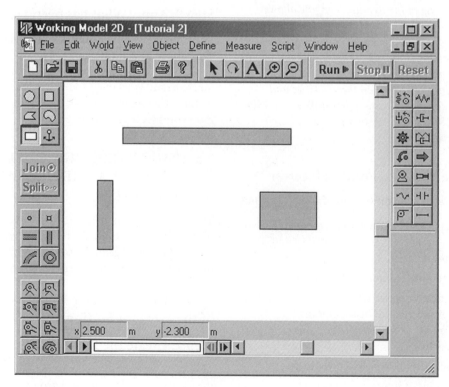

FIGURE 2.8 Three links sketched using the rectangle tool.

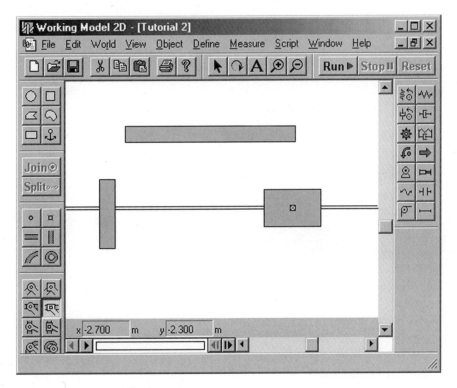

FIGURE 2.9 Point and slot location.

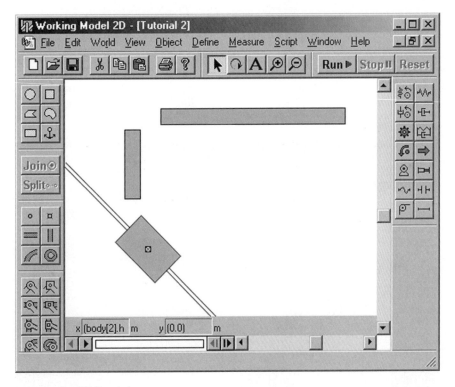

FIGURE 2.10 Sliding joint.

7. Select the pointer tool.

8. With the pointer tool selected, click on one point that will be connected with a pin joint. Then, holding down the shift key, click on the second point that will form a pin joint. Notice that the two points should now be highlighted (darkened).

9. Click on the "Join" button in the toolbar, merging the two points into a pin joint.
   The SmartEditor creates a pin joint between the two selected points, moving the unanchored link into place. The moved link may no longer be vertical. This will be fixed in a moment.

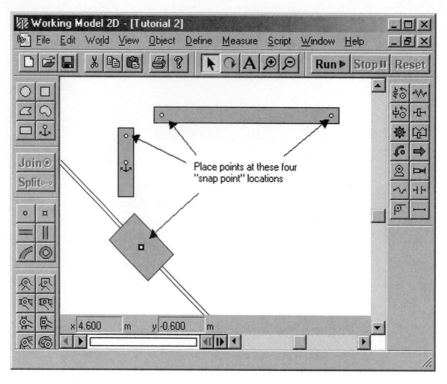

FIGURE 2.11 Placing points on the other links.

10. Perform steps 8 and 9 for the other two points that will create another pin joint. The screen will appear similar to Figure 2.12.

    Once again, the vertical link remains in this original position, and the SmartEditor moves the vertical link to create the pin joint.

11. Click on the vertical link.
    Four black boxes appear around the link, indicating that it has been selected.

12. Select the "Move to front" option in the "Object" menu.
    This places the vertical link in front of the connecting link, making the anchor visible.

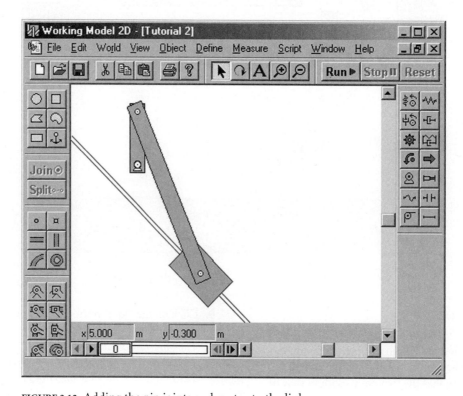

FIGURE 2.12 Adding the pin joints and motor to the linkage.

13. Select the anchor, which is used to keep the vertical link in position during building, and press the delete key to remove it.
    The anchor is no longer needed and should be removed.

Step 5: Add a Motor to the Linkage

This step adds the motor to one of the links to drive the linkage.

1. Click on the motor tool in the toolbox. This tool appears as a circle, sitting on a base with a point in its center.
   The motor tool becomes shaded, indicating that it has been selected. The cursor should now look like a small motor.

2. Place the cursor over the "snap point" on the vertical link. Click the mouse.
   A motor appears on the slider-crank linkage, as shown in Figure 2.12. Similar to a pin joint, a motor has two attachment points. A motor automatically connects the top two bodies. If only one body were to lay beneath the motor, the motor would join the body to the background. The motor then applies a torque between the two bodies to which it is pinned.

3. Click on "Run" in the toolbar.
   The slider-crank linkage begins slowly cranking through its range of motion.

4. Click on "Reset" in the toolbar.
   The simulation resets to frame 0.

5. Double-click on the motor to open the "Properties" box.
   This can also be accomplished by selecting the motor and choosing "Properties" from the "Window" menu to open the "Properties" box.

6. Increase the velocity of the motor to −300 deg/s by typing this value in the "Properties" box.
   Users can define a motor to apply a certain torque, to move to a given rotational position, or to turn at a given velocity or acceleration. Rotation, velocity, and acceleration motors have built-in control systems that automatically calculate the torque needed. In this demo, we use the velocity motor.

7. Click on "Run" in the toolbar.
   The slider-crank linkage once again begins cranking, this time at a much higher velocity.

Step 6: Apply What Has Been Learned

The student is encouraged to experiment with this simulation or to create an original mechanism. Working Model has an incredible array of features that allows for the creation of a model to analyze most complex mechanical devices.

PROBLEMS

Use Working Model software to generate a model of a four-bar mechanism. Use the following values:

2–1. frame = 9 in.;   crank = 1 in.;   coupler = 10 in.; follower = 3.5 in.; crank speed = 200 rad/s

2–2. frame = 100 mm;   crank = 12 mm;   coupler = 95 mm ; follower = 24 mm; crank speed = 30 rad/s

2–3. frame = 2 ft;   crank = 0.5 ft;   coupler = 2.1 ft; follower = 0.75 ft; crank speed = 25 rpm

Use the Working Model software to generate a model of a slider-crank mechanism. Use the following values:

2–4. offset = 0 in.; crank = 1.45 in.; coupler = 4.5 in.; crank speed = 200 rad/s

2–5. offset = 0 mm; crank = 95 mm; coupler = 350 mm; crank speed = 200 rad/s

2–6. offset = 50 mm;   crank = 95 mm;   coupler = 350 mm; crank speed = 200 rad/s

2–7. Figure P2.7 shows a mechanism that operates the landing gear in a small airplane. Use the Working Model software to generate a model of this linkage. The motor operates clockwise at a constant rate of 20 rpm.

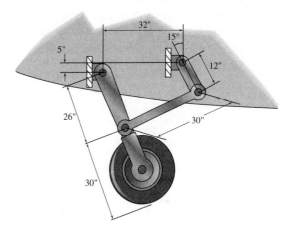

FIGURE P2.7 Problem 7.

2–8. Figure P2.8 shows a mechanism that operates a coin-operated child's amusement ride. Use the Working Model software to generate a model of this

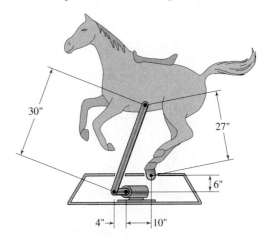

FIGURE P2.8 Problem 8.

linkage. The motor operates counterclockwise at a constant rate of 60 rpm.

2–9. Figure P2.9 shows a transfer mechanism that lifts crates from one conveyor to another. Use the Working Model software to generate a model of this linkage. The motor operates counterclockwise at a constant rate of 20 rpm.

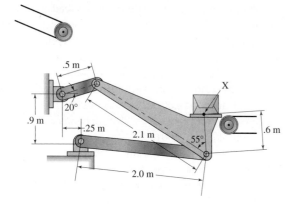

FIGURE P2.9 Problem 9.

2–10. Figure P2.10 shows another transfer mechanism that pushes crates from one conveyor to another. Use the Working Model software to generate a model of this linkage. The motor operates clockwise at a constant rate of 40 rpm.

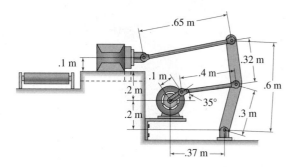

FIGURE P2.10 Problem 10.

2–11. Figure P2.11 shows yet another transfer mechanism that lowers crates from one conveyor to another. Use the Working Model software to generate a model of

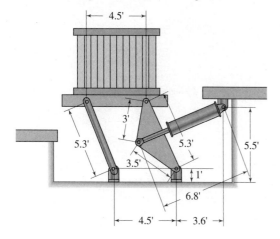

FIGURE P2.11 Problem 11.

this linkage. The cylinder extends at a constant rate of 1 fpm.

2–12. Figure P2.12 shows a mechanism that applies labels to packages. Use the Working Model software to generate a model of this linkage. The motor operates counterclockwise at a constant rate of 300 rpm.

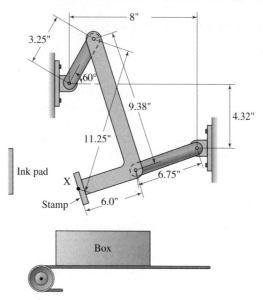

FIGURE P2.12 Problem 12.

CASE STUDY

2–1. The mechanism shown in Figure C2.1 is a top view of a fixture in a machining operation. Carefully examine the configuration of the components in the mechanism. Then answer the following leading questions to gain insight into the operation of the mechanism.

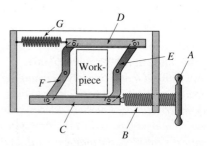

FIGURE C2.1 Mechanism for Case Study 2.1.

1. As the handle *A* is turned, moving the threaded rod *B* to the left, describe the motion of grip *C*.
2. As the handle *A* is turned, moving the threaded rod *B* to the left, describe the motion of grip *D*.
3. What is the purpose of this mechanism?
4. What action would cause link *D* to move upward?
5. What is the purpose of spring *G*?
6. Discuss the reason for the odd shape to links *E* and *F*.
7. What would you call such a device?
8. Describe the rationale behind using a rounded end for the threaded rod *B*.

VECTORS

3.1 INTRODUCTION

Mechanism analysis involves manipulating vector quantities. Displacement, velocity, acceleration, and force are the primary performance characteristics of a mechanism, and are all vectors. Prior to working with mechanisms, a thorough introduction to vectors and vector manipulation is in order. In this chapter, both graphical and analytical solution techniques are presented. Students who have completed a mechanics course may omit this chapter or use it as a reference to review vector manipulation.

3.2 SCALARS AND VECTORS

In the analysis of mechanisms, two types of quantities need to be distinguished. A *scalar* is a quantity that is sufficiently defined by simply stating a magnitude. For example, by saying "a dozen donuts," one describes the quantity of donuts in a box. Because the number "12" fully defines the

amount of donuts in the box, the amount is a scalar quantity. The following are some more examples of scalar quantities: a board is 8 ft long, a class meets for 50 min, or the temperature is 78°F—length, time, and temperature are all scalar quantities.

In contrast, a *vector* is not fully defined by stating only a magnitude. Indicating the direction of the quantity is also required. Stating that a golf ball traveled 200 yards does not fully describe its path. Neglecting to express the direction of travel hides the fact that the ball has landed in a lake. Thus, the direction must be included to fully describe such a quantity. Examples of properly stated vectors include "the crate is being pulled to the right with 5 lb" or "the train is traveling at a speed of 50 mph in a northerly direction." Displacement, force, and velocity are vector quantities.

Vectors are distinguished from scalar quantities through the use of boldface type (**v**). The common notation used to graphically represent a vector is a line segment having an arrowhead placed at one end. With a graphical approach to analysis, the length of the line segment is drawn proportional to the magnitude of the quantity that the vector describes. The direction is defined by the arrowhead and the incline of the line with respect to some reference axis. The direction is always measured at its root, not at its head. Figure 3.1 shows a fully defined velocity vector.

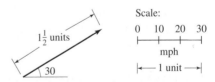

FIGURE 3.1 A 45 mph velocity vector.

3.3 GRAPHICAL VECTOR ANALYSIS

Much of the work involved in the study of mechanisms and analysis of vectors involves geometry. Often, graphical methods are employed in such analyses because the motion of a mechanism can be clearly visualized. For more complex mechanisms, analytical calculations involving vectors also become laborious.

A graphical approach to analysis involves drawing scaled lines at specific angles. To achieve results that are consistent with analytical techniques, accuracy must be a major objective. For several decades, accuracy in mechanism analyses was obtained with attention to precision and proper drafting equipment. Even though they were popular, many scorned graphical techniques as being imprecise. However, the development of computer-aided design (CAD) with its accurate geometric constructions has allowed graphical techniques to be applied with precision.

3.4 DRAFTING TECHNIQUES REQUIRED IN GRAPHICAL VECTOR ANALYSIS

The methods of graphical mechanism and vector analysis are identical, whether using drafting equipment or a CAD package. Although it may be an outdated mode in industrial analyses, drafting can be successfully employed to learn and understand the techniques.

For those using drafting equipment, fine lines and circular arcs are required to produce accurate results. Precise linework is needed to accurately determine intersection points. Thus, care must be taken in maintaining sharp drawing equipment.

Accurate measurement is as important as line quality. The length of the lines must be drawn to a precise scale, and linear measurements should be made as accurately as possible. Therefore, using a proper engineering scale with inches divided into 50 parts is desired. Angular measurements must be equally precise.

Lastly, a wise choice of a drawing scale is also a very important factor. Typically, the larger the construction, the more accurate the measured results are. Drawing precision to 0.05 in. produces less error when the line is 10 in. long as opposed to 1 in. Limits in size do exist in that very large constructions require special equipment. However, an attempt should be made to create constructions as large as possible.

A drawing textbook should be consulted for the details of general drafting techniques and geometric constructions.

3.5 CAD KNOWLEDGE REQUIRED IN GRAPHICAL VECTOR ANALYSIS

As stated, the methods of graphical mechanism and vector analysis are identical, whether using drafting equipment or a CAD package. CAD allows for greater precision. Fortunately, only a limited level of proficiency on a CAD system is required to properly complete graphical vector analysis. Therefore, utilization of a CAD system is preferred and should not require a large investment on a "learning curve."

As mentioned, the graphical approach of vector analysis involves drawing lines at precise lengths and specific angles. The following list outlines the CAD abilities required for vector analysis. The user should be able to

- Draw lines at a specified length and angle;
- Insert lines, perpendicular to existing lines;
- Extend existing lines to the intersection of another line;
- Trim lines at the intersection of another line;
- Draw arcs, centered at a specified point, with a specified radius;
- Locate the intersection of two arcs;
- Measure the length of existing lines;
- Measure the included angle between two lines.

Of course, proficiency beyond these items facilitates more efficient analysis. However, familiarity with CAD commands that accomplish these actions is sufficient to accurately complete vector analysis.

3.6 TRIGONOMETRY REQUIRED IN ANALYTICAL VECTOR ANALYSIS

In the analytical analysis of vectors, knowledge of basic trigonometry concepts is required. Trigonometry is the study of the properties of triangles. The first type of triangle examined is the right triangle.

3.6.1 Right Triangle

In performing vector analysis, the use of the basic trigonometric functions is vitally important. The basic trigonometric functions apply only to right triangles. Figure 3.2 illustrates a right triangle with sides denoted as a, b, and c and interior angles as A, B, and C. Note that angle C is a 90° right angle. Therefore, the triangle is called a right triangle.

The basic trigonometric relationships are defined as:

$$\text{sine} \angle A = \sin \angle A = \frac{\text{side opposite}}{\text{hypotenuse}} = \frac{a}{c} \quad (3.1)$$

$$\text{cosine} \angle A = \cos \angle A = \frac{\text{side adjacent}}{\text{hypotenuse}} = \frac{b}{c} \quad (3.2)$$

$$\text{tangent} \angle A = \tan \angle A = \frac{\text{side opposite}}{\text{side adjacent}} = \frac{a}{b} \quad (3.3)$$

These definitions can also be applied to angle B:

$$\sin \angle B = \frac{b}{c}$$

$$\cos \angle B = \frac{a}{c}$$

$$\tan \angle B = \frac{b}{a}$$

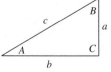

FIGURE 3.2 The right triangle.

The Pythagorean theorem gives the relationship of the three sides of a right triangle. For the triangle shown in Figure 3.2, it is defined as

$$a^2 + b^2 = c^2 \qquad (3.4)$$

Finally, the sum of all angles in a triangle is 180°. Knowing that angle C is 90°, the sum of the other two angles must be

$$\angle A + \angle B = 90° \qquad (3.5)$$

EXAMPLE PROBLEM 3.1

Figure 3.3 shows a front loader with cylinder BC in a vertical position. Use trigonometry to determine the required length of the cylinder to orient arm AB in the configuration shown.

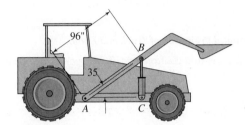

FIGURE 3.3 Front loader for Example Problem 3.1.

SOLUTION: 1. ***Determine Length* BC**

Focus on the triangle formed by points A, B, and C in Figure 3.3. The triangle side BC can be found using equation (3.1).

$$\sin\angle A = \frac{\text{opposite side}}{\text{hypotenuse}}$$

$$\sin 35° = \frac{BC}{(96 \text{ in.})}$$

solving:

$$BC = (96 \text{ in.}) \sin 35° = 55.06 \text{ in.}$$

2. ***Determine Length* AC**

Although not required, notice that the distance between A and C can similarly be determined using equation (3.2). Thus

$$\cos\angle A = \frac{\text{adjacent side}}{\text{hypotenuse}}$$

$$\cos 35° = \frac{AC}{(96 \text{ in.})}$$

solving:

$$AC = (96 \text{ in.}) \cos 35° = 78.64 \text{ in.}$$

EXAMPLE PROBLEM 3.2

Figure 3.4 shows a tow truck with an 8-ft boom, which is inclined at a 25° angle. Use trigonometry to determine the horizontal distance that the boom extends from the truck.

SOLUTION: 1. ***Determine the Horizontal Projection of the Boom***

The horizontal projection of the boom can be determined from equation (3.2):

$$\cos 25° = \frac{\text{horizontal projection}}{(8 \text{ ft})}$$

$$\text{horizontal projection} = (8 \text{ ft})\cos 25° = 7.25 \text{ ft}$$

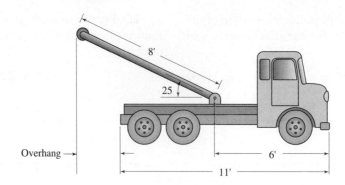

FIGURE 3.4 Tow truck for Example Problem 3.2.

2. ***Determine Horizontal Projection of Truck and Boom***

    The horizontal distance from the front end of the truck to the end of the boom is

$$6 \text{ ft} + 7.25 \text{ ft} = 13.25 \text{ ft}$$

3. ***Determine the Overhang***

    Because the overall length of the truck is 11 ft, the horizontal distance that the boom extends from the truck is

$$13.25 \text{ ft} - 11 \text{ ft} = 2.25 \text{ ft}$$

3.6.2 Oblique Triangle

In the previous discussion, the analysis was restricted to right triangles. An approach to general, or oblique, triangles is also important in the study of mechanisms. Figure 3.5 shows a general triangle. Again, *a*, *b*, and *c* denote the length of the sides and $\angle A$, $\angle B$, and $\angle C$ represent the interior angles.

For this general case, the basic trigonometric functions described in the previous section are not applicable. To analyze the general triangle, the law of sines and the law of cosines have been developed.

The *law of sines* can be stated as

$$\frac{a}{\sin \angle A} = \frac{b}{\sin \angle B} = \frac{c}{\sin \angle C} \qquad (3.6)$$

The *law of cosines* can be stated as:

$$c^2 = a^2 + b^2 - 2ab \ \cos \angle C \qquad (3.7)$$

In addition, the sum of all interior angles in a general triangle must total 180°. Stated in terms of Figure 3.4 the equation would be

$$\angle A + \angle B + \angle C = 180° \qquad (3.8)$$

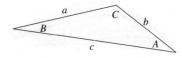

FIGURE 3.5 The oblique triangle.

Problems involving the solution of a general triangle fall into one of four cases:

Case 1: Given one side (*a*) and two angles ($\angle A$ and $\angle B$).

To solve a problem of this nature, equation (3.8) can be used to find the third angle:

$$\angle C = 180° - \angle A - \angle B$$

The law of sines can be rewritten to find the remaining sides.

$$b = a\left\{ \frac{\sin \angle B}{\sin \angle A} \right\}$$

$$c = a\left\{ \frac{\sin \angle C}{\sin \angle A} \right\}$$

Case 2: Given two sides (*a* and *b*) and the angle opposite to one of the sides ($\angle A$).

To solve a Case 2 problem, the law of sines can be used to find the second angle. Equation (3.6) is rewritten as

$$\angle B = \sin^{-1}\left\{ \left(\frac{b}{a} \right) \sin \angle A \right\}$$

Equation (3.8) can be used to find the third angle:

$$\angle C = 180° - \angle A - \angle B$$

The law of cosines can be used to find the third side. Equation (3.7) is rewritten as:

$$c = \sqrt{\{a^2 + b^2 - 2ab \ \cos \angle C\}}$$

Case 3 Given two sides (*a* and *b*) and the included angle (∠C).

To solve a Case 3 problem, the law of cosines can be used to find the third side:

$$c = \sqrt{a^2 + b^2 - 2ab\cos\angle C}$$

The law of sines can be used to find a second angle. Equation (3.6) is rewritten as

$$\angle A = \sin^{-1}\left\{\left(\frac{a}{c}\right)\sin\angle C\right\}$$

Equation (3.8) can be used to find the third angle:

$$\angle B = 180° - \angle A - \angle C$$

Case 4 Given three sides.

To solve a Case 4 problem, the law of cosines can be used to find an angle. Equation (3.7) is rewritten as

$$\angle C = \cos^{-1}\left(\frac{a^2 + b^2 - c^2}{2ab}\right)$$

The law of sines can be used to find a second angle. Equation (3.6) is rewritten as

$$\angle A = \sin^{-1}\left\{\left(\frac{a}{c}\right)\sin\angle C\right\}$$

Equation (3.8) can be used to find the third angle:

$$\angle B = 180° - \angle A - \angle C$$

Once familiarity in solving problems involving general triangles is gained, referring to the specific cases will be unnecessary.

EXAMPLE PROBLEM 3.3

Figure 3.6 shows a front loader. Use trigonometry to determine the required length of the cylinder to orient arm *AB* in the configuration shown.

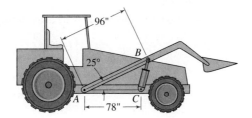

FIGURE 3.6 Front loader for Example Problem 3.3.

SOLUTION: 1.   ***Determine Length* BC**

By focusing on the triangle created by points *A*, *B*, and *C*, it is apparent that this is a Case 3 problem. The third side can be found by using the law of cosines:

$$BC = \sqrt{AC^2 + AB^2 - 2(AC)(AB)\cos\angle BAC}$$
$$= \sqrt{(78\,\text{in.})^2 + (96\,\text{in.})^2 - 2(78\,\text{in.})(96\,\text{in.})\cos 25°}$$
$$= 41.55\,\text{in.}$$

Because determining the remaining angles was not required, the procedure described for Case 3 problems will not be completed.

EXAMPLE PROBLEM 3.4

Figure 3.7 shows the drive mechanism for an engine system. Use trigonometry to determine the crank angle as shown in the figure.

SOLUTION: 1.   ***Determine Angle* BAC**

By focusing on the triangle created by points *A, B*, and *C*, it is apparent that this is a Case 4 problem. Angle *BAC* can be determined by redefining the variables in the law of cosines:

$$\angle BAC = \cos^{-1}\left\{\frac{AC^2 + AB^2 - BC^2}{2(AC)(AB)}\right\}$$
$$= \cos^{-1}\left\{\frac{(5.3\,\text{in.})^2 + (1\,\text{in.})^2 - (5\,\text{in.})^2}{2(5.3\,\text{in.})(1\,\text{in.})}\right\} = 67.3°$$

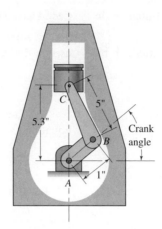

FIGURE 3.7 Engine linkage for Example Problem 3.4.

2. **Determine the Crank Angle**

   Angle *BAC* is defined between side *AC* (the vertical side) and leg *AB*. Because the crank angle is defined from a horizontal axis, the crank angle can be determined by the following:

   $$\text{Crank angle} = 90° - 67.3° = 22.7°$$

3. **Determine the Other Interior Angles**

   Although not required in this problem, angle *ACB* can be determined by

   $$\angle ACB = \sin^{-1}\left\{\left(\frac{AB}{BC}\right)\sin\angle BAC\right\}$$

   $$= \sin^{-1}\left\{\left(\frac{1\text{ in.}}{5\text{ in.}}\right)\sin 67.3°\right\} = 10.6°$$

   Finally, angle *ABC* can be found by

   $$\angle ABC = 180° - 67.3° - 10.6° = 102.1°$$

3.7 VECTOR MANIPULATION

Throughout the analysis of mechanisms, vector quantities (e.g., displacement or velocity) must be manipulated in different ways. In a similar manner to scalar quantities, vectors can be added and subtracted. However, unlike scalar quantities, these are not simply algebraic operations. Because it is also required to define a vector, direction must be accounted for during mathematical operations. Vector addition and subtraction are explored separately in the following sections.

Adding vectors is equivalent to determining the combined, or net, effect of two quantities as they act together. For example, in playing a round of golf, the first shot off the tee travels 200 yards but veers off to the right. A second shot then travels 120 yards but to the left of the hole. A third shot of 70 yards places the golfer on the green. As this golfer looks on the score sheet, she notices that the hole is labeled as 310 yards; however, her ball traveled 390 yards (200 + 120 + 70 yards).

As repeatedly stated, the direction of a vector is just as important as the magnitude. During vector addition, 1 + 1 does not always equal 2; it depends on the direction of the individual vectors.

3.8 GRAPHICAL VECTOR ADDITION (+>)

Graphical addition is an operation that determines the net effect of vectors. A graphical approach to vector addition involves drawing the vectors to scale and at the proper orientation. These vectors are then relocated, maintaining the scale and orientation. The tail of the first vector is designated as the origin (point *O*). The second vector is relocated so that its tail is placed on the tip of the first vector. The process then is repeated for all remaining vectors. This technique is known as the *tip-to-tail* method of vector addition. This name is derived from viewing a completed vector polygon. The tip of one vector runs into the tail of the next.

The combined effect is the vector that extends from the tail of the first vector in the series to the tip of the last vector in the series. Mathematically, an equation can be written that represents the combined effect of vectors:

$$\mathbf{R} = \mathbf{A} +> \mathbf{B} +> \mathbf{C} +> \mathbf{D} +> \ldots$$

Vector **R** is a common notation used to represent the resultant of a series of vectors. A *resultant* is a term used to describe the combined effect of vectors. Also note that the symbol +> is used to identify vector addition and to differentiate it from algebraic addition [Ref. 5].

It should be noted that vectors follow the commutative law of addition; that is, the order in which the vectors are added does not alter the result. Thus,

$$R = (A +> B +> C) = (C +> B +> A) =$$
$$(B +> A +> C) = \ldots$$

The process of combining vectors can be completed graphically, using either manual drawing techniques or CAD software. Whatever method is used, the underlying concepts are identical. The following example problems illustrate this concept.

EXAMPLE PROBLEM 3.5

Graphically determine the combined effect of velocity vectors **A** and **B**, as shown in Figure 3.8.

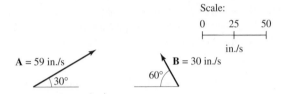

FIGURE 3.8 Vectors for Example Problem 3.5.

SOLUTION:

1. *Construct Vector Diagrams*

   To determine the resultant, the vectors must be relocated so that the tail of **B** is located at the tip of **A**. To verify the commutative law, the vectors were redrawn so that the tail of **A** is placed at the tip of **B**. The resultant is the vector drawn from the tail of the first vector, the origin, to the tip of the second vector. Both vector diagrams are shown in Figure 3.9.

2. *Measure the Resultant*

   The length vector **R** is measured as 66 in./s. The direction is also required to fully define vector **R**. The angle from the horizontal to vector **R** is measured as 57°. Therefore, the proper manner of presenting the solution is as follows:

   $$R = 66 \text{ in./s} \diagup 57°$$

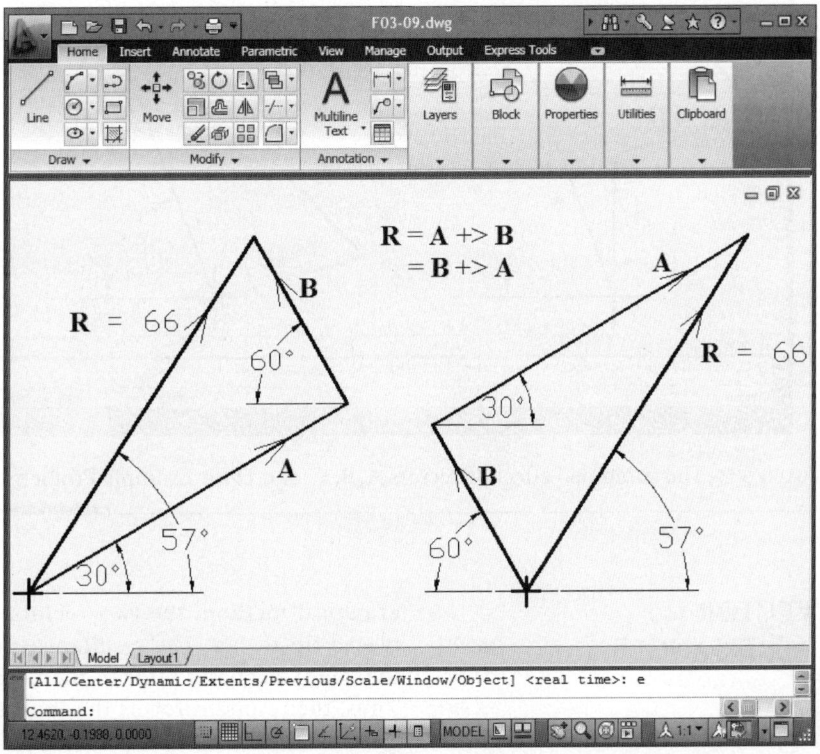

FIGURE 3.9 The combined effect of vectors **A** and **B** for Example Problem 3.5.

EXAMPLE PROBLEM 3.6

Graphically determine the combined effect of force vectors **A, B, C,** and **D,** as shown in Figure 3.10.

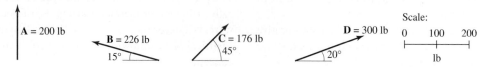

FIGURE 3.10 Vectors for Example Problem 3.6.

SOLUTION:

1. **Construct Vector Diagrams**

   To determine the resultant, the vectors must be relocated so that the tail of **B** is located at the tip of **A**. Then the tail of **C** is placed on the tip of **B**. Finally, the tail of **D** is placed on the tip of **C**. Again, the ordering of vectors is not important, and any combination could be used. As an illustration, another arbitrary combination is used in this example. The resultant is the vector drawn from the tail of the first vector, at the origin, to the tip of the fourth vector. The vector diagrams are shown in Figure 3.11.

2. **Measure the Resultant**

   The length vector **R** is measured as 521 lb. The direction is also required to fully define the vector **R**. The angle from the horizontal to vector **R** is measured as 68°. Therefore, the proper manner of presenting the solution is as follows:

$$\mathbf{R} = 521 \text{ in./s} \diagup 68°$$

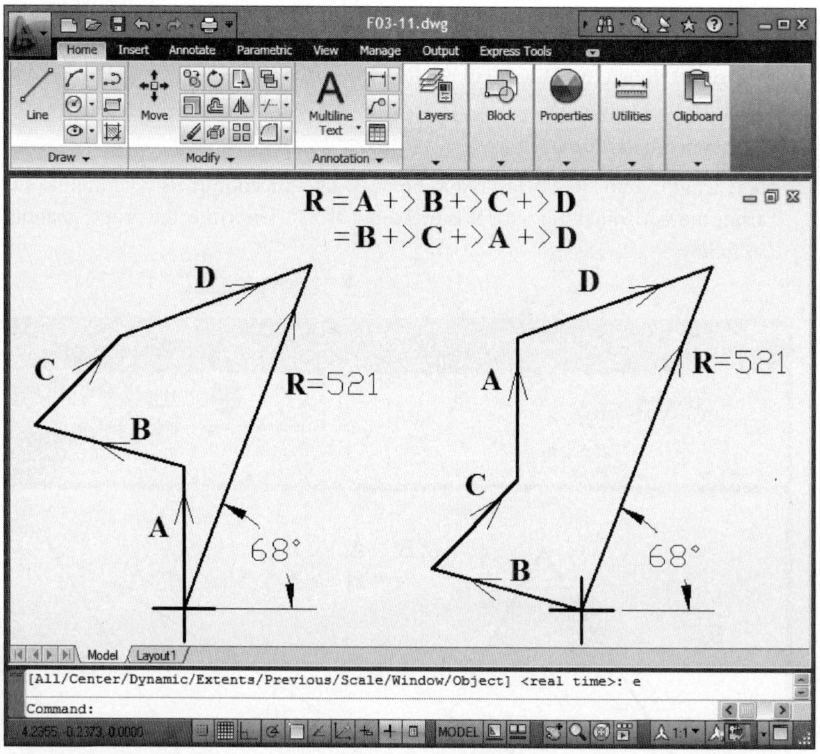

FIGURE 3.11 The combined effect of vectors **A, B, C,** and **D** for Example Problem 3.6.

3.9 ANALYTICAL VECTOR ADDITION (+>): TRIANGLE METHOD

Two analytical methods can be used to determine the net effect of vectors. The first method is best suited when the resultant of only two vectors is required. As with the graphical method, the two vectors to be combined are placed tip-to-tail. The resultant is found by connecting the tail of the first vector to the tip of the second vector. Thus, the resultant forms the third side of a triangle. In general, this is an oblique triangle, and the laws described in Section 3.6.2 can be applied. The length of the third side and a reference angle must be determined through the laws of sines and cosines to fully define the resultant vector. This method can be illustrated through an example problem. To clearly distinguish quantities, vectors are shown in boldface type (**D**) while the magnitude of the vector is shown as non-bold, italic type (*D*).

EXAMPLE PROBLEM 3.7

Analytically determine the resultant of two acceleration vectors as shown in Figure 3.12.

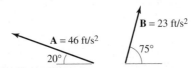

FIGURE 3.12 Vectors for Example Problem 3.7.

SOLUTION:

1. *Sketch a Rough Vector Diagram*

   The vectors are placed tip-to-tail as shown in Figure 3.13. Note that only a rough sketch is required because the resultant is analytically determined.

   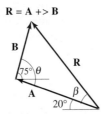

   FIGURE 3.13 Combined effect of vectors **A** and **B** for Example Problem 3.7.

2. *Determine an Internal Angle*

   The angle between **A** and the horizontal is 20°. By examining Figure 3.13, the angle between vectors **A** and **B** is:

   $$\theta = 20° + 75° = 95°$$

   Therefore, the problem of determining the resultant of two vectors is actually a general triangle situation as described in Section 3.6.2 (Case 3).

3. *Determine Resultant Magnitude*

   By following the procedure outlined for a Case 3 problem, the law of cosines is used to find the magnitude of the resultant:

   $$R = \sqrt{A^2 + B^2 - 2AB\cos\theta}$$

   $$= \sqrt{(46\,\text{ft/s}^2)^2 + (23\,\text{ft/s}^2)^2 - 2(46\,\text{ft/s}^2)(23\,\text{ft/s}^2)\{\cos 95°\}} = 53.19 \text{ ft/s}^2$$

4. *Determine Magnitude Direction*

   The law of sines can be used to find the angle between vectors **A** and **R**:

   $$\beta = \sin^{-1}\left\{\left(\frac{B}{R}\right)\sin\theta\right\}$$

   $$= \sin^{-1}\left\{\frac{(23\,\text{ft/s}^2)}{(53.19\,\text{ft/s}^2)\sin 95°}\right\} = 25.5°$$

5. *Fully Specify Resultant*

   The angle from the horizontal is 20° + 25.5° = 45.5°. The resultant can be properly written as:

   $$\mathbf{R} = 53.19 \text{ ft./s}^2 \ \underline{45.5°}\nwarrow$$

   or

   $$\mathbf{R} = 53.19 \text{ ft./s}^2 \ \diagup\underline{134.5°}$$

3.10 COMPONENTS OF A VECTOR

The second method for analytically determining the resultant of vectors is best suited for problems where more than two vectors are to be combined. This method involves resolving vectors into perpendicular components.

Resolution of a vector is the reverse of combining vectors. A single vector can be broken into two separate vectors, along convenient directions. The two vector components have the same effect as the original vector.

In most applications, it is desirable to concentrate on a set of vectors directed vertically and horizontally; therefore, a typical problem involves determining the horizontal and vertical components of a vector. This problem can be solved by using the tip-to-tail approach, but in reverse. To explain the method, a general vector, **A**, is shown in Figure 3.14.

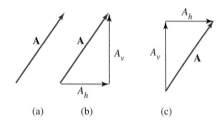

(a)   (b)   (c)

FIGURE 3.14 Components of a vector.

Two vectors can be drawn tip-to-tail along the horizontal and vertical that have the net effect of the original. The tail of the horizontal vector is placed at the tail of the original, and the tip of the vertical vector is placed at the tip of the original vector. This vector resolution into a horizontal component, A_h, and the vertical component, A_v, is shown in Figure 3.14b. Recall that the order of vector addition is not

important. Therefore, it is irrelevant whether the horizontal or vertical vector is drawn first. Figure 3.14c illustrates the components of a general vector in the opposite order.

Notice that the magnitude of the components can be found from determining the sides of the triangles shown in Figure 3.14. These triangles are always right triangles, and the methods described in Section 3.3 can be used. The directions of the components are taken from sketching the vectors as in Figure 3.14b or 3.14c. Standard notation consists of defining horizontal vectors directed toward the right as positive. All vertical vectors directed upward are also defined as positive. In this fashion, the direction of the components can be determined from the algebraic sign associated with the component.

An alternative method to determine the rectangular components of a vector is to identify the vector's angle with the positive x-axis of a conventional Cartesian coordinate system. This angle is designated as θ. The magnitude of the two components can be computed from the basic trigonometric relations as

$$A_h = A \cos \theta_x \qquad (3.9)$$

$$A_v = A \sin \theta_x \qquad (3.10)$$

The importance of this method lies in the fact that the directions of the components are evident from the sign that results from the trigonometric function. That is, a vector that points into the second quadrant of a conventional Cartesian coordinate system has an angle, θ_x, between 90° and 180°. The cosine of such an angle results in a negative value, and the sine results in a positive value. Equations (3.9) and (3.10) imply that the horizontal component is negative (i.e., toward the left in a conventional coordinate system) and the vertical component is positive (i.e., upward in a conventional system).

EXAMPLE PROBLEM 3.8

A force, **F**, of 3.5 kN is shown in Figure 3.15. Using the analytical triangle method, determine the horizontal and vertical components of this force.

FIGURE 3.15 Force vector for Example Problem 3.8.

SOLUTION: 1.   **Sketch the Vector Components**

The horizontal vector (component) is drawn from the tail of vector **F**. A vertical vector (component) is drawn from the horizontal vector to the tip of the original force vector. These two components are shown in Figure 3.16.

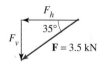

FIGURE 3.16 Force components for Example Problem 3.8.

2. *Use Triangle Method*

Working with the right triangle, an expression for both components can be written using trigonometric functions:

$$\sin 35° = \frac{\text{opposite side}}{\text{hypotenuse}} = \frac{F_v}{3.5 \text{ kN}}$$

$$\cos 35° = \frac{\text{adjacent side}}{\text{hypotenuse}} = \frac{F_h}{3.5 \text{ kN}}$$

Both these expressions can be solved in terms of the magnitude of the desired components:

$$F_h = (3.5 \text{ kN}) \cos 35° = 2.87 \text{ kN} \leftarrow$$
$$= -2.87 \text{ kN}$$
$$\mathbf{F}_v = (3.5 \text{ kN}) \sin 35° = 2.00 \text{ kN} \downarrow$$
$$= -2.00 \text{ kN}$$

3. *Use x-axis, Angle Method*

An alternative solution is obtained by using equations (3.9) and (3.10). The angle θ_x from the positive x-axis to the vector $\mathbf{F}$ is 215°. The components are computed as follows:

$$F_h = F \cos\theta_x = (3.5 \text{ kN}) \cos 215° = -2.87 \text{ kN}$$
$$= 2.87 \text{ kN} \leftarrow$$
$$F_v = F \sin\theta_x = (3.5 \text{ kN}) \sin 215° = -2.0 \text{ kN}$$
$$= 2.0 \text{ kN} \downarrow$$

3.11 ANALYTICAL VECTOR ADDITION (+>): COMPONENT METHOD

The components of a series of vectors can be used to determine the net effect of the vectors. As mentioned, this method is best suited when more than two vectors need to be combined. This method involves resolving each individual vector into horizontal and vertical components. It is standard to use the algebraic sign convention for the components as described previously.

All horizontal components may then be added into a single vector component. This component represents the net horizontal effect of the series of vectors. It is worth noting that the component magnitudes can be simply added together because they all lay in the same direction. These components are treated as scalar quantities. A positive or negative sign is used to denote the sense of the component. This concept can be summarized in the following equation:

$$R_h = A_h + B_h + C_h + D_h + \ldots \quad (3.11)$$

Similarly, all vertical components may be added together into a single vector component. This component represents the net vertical effect of the series of vectors:

$$R_v = A_v + B_v + C_v + D_v + \ldots \quad (3.12)$$

The two net components may then be added vectorally into a resultant. Trigonometric relationships can be used to produce the following equations:

$$R = \sqrt{R_h^2 + R_v^2} \quad (3.13)$$

$$\theta_x = \tan^{-1}\left(\frac{R_v}{R_h}\right) \quad (3.14)$$

This resultant is the combined effect of the entire series of vectors. This procedure can be conducted most efficiently when the computations are arranged in a table, as demonstrated in the following example problem.

EXAMPLE PROBLEM 3.9

Three forces act on a hook as shown in Figure 3.17. Using the analytical component method, determine the net effect of these forces.

SOLUTION: 1. *Use x-axis, Angle Method to Determine Resultant Components*

The horizontal and vertical components of each force are determined by trigonometry and shown in Figure 3.18. Also shown are the vectors rearranged in a tip-to-tail fashion. The components are organized in Table 3.1.

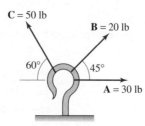

FIGURE 3.17 Forces for Example Problem 3.9.

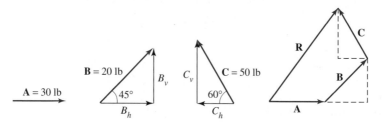

FIGURE 3.18 Components of vectors in Example Problem 3.9.

TABLE 3.1	Vector Components for Example Problem 3.9.		
Vector	Reference Angle θ_x	**h component (lb)** $F_h = F \cos \theta_x$	**v component (lb)** $F_v = F \sin \theta_x$
A	0°	$A_h = (30)\cos 0° = +30$ lb	$A_v = (30)\sin 0° = 0$
B	45°	$B_h = (20)\cos 45° = +14.14$ lb	$B_v = (20)\sin 45° = +14.14$ lb
C	120°	$C_h = (50)\cos 120° = -25$ lb	$C_v = (50)\sin 120° = +43.30$ lb
		$R_h = 19.14$	$R_v = 57.44$

Notice in Figure 3.18 that adding the magnitudes of the horizontal components is tracking the total "distance" navigated by the vectors in the horizontal direction. The same holds true for adding the magnitudes of the vertical components. This is the logic behind the component method of combining vectors. For this problem, adding the individual horizontal and vertical components gives the components of the resultant as follows:

$$R_h = 19.14 \text{ lb.}$$

and

$$R_v = 57.44 \text{ lb.}$$

2. ***Combine the Resultant Components***

The resultant is the vector sum of two perpendicular vectors, as shown in Figure 3.19.

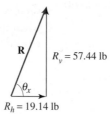

FIGURE 3.19 Resultant vector for Example Problem 3.9.

The magnitude of the resultant can be found from equation (2.13):

$$R = \sqrt{R_h^2 + R_v^2}$$
$$= \sqrt{(19.14 \text{ lb})^2 + (57.44 \text{ lb})^2} = 60.54 \text{ lb}$$

The angle of the resultant can be found:

$$\theta_x = \tan^{-1}\left(\frac{R_v}{R_h}\right) = \tan^{-1}\left(\frac{57.44 \text{ lb}}{19.14 \text{ lb}}\right) = 71.6°$$

Thus the resultant of the three forces can be formally stated as

$$\mathbf{R} = 60.54 \text{ lb.} \; \diagup \underline{71.6°}$$

3.12 VECTOR SUBTRACTION ($->$)

In certain cases, the difference between vector quantities is desired. In these situations, the vectors need to be subtracted. The symbol $->$ denotes vector subtraction, which differentiates it from algebraic subtraction [Ref. 5]. Subtracting vectors is accomplished in a manner similar to adding them. In effect, subtraction adds the negative, or opposite, of the vector to be subtracted. The negative of a vector is equal in magnitude, but opposite in direction. Figure 3.20 illustrates a vector **A** and its negative, $-> \mathbf{A}$.

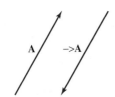

FIGURE 3.20 Negative vector.

Whether a graphical or analytical method is used, a vector diagram should be drawn to understand the procedure. Consider a general problem where vector **B** must be subtracted from **A**, as shown in Figure 3.21a.

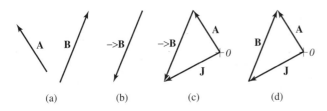

FIGURE 3.21 Vector subtraction.

This subtraction can be accomplished by first drawing the negative of vector **B**, $->\mathbf{B}$. This is shown in Figure 3.21b. Then, vector $->\mathbf{B}$ can be added to vector **A**, as shown in Figure 3.21c. This subtraction can be stated mathematically as

$$\mathbf{J} = \mathbf{A} -> \mathbf{B} = \mathbf{A} +> (->\mathbf{B})$$

Notice that this expression is identical to the subtraction of scalar quantities through basic algebraic methods. Also, the outcome of the vector subtraction has been designated **J**. The notation **R** is typically reserved to represent the result of vector addition.

Figure 3.21d shows that the same vector subtraction result by placing the vector **B** onto vector **A**, but opposite to the tip-to-tail orientation. This method is usually preferred, after some confidence has been established, because it eliminates the need to redraw a negative vector. Generally stated, vectors are added in a tip-to-tail format whereas they are subtracted in a tip-to-tip format. This concept is further explored as the individual solution methods are reviewed in the following example problems.

3.13 GRAPHICAL VECTOR SUBTRACTION ($->$)

As discussed, vector subtraction closely parallels vector addition. To graphically subtract vectors, they are relocated to scale to form a tip-to-tip vector diagram. The vector to be subtracted must be treated in the manner discussed in Section 3.12.

Again, the process of subtracting vectors can be completed graphically, using either manual drawing techniques or CAD software. Whatever method is used, the underlying concepts are identical. The specifics of the process are shown in the following examples.

EXAMPLE PROBLEM 3.10

Graphically determine the result of subtracting the velocity vector **B** from **A**, $\mathbf{J} = \mathbf{A} -> \mathbf{B}$, as shown in Figure 3.22.

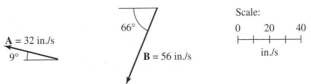

FIGURE 3.22 Vectors for Example Problem 3.10.

SOLUTION: 1.   ***Construct the Vector Diagram***

To determine the result, the vectors are located in the tip-to-tail form, but vector **B** points toward vector **A**. Again, this occurs because **B** is being subtracted (opposite to addition). The vector diagram is shown in Figure 3.23.

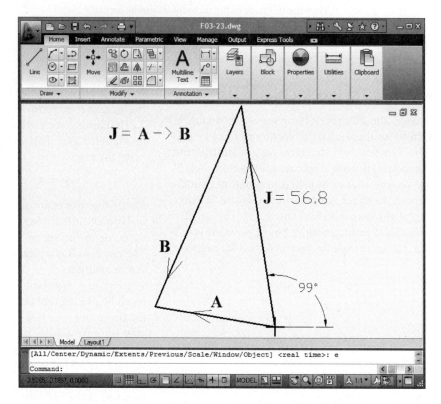

FIGURE 3.23 **J** = **A** −> **B** for Example Problem 3.10.

2.   ***Measure the Result***

The resultant extends from the tail of **A**, the origin, to the tail of **B**. The length vector **J** is measured as 56.8 in./s.

The direction is also required to fully define the vector **J**. The angle from the horizontal to vector **J** is measured as 99°. Therefore, the proper manner of presenting the solution is as follows:

$$\textbf{J} = 56.8 \text{ in./s} \ \diagup \underline{81°}$$

or

$$\textbf{J} = 56.8 \text{ in./s} \ \underline{99°}\diagdown$$

EXAMPLE PROBLEM 3.11

Graphically determine the result, **J** = **A** −> **B** −> **C** +> **D**, of the force vectors as shown in Figure 3.24.

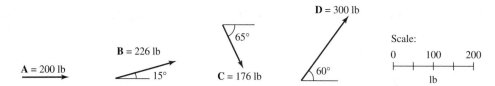

FIGURE 3.24 Vectors for Example Problem 3.11.

SOLUTION: 1.   *Construct the Vector Diagram*

To determine the result, $J = A \rightarrow B \rightarrow C +\!\!> D$, the vectors must be relocated tip-to-tail or tip-to-tip, depending on whether they are added or subtracted. Vector **B** must be drawn pointing toward vector **A** because **B** is being subtracted. A similar approach is taken with vector **C**. The tail of vector **D** is then placed on the tail of **C** because **D** is to be added to the series of previously assembled vectors. The completed vector diagram is shown in Figure 3.25.

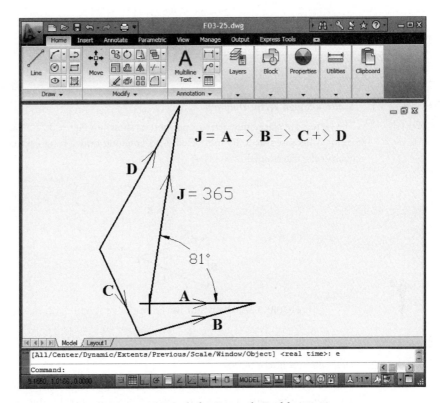

FIGURE 3.25   Result for Example Problem 3.11.

From viewing the vector polygon in Figure 3.25, it appears that vectors **B** and **C** were placed in backward, which occurs with vector subtraction.

2.   *Measure the Result*

The length of vector **J** is measured as 365 lb. The angle from the horizontal to vector **J** is measured as 81°. Therefore, the proper manner of presenting the solution is as follows:

$$J = 365 \text{ lb } \angle 81°$$

3.14 ANALYTICAL VECTOR SUBTRACTION ($\rightarrow$): TRIANGLE METHOD

As in analytically adding vectors, the triangle method is best suited for manipulation of only two vectors. A vector diagram should be sketched using the logic as described in the previous section. Then the triangle laws can be used to determine the result of vector subtraction. This general method is illustrated through the following example problem.

EXAMPLE PROBLEM 3.12

Analytically determine the result of the vectors **J** = **A** –> **B** shown in Figure 3.26.

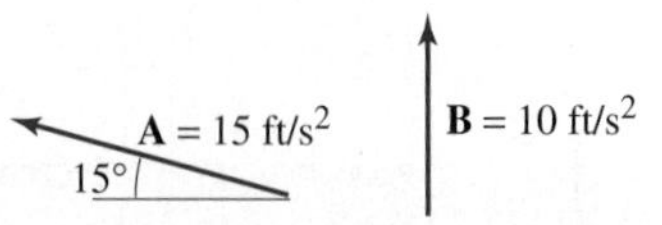

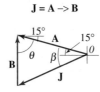

A = 15 ft/s²   15°   B = 10 ft/s²

FIGURE 3.26 Vectors for Example Problem 3.12.

SOLUTION:

1. ***Sketch a Rough Vector Diagram***

   The vectors are placed into a vector polygon, as shown in Figure 3.27. Again, vector **B** is placed pointing toward vector **A** because it is to be subtracted. Also note that only a rough sketch is required because the resultant is analytically determined.

   J = A –> B

   FIGURE 3.27 The result for Example Problem 3.12.

2. ***Determine an Internal Angle***

   Because the angle between **A** and the horizontal is 15°, the angle between **A** and the vertical is 75°. Notice that the angle between the vertical and **A** is the same as the angle labeled θ; thus, $\theta = 75°$.

   The problem of determining the result of **A** –> **B** is actually a general triangle situation as described in Section 3.6.2 (Case 3).

3. ***Determine the Magnitude of the Result***

   Following the procedure outlined for a Case 3 problem, the law of cosines is used to find the magnitude of the resultant:

   $$J = \sqrt{A^2 + B^2 - 2AB \cos\theta}$$
   $$= \sqrt{(15 \text{ ft/s}^2)^2 + (10 \text{ ft/s}^2)^2 - 2(15 \text{ ft/s}^2)(10 \text{ ft/s}^2) \cos 75°} = 15.73 \text{ ft/s}^2$$

4. ***Determine the Direction of the Result***

   The law of sines can be used to find the angle between vectors **A** and **J**:

   $$\beta = \sin^{-1}\left\{\left(\frac{B}{J}\right)\sin\theta\right\}$$
   $$= \sin^{-1}\left\{\frac{10 \text{ ft /s}^2}{15.73 \text{ ft /s}^2}\sin 75\right\} = 37.9°$$

5. ***Fully Specify the Result***

   From examining Figure 3.27, the angle that **J** makes with the horizontal is 37.9° − 15° = 22.9°. The solution can be properly written as

   $$\mathbf{J} = 15.73 \text{ ft/s}^2 \ \ \overline{22.9°}$$

3.15 ANALYTICAL VECTOR SUBTRACTION (−>): COMPONENT METHOD

The component method can be best used to analytically determine the result of the subtraction of a series of vectors. This is done in the same manner as vector addition. Consider the general problem of vector subtraction defined by the following equation:

$$J = A +> B −> C +> D +> \dots$$

The horizontal and vertical components of each vector can be determined (as in Section 3.10). Also, a sign convention to denote the sense of the component is required. The convention that was used in Section 3.10 designated components that point either to the right or upward with a positive algebraic sign.

Because they are scalar quantities, the individual components can be algebraically combined by addition or subtraction. For the general problem stated here, the horizontal and vertical components of the result can be written as follows:

$$J_h = A_h + B_h − C_h + D_h + \dots$$
$$J_v = A_v + B_v − C_v + D_v + \dots$$

Notice the components of **C** are subtracted from all the other components. This is consistent with the desired vector subtraction. Using equations (3.13) and (3.14), the two result components may then be combined vectorally into a resultant. This resultant is the result of the vector manipulation of the entire series of vectors. Again, the procedure can be conducted most efficiently when the computations are arranged in a table.

EXAMPLE PROBLEM 3.13

Analytically determine the result $J = A −> B +> C +> D$ for the velocity vectors shown in Figure 3.28.

FIGURE 3.28 Forces for Example Problem 3.13.

SOLUTION:

1. **Sketch a Rough Vector Diagram**

    The horizontal and vertical components of each velocity are determined by trigonometry using equations (3.9) and (3.10) and shown in Figure 3.29. Also shown are the vectors rearranged in a series: tip-to-tail for addition and tail-to-tip for subtraction.

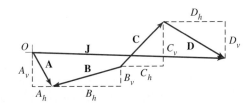

FIGURE 3.29 Result for Example Problem 3.13.

2. **Use x-axis, Angle Method to Determine Components**

    The values of the component are listed in Table 3.2.

Vector	Reference Angle θ_x	h component (ft/s) $V_h = V \cos \theta_x$	v component (ft/s) $V_v = V \sin \theta_x$
A	300°	+3.00	−5.19
B	195°	−11.59	−3.11
C	45°	+5.66	+5.66
D	330°	+8.66	−5.00

TABLE 3.2 **Component Values for Example Problem 3.13.**

3. **Determine the Components of the Solution**

Algebraic manipulation of the individual horizontal and vertical components gives the components of the resultant:

$$J_h = A_h - B_h + C_h + D_h$$
$$= (+3.0) - (-11.59) + (+5.66) + (+8.66) = +28.91 \text{ ft/s}$$
$$J_v = A_v - B_v + C_v + D_v$$
$$= (-5.19) - (-3.11) + (+5.66) + (-5.00) = -1.42 \text{ ft/s}$$

4. **Combine the Components of the Solution**

The magnitude and direction of the resultant may be determined by vectorally adding the components (Figure 3.30).

FIGURE 3.30 Resultant vector for Example Problem 3.13.

The magnitude of the solution can be determined from equation (3.13):

$$J = \sqrt{J_h^2 + J_v^2}$$
$$= \sqrt{(28.91 \text{ ft/s})^2 + (-1.42 \text{ ft/s})^2} = 28.94 \text{ ft/s}$$

The angle of the solution can be found from the tangent function:

$$\theta_x = \tan^{-1}\left(\frac{J_v}{J_h}\right) = \tan^{-1}\left(\frac{-1.42 \text{ ft/s}}{28.91 \text{ ft/s}}\right) = -2.8°$$

Thus, the solution can be formally stated as

$$J = 28.94 \text{ ft/s} \diagdown 2.8°$$

3.16 VECTOR EQUATIONS

As already shown in Section 3.8, vector operations can be expressed in equation form. The expression for subtracting two vectors, $J = A \rightarrow B$, is actually a vector equation. Vector equations can be manipulated in a manner similar to algebraic equations. The terms can be transposed by changing their signs. For example, the equation

$$A +> B -> C = D$$

can be rearranged as:

$$A +> B = C +> D$$

The significance of vector equations has been seen with vector addition and subtraction operations. In vector addition, vectors are placed tip-to-tail, and the resultant is a vector that extends from the origin of the first vector to the end of the final vector. Figure 3.31a illustrates the vector diagram for the following:

$$R = A +> B +> C$$

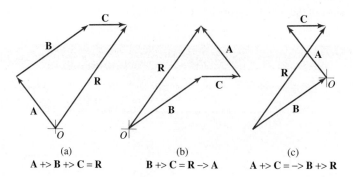

| (a) | (b) | (c) |
| A +> B +> C = R | B +> C = R -> A | A +> C = -> B +> R |

FIGURE 3.31 Vector equations.

The equation can be rewritten as

$$B +> C = R -> A$$

The vector diagram shown in Figure 3.31b illustrates this form of the equation. Notice that because vector **A** is subtracted from **R**, vector **A** must point toward **R**. Recall that this is the opposite of the tip-to-tail method because subtraction is the opposite of addition.

Notice that as the diagram forms a closed polygon, the magnitude and directions for all vectors are maintained.

This verifies that vector equations can be manipulated without altering their meaning. The equation can be rewritten once again as (Figure 3.31c):

$$-> B +> R = A +> C$$

As illustrated in Figure 3.31, a vector equation can be rewritten into several different forms. Although the vector polygons created by the equations have different shapes, the individual vectors remain unaltered. By using this principle, a vector equation can be written to describe a vector diagram.

EXAMPLE PROBLEM 3.14

Write a vector equation for the arrangement of vectors shown in Figure 3.32.

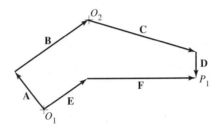

FIGURE 3.32 Vector diagram for Example Problem 3.14.

SOLUTION: 1. ***Write an Equation to Follow the Two Paths from O_1 to P_1***

Use point O_1 as the origin for the vector equation and follow the paths to point P_1:

The upper path states: $A +> B +> C +> D$
The lower path states: $E +> F$

Because they start at a common point and end at a common point, both paths must be vectorally equal. Thus, the following equation can be written as:

$$O_1P_1 = A +> B +> C +> D = E +> F$$

2. ***Write an Equation to Follow the Two Paths from O_2 to P_1***

Another equation can be written by using point O_2 as the origin and following the paths to point P_1:

The upper path states: $C +> D$
The lower path states: $-> B -> A +> E -> F$

Thus, the equation can be written as follows:

$$O_2P_1 = C +> D = -> A -> B +> E +> F$$

Notice that these are two forms of the same equation.

EXAMPLE PROBLEM 3.15

Write a vector equation for the arrangement of vectors shown in Figure 3.33. Then rewrite the equation to eliminate the negative terms and draw the associated vector diagram.

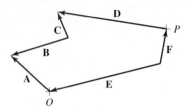

FIGURE 3.33 Vector diagram for Example Problem 3.15.

SOLUTION:

1. ***Write an Equation to Follow the Two Paths from O_1 to P_1***

   Use point O_1 as the origin for the vector equation and follow the paths to the point P_1:

   The upper path states: $\mathbf{A} -> \mathbf{B} +> \mathbf{C} -> \mathbf{D}$
   The lower path states: $-> \mathbf{E} +> \mathbf{F}$

   Thus, the following equation can be written as

   $$\mathbf{O_1P_1} = \mathbf{A} -> \mathbf{B} +> \mathbf{C} -> \mathbf{D} = -> \mathbf{E} +> \mathbf{F}$$

2. ***Rewrite the Equation***

   To eliminate the negative terms, vectors **B, D,** and **E** all must be transposed to their respective opposite sides of the equation. This yields the following equation:

   $$\mathbf{A} +> \mathbf{C} +> \mathbf{E} = \mathbf{B} +> \mathbf{D} +> \mathbf{F}$$

   Note that the order of addition is not significant. Rearranging the vectors into a new diagram is shown in Figure 3.34.

   Familiarity with vector equations should be gained, as they are used extensively in mechanism analysis. For example, determining the acceleration of even simple mechanisms involves vector equations with six or more vectors.

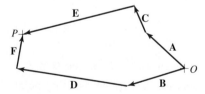

FIGURE 3.34 Rearranged diagram for Example Problem 3.15.

3.17 APPLICATION OF VECTOR EQUATIONS

Each vector in an equation represents two quantities: a magnitude and a direction. Therefore, a vector equation actually represents two constraints: The combination of the vector magnitudes and the directions must be equivalent. Therefore, a vector equation can be used to solve for two unknowns. In the addition and subtraction problems previously discussed, the magnitude and direction of the resultant were determined.

A common situation in mechanism analysis involves determining the magnitude of two vectors when the direction of all vectors is known. Like the addition of vectors, this problem also involves two unknowns. Therefore, one vector equation is sufficient to complete the analysis.

3.18 GRAPHICAL DETERMINATION OF VECTOR MAGNITUDES

For problems where the magnitude of two vectors in an equation must be determined, the equation should be rearranged so that one unknown vector is the last term on each side of the equation. To illustrate this point, consider the case where the magnitudes of vectors **A** and **B** are to be found. The vector equation consists of the following:

$$\mathbf{A} +> \mathbf{B} +> \mathbf{C} = \mathbf{D} +> \mathbf{E}$$

and should be rearranged as

$$\mathbf{C} +> \mathbf{B} = \mathbf{D} +> \mathbf{E} +> \mathbf{A}$$

Notice that both vectors with unknown magnitudes, **A** and **B**, are the last terms on both sides of the equation.

To graphically solve this problem, the known vectors on each side of the equation are placed tip-to-tail (or tip-to-tip if the vectors are subtracted) starting from a common origin. Of course, both sides of the equation must end at the same point. Therefore, lines at the proper direction should be inserted into the vector polygon. The intersection of these two lines represents the equality of the governing equation and solves the problem. The lines can be measured and scaled to determine the magnitudes of the unknown vectors. The sense of the unknown vector is also discovered.

This process of determining vector magnitudes can be completed graphically, using either manual drawing techniques or CAD software. Whatever method is used, the underlying strategy is identical. The solution strategy can be explained through example problems.

EXAMPLE PROBLEM 3.16

A vector equation can be written as

$$\mathbf{A} +> \mathbf{B} +> \mathbf{C} = \mathbf{D} +> \mathbf{E}$$

The directions for vectors **A**, **B**, **C**, **D**, and **E** are known, and the magnitudes of vectors **B**, **C**, and **D** are also known (Figure 3.35). Graphically determine the magnitudes of vectors **A** and **E**.

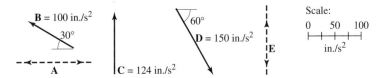

FIGURE 3.35 Vectors for Example Problem 3.16.

SOLUTION: 1.   *Rewrite Vector Equations*

First, the equation is rewritten so that the unknown magnitudes appear as the last term on each side of the equation:

$$\mathbf{B} +> \mathbf{C} +> \mathbf{A} = \mathbf{D} +> \mathbf{E}$$

2.   *Place All Fully Known Vectors into the Diagram*

Using point *O* as the common origin, vectors **B** and **C** can be drawn tip-to-tail. Because it is on the other side of the equation, vector **D** should be drawn from the origin (Figure 3.36a).

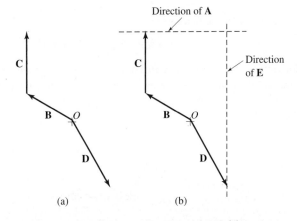

FIGURE 3.36 Vector diagrams for Example Problem 3.16.

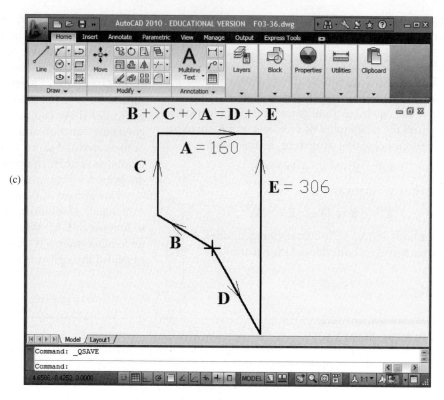

FIGURE 3.36 Continued

3. **Place Directional Lines for Unknown Vectors**

Obviously, vectors **A** and **E** close the gap between the end of vectors **C** and **D**. A line that represents the direction of vector **A** can be placed at the tip of **C**. This is dictated by the left side of the vector equation. Likewise, a line that represents the direction of vector **E** can be placed at the tip of **D** (Figure 3.36b).

4. **Trim Unknown Vectors at the Intersection and Measure**

The point of intersection of the two lines defines both the magnitude and sense of vectors **A** and **E**. A complete vector polygon can be drawn as prescribed by a vector equation (Figure 3.36c).

The following results are obtained by measuring vectors **A** and **E**:

$$\mathbf{A} = 160 \text{ in./s}^2 \rightarrow$$

$$\mathbf{E} = 306 \text{ in./s}^2 \uparrow$$

EXAMPLE PROBLEM 3.17

A vector equation can be written as follows:

$$\mathbf{A} +> \mathbf{B} -> \mathbf{C} +> \mathbf{D} = \mathbf{E} +> \mathbf{F}$$

The directions for vectors **A, B, C, D, E,** and **F** are known, and the magnitudes of vectors **B, C, E,** and **F** are also known, as shown in Figure 3.37. Graphically solve for the magnitudes of vectors **A** and **D**.

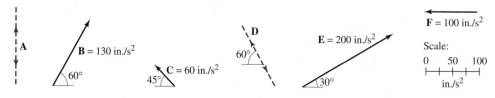

FIGURE 3.37 Vectors for Example Problem 3.17.

SOLUTION:

1. *Rewrite the Vector Equation*

   The equation is first rewritten so that the unknown magnitudes appear as the last term on each side of the equation:

   $$B -> C +> A = E +> F -> D$$

2. *Place All Fully Known Vectors into the Diagram*

   Using point O as the common origin, vectors **B** and **C** can be drawn tip-to-tip (because **C** is being subtracted). Because they are on the other side of the equation, vectors **E** and **F** are placed tip-to-tail starting at the origin (Figure 3.38a).

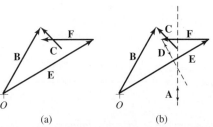

   (a)   (b)

(c)

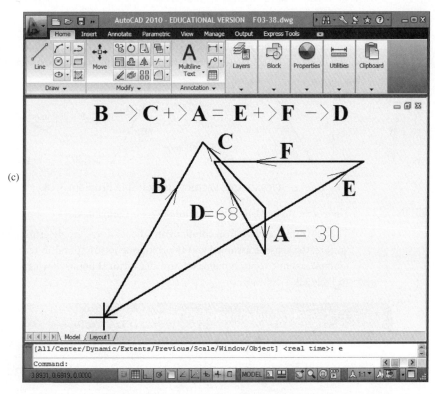

FIGURE 3.38 Vector diagrams for Example Problem 3.17.

3. *Place Directional Lines for the Unknown Vectors*

   As in Example Problem 3.16, vectors **A** and **D** must close the gap between the end of vectors **C** and **D**. A line that represents the direction of vector **A** can be placed at the tip of **C**. This is dictated by the left side of the vector equation. Likewise, a line that represents the direction of vector **D** can be placed at the tip of **F** (Figure 3.38b).

4. *Trim the Unknown Vectors at the Intersection and Measure*

   The point of intersection of the two lines defines both the magnitude and sense of vectors **A** and **D**. The sense of **D** is chosen in a direction that is consistent with its being subtracted from the right side of the equation. The complete vector polygon can be drawn as prescribed by the vector equation (Figure 3.38c).

   The following results are obtained by measuring vectors **A** and **D**:

   $$A = 30 \text{ in./s}^2 \downarrow$$
   $$D = 68 \text{ in./s}^2 \ \underline{60°} \nwarrow$$

3.19 ANALYTICAL DETERMINATION OF VECTOR MAGNITUDES

An analytical method can also be used to determine the magnitude of two vectors in an equation. In these cases, the horizontal and vertical components of all vectors should be determined as in Section 3.10. Components of the vectors with unknown magnitudes can be written in terms of the unknown quantity. As in the previous component methods, an algebraic sign convention must be followed while computing components. Therefore, at this point, a sense must be arbitrarily assumed for the unknown vectors.

The horizontal components of the vectors must adhere to the original vector equation. Likewise, the vertical components must adhere to the vector equation. Thus, two algebraic equations are formed and two unknown magnitudes must be determined. Solving the two equations simultaneously yields the desired results. When either of the magnitudes determined has a negative sign, the result indicates that the assumed sense of the vector was incorrect. Therefore, the magnitude determined and the opposite sense fully define the unknown vector.

This method is illustrated in the example problem below.

EXAMPLE PROBLEM 3.18

A vector equation can be written as follows:

$$\mathbf{A} +> \mathbf{B} -> \mathbf{C} +> \mathbf{D} = \mathbf{E} +> \mathbf{F}$$

The directions for vectors **A, B, C, D, E,** and **F** are known, and the magnitudes of vectors **B, C, E,** and **F** are also known, as shown in Figure 3.39. Analytically solve for the magnitudes of vectors **A** and **D**.

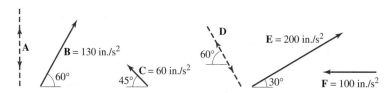

FIGURE 3.39 Vectors for Example Problem 3.18.

SOLUTION: 1. *Use x-axis, Angle Method to Determine Vector Components*

The horizontal and vertical components of each force are determined by trigonometry. For the unknown vectors, the sense is assumed and the components are found in terms of the unknown quantities. For this example, assume vector **A** points upward and vector **D** points down and to the right. The components are given in Table 3.3.

	TABLE 3.3 **Vector Components for Example Problem 3.18.**		
Vector	Reference Angle θ_x	h component (in./s²) $a_h = a \cos \theta_x$	v component (in./s²) $a_v = a \sin \theta_x$
A	90°	0	+A
B	60°	+65.0	+112.6
C	135°	−42.4	+42.4
D	300°	+.500D	−.866D
E	30°	+173.2	+100
F	180°	−100	0

2. *Use the Vector Equations to Solve for Unknown Magnitudes*

The components can be used to generate algebraic equations that are derived from the original vector equation:

$$\mathbf{A} +> \mathbf{B} -> \mathbf{C} +> \mathbf{D} = \mathbf{E} +> \mathbf{F}$$

horizontal components:

$$A_h + B_h - C_h + D_h = E_h + F_h$$

$$(0) + (+65.0) - (-42.4) + (+0.500\,D) = (+173.2) + (-100.0)$$

vertical components:

$$A_v + B_v - C_v + D_v = E_v + F_v$$

$$(+A) + (+112.6) - (42.4) + (-0.866D) = (+100.0) + (0)$$

In this case, the horizontal component equation can be solved independently for D. In general, both equations are coupled and need to be solved simultaneously. In this example, the horizontal component equation can be solved to obtain the following:

$$D = -68.4 \text{ in./s}^2$$

Substitute this value of D into the vertical component equation to obtain:

$$A = -29.4 \text{ in./s}^2$$

3. **Fully Specify the Solved Vectors**

Because both values are negative, the original directions assumed for the unknown vectors were incorrect. Therefore, the corrected results are

$$\mathbf{A} = 29.4 \text{ in./s}^2 \downarrow$$

$$\mathbf{D} = 68.4 \text{ in./s}^2 \, \underline{60°} \, \searrow$$

PROBLEMS

Although manual drafting techniques are instructive for problems that require graphical solution, use of a CAD system is highly recommended.

Working with Triangles

3–1. Analytically determine the angle θ in Figure P3.1.

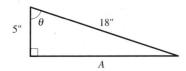

FIGURE P3.1 Problems 1 and 2.

3–2. Analytically determine the length of side A in Figure P3.1.

3–3. Analytically determine the length of side X in Figure P3.3.

3–4. Calculate the angle θ and the hypotenuse R in Figure P3.3.

3–5. Calculate the angle θ and the hypotenuse R for all triangles in Figure P3.5.

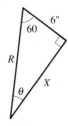

FIGURE P3.3 Problems 3 and 4.

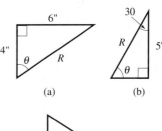

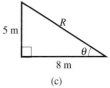

FIGURE P3.5 Problem 5.

3–6. Determine the angle, β, and the length, s, of the two identical support links in Figure P3.6 when $x = 150$ mm and $y = 275$ mm.

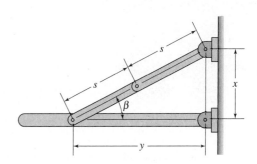

FIGURE P3.6 Problems 6 to 9.

3–7. Determine the distance, x, and the length, s, of the two identical support links in Figure P3.6 when $\beta = 35°$ and $y = 16$ in.

3–8. For the folding shelf in Figure P3.6, with $\beta = 35°$ and $s = 10$ in., determine the distances x and y.

3–9. A roof that has an 8-on-12 pitch slopes upward 8 vertical in. for every 12 in. of horizontal distance. Determine the angle with the horizontal of such a roof.

3–10. For the swing-out window in Figure P3.10, determine the length, s, of the two identical support links when $x = 850$ mm, $d = 500$ mm, and $\beta = 35°$.

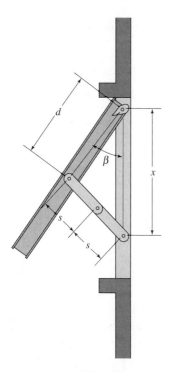

FIGURE P3.10 Problems 10 and 11.

3–11. For the swing-out window in Figure P3.10, determine the angle β when $x = 24$ in., $d = 16$ in., and $s = 7$ in.

3–12. If the height, h, of the trailer shown in Figure P3.12 is 52 in., determine the length of ramp needed to maintain an angle, β, of 30°.

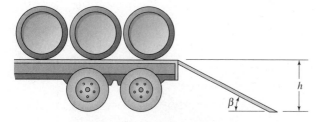

FIGURE P3.12 Problems 12 and 13.

3–13. For the ramp shown in Figure P3.12, determine the angle with the ground, β. The trailer height is 1.5 m and the ramp is 4 m long.

3–14. The length of the ladder shown in Figure P3.14 is 12 ft and the angle with the ground, β, is 70°. Determine the vertical distance on the wall where the ladder is resting.

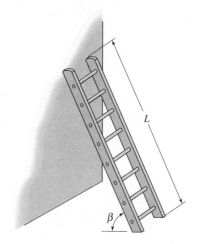

FIGURE P3.14 Problems 14 and 15.

3–15. For the ladder shown in Figure P3.14, determine the angle with the ground. The ladder is 7 m long and rests on the ground 2 m from the wall.

3–16. For the farm conveyor shown in Figure P3.16, determine the required length of the support rod. The angle is $\beta = 28°$ and the distances are $x = 20$ ft and $d = 16$ ft. Also determine the vertical height of the end of the conveyor when $L = 25$ ft.

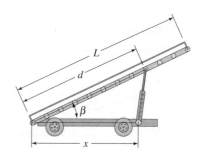

FIGURE P3.16 Problems 16 and 17.

3–17. For the farm conveyor shown in Figure P3.16, determine the angle β when a vertical height of 8 m is required at the end of the conveyor and $x = 8$ m, $d = 10$ m, and $L = 13$ m.

3–18. Determine the vertical height of the basket in Figure P3.18 when $a = 24$ in., $b = 36$ in., $c = 30$ in., $d = 60$ in., $e = 6$ ft, and $f = 10$ ft.

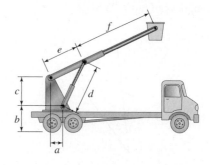

FIGURE P3.18 Problems 18 and 19.

3–19. For the lift described in Problem 3–18, determine the vertical height of the basket when the hydraulic cylinder is shortened to 50 in.

Graphical Vector Addition

3–20. For the vectors shown in Figure P3.20, graphically determine the resultant, **R** = **A** +> **B**.

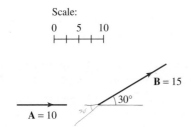

FIGURE P3.20 Problems 20, 26, 32, 33, 38, 39.

3–21. For the vectors shown in Figure P3.21, graphically determine the resultant, **R** = **A** +> **B**.

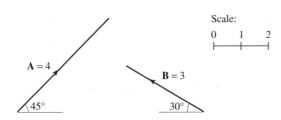

FIGURE P3.21 Problems 21, 27, 34, 35, 40, 41.

3–22. For the vectors shown in Figure P3.22, graphically determine the resultant, **R** = **A** +> **B**.

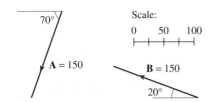

FIGURE P3.22 Problems 22, 28, 36, 37, 42, 43.

3–23. For the vectors shown in Figure P3.23, graphically determine the resultant, **R** = **A** +> **B** +> **C**.

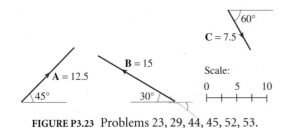

FIGURE P3.23 Problems 23, 29, 44, 45, 52, 53.

3–24. For the vectors shown in Figure P3.24, graphically determine the resultant, **R** = **A** +> **B** +> **C** +> **D**.

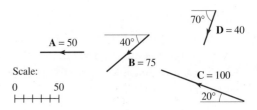

FIGURE P3.24 Problems 24, 30, 46, 47, 54, 55.

3–25. For the vectors shown in Figure P3.25, graphically determine the resultant, **R** = **A** +> **B** +> **C** +> **D** +> **Es**

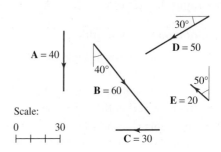

FIGURE P3.25 Problems 25, 31, 48, 49, 56, 57.

Analytical Vector Addition

3–26. For the vectors shown in Figure P3.20, analytically determine the resultant, **R** = **A** +> **B**.

3–27. For the vectors shown in Figure P3.21, analytically determine the resultant, **R** = **A** +> **B**.

3–28. For the vectors shown in Figure P3.22, analytically determine the resultant, **R** = **A** +> **B**.

3–29. For the vectors shown in Figure P3.23, analytically determine the resultant, **R** = **A** +> **B** +> **C**.

3–30. For the vectors shown in Figure P3.24, analytically determine the resultant, **R** = **A** +> **B** +> **C** +> **D**.

3–31. For the vectors shown in Figure P3.25, analytically determine the resultant, **R** = **A** +> **B** +> **C** +> **D** +> **E**.

Graphical Vector Subtraction

3–32. For the vectors shown in Figure P3.20, graphically determine the vector, **J** = **A** –> **B**.

3–33. For the vectors shown in Figure P3.20, graphically determine the vector, **K** = **B** –> **A**.

3–34. For the vectors shown in Figure P3.21, graphically determine the vector, **J** = **A** –> **B**.

3–35. For the vectors shown in Figure P3.21, graphically determine the vector, **K** = **B** –> **A**.

3–36. For the vectors shown in Figure P3.22, graphically determine the vector, **J** = **A** –> **B**.

3–37. For the vectors shown in Figure P3.22, graphically determine the vector, **K** = **B** –> **A**.

Analytical Vector Subtraction

3–38. For the vectors shown in Figure P3.20, analytically determine the vector, $J = A \rightarrow B$.

3–39. For the vectors shown in Figure P3.20, analytically determine the vector, $K = B \rightarrow A$.

3–40. For the vectors shown in Figure P3.21, analytically determine the vector, $J = A \rightarrow B$.

3–41. For the vectors shown in Figure P3.21, analytically determine the vector, $K = B \rightarrow A$.

3–42. For the vectors shown in Figure P3.22, analytically determine the vector, $J = A \rightarrow B$.

3–43. For the vectors shown in Figure P3.22, analytically determine the vector, $K = B \rightarrow A$.

General Vector Equations (Graphical)

3–44. For the vectors shown in Figure P3.23, graphically determine the vector, $J = C +> A \rightarrow B$.

3–45. For the vectors shown in Figure P3.23, graphically determine the vector, $K = B \rightarrow A \rightarrow C$.

3–46. For the vectors shown in Figure P3.24, graphically determine the vector, $J = C +> A \rightarrow B +> D$.

3–47. For the vectors shown in Figure P3.24, graphically determine the vector, $K = B \rightarrow D +> A \rightarrow C$.

3–48. For the vectors shown in Figure P3.25, graphically determine the vector, $J = C +> A \rightarrow B +> D \rightarrow E$.

3–49. For the vectors shown in Figure P3.25, graphically determine the vector, $K = B \rightarrow D +> A \rightarrow C +> E$.

3–50. Using the vector diagram in Figure P3.50:

    a. Generate an equation that describes the vector diagram.

    b. Rewrite the equations to eliminate the negative terms.

    c. Roughly scale the vectors and rearrange them according to the equation generated in part b.

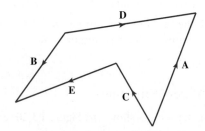

FIGURE P3.50 Problem 50.

3–51. Using the vector diagram in Figure P3.51:

    a. Generate an equation that describes the vector diagram.

    b. Rewrite the equations to eliminate the negative terms.

    c. Roughly scale the vectors and rearrange them according to the equation generated in part b.

3–52. For the vectors shown in Figure P3.23, analytically determine the vector, $J = C +> A \rightarrow B$.

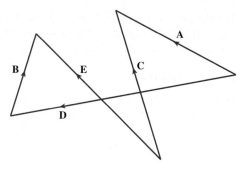

FIGURE P3.51 Problem 51.

3–53. For the vectors shown in Figure P3.23, analytically determine the vector, $K = B \rightarrow A \rightarrow C$.

3–54. For the vectors shown in Figure P3.24, analytically determine the vector, $J = C +> A \rightarrow B +> D$.

3–55. For the vectors shown in Figure P3.24, analytically determine the vector, $K = B \rightarrow D +> A \rightarrow C$.

3–56. For the vectors shown in Figure P3.25, analytically determine the vector, $J = C +> A \rightarrow B +> D \rightarrow E$.

3–57. For the vectors shown in Figure P3.25, analytically determine the vector, $K = B \rightarrow D +> A \rightarrow C +> E$.

Solving for Vector Magnitudes (Graphical)

3–58. A vector equation can be written as $A +> B \rightarrow C = D \rightarrow E$. The directions of all vectors and magnitudes of A, B, and D are shown in Figure P3.58. Graphically (using either manual drawing techniques or CAD) determine the magnitudes of vectors C and E.

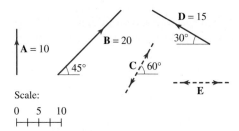

FIGURE P3.58 Problems 58 and 61.

3–59. A vector equation can be written as $A +> B + C \rightarrow D = E + F$. The directions of all vectors and magnitudes of A, B, C, and E are shown in Figure P3.59. Graphically (using either manual drawing techniques or CAD) determine the magnitudes of vectors D and F.

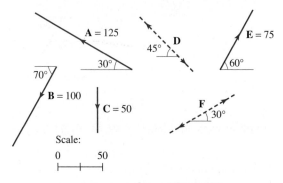

FIGURE P3.59 Problems 59 and 62.

3–60. A vector equation can be written as $\mathbf{A} +> \mathbf{B} + \mathbf{C} -> \mathbf{D} +> \mathbf{E}$. The directions of all vectors and magnitudes of **A, D, E,** and **F** are shown in Figure P3.60. Graphically (using either manual drawing techniques or CAD) determine the magnitudes of vectors **B** and **C**.

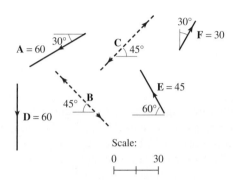

FIGURE P3.60 Problems 60 and 63.

Solving for Vector Magnitudes (Analytical)

3–61. Analytically determine vectors **C** and **E** from Problem 3–58.

3–62. Analytically determine vectors **D** and **F** from Problem 3–59.

3–63. Analytically determine vectors **B** and **C** from Problem 3–60.

CASE STUDIES

3–1. Figure C3.1 shows two of many keys from an adding machine that was popular several years ago. End views are also shown to illustrate the configuration at keys 1 and 2. Carefully examine the configuration of the components in the mechanism. Then, answer the following leading questions to gain insight into the operation of the mechanism.

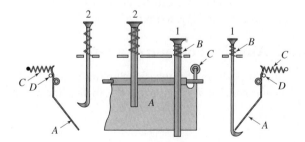

FIGURE C3.1 (Courtesy, Industrial Press)

1. As key 2 is pressed, what happens to rocker plate *A*?
2. What is the purpose of spring *C*?
3. What is the purpose of spring *B*?
4. As button 2 is pressed, what happens to button 1?
5. What is the purpose of this device?
6. Because force is a vector, its direction is important. In what direction must the force applied to button 1 by the spring *B* act?

7. In what direction must the force applied to plate *A* by spring *C* act?
8. List other machines, other than an adding machine, that could use this device.
9. What is the function of pin *D*?

3–2. An automatic machine that forms steel wire occasionally jams when the raw material is oversized. To prevent serious damage to the machine, it was necessary for the operator to cut off power immediately when the machine became jammed. However, the operator is unable to maintain a close watch over the machine to prevent damage. Therefore, the following mechanism has been suggested to solve the problem.

3. Figure C3.2 shows that gear *C* drives a mating gear (not shown) that operates the wire-forming machine. Driveshaft *A* carries collar *B*, which is keyed to it. Gear *C* has a slip fit onto shaft *A*. Two pins, *G* and *E*, attach links *F* and *D*, respectively, to gear *C*. An additional pin on gear *C* is used to hold the end of spring *H*. Carefully examine the configuration of the components in the mechanism. Then, answer the following leading questions to gain insight into the operation of the mechanism.

1. As driveshaft *A* turns clockwise, what is the motion of collar *B*?

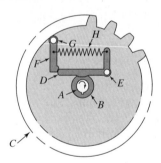

FIGURE C3.2 (Courtesy, Industrial Press)

2. If gear *C* is not fixed to collar *B*, how can the clockwise motion from the shaft rotate the gear?
3. What happens to the motion of gear *C* if link *D* is forced upward?
4. What action would cause link *D* to move upward?
5. What resistance would link *D* have to move upward?
6. What is the purpose of this device?
7. What would you call such a device?
8. How does this device aid the automatic wire-forming machine described here?
9. This device must be occasionally "reset." Why and how will that be accomplished?
10. Because force is a vector, its direction is important. In what direction must the forces applied by the spring *H* act?
11. List other machines, other than the wire-forming one, that could use this device.

POSITION AND DISPLACEMENT ANALYSIS

OBJECTIVES

Upon completion of this chapter, the student will be able to:

1. Define position and displacement of a point.

2. Graphically and analytically determine the position of all links in a mechanism as the driver links are displaced.

3. Graphically and analytically determine the limiting positions of a mechanism.

4. Graphically and analytically determine the position of all links for an entire cycle of mechanism motion.

5. Plot a displacement diagram for various points on a mechanism as a function of the motion of other points on the mechanism.

4.1 INTRODUCTION

For many mechanisms, the sole purpose of analysis is to determine the location of all links as the driving link(s) of the mechanism is moved into another position. Consider a machining clamp, as shown in Figure 4.1. If such a clamp is integrated into a machine, it is essential to understand the motion of the various links. One investigation might be to determine the motion of the handle that is required to close the jaw. This is a repeated motion that will be required from machine operators. Access, the effort required to operate, and other "human factors" criteria must be considered in using the clamp. Position analysis often involves repositioning the links of a mechanism between two alternate arrangements.

Another investigation might be to understand the path of the different components during the clamping process. Proper clearances with other machine components must be ensured. Position analysis is commonly repeated at several intervals of mechanism movement to determine the location of all links at various phases of the operation cycle. The focus of this chapter is on these types of position and displacement analyses.

4.2 POSITION

Position refers to the location of an object. The following sections will address the position of points and links.

4.2.1 Position of a Point

The *position* of a point on a mechanism is the spatial location of that point. It can be defined with a *position vector*, **R**, from a reference origin to the location of the point. Figure 4.2 illustrates a position vector, $\mathbf{R}_P$, defining the planar position of point P. As with all vectors, the planar position of a point can be specified with a distance from the origin (vector magnitude) and angle from a reference axis (orientation).

An alternative practice used to identify the position of a point is with rectangular components of the position vector in a reference coordinate system. Notice that the position of point P in Figure 4.2 can be defined with its x and y components, $\mathbf{R}^x_P$ and $\mathbf{R}^y_P$, respectively.

4.2.2 Angular Position of a Link

The angular position of a link is also an important quantity. An *angular position*, θ, is defined as the angle a line between

FIGURE 4.1 Machining clamp. (*Courtesy of Carr Lane Mfg.*)

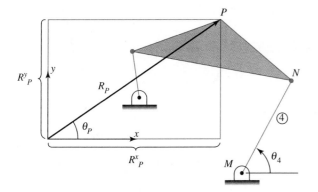

FIGURE 4.2 Position vector for point *P*.

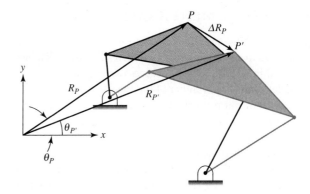

FIGURE 4.3 Displacement vector for point *P*.

$$\Delta R_P = R_P{}' - R_P$$

two points on that link forms with a reference axis. Referring to Figure 4.2, line *MN* lies on link 4. The angular position of link 4 is defined by θ_4, which is the angle between the *x*-axis and line *MN*. For consistency, angular position is defined as positive if the angle is measured counterclockwise from the reference axis and negative if it is measured clockwise.

4.2.3 Position of a Mechanism

The primary purpose in analyzing a mechanism is to study its motion. Motion occurs when the position of the links and the reference points that comprise the mechanism are changed. As the position of the links is altered, the mechanism is forced into a different configuration, and motion proceeds.

Recall from Chapter 1 that an important property of a mechanism is the mobility or degrees of freedom. For linkages with one degree of freedom, the position of one link or point can precisely determine the position of all other links or points. Likewise, for linkages with two degrees of freedom, the position of two links can precisely determine the position of all other links.

Therefore, the position of all points and links in a mechanism is not arbitrary and independent. The independent parameters are the positions of certain "driver" links or "driver" points. The primary goal of position analysis is to determine the resulting positions of the points on a mechanism as a function of the position of these "driver" links or points.

4.3 DISPLACEMENT

Displacement is the end product of motion. It is a vector that represents the distance between the starting and ending positions of a point or link. There are two types of displacements that will be considered: linear and angular.

4.3.1 Linear Displacement

Linear displacement, $\Delta \mathbf{R}$, is the straight line distance between the starting and ending position of a point during a time interval under consideration. Figure 4.3 illustrates a point *P* on a mechanism that is displaced to position *P'*.

The linear displacement of point *P* is denoted as ΔR_P and is calculated as the vectoral difference between the initial position and the final position. Given in equation form:

$$\Delta \mathbf{R}_P = \mathbf{R}_{P'} \text{--}> \mathbf{R}_P \qquad (4.1)$$

Notice that linear displacement is not the distance traveled by the point during motion.

The magnitude of the displacement vector is the distance between the initial and final position during an interval. This magnitude will be in linear units (inches, feet, millimeters, etc.). The direction can be identified by an angle from a reference axis to the line that connects the two positions. The sense of the vector is from the initial position and pointing toward the final position.

4.3.2 Angular Displacement

Angular displacement, $\Delta \theta$, is the angular distance between two configurations of a rotating link. It is the difference between the starting and ending angular positions of a link, as shown in Figure 4.4. While possessing a magnitude and direction (clockwise or counterclockwise), angular displacement is not technically a vector since it does not adhere to commutative and associative laws of vector addition.

The angular displacement of a link, say link 3, is denoted as $\Delta \theta_3$ and determined with equation (4.2).

$$\Delta \theta_3 = \theta_{3'} - \theta_3 \qquad (4.2)$$

The magnitude of the angular displacement is the angle between the initial and final configuration of a link during an interval. This magnitude will be in rotational units (e.g., degrees, radians, and revolutions), and denoting either clockwise or counterclockwise specifies the direction.

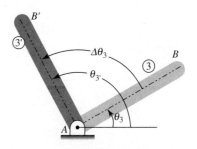

FIGURE 4.4 Angular displacement.

4.4 DISPLACEMENT ANALYSIS

A common kinematic investigation is locating the position of all links in a mechanism as the driver link(s) is displaced. As stated in Section 4.2, the degrees of freedom of a mechanism determine the number of independent driver links. For the most common mechanisms, those with one degree of freedom, displacement analysis consists of determining the position of all links as one link is displaced. The positions of all links are called the *configuration* of the mechanism.

Figure 4.5 illustrates this investigation. The mechanism shown has four links, as numbered. Recall that the fixed link, or frame, must always be included as a link. The mechanism also has four revolute, or pin, joints.

From equation (1.1), the degrees of freedom can be calculated as follows:

$$M = 3(4 - 1) - 2(4) = 1$$

With one degree of freedom, moving one link precisely positions all other links in the mechanism. Therefore, a typical displacement analysis problem involves determining the position of links 3 and 4 in Figure 4.5 as link 2 moves to a specified displacement. In this example, the driving displacement is angular, $\Delta\theta_2 = 15°$ clockwise.

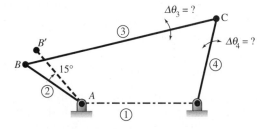

FIGURE 4.5 Typical position analysis.

Nearly all linkages exhibit alternate configurations for a given position of the driver link(s). Two configurations for the same crank position of a four-bar mechanism are shown in Figure 4.6. These alternate configurations are called *geometric inversions*. It is a rare instance when a linkage can move from one geometric inversion to a second without

FIGURE 4.6 Two geometric inversions of a four-bar mechanism.

disassembling the mechanism or traveling through dead points. Thus, when conducting a displacement analysis, inspecting the original configuration of the mechanism is necessary to determine the geometric inversion of interest.

4.5 DISPLACEMENT: GRAPHICAL ANALYSIS

4.5.1 Displacement of a Single Driving Link

In placing a mechanism in a new configuration, it is necessary to relocate the links in their respective new positions. Simple links that rotate about fixed centers can be relocated by drawing arcs, centered at the fixed pivot, through the moving pivot, at the specified angular displacement. This was illustrated in Figure 4.5 as link 2 was rotated 15° clockwise.

In some analyses, complex links that are attached to the frame also must be rotated. This can be done using several methods. In most cases, the simplest method begins by relocating only one line of the link. The other geometry that describes the link can then be reconstructed, based on the position of the line that has already been relocated.

Figure 4.7 illustrates the process of rotating a complex link. In Figure 4.7a, line *AB* of the link is displaced to its desired position, $\Delta\theta_2 = 80°$ clockwise. Notice that the relocated position of point *B* is designated as *B'*.

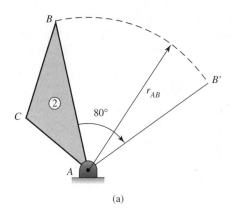

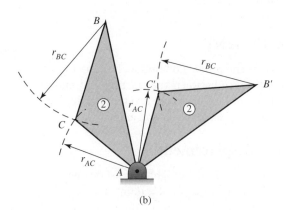

(a)   (b)

FIGURE 4.7 Rotating a complex link.

The next step is to determine the position of the relocated point C, which is designated as C'. Because the complex link is rigid and does not change shape during movement, the lengths of lines AC and BC do not change. Therefore, point C' can be located by measuring the lengths of AC and BC and striking arcs from points A and B', respectively (Figure 4.7b).

A second approach can be employed on a CAD system. The lines that comprise the link can be duplicated and rotated to yield the relocated link. All CAD systems have a command that can easily rotate and copy geometric entities. This command can be used to rotate all lines of a link about a designated point, a specified angular displacement. It is convenient to display the rotated link in another color and place it on a different layer.

4.5.2 Displacement of the Remaining Slave Links

Once a driver link is repositioned, the position of all other links must be determined. To accomplish this, the possible paths of all links that are connected to the frame should be constructed. For links that are pinned to the frame, all points on the link can only rotate relative to the frame. Thus, the possible paths of those points are circular arcs, centered at the pin connecting the link to the frame.

Figure 4.8 illustrates a kinematic diagram of a mechanism. Links 2, 4, and 6 are all pinned to the frame. Because points B, C, and E are located on links 2, 4, and 6, respectively, their constrained paths can be readily constructed. The constrained path of point B is a circular arc, centered at point A, which is the pin that connects link 2 to the frame. The constrained paths of C and E can be determined in a similar manner.

The constrained path of a point on a link that is connected to the frame with a slider joint can also be easily determined. All points on this link move in a straight line, parallel to the direction of the sliding surface.

After the constrained paths of all links joined to the frame are constructed, the positions of the connecting links can be determined. This is a logical process that stems from the fact that all links are rigid. Rigidity means that the links do not change length or shape during motion.

In Figure 4.5, the positions of links 3 and 4 are desired as link 2 rotates 15° clockwise. Using the procedures described

in Section 4.5.1, Figure 4.9 shows link 2 relocated to its displaced location, which defines the position of point B'. The constrained path of point C has also been constructed and shown in Figure 4.9.

Because of its rigidity, the length of link 3 does not change during motion. Although link 2 has been repositioned, the length between points B and C (r_{BC}) does not change. To summarize the facts of this displacement analysis, the following is known:

1. Point B has been moved to B'
2. Point C must always lay on its constrained path (length r_{CD} from D) and
3. The length between B and C must stay constant (C' must be a length r_{BC} from B').

From these facts, the new position of link 3 can be constructed. The length of line BC should be measured. Because point B has been moved to B', an arc of length r_{BC} is constructed with its center at B'. By sweeping this arc, the feasible path of point C' has been determined. However, point C must also lay on its constrained path, as shown in Figure 4.9. Therefore, point C' must be located at the intersection of the two arcs. This process is illustrated in Figure 4.10. Note that the two arcs will also intersect at a second point. This second point of intersection is a considerable distance from C and represents a second geometric inversion for this linkage. The linkage must be disassembled and reassembled to achieve this alternate configuration, so that intersection can be ignored.

It is possible that the two arcs do not intersect at all. Cases where the constrained path and feasible path do not intersect indicate that length of the individual links prevents the driver link from achieving the specified displacement.

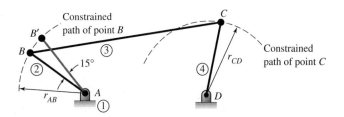

FIGURE 4.9 Constructing the constrained path of C.

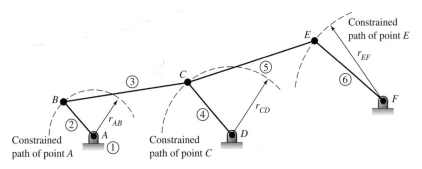

FIGURE 4.8 Constrained paths of points on a link pinned to the frame.

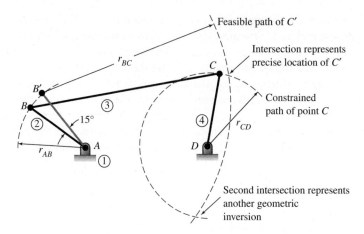

FIGURE 4.10 Locating the position of C'.

Once C' has been located, the position of links 3 and 4 can be drawn. Thus, the configuration of the mechanism as the driver link was repositioned has been determined

This section presents the logic behind graphical position analysis—that is, locating a displaced point as the intersection of the constrained and feasible paths. This logic is merely repeated as the mechanisms become more complex. The actual solution can be completed using manual drawing techniques (using a protractor and compass) or can be completed on a CAD system (using a rotate and copy command). The logic is identical; however, the CAD solution is not susceptible to the limitations of drafting accuracy. Regardless of the method used, the underlying concepts of graphical position analysis can be further illustrated and expanded through the following example problems.

EXAMPLE PROBLEM 4.1

Figure 4.11 shows a kinematic diagram of a mechanism that is driven by moving link 2. Graphically reposition the links of the mechanism as link 2 is displaced 30° counterclockwise. Determine the resulting angular displacement of link 4 and the linear displacement of point E.

SOLUTION: 1.   *Calculate Mobility*

To verify that the mechanism is uniquely positioned by moving one link, its mobility can be calculated. Six links are labeled. Notice that three of these links are connected at point C. Recall from Chapter 1 that this arrangement must be counted as two pin joints. Therefore, a total of six pin joints are tallied. One sliding joint connects links 1 and 6. No gear or cam joints exist:

$$n = 6 \quad j_p = (6\,\text{pins} + 1\,\text{sliding}) = 7 \quad j_h = 0$$

and

$$M = 3(n - 1) - 2j_p - j_h = 3(6 - 1) - 2(7) - 0 = 15 - 14 = 1$$

With one degree of freedom, moving one link uniquely positions all other links of the mechanism.

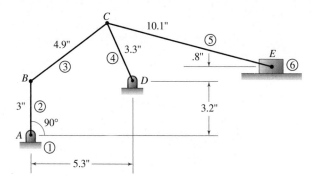

FIGURE 4.11 Kinematic diagram for Example Problem 4.1.

2. *Reposition the Driving Link*

Link 2 is graphically rotated 30° counterclockwise, locating the position of point B'. This is shown in Figure 4.12a

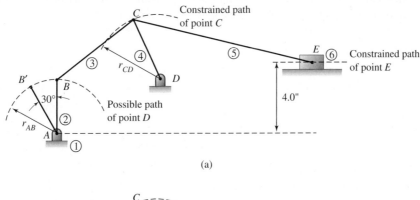

(a)

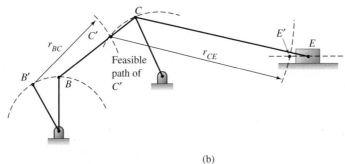

(b)

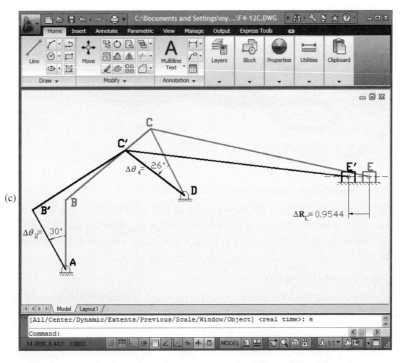

(c)

FIGURE 4.12 Displacement constructions for Example Problem 4.1.

3. *Determine the Paths of All Links Directly Connected to the Frame*

To reposition the mechanism, the constrained paths of all the points on links that are connected to the frame (B,C, and E) are drawn. This is also shown in Figure 4.12a.

4. *Determine the Precise Position of Point C'*

Being rigid, the shape of link 3 cannot change, and the distance between points B and C (r_{BC}) remains constant. Because point B has been moved to B', an arc can be drawn of length r_{BC}, centered at B'. This arc represents the feasible path of point C'. The intersection of this arc with the constrained path of C yields the position of C'. This is shown in Figure 4.12b.

5. ***Determine the Precise Position of Point E'***

   This same logic can be used to locate the position of point E'. The shape of link 5 cannot change, and the distance between points C and E (r_{CE}) remains constant. Because point C has been moved to C', an arc can be drawn of length r_{CE}, centered at C'. This arc represents the feasible path of point E'. The intersection of this arc with the constrained path of E yields the position of E' (Figure 4.12b).

6. ***Measure the Displacement of Link 4 and Point E'***

   Finally, with the position of C' and E' determined, links 3 through 6 can be drawn. This is shown in Figure 4.12c. The displacement of link 4 is the angular distance between the new and original position and measured as

$$\Delta\theta_4 = 26°, \text{counterclockwise}$$

   The displacement of point E is the linear distance between the new and original position of point E. The distance between E and E' is measured and adjusted for the drawing scale.

$$\Delta\mathbf{R}_E = .9544 \text{ in. } \leftarrow$$

EXAMPLE PROBLEM 4.2

Compound-lever snips, as shown in Figure 4.13, are often used in place of regular tinner snips when large cutting forces are required. Using the top handle as the frame, graphically reposition the components of the snips when the jaw is opened 15°. Determine the resulting displacement of the lower handle.

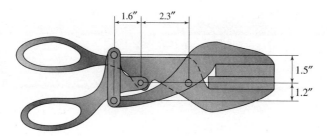

FIGURE 4.13 Cutting snips for Example Problem 4.2.

SOLUTION:

1. ***Draw the Kinematic Diagram and Calculate Mobility***

   The kinematic diagram for the snips is given in Figure 4.14a. The top handle has been designated as the frame, and points of interest were identified at the tip of the upper cutting jaw (X) and at the end of the bottom handle (Y). Notice that this is the familiar four-bar mechanism, with one degree of freedom. Moving one link, namely the jaw, uniquely positions all other links of the mechanism.

2. ***Reposition the Driving Link***

   To reposition the mechanism, the top cutting jaw, link 2, is rotated 15° counterclockwise. This movement corresponds to an open position. The point of interest, X, also needs to be rotated with link 2.

3. ***Determine the Precise Position of Point C'***

   Because this is a four-bar mechanism, the position of point C' is the intersection of its constrained path and feasible path. Figure 4.14b shows the constructions necessary to determine the position of C'.

4. ***Determine the Precise Position of Point Y'***

   Finally, the location of point of interest Y must be determined. Link 4 is rigid and its shape is not altered. Because the side $C'D$ has been located, point Y' can be readily found.

   Similar to the procedure described in Figure 4.7b, the length of side DY does not change. Therefore, the path of point Y can be constructed from point D. Also, the length of side CY will not change. However, point C has been relocated to C'. Another feasible path for Y' can be constructed from C'. The intersection of these two paths gives the final location of Y'. This construction is shown in Figure 4.14c.

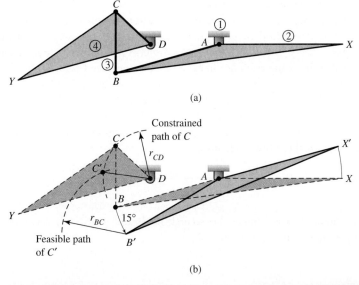

(a)

(b)

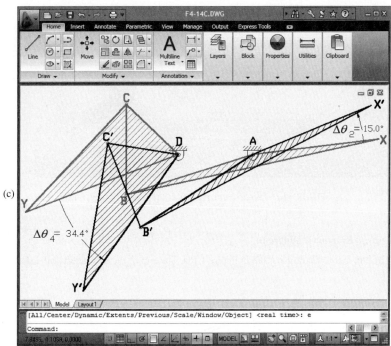

(c)

FIGURE 4.14 Constructions for Example Problem 4.2.

5. ***Measure the Displacement of Link 4***

The displacement required from the bottom handle in order to open the jaw 15° can be measured. From Figure 4.14c, the bottom handle, link 4, must be displaced:

$$\Delta\theta_4 = 35°, \text{counterclockwise}$$

4.6 POSITION: ANALYTICAL ANALYSIS

Generally speaking, analytical methods can be used in position analysis to yield results with a high degree of accuracy. This accuracy comes with a price in that the methods often become numerically intensive. Methods using complex

notation, involving higher-order math, have been developed for position analysis [Refs. 4, 9, 11, 12].

For design situations, where kinematic analysis is not a daily task, these complex methods can be difficult to understand and implement. A more straightforward method of position analysis involves using the trigonometric laws for triangles. Admittedly, this "brute-force" technique is not efficient for those involved in kinematic research. However, for the typical design engineer, the simplicity far outweighs all inefficiencies. Thus, this triangle method of position analysis will be used in this text.

In general, this method involves inserting reference lines within a mechanism and analyzing the triangles. Laws of general and right triangles are then used to determine the lengths of the triangle sides and the magnitude of the interior angles. As details about the geometry of the triangles are determined, this information is assembled to analyze the entire mechanism.

A substantial benefit of analytical analysis is the ability to alter dimensions and quickly recalculate a solution. During the design stages, many machine configurations and dimensions are evaluated. Graphical analysis must be completely repeated for each evaluation. Analytical methods, specifically when implemented with spreadsheets or other computer-based tools, can update solutions quickly.

The analytical method of position analysis can best be seen through the following examples.

EXAMPLE PROBLEM 4.3

Figure 4.15 shows a toggle clamp used to securely hold parts. Analytically determine the displacement of the clamp surface as the handle rotates downward, 15°.

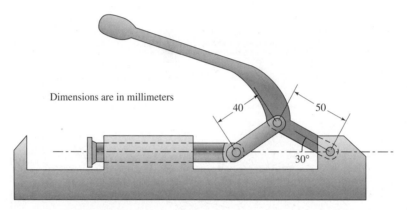

FIGURE 4.15 Toggle clamp for Example Problem 4.3.

SOLUTION: 1.   ***Draw a Kinematic Diagram***

The kinematic diagram is given in Figure 4.16a. The end of the handle was labeled as point of interest *X*.

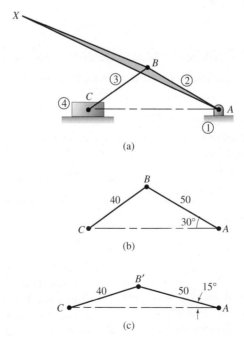

FIGURE 4.16 Mechanism for Example Problem 4.3.

2. **Analyze the Geometry in the Original Configuration**

For this slider-crank mechanism, a triangle is naturally formed between pin joints A, B, and C. This triangle is shown in Figure 4.16b.

Prior to observing the mechanism in a displaced configuration, all properties of the original configuration must be determined. The internal angle at joint C, $\angle BCA$, can be determined from the law of sines, equation (3.6):

$$\frac{sin \angle BAC}{(BC)} = \frac{sin \angle BCA}{(AB)}$$

$$\angle BCA = sin^{-1}\left[\left(\frac{AB}{BC}\right)sin \angle BAC\right] = sin^{-1}\left[\left(\frac{50 \text{ mm}}{40 \text{ mm}}\right)sin 30°\right] = 38.68°$$

The interior angle at joint B, $\angle ABC$, can be found because the sum of all interior angles in any triangle must total 180°:

$$\angle ABC = 180° - (30° + 38.68°) = 111.32°$$

The length side AC represents the original position of the slider and can be determined from the law of cosines, equation (3.7):

$$AC = \sqrt{AB^2 + BC^2 - 2(AB)(BC)\cos \angle ABC}$$

$$= \sqrt{(50 \text{ mm})^2 + (40 \text{ mm})^2 - 2(50 \text{ mm})(40 \text{ mm})\{\cos 111.32°\}}$$

$$= 74.52 \text{ mm}$$

3. **Analyze the Geometry in the Displaced Configuration**

The displaced configuration is shown in Figure 4.16c when the handle is rotated downward 15°. Note that this displacement yields an interior angle at joint A, $\angle C'AB'$, of 15°. The law of sines can be used to find the interior angle at joint C', $\angle B'C'A$:

$$\angle B'C'A = sin^{-1}\left[\left(\frac{AB'}{B'C'}\right)sin \angle C'AB'\right] = sin^{-1}\left[\left(\frac{50 \text{ mm}}{40 \text{ mm}}\right)sin 15°\right] = 18.88°$$

Again, the interior angle at joint B', $\angle AB'C'$, can be found because the sum of all interior angles in any triangle must total 180°:

$$\angle AB'C' = 180° - (15° + 18.88°) = 146.12°$$

The length side AC' represents the displaced position of the slider. As before, it can be determined from the law of cosines:

$$AC' = \sqrt{AB'^2 + B'C'^2 - 2(AB')(B'C')\cos \angle AB'C'}$$

$$= \sqrt{(50 \text{ mm})^2 + (40 \text{ mm})^2 - 2(50 \text{ mm})(40 \text{ mm})\cos(146.12°)} = 86.14 \text{ mm}$$

$$= 86.14 \text{ mm}$$

4. **Calculate the Desired Displacement**

The displacement of point C during this motion can be found as the difference of the triangle sides AC' and AC:

$$\Delta \mathbf{R}_C = AC' - AC = 86.14 - 74.52 = 11.62 \text{ mm} \leftarrow$$

4.6.1 Closed-Form Position Analysis Equations for an In-Line Slider-Crank

The clamp mechanism in Example Problem 4.3 is a slider-crank linkage. Specifically, it is termed an in-line slider-crank mechanism because the constrained path of the pin joint on the slider extends through the center of the crank rotation. Figure 4.17 illustrates the basic configuration of an in-line slider-crank linkage.

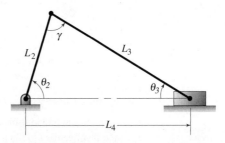

FIGURE 4.17 In-line slider-crank mechanism.

Because this is a common mechanism, the results from the previous problem can be generalized [Ref. 12]. A typical analysis involves locating the position of the links, given their lengths (L_2 and L_3) and the crank angle (θ_2). Specifically, the position of the slider (L_4) and the interior joint angles (θ_3 and θ) must be determined.

The equations used in Example Problem 4.3 are summarized in terms of L_2, L_3, and θ_2:

$$\theta_3 = \sin^{-1}\left[\frac{L_2}{L_3}\sin\theta_2\right] \tag{4.3}$$

$$\gamma = 180° - (\theta_2 + \theta_3) \tag{4.4}$$

$$L_4 = \sqrt{L_2^2 + L_3^2 - 2(L_2)(L_3)\cos\gamma} \tag{4.5}$$

These equations can be used to determine the position of the links in any configuration of an in-line slider-crank mechanism.

EXAMPLE PROBLEM 4.4

Figure 4.18 shows a concept for a hand pump used for increasing oil pressure in a hydraulic line. Analytically determine the displacement of the piston as the handle rotates 15° counterclockwise.

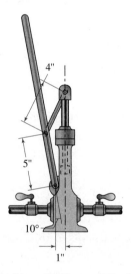

FIGURE 4.18 Toggle clamp for Example Problem 4.4.

SOLUTION: 1. ***Draw a Kinematic Diagram***

The kinematic diagram is given in Figure 4.19a. The end of the handle was labeled as point of interest X.

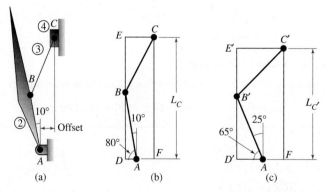

FIGURE 4.19 Mechanism diagrams for Example Problem 4.4.

2. ***Analyze the Geometry in the Original Configuration***

In contrast to the previous problem, this mechanism is an offset slider-crank mechanism. For this type of mechanism, it is convenient to focus on two right triangles. These triangles are shown in Figure 4.19b. Notice that the 10° angle and its 80° complement are shown.

Prior to observing the mechanism in a displaced configuration, all properties of the original configuration must be determined. Focusing on the lower right triangle, the sides AD and BD can be determined from the following trigonometric functions:

$$\cos \angle BAD = \frac{AD}{AB}$$

$$AD = (AB)\cos \angle BAD = (5\,\text{in.})\ \{\cos 80°\} = 0.87\,\text{in.}$$

$$\sin \angle BAD = \frac{BD}{AB}$$

$$BD = (AB)\sin \angle BAD = (5\,\text{in.})\ \{\sin 80°\} = 4.92\,\text{in.}$$

By focusing on the top triangle, the length of side CE can be found as the sum of the offset distance and the length of side AD from the lower triangle:

$$CE = \text{offset} + AD = 1.0 + 0.87 = 1.87\,\text{in.}$$

Use the Pythagorean theorem, equation (3.4), to determine side BE:

$$BE = \sqrt{BC^2 - CE^2}$$
$$= \sqrt{(4)^2 - (1.87)^2} = 3.54\,\text{in.}$$

The original position of the piston, point C, can be determined by summing BD and BE:

$$L_C = BD + BE = 4.92 + 3.54 = 8.46\,\text{in.}$$

Although not required in this problem, the angle that defines the orientation of link 3 is often desired. The angle $\angle BCE$ can be determined with the inverse cosine function:

$$\angle BCE = \cos^{-1}\left(\frac{CE}{BC}\right) = \cos^{-1}\left(\frac{1.87\,\text{in.}}{4\,\text{in.}}\right) = 62.13°$$

3. **Analyze the Geometry in the Displaced Configuration**

   The displaced configuration is shown in Figure 4.19c with the handle rotated downward 15°. Note that this displacement yields an angle at joint A of 25°, and its complement, 65°, is also shown. Focusing on the lower right triangle, the sides AD' and $B'D'$ can be determined from the following trigonometric functions:

   $$AD' = (AB')\cos \angle B'AD' = (5\,\text{in.})\ \{\cos 65°\} = 2.11\,\text{in.}$$

   $$B'D' = (AB')\sin \angle B'AD' = (5\,\text{in.})\ \{\sin 65°\} = 4.53\,\text{in.}$$

   Focusing on the top triangle, the length of side $C'E'$ can be found as the sum of the offset distance (AF) and the length of side AD' from the lower triangle:

   $$C'E' = AF + AD'$$
   $$= 1.0 + 2.11 = 3.11\,\text{in.}$$

   Side $B'E'$ can then be determined:

   $$B'E' = \sqrt{(B'C')^2 - (C'E')^2} = \sqrt{(4\,\text{in.})^2 - (3.11\,\text{in.})^2} = 2.52\,\text{in.}$$

   The displaced position of the piston can be determined by summing $B'D'$ and $B'E'$:

   $$L_C' = B'D' + B'E' = 4.53 + 2.52 = 7.05\,\text{in.}$$

4. **Calculate the Resulting Displacement**

   The displacement of the piston, point C, during this motion can be found by subtracting the length L_C' from L_C:

   $$\Delta \mathbf{R}_C = 8.46 - 7.05 = 1.41\,\text{in.} \downarrow$$

4.6.2 Closed-Form Position Analysis Equations for an Offset Slider-Crank

The mechanism in Example Problem 4.4 is an offset slider-crank mechanism because the constrained path of the pin joint on the slider does not extend through the center of the crank rotation. Figure 4.20 illustrates the basic configuration of an offset slider-crank linkage.

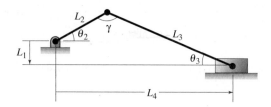

FIGURE 4.20 Offset slider-crank mechanism.

Because this is also a common mechanism, the results from the previous problem can be generalized [Ref. 12].

A typical analysis involves locating the position of the links, given the lengths (L_1, L_2, and L_3) and a crank angle (θ_2). Specifically, the position of the slider (L_4) and the interior joint angles (θ_3 and θ) must be determined.

The generalized equations are given as

$$\theta_3 = \sin^{-1}\left[\frac{L_1 + L_2 \sin \theta_2}{L_3}\right] \qquad (4.6)$$

$$L_4 = L_2 \cos \theta_2 + L_3 \cos \theta_3 \qquad (4.7)$$

$$\gamma = 180° - (\theta_2 + \theta_3) \qquad (4.8)$$

These equations can be used to determine the position of the links in any mechanism configuration. Recall, however, that these equations are only applicable to an offset slider-crank mechanism. The equations also apply when the offset distance is in the opposite direction as shown in Figure 4.20. For these cases, L_1 in equation (4.6) should be substituted as a negative value.

EXAMPLE PROBLEM 4.5

Figure 4.21 shows a toggle clamp used for securing a workpiece during a machining operation. Analytically determine the angle that the handle must be displaced in order to lift the clamp arm 30° clockwise.

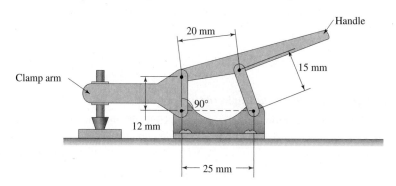

FIGURE 4.21 Clamp for Example Problem 4.5.

SOLUTION:

1. **Draw a Kinematic Diagram**

   The kinematic diagram for the clamp is given in Figure 4.22a. The end of the handle was labeled as point of interest X. The clamp nose was identified as point of interest Y.

2. **Analyze the Geometry in the Original Configuration**

   This mechanism is the common four-bar linkage. In order to more closely analyze the geometry, Figure 4.22b focuses on the kinematic chain ABCD. A diagonal is created by connecting B and D, forming two triangles.

   Prior to observing the mechanism in a displaced configuration, all properties of the original configuration must be determined. Notice that the lower triangle, ABD, is a right triangle. The length of BD can be found using the Pythagorean theorem introduced in equation (3.4).

   $$BD = \sqrt{(AB)^2 + (AD)^2} = \sqrt{(12)^2 + (25)^2} = 27.73 \text{ mm}$$

   The internal angles, $\angle ABD$ and $\angle BDA$, can be determined from the following basic trigonometric functions:

   $$\angle ABD = \sin^{-1}\left(\frac{25 \text{ mm}}{27.73 \text{ mm}}\right) = 64.4°$$

   $$\angle BDA = \cos^{-1}\left(\frac{25 \text{ mm}}{27.73 \text{ mm}}\right) = 25.6°$$

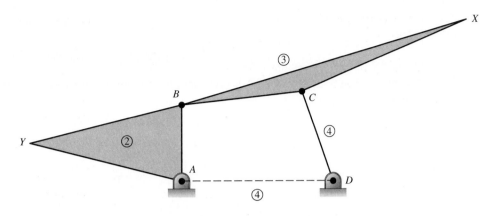

(a) Kinematic diagram

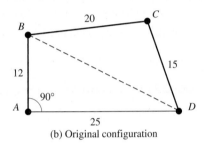

(b) Original configuration

(c) Displaced configuration

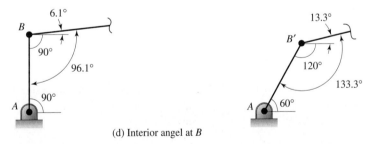

(d) Interior angel at B

FIGURE 4.22 Mechanism for Example Problem 4.5.

Focusing on the top triangle, the internal angle $\angle BCD$ can be found from the law of cosines, introduced in equation (3.7):

$$\angle BCD = \cos^{-1}\left(\frac{BC^2 + CD^2 - BD^2}{2(BC)(CD)}\right)$$

$$= \cos^{-1}\left(\frac{(20 \text{ mm})^2 + (15 \text{ mm})^2 - (27.73 \text{ mm})^2}{2(20 \text{ mm})(15 \text{ mm})}\right) = 103.9°$$

The internal angle $\angle CBD$ can be determined from the law of sines:

$$\angle CBD = \sin^{-1}\left[\left(\frac{CD}{BD}\right)\sin \angle BCD\right]$$

$$= \sin^{-1}\left[\left(\frac{15 \text{ mm}}{27.73 \text{ mm}}\right)\sin 103.9°\right] = 31.7°$$

The interior angle at $\angle BDC$ can be found because the sum of all interior angles in any triangle must total 180°. Thus

$$\angle BDC = 180° - (103.9° + 31.7°) = 44.4°$$

The total mechanism angles at joint B (between links 2 and 3) and at joint D (between links 1 and 4) can be determined.

At joint B:

$$\angle ABC = \angle ABD + \angle CBD = 64.4° + 31.7° = 96.1°$$

At joint D:

$$\angle CDA = \angle BDC + \angle BDA = 44.4° + 25.6° = 70.0°$$

3. **Analyze the Geometry in the Displaced Configuration**

    The displaced configuration is shown in Figure 4.22c with the clamp nose, link 2, rotated clockwise 30°. Notice that this leaves the interior angle at joint A, $\angle DAB'$, as 60°. Also, the lower triangle is no longer a right triangle.

    The length of diagonal $B'D$ can be found by using the lower triangle, ΔABD, and the law of cosines:

    $$B'D = \sqrt{(12 \text{ mm})^2 + (25 \text{ mm})^2 - 2(12 \text{ mm})(25 \text{ mm}) \cos 60°} = 21.66 \text{ mm}$$

    The internal angle $\angle AB'D$ can also be determined from the law of cosines:

    $$\angle AB'D = \cos^{-1} \frac{(AB)^2 + (B'D)^2 - (AD)^2}{2(AB')(B'D)}$$

    $$= \cos^{-1}\left[\frac{(12)^2 + (21.66)^2 - (25)^2}{2(12)(21.66)}\right] = 91.3°$$

    The total of the interior angles of any triangle must be 180°. Therefore, angle $\angle B'DA$ can be readily determined:

    $$\angle B'DA = 180° - (\angle DAB' + \angle AB'D)$$

    $$= 180° - (60° + 91.3°) = 28.7°$$

    Focusing on the top triangle, the internal angle $\angle B'C'D$ can be found from the law of cosines:

    $$\angle B'C'D = \cos^{-1}\left[\frac{(B'C')^2 + (C'D)^2 - (B'D)^2}{2(B'C')(C'D)}\right]$$

    $$= \cos^{-1}\left[\frac{(20 \text{ mm})^2 + (15 \text{ mm})^2 - (21.66 \text{ mm})^2}{2(20 \text{ mm})(15 \text{ mm})}\right] = 74.9°$$

    The internal angle $\angle C'B'D$ can be determined from the law of sines:

    $$\angle C'B'D = \sin^{-1}\left[\left(\frac{C'D}{B'D}\right)\sin \angle B'C'D\right]$$

    $$= \sin^{-1}\left[\left(\frac{15 \text{ mm}}{21.66 \text{ mm}}\right)\sin 74.9°\right] = 42.0°$$

    The final interior angle, $\angle B'DC'$, of the upper triangle can be found by the following:

    $$\angle B'DC' = 180° - (\angle C'B'D + \angle B'C'D) = 180° - (42.0° + 74.9°) = 63.1°$$

    The total mechanism angles at joint B (between links 2 and 3) and at joint D (between links 1 and 4) can be determined by the following:

    At joint B':

    $$\angle AB'C' = \angle AB'D + \angle C'B'D = 91.3° + 42.0° = 133.3°$$

    At joint D:

    $$\angle C'DA = \angle B'DC' + \angle B'DA = 63.1° + 28.7° = 91.8°$$

4.   ***Calculate the Resulting Displacement***

The angular displacement of the handle, link 3, can be determined by focusing on joint B, as shown in Figure 4.22d. For the original configuration, the angle of link 3 above the horizontal is expressed as

$$\angle ABC - 90° = 96.1° - 90.0° = 6.1°$$

For the displaced configuration, the angle of link 3 above the horizontal is expressed as

$$\angle AB'C' - 120° = 133.3° - 120.0° = 13.3°$$

Finally, the angular displacement of link 3 is determined by

$$\Delta \theta_3 = 13.3° - 6.1° = 7.2°, \text{counterclockwise}$$

4.6.3 Closed-Form Position Equations for a Four-Bar Linkage

The four-bar mechanism is another very common linkage. Figure 4.23 illustrates a general four-bar linkage.

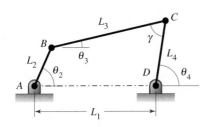

FIGURE 4.23 The four-bar mechanism.

The specific equations used in Example Problem 4.5 can be generalized [Ref. 12]. A typical analysis involves determining the interior joint angles (θ_3, θ_4, and γ) for known links (L_1, L_2, L_3, and L_4) at a certain crank angle (θ_2). Specifically, the interior joint angles (θ_3, θ_4, and γ) must be determined.

$$BD = \sqrt{L_1^2 + L_2^2 - 2(L_1)(L_2)\cos(\theta_2)} \quad \textbf{(4.9)}$$

$$\gamma = \cos^{-1}\left[\frac{(L_3)^2 + (L_4)^2 - (BD)^2}{2(L_3)(L_4)}\right] \quad \textbf{(4.10)}$$

$$\theta_3 = 2\tan^{-1}\left[\frac{-L_2\sin\theta_2 + L_4\sin\gamma}{L_1 + L_3 - L_2\cos\theta_2 - L_4\cos\gamma}\right] \quad \textbf{(4.11)}$$

$$\theta_4 = 2\tan^{-1}\left[\frac{L_2\sin\theta_2 - L_3\sin\gamma}{L_2\cos\theta_2 + L_4 - L_1 - L_3\cos\gamma}\right] \quad \textbf{(4.12)}$$

These equations can be used to determine the position of the links in any mechanism configuration. The equations are applicable to any four-bar mechanism assembled as shown in Figure 4.23.

4.6.4 Circuits of a Four-Bar Linkage

For four-bar mechanisms classified as crank-rockers (as described in Section 1.10), there are two regions of possible motion corresponding with the two geometric inversions. These regions are termed *assembly circuits*. A mechanism is unable to move between assembly circuits without being

disassembled. The mechanism shown in Figure 4.23 operates in the first circuit (Figure 4.24a).

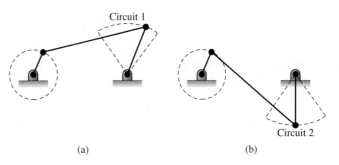

(a)   (b)

FIGURE 4.24 Circuits of a four-bar mechanism.

By physically disconnecting joint C, the links can be reoriented and reassembled into the configuration shown in Figure 4.24b. As this mechanism is operated, it exhibits motion in the second circuit. Although the motion of the mechanism appears to be different, depending on the circuit of operation, the relative motion between the links does not change. However, the circuit in which the mechanism is assembled must be specified to understand the absolute motion and operation of the mechanism.

For four-bar mechanisms operating in the second circuit, equation (4.11) must be slightly altered as follows:

$$\theta_3 = 2\tan^{-1}\left[\frac{-L_2\sin\theta_2 - L_4\sin\gamma}{L_1 + L_3 - L_2\cos\theta_2 - L_4\cos\gamma}\right] \textbf{(4.13)}$$

$$\theta_4 = 2\tan^{-1}\left[\frac{L_2\sin\theta_2 + L_3\sin\gamma}{L_2\cos\theta_2 + L_4 - L_1 - L_3\cos\gamma}\right]\textbf{(4.14)}$$

4.7 LIMITING POSITIONS: GRAPHICAL ANALYSIS

The configuration of a mechanism that places one of the follower links in an extreme location is called a *limiting position*. Many machines have linkages that continually oscillate between two limiting positions. Figure 4.25 illustrates the limiting positions of an offset slider-crank mechanism.

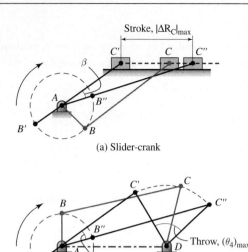

(a) Slider-crank

(b) Four-bar

FIGURE 4.25 Limiting positions.

The displacement of the follower link from one limiting position to the other defines the *stroke* of the follower. For translating links, as shown in Figure 4.25a, the stroke is a linear measurement. For links that exhibit pure rotation, the stroke is an angular quantity, and is also called *throw*, as shown in Figure 4.25b. The configuration of links that place a follower in a limiting position is associated with the crank and coupler becoming collinear. Figure 4.25 illustrates the limiting configurations for a slider-crank and a four-bar linkage. An imbalance angle β is defined as the angle between the coupler configuration at the two limiting positions. The imbalance angle influences the timing of the inward and outward stroke and will be extensively utilized in Chapter 5. The position of a driver, or actuated link, that places a follower link in an extreme, or limiting, position is often desired. In addition, the motion of a linkage is commonly referenced from the actuator position that places the follower in a limiting position.

The logic used on solving such a problem is similar to the position analysis just discussed. The following examples illustrate this analysis.

EXAMPLE PROBLEM 4.6

The mechanism shown in Figure 4.26 is the driving linkage for a reciprocating saber saw. Determine the configurations of the mechanism that places the saw blade in its limiting positions.

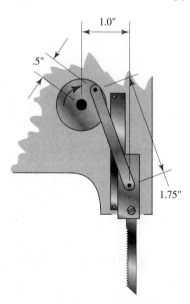

FIGURE 4.26 Saber saw mechanism for Example Problem 4.6.

SOLUTION:

1. ***Draw a Kinematic Diagram***

   The kinematic diagram for the reciprocating saw mechanism is given in Figure 4.27a. Notice that this is a slider-crank mechanism as defined in Chapter 1. The slider-crank has one degree of freedom.

2. ***Construct the Extended Limiting Position***

   The saw blade, link 4, reaches its extreme downward position as links 2 and 3 move into a collinear alignment. This configuration provides the maximum distance between points A and C. To determine this maximum distance, the lengths of links 2 and 3 must be combined. Adding these lengths,

   $$L_2 + L_3 = 0.5 \text{ in.} + 1.75 \text{ in.} = 2.25 \text{ in.}$$

   Once the combined length of lines 2 and 3 is determined, an arc should be constructed of this length, centered at point A. As shown in Figure 4.29b, the intersection of this arc and the possible path of point determines the limiting extended position of C, denoted C. Links 2 and 3 can be drawn, and point B' can be determined. This is shown in Figure 4.29c.

3. ***Construct the Retracted Limiting Position***

Next, the configuration that places the saw blade, link 4, in its extreme upper position must be determined. In this configuration, links 2 and 3 are again collinear but overlapped. This provides the minimum distance between points A and C. Thus, this minimum distance is the difference between the lengths of links 3 and 2. Subtracting the link lengths gives

$$L_3 - L_2 = 1.75 \text{ in.} - 0.5 \text{ in.} = 1.25 \text{ in.}$$

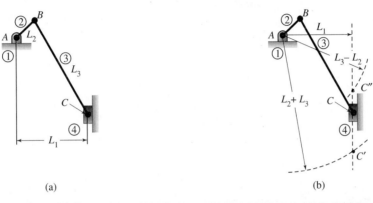

(a)   (b)

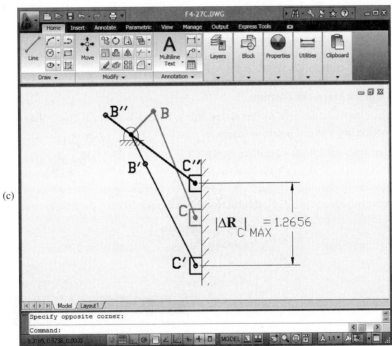

FIGURE 4.27 Extreme positions for Example Problem 4.6.

This retracted limiting position can be determined using a technique similar to determining the extended position. Recall that the distance between A and C' in Figure 4.27b represents the combined length of links 2 and 3. Similarly, the distance between points A and C'' represents the difference between links 3 and 2.

Using the $L_3 - L_2$ distance, the position of point C at its extreme upward position, denoted as C'', can be determined (Figure 4.27b). Finally, links 2 and 3 can be drawn and the position of point B'' is located.

4. ***Measure the Stroke of the Follower Link***

As shown in Figure 4.27c, the stroke of the saw blade can be measured as the extreme displacement of point C. Scaling this from the kinematic diagram yields the following result:

$$|\Delta R_C|_{max} = 1.27 \text{ in.}$$

EXAMPLE PROBLEM 4.7

Figure 4.28 illustrates a linkage that operates a water nozzle at an automatic car wash. Determine the limiting positions of the mechanism that places the nozzle in its extreme positions.

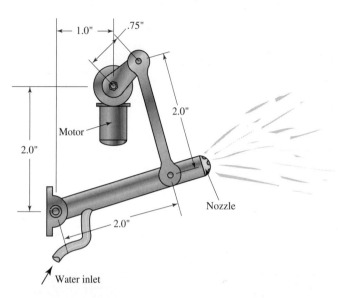

FIGURE 4.28 Water nozzle linkage for Example Problem 4.7.

SOLUTION: 1. ***Draw the Kinematic Diagram***

The kinematic diagram for the water nozzle linkage is given in Figure 4.29. Notice that this is a four-bar mechanism with one degree of freedom.

2. ***Construct the Extended Limiting Position***

The analysis in this example is very similar to Example Problem 4.6. The nozzle, link 4, reaches its extreme downward position as links 2 and 3 become collinear. This configuration provides the maximum distance between points A and C. To determine this maximum distance, the lengths of links 2 and 3 must be combined. Adding these lengths gives

$$L_2 + L_3 = 0.75 \text{ in.} + 2.00 \text{ in.} = 2.75 \text{ in.}$$

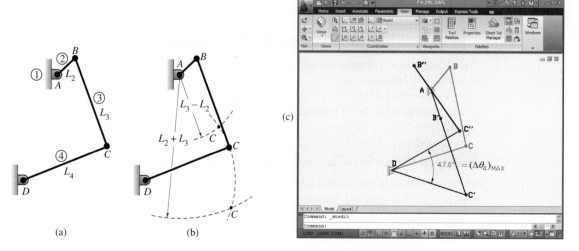

FIGURE 4.29 Extreme positions for Example Problem 4.7.

Once the combined length of lines 2 and 3 is determined, an arc should be constructed of this length, centered at point A. As shown in Figure. 4.28b, the intersection of this arc and the possible path of point C determines the extreme downward position of C, denoted C'. Links 2 and 3 can be drawn, and point B' can be determined. This is shown in Figure 4.29c.

3. ***Construct the Retracted Limiting Position***

    Next, the configuration that places the nozzle, link 4, in its extreme upper position must be determined. Similar to the slider-crank discussed in Example Problem 4.6, the retracted configuration occurs when links 2 and 3 are collinear but overlapped. This produces the minimum distance between points A and C. Thus, this minimum distance is the difference between the lengths of links 3 and 2. Subtracting the link lengths gives

$$L_3 - L_2 = 2.00 \text{ in.} - .75 \text{ in.} = 1.25 \text{ in.}$$

    This minimum distance can be constructed similar to the technique for the maximum distance. Recall that the distance between A and C' in Figure 4.29c represents the combined length of links 2 and 3. Similarly, the distance between points A and C'' represents the difference between links 3 and 2.

    Using the $L_3 - L_2$ distance, the position of point C at its extreme upward position, denoted as C'', can be determined. This is shown in Figure 4.29b. Finally, links 2 and 3 can be drawn and the position of point B'' is located.

4. ***Measure the Stroke of the Follower Link***

    As shown in Figure 4.29c, the stroke of the nozzle can be measured as the extreme angular displacement of link 4. Measuring this from the graphical layout yields the following:

$$|\Delta\theta_4|_{\text{max}} = 47.0°$$

4.8 LIMITING POSITIONS: ANALYTICAL ANALYSIS

Analytical determination of the limiting positions for a mechanism is a combination of two concepts presented earlier in this chapter:

I. The logic of configuring the mechanism into a limiting configuration. This was incorporated in the graphical method for determining the limiting positions, as presented in Section 4.7.

II. The method of breaking a mechanism into convenient triangles and using the laws of trigonometry to determine all mechanism angles and lengths, as presented in Section 4.6.

Combining these two concepts to determine the position of all links in a mechanism at a limiting position is illustrated through Example Problem 4.8.

EXAMPLE PROBLEM 4.8

Figure 4.30 shows a conveyor transfer mechanism. Its function is to feed packages to a shipping station at specific intervals. Analytically determine the extreme positions of the lifting conveyor segment.

SOLUTION:

1. ***Draw the Kinematic Diagram***

    The kinematic diagram for the mechanism is given in Figure 4.31a. The end of the conveyor segment is labeled as point of interest X.

2. ***Analyze the Geometry at the Extended Limiting Position***

    This mechanism is another four-bar linkage. As seen in Example Problem 4.7, the follower of a four-bar is in the extended limiting position when links 2 and 3 become collinear. Figure 4.31b illustrates this mechanism with the follower in its upper position. Notice that the links form a general triangle, $\Delta AC'D$. Also note that the length of AC' is 20 in. (16 + 4).

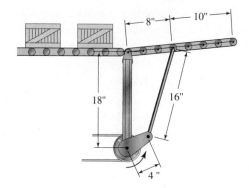

FIGURE 4.30 Conveyor feed for Example Problem 4.8.

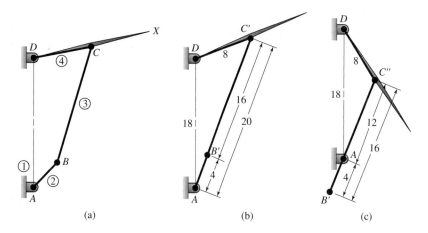

FIGURE 4.31 Mechanism for Example Problem 4.8.

This upper-limiting position is fully defined by determination of the internal angles. The internal angle at joint A, $\angle C'AD$, can be found using the law of cosines:

$$\angle C'AD = \cos^{-1}\left[\frac{AD^2 + AC' - C'D^2}{2(AD)(AC')}\right]$$

$$= \cos^{-1}\left[\frac{(18 \text{ in.})^2 + (20 \text{ in.})^2 - (8 \text{ in.})^2}{2(18 \text{ in.})(20 \text{ in.})}\right] = 23.6°$$

The law of sines can be used to find either of the remaining internal angles. However, the law of sines may present some confusion with angles between 90° and 180° because

$$\sin \theta = \sin(180° - \theta)$$

When the inverse sine function is used on a calculator, an angle is between 0° and 90°. However, the desired result may be an angle between 90° and 180°. To minimize this confusion, it is recommended to draw the triangles to an approximate scale and verify numerical results. Also, it is best to use the law of sines with angles that are obviously in the range of 0° to 90°.

Using that approach, the internal angle at joint C', $\angle AC'D$, is determined using the law of sines because it is obviously smaller than 90°.

$$\angle AC'D = \sin^{-1}\left[\left(\frac{AD}{DC'}\right)\sin\angle C'AD\right]$$

$$= \sin^{-1}\left[\left(\frac{18 \text{ in.}}{8 \text{ in.}}\right)\sin 23.6°\right] = 64.1°$$

The internal angle at joint D, $\angle ADC'$, can be determined:

$$\angle ADC' = 180° - (\angle C'AD + \angle ADC')$$

$$= 180° - (23.6° + 64.1°) = 92.3°$$

3. **Analyze the Geometry at the Retracted Limiting Position**

Figure 4.31c illustrates this mechanism with the follower in its lower position. Again, the links form a general triangle, $\triangle AC''D$. Notice that the length of AC'' is 12 in. $(16 - 4)$.

To fully define this configuration, the internal angles are determined through a procedure identical to the one just described.

For the internal angle at joint A, $\angle C''AD$:

$$\angle C''AD = \cos^{-1}\left[\frac{AD^2 + AC'^2 - C''D^2}{2(AD)(AC'')}\right]$$

$$= \cos^{-1}\left[\frac{(18\text{ in.})^2 + (12\text{ in.})^2 - (8\text{ in.})^2}{2(18\text{ in.})(12\text{ in.})}\right] = 20.7°$$

The internal angle at D is in the range of $0°$ to $90°$. Therefore, for the internal angle at joint D, $\angle ADC''$:

$$\angle ADC'' = \sin^{-1}\left[\left(\frac{AC'}{DC'}\right)\sin\angle C''AD\right]$$

$$= \sin^{-1}\left[\left(\frac{12\text{ in.}}{8\text{ in.}}\right)\sin 20.7°\right] = 32.1°$$

Finally, the internal angle at joint C'', $\angle AC''D$, can be determined by the following:

$$\angle AC''D = 180° - (\angle C''AD + \angle ADC'')$$

$$= 180° - (20.7° + 32.1°) = 127.2°$$

4. **Measure the Stroke of the Follower Link**

To summarize, the conveyor segment (internal angle at joint D, $\angle ADC$) cycles between $92.3°$ and $32.1°$, as measured upward from the vertical:

$$32.1° < \theta_4 < 92.3°$$

and the stroke is

$$|\Delta\theta_4|_{max} = 92.3° - 32.1° = 60.2°$$

4.9 TRANSMISSION ANGLE

The *mechanical advantage* of a mechanism is the ratio of the output force (or torque) divided by the input force (or torque). In a linkage, the *transmission angle* γ quantifies the force transmission through a linkage and directly affects the mechanical efficiency. Clearly, the definitions of transmission angle depend on the choice of driving link. The transmission angle for slider-crank and four-bar mechanisms driven by the crank is shown in Figure 4.32. In these linkages, the mechanical advantage is proportional to the sine of the angle γ. As the linkage moves, the transmission angle, along with all other joint angles, and the mechanical advantage constantly change. Often, the extreme transmission angle values are desired.

In the slider-crank, the transmission angle is measured between the coupler and a line normal to the sliding direction. The values for the minimum and maximum transmission angles can be determined by geometrically constructing the configurations as shown in Figure 4.32a. Alternatively, the minimum and maximum transmission angles for a slider-crank can be calculated from

$$\gamma_{min} = \cos^{-1}\left[\frac{L_1 + L_2}{L_3}\right] \tag{4.15}$$

$$\gamma_{max} = \cos^{-1}\left[\frac{L_1 - L_2}{L_3}\right] \tag{4.16}$$

In the four-bar, the transmission angle is measured between the output link and the coupler. As with the slider-crank, the values for the minimum and maximum transmission angles can be determined by geometrically constructing the configurations as shown in Figure 4.32b. Alternatively, the minimum and maximum transmission angles can be calculated from

$$\gamma_{min} = \cos^{-1}\left[\frac{L_3^2 + L_4^2 - (L_1 - L_2)^2}{2L_3L_4}\right] \tag{4.17}$$

$$\gamma_{max} = \cos^{-1}\left[\frac{L_3^2 + L_4^2 - (L_1 + L_2)^2}{2L_3L_4}\right] \tag{4.18}$$

The transmission angle is one measure of the quality of force transmission in the mechanism. Ordinarily, the

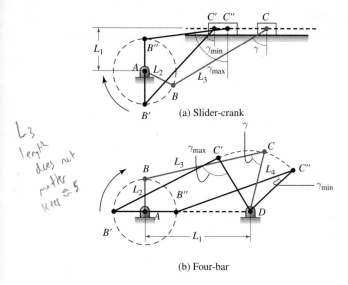

(a) Slider-crank

(b) Four-bar

FIGURE 4.32 Transmission angles.

L_3
length
dues not
matter ≥ 5
well

coupler is a tension or compression link. Thus, it is only able to push or pull along the line that connects the two pins. As a torque applied to the output pivot, optimal force transmission occurs when the transmission angle is 90°. As the transmission angle deviates from 90°, only a component of the coupler force is converted to torque at the pivot. Thus, the transmission angle influences mechanical advantage of a mechanism. The configurations of slider-crank and four-bar mechanisms that produce the maximum and

minimum transmission angles are also shown in Figure 4.32. A common rule of thumb is that the transmission angles should remain between 45° and 135°. Further details are given during the discussion of linkage design in Chapter 5.

4.10 COMPLETE CYCLE: GRAPHICAL POSITION ANALYSIS

The configuration of a mechanism at a particular instant is also referred to as the *phase* of the mechanism. Up to this point, the position analyses focused on determining the phase of a mechanism at a certain position of an input link. Cycle analysis studies the mechanism motion from an original phase incrementally through a series of phases encountered during operation. The assignment of an original phase is used as a reference for the subsequent phases. Any convenient configuration can be selected as the original phase. It is common to use a limiting position as the original, or reference, phase.

To complete a position analysis for an entire cycle, the configuration of the mechanism must be determined at interval phases of its cycle. The procedure, whether graphical or analytical, is exactly the same as detailed in the previous sections. The only adaptation is that these procedures are repeated at set intervals of the input displacement. The following example problems illustrate the position analysis for a full cycle.

EXAMPLE PROBLEM 4.9

Figure 4.33 shows the driving mechanism of handheld grass shears. The mechanism operates by rotating the large disc as shown. Graphically determine the position of the driving mechanism at several phases of its operating cycle.

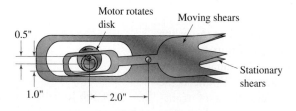

FIGURE 4.33 Grass shears for Example Problem 4.9.

SOLUTION: 1. **Draw the Kinematic Diagram and Calculate Mobility**

The kinematic diagram is given in Figure 4.34a. The end of the middle cutting blade is labeled as point of interest X.

The mobility of the mechanism is calculated as:

$$n = 4 \; j_p = (3 \text{ pins} + 1 \text{ sliding}) = 4 \; j_h = 0$$

and

$$M = 3(n - 1) - 2j_p - j_h$$
$$= 3(4 - 1) - 2(4) - 0 = 1$$

Therefore, the one input link will be moved to operate the shears.

2. ***Designate the Reference Phase***

To assign a reference phase, only the position of the input link must be specified. Arbitrarily select the configuration when the drive disk, link 2, is in a vertical position, with joint B directly below joint A.

3. ***Construct Interval Phases***

Drawing the mechanism in several phases of its cycle is identical to the previous position analysis, but repetitive. While drawing the different phases using graphical methods, the kinematic diagram can become cluttered very

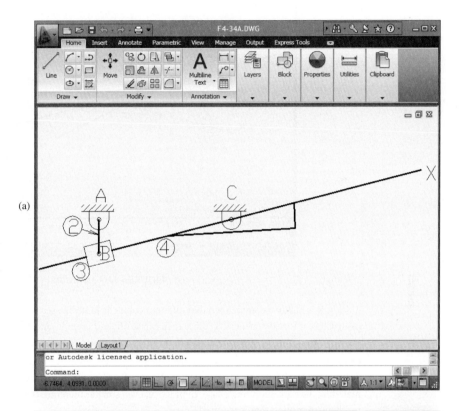

(a)

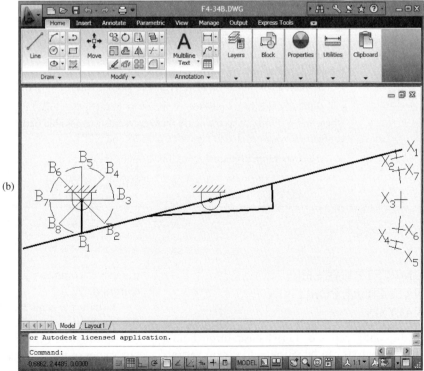

(b)

FIGURE 4.34 Mechanism phases for Example Problem 4.9.

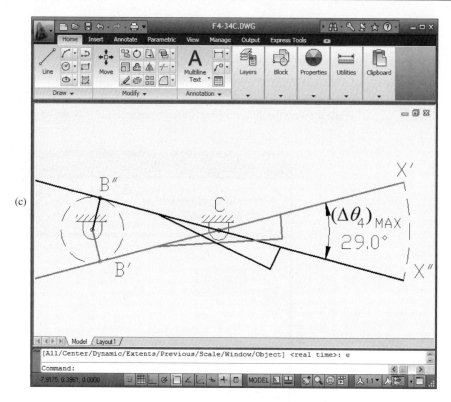

FIGURE 4.34 Continued

quickly. It is highly recommended that different colors or fonts be used to represent each phase of the cycle. When using CAD, it is also beneficial to place each phase on a different layer, which can be rapidly displayed or hidden.

For this problem, the driving link, link 2, is positioned at 45° intervals throughout its cycle. Therefore, eight phases of the mechanism are constructed. The phases are designated as phase 1 through 8. The eight positions of points B and X are shown in Figure 4.34b. Notice that the points are identified by using a subscript from 1 to 8, matching the corresponding phase. In practice, even smaller increments are used depending on the details of the mechanism motion that is required.

4. ***Construct the Limiting Positions***

   The phases associated with the limiting positions should also be determined. The shear blade reaches its upwardmost position when link 4 rotates to the greatest angle possible. This occurs when link 4 is tangent to the circle that represents the possible positions of point B. The point of tangency is denoted as B', and the corresponding position of the blade is denoted as X'. This is shown in Figure 4.34c.

   Similarly, the lowest position of the blade occurs when link 4 dips to its lowest angle. Again, this occurs when link 4 is tangent to the circle that represents the possible paths of B. The points related to this lowest configuration are denoted in Figure 4.34c as B'' and X''.

   The maximum displacement of link 4 can be measured from the kinematic construction:

$$|\Delta\theta_4|_{\max} = 29.0°$$

4.11 COMPLETE CYCLE: ANALYTICAL POSITION ANALYSIS

To generate the configuration of a mechanism throughout a cycle, analytical analysis can be repeated to obtain various phases. This can be an extremely repetitive process and the use of computer programs as discussed in Chapter 8 is common.

Equations, generated from triangles defined in part by the mechanism links, can be summarized as in equations (4.1) through (4.12). These equations can be solved for various values of the driver position. Computer spreadsheets as discussed in Chapter 8 are ideal for such analyses.

EXAMPLE PROBLEM 4.10

Figure 4.35 shows a mechanism that is designed to push parts from one conveyor to another. During the transfer, the parts must be rotated as shown. Analytically determine the position of the pusher rod at several phases of its motion.

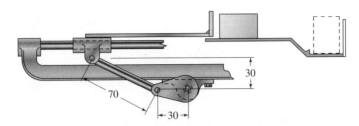

FIGURE 4.35 Conveyor feed for Example Problem 4.10.

SOLUTION:

1. **Draw the Kinematic Diagram**

   The kinematic diagram for this mechanism is shown in Figure 4.36. Notice that it is an offset slider-crank mechanism having one degree of freedom.

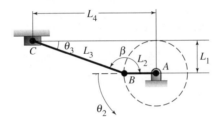

FIGURE 4.36 Kinematic diagram for Example Problem 4.10.

2. **Designate the Original Phase**

   Arbitrarily, the original phase is selected to be when the crank is horizontal, placing joint B directly left of joint A.

3. **Construct the Interval Phases**

   Recall that equations (4.6), (4.7), and (4.8) describe the position of an offset slider-crank mechanism. These can be used in a full-cycle analysis. The equations were used in conjunction with a spreadsheet, yielding the results shown in Figure 4.37. If you are not familiar with using spreadsheets, refer to Chapter 8.

4. **Identify the Limiting Positions**

   Focusing on the position of link 4, the slider oscillation can be approximated as

   $$26.51 \text{ mm} < L_4 < 93.25 \text{ mm}$$

   and the maximum displacement as

   $$|\Delta \mathbf{R}_4|_{\max} = (L_4)_{\max} - (L_4)_{\min} \cong 93.25 - 26.51 = 66.74 \text{ mm}$$

This is only an approximation because, at 15° increments, the limiting position cannot be precisely detected. When exact information on the limiting position is required, the techniques discussed in Section 4.8 can be used.

Confusion may arise when observing the value of the angle β at the crank angle, θ_2, of 360°. The value should be identical to the initial value at the 0° crank angle. Note that the values differ by 360°. One is measuring the inner angle, and the other is measuring the outer angle. This illustrates the need to verify the information obtained from the equations with the physical mechanism.

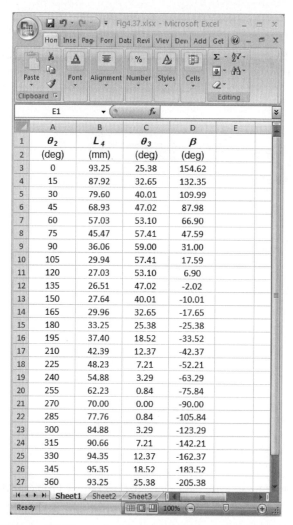

FIGURE 4.37 Positions of pusher rod for Example Problem 4.10.

4.12 DISPLACEMENT DIAGRAMS

Once a full-cycle position analysis is completed, it is insightful to plot the displacement of one point corresponding to the displacement of another point. It is most common to plot the displacement of a point on the follower relative to the displacement of a point on the driver.

Typically, the displacement of the driver is plotted on the horizontal. In the case of a crank, the driver displacement consists of one revolution. The corresponding displacement of the follower is plotted along the vertical. The displacement plotted on the vertical axis may be linear or angular depending on the motion obtained from the specific mechanism.

EXAMPLE PROBLEM 4.11

Figure 4.38 shows the driving mechanism of a reciprocating compressor. Plot a displacement diagram of the piston displacement relative to the crankshaft rotation.

SOLUTION: 1. **Draw the Kinematic Diagram**

After close examination, the compressor mechanism is identified as a slider-crank. Recall that this mechanism has one degree of freedom and can be operated by rotating the crank. The kinematic diagram with the appropriate dimensions is shown in Figure 4.39.

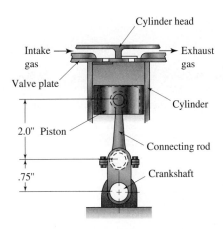

FIGURE 4.38 Compressor for Example Problem 4.11.

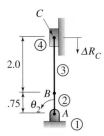

FIGURE 4.39 Kinematic diagram for Example Problem 4.11.

2. **Designate the Reference Phase**

   As shown in Figure 4.39, the reference phase is arbitrarily selected as the crank is vertical, placing joint *B* directly above joint *A*. The position of the piston (point *C* will be measured from this reference position.

3. **Construct the Interval Phases**

   The actual displacements can be determined either analytically or graphically, using the methods presented in the previous sections. For this slider-crank mechanism, the displacements were obtained analytically using equations (4.3) through (4.5). Using a spreadsheet, the results were obtained as shown in Figure 4.40. The crank displacement (θ_2) is in degrees and the piston displacement (ΔR_C) is in inches.

4. **Identify the Limiting Positions**

   Focusing on the position of the piston, the oscillation can be approximated as

$$|\Delta \mathbf{R}_c|_{max} \cong 1.50 \text{ in.}$$

   As stated in the previous problem, this is only an approximation because, at 30° increments, the limiting position will not be precisely detected. However, for the in-line slider-crank mechanism, inspecting the geometry reveals that the limiting positions occur at crank angles of 0° and 180°. Therefore, the stroke is exactly 1.50 in.

5. **Plot the Displacement Diagram**

   The values calculated in the spreadsheet and tabulated in Figure 4.40 were plotted in Figure 4.41 to form a displacement diagram.

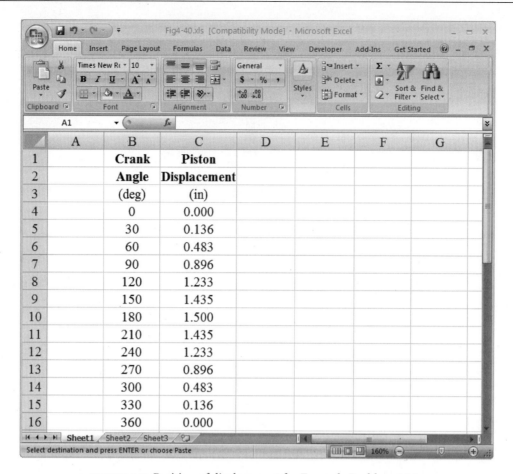

FIGURE 4.40 Position of displacement for Example Problem 4.11.

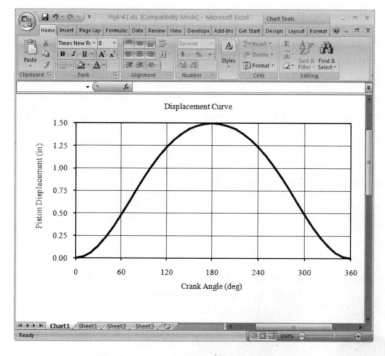

FIGURE 4.41 Displacement diagram for Example Problem 4.11.

4.13 COUPLER CURVES

Often, the function of a mechanism is to guide a part along a particular path. The paths generated by points on a connecting rod, or coupler, of a four-bar mechanism can often achieve the complex motion desired. The *trace* of a point is the path that the point follows as the mechanism moves through its cycle. The path traced by any point on the coupler is termed a *coupler curve*. Two coupler curves—namely, those traced by the pin connections of the coupler—are simple arcs, centered at the two fixed pivots. However, other points on the coupler trace rather complex curves. Figure 4.42 illustrates a four-bar mechanism. The coupler curves of a few select points are displayed.

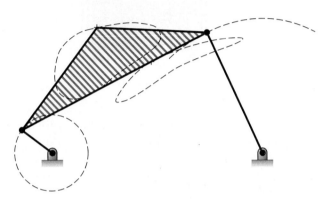

FIGURE 4.42 Coupler curves.

The methods in this chapter can be used to construct the trace of the motion of certain points on a mechanism. Section 4.10 introduces the concept of constructing the configuration at several phases of its cycle. As these phases are constructed, the position of a certain point can be retained. The curve formed by joining the position of this point at several phases of the mechanism forms the trace of that point. If the point resides on a floating link, the resulting trace, or coupler curve, is a complex shape. These traces can be used to determine the spatial requirements of a mechanism.

PROBLEMS

Although manual drafting techniques are instructive for problems that require graphical solution, use of a CAD system is highly recommended.

General Displacement

4–1. The device shown in Figure P4.1 is a scotch yoke mechanism. The horizontal position of link 4 can be described as $x = 3 \cos (50t + 40°)$. Determine the displacement of link 4 during the interval of 0.10 to 1.50 s.

4–2. For the scotch yoke mechanism shown in Figure P4.1, the horizontal position of link 4 can be described as $x = 3 \cos (50t + 40°)$. Determine the displacement of link 4 during the interval of 3.8 to 4.7 s.

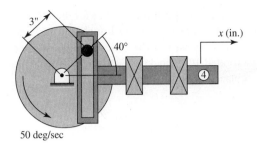

FIGURE P4.1 Problems 1 and 2.

Graphical Displacement Analysis

4–3. Graphically determine the displacement of points P and Q as the link shown in Figure P4.3 is displaced 25° counterclockwise. Use $\beta = 55°$ and $\gamma = 30°$.

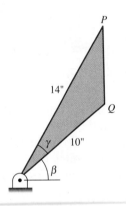

FIGURE P4.3 Problems 3, 4, 38, 39.

4–4. Graphically determine the displacement of points P and Q as the link shown in Figure P4.3 is displaced 35° clockwise. Use $\beta = 65°$ and $\gamma = 15°$.

4–5. Graphically position the links for the compressor linkage in the configuration shown in Figure P4.5. Then reposition the links as the 45-mm crank is rotated 90° counterclockwise. Determine the resulting displacement of the piston.

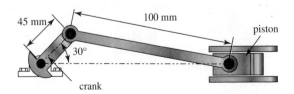

FIGURE P4.5 Problems 5, 6, 40, 56, 63, 70, 76, 82.

4–6. Graphically position the links for the compressor linkage in the configuration shown in Figure P4.5. Then reposition the links as the 45-mm crank is rotated 120° clockwise. Determine the resulting displacement of the piston.

4–7.   Graphically position the links for the shearing mechanism in the configuration shown in Figure P4.7. Then reposition the links as the 0.75-in. crank is rotated 100° clockwise. Determine the resulting displacement of the blade.

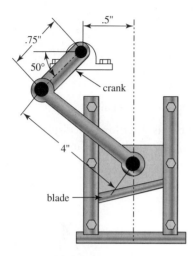

FIGURE P4.7   Problems 7, 8, 41, 57, 64, 71, 77, 83.

4–8.   Graphically position the links for the shearing mechanism in the configuration shown in Figure P4.7. Then reposition the links as the blade is lowered 0.2 in. Determine the resulting angular displacement of the crank.

4–9.   Graphically position the links for the embossing mechanism in the configuration shown in Figure P4.9. Then reposition the links as the handle is rotated 15° clockwise. Determine the resulting displacement of the stamp and the linear displacement of the handle end.

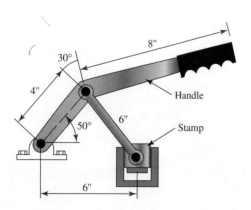

FIGURE P4.9   Problems 9, 10, 42.

4–10.   Graphically position the links for the embossing mechanism in the configuration shown in Figure P4.9. Then reposition the links as the handle is rotated 10° counterclockwise. Determine the resulting displacement of the stamp and the linear displacement of the handle end.

4–11.   Graphically position the links for the furnace door in the configuration shown in Figure P4.11. Then reposition the links as the handle, which is originally set at 10°, is rotated counterclockwise to 40°. Determine the resulting displacement of the door.

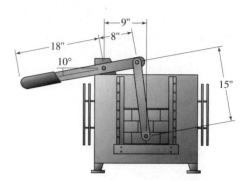

FIGURE P4.11   Problems 11, 12, 43.

4–12.   Graphically position the links for the furnace door in the configuration shown in Figure P4.11. Then reposition the links as the door is raised 3 in. Determine the angular displacement of the handle required to raise the door 3 in.

4–13.   A rock-crushing mechanism is given in Figure P4.13. Graphically position the links for the configuration shown. Then reposition the links as the crank is rotated 30° clockwise. Determine the resulting angular displacement of the crushing ram.

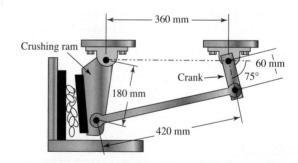

FIGURE P4.13   Problems 13, 14, 44, 58, 65, 72, 78, 84.

4–14.   A rock-crushing mechanism is given in Figure P4.13. Graphically position the links for the configuration shown. Then reposition the links as the crank is rotated 150° counterclockwise. Determine the resulting angular displacement of the crushing ram.

4–15.   Graphically position the links for the rear-wiper mechanism shown in Figure P4.15. Then reposition the links as the 2-in. crank is rotated 50° clockwise. Determine the resulting angular displacement of the wiper arm and the linear displacement at the end of the wiper blade.

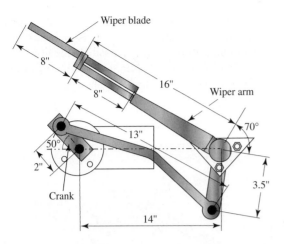

FIGURE P4.15 Problems 15, 16, 45, 59, 66, 73, 79, 85.

4–16. Graphically position the links for the rear-wiper mechanism shown in Figure P4.15. Then reposition the links as the 2-in. crank is rotated 110° clockwise. Determine the resulting angular displacement of the wiper arm and the linear displacement at the end of the wiper blade.

4–17. Graphically position the links for the vise grips shown in Figure P4.17. Then reposition the links as the top jaw is opened 40° from the orientation shown, while the lower jaw remains stationary. Determine the resulting angular displacement of the top handle.

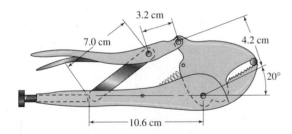

FIGURE P4.17 Problems 17, 18, 19, 46.

4–18. Graphically position the links for the vise grips shown in Figure P4.17. Then reposition the links as the top jaw is opened 20° from the orientation shown, while the lower jaw remains stationary. Determine the resulting angular displacement of the top handle.

4–19. When the thumbscrew in the vise grips shown in Figure P4.17 is rotated, the effective pivot point of the 7.0-cm link is moved. During this motion, the spring prevents the jaws from moving. Graphically position the links as the effective pivot point is moved 2 cm to the right. Then reposition the links as the top jaw is opened 40° from the new orientation, while the lower jaw remains stationary. Determine the resulting angular displacement of the top handle.

4–20. Graphically position the links for the small aircraft nosewheel actuation mechanism shown in Figure P4.20. Then reposition the links as the 12-in. crank is rotated 60° clockwise from the orientation shown. Determine the resulting angular displacement of the wheel assembly.

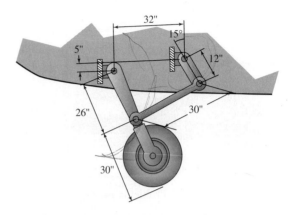

FIGURE P4.20 Problems 20, 21, 47, 60, 67, 74, 80, 86.

4–21. Graphically position the links for the small aircraft nosewheel actuation mechanism shown in Figure P4.20. Then reposition the links as the 12-in. crank is rotated 110° clockwise from the orientation shown. Determine the resulting angular displacement of the wheel assembly.

4–22. Graphically position the links for the foot-operated air pump shown in Figure P4.22. Then reposition the links as the foot pedal is rotated 25° counterclockwise from the orientation shown. Determine the resulting linear displacement of point X and the amount that the air cylinder retracts. Also, with the diameter of the cylinder at 25 mm, determine the volume of air displaced by this motion.

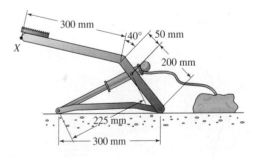

FIGURE P4.22 Problems 22, 23, 48.

4–23. Graphically position the links for the foot-operated air pump shown in Figure P4.22. Then reposition the links as the air cylinder retracts to 175 mm. Determine the resulting angular displacement of the foot pedal and the linear displacement of point X.

4–24. Graphically position the links for the microwave oven lift, which assists people in wheelchairs, as shown in Figure P4.24. Then reposition the links as the linear actuator is retracted to a length of 400 mm.

Determine the resulting angular displacement of the front support link and the linear displacement of any point on the oven carrier.

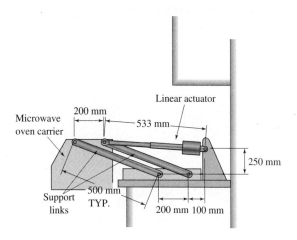

FIGURE P4.24 Problems 24, 25, 49.

4–25. Graphically position the links for the microwave oven lift, which assists people in wheelchairs, as shown in Figure P4.24. Then reposition the links as the front support link is raised 45° from the orientation shown. Determine the distance that the linear actuator needs to retract.

4–26. Graphically position the links for the box truck, used to load supplies onto airplanes, as shown in Figure P4.26. Then reposition the links as the lower sliding pin moves 0.5 m toward the cab. Determine the resulting linear displacement of any point on the cargo box.

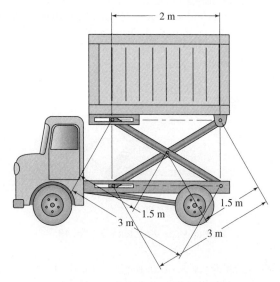

FIGURE P4.26 Problems 26, 27, 50.

4–27. Graphically position the links for the box truck, used to load supplies onto airplanes, as shown in Figure P4.26. Then reposition the links as the lower sliding pin moves 0.75 m away from the cab. Determine the resulting linear displacement of any point on the cargo box.

4–28. Graphically position the links for the lift platform shown in Figure P4.28. Determine the length of the hydraulic cylinder. Then reposition the links as the platform is raised to 40 in. Determine the amount that the hydraulic cylinder must extend to accomplish this movement.

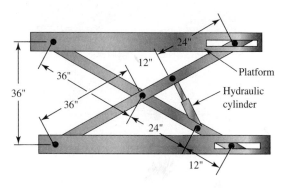

FIGURE P4.28 Problems 28, 29, 51.

4–29. Graphically position the links for the lift platform shown in Figure P4.28. Determine the length of the hydraulic cylinder. Then reposition the links as the platform is lowered to 30 in. Determine the amount that the hydraulic cylinder must contract to accomplish this movement.

4–30. The mechanism shown in Figure P4.30 is used to advance the film in movie-quality projectors. Graphically position the links for the configuration shown. Then reposition the links as the crank is rotated 90° clockwise. Determine the resulting displacement of the advancing claw.

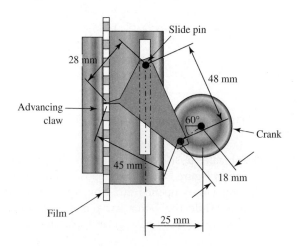

FIGURE P4.30 Problems 30, 31, 52, 61, 68.

4–31. The mechanism shown in Figure P4.30 is used to advance the film in movie-quality projectors. Graphically position the links for the configuration shown. Then reposition the links as the crank is rotated 130° clockwise. Determine the resulting displacement of the advancing claw.

4–32. Graphically position the links in the automotive front suspension mechanism shown in Figure P4.32.

Then reposition the links as the upper control arm is rotated 20° clockwise. Determine the resulting displacement of the bottom of the tire. Also determine the change of length of the spring.

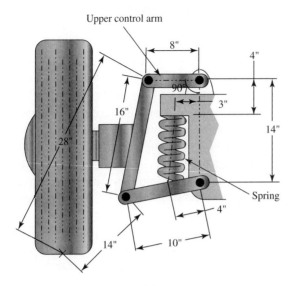

FIGURE P4.32 Problems 32, 33, 53.

4–33. Graphically position the links in the automotive front suspension mechanism shown in Figure P4.32. Then reposition the links as the upper control arm is rotated 10° counterclockwise. Determine the resulting displacement of the bottom of the tire. Also determine the change of length of the spring.

4–34. Graphically position the links for the rock-crushing mechanism shown in Figure P4.34. Then reposition the links as the crank is rotated 120° clockwise. Determine the resulting angular displacement of the crushing ram.

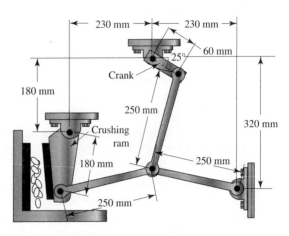

FIGURE P4.34 Problems 34, 35, 54, 62, 69, 75, 81, 87.

4–35. Graphically position the links for the rock-crushing mechanism as shown in Figure P4.34. Then reposition the links as the crank is rotated 75° clockwise. Determine the resulting angular displacement of the crushing ram.

4–36. Graphically position the links for the dump truck shown in Figure P4.36. Then reposition the links as the cylinder is shortened 0.15 m. Determine the

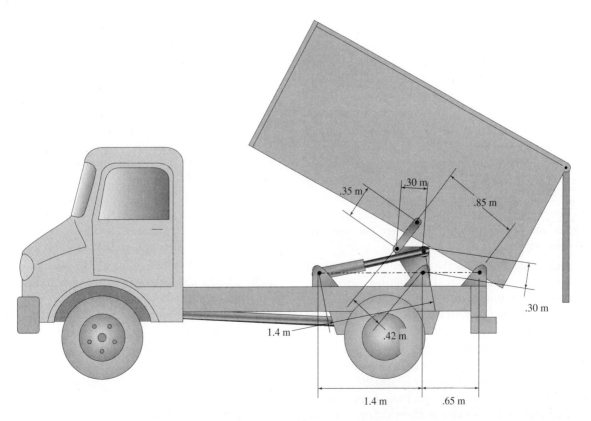

FIGURE P4.36 Problems 36, 37, 55.

resulting angular displacement of any point on the bed.

4–37. Graphically position the links for the dump truck shown in Figure P4.36. Then reposition the links as the cylinder is lengthened 0.2 m. Determine the resulting angular displacement of any point on the bed.

Analytical Displacement Analysis

4–38. Analytically determine the displacement of points P and Q, as the link shown in Figure P4.3 is displaced 30° counterclockwise. Use $\beta = 55°$ and $\gamma = 30°$.

4–39. Analytically determine the displacement of points P and Q, as the link shown in Figure P4.3 is displaced 40° clockwise. Use $\beta = 65°$ and $\gamma = 15°$.

4–40. Analytically determine the linear displacement of the piston in the compressor linkage shown in Figure P4.5 as the 45-mm crank is rotated from its current position 90° counterclockwise.

4–41. Analytically determine the linear displacement of the blade in the shearing mechanism shown in Figure P4.7 as the 0.75-in. crank is rotated from its current position 50° counterclockwise.

4–42. Analytically determine the linear displacement of the stamp in the mechanism shown in Figure P4.9 as the handle is rotated from its current position 20° clockwise.

4–43. Analytically determine the linear displacement of the furnace door in the mechanism shown in Figure P4.11 as the 26-in. handle is rotated 25° counterclockwise from its current position.

4–44. Analytically determine the angular displacement of the ram in the rock-crushing mechanism shown in Figure P4.13 as the 60-mm crank is rotated from its current position 40° clockwise.

4–45. Analytically determine the angular displacement of the wiper arm in the rear-wiper mechanism shown in Figure P4.15 as the 2-in. crank is rotated from its current position 100° clockwise.

4–46. Analytically determine the angular displacement of the top handle in the vise grips shown in Figure P4.17 as the top jaw is opened 25° from its current position, while the lower jaw remains stationary.

4–47. Analytically determine the angular displacement of the wheel assembly in the nosewheel actuation mechanism shown in Figure P4.20 as the 12-in. crank is rotated from its current position 60° counterclockwise.

4–48. Analytically determine the distance that the air cylinder in the foot pump shown in Figure P4.22 retracts when the foot pedal is rotated 20° counterclockwise from its current position. Also, with the diameter of the cylinder at 25 mm, determine the volume of air displaced during this motion.

4–49. Analytically determine the angular displacement of the front support link of the microwave lift shown in Figure P4.24 as the linear actuator is retracted to a length of 425 mm.

4–50. Analytically determine the vertical distance that the box truck in Figure P4.26 lowers if the bottom pins are separated from 2.0 m to 1.5 m.

4–51. Analytically determine the extension required from the hydraulic cylinder to raise the platform shown in Figure P4.28 to a height of 45 in.

4–52. Analytically determine the displacement of the claw in the film-advancing mechanism shown in Figure P4.30 as the crank is rotated 100° clockwise.

4–53. Analytically determine the displacement of the bottom of the tire in the automotive suspension mechanism shown in Figure P4.32 as the upper control arm is rotated 15° clockwise.

4–54. Analytically determine the angular displacement of the crushing ram in the rock-crushing mechanism shown in Figure P4.34 as the crank is rotated 95° clockwise.

4–55. Analytically determine the angular displacement of the dump truck bed shown in Figure P4.36 as the cylinder is shortened 0.1 m.

Limiting Positions—Graphical

4–56. Graphically position the links for the compressor mechanism shown in Figure P4.5 into the configurations that place the piston in its limiting positions. Determine the maximum linear displacement (stroke) of the piston.

4–57. Graphically position the links for the shearing mechanism shown in Figure P4.7 into the configurations that place the blade in its limiting positions. Determine the maximum linear displacement (stroke) of the blade.

4–58. Graphically position the links for the rock-crushing mechanism shown in Figure P4.13 into the configurations that place the ram in its limiting positions. Determine the maximum angular displacement (throw) of the ram.

4–59. Graphically position the links for the windshield wiper mechanism shown in Figure P4.15 into the configurations that place the wiper in its limiting positions. Determine the maximum angular displacement (throw) of the wiper.

4–60. Graphically position the links for the wheel actuator mechanism shown in Figure P4.20 into the configurations that place the wheel assembly in its limiting positions. Determine the maximum angular displacement (throw) of the wheel assembly.

4–61. Graphically position the links for the film-advancing mechanism shown in Figure P4.30 into the configurations that place the slide pin in its limiting positions. Determine the maximum linear displacement (stroke) of the slide pin.

4–62. Graphically position the links for the rock-crushing mechanism shown in Figure P4.34 into the configurations that place the ram in its limiting positions. Determine the maximum angular displacement (throw) of the crushing ram.

Limiting Positions—Analytical

4–63. Analytically calculate the maximum linear displacement (stroke) of the piston for the compressor mechanism shown in Figure P4.5.

4–64. Analytically calculate the maximum linear displacement (stroke) of the blade for the shearing mechanism shown in Figure P4.7.

4–65. Analytically calculate the maximum angular displacement (throw) of the ram for the rock-crushing mechanism shown in Figure P4.13.

4–66. Analytically calculate the maximum angular displacement (throw) of the wiper for the windshield wiper mechanism shown in Figure P4.15.

4–67. Analytically calculate the maximum angular displacement (throw) of the wheel assembly for the wheel actuator mechanism shown in Figure P4.20.

4–68. Analytically calculate the maximum linear displacement (stroke) of the slide pin for the film-advancing mechanism shown in Figure P4.30.

4–69. Analytically calculate the maximum angular displacement (throw) of the ram for the rock-crushing mechanism shown in Figure P4.34.

Displacement Diagrams—Graphical

4–70. For the compressor mechanism shown in Figure P4.5, graphically create a displacement diagram for the position of the piston as the crank rotates a full revolution clockwise.

4–71. For the shearing mechanism shown in Figure P4.7, graphically create a displacement diagram for the position of the blade as the crank rotates a full revolution clockwise.

4–72. For the rock-crushing mechanism shown in Figure P4.13, graphically create a displacement diagram of the angular position of the ram as the crank rotates a full revolution clockwise.

4–73. For the windshield wiper mechanism shown in Figure P4.15, graphically create a displacement diagram of the angular position of the wiper as the crank rotates a full revolution clockwise.

4–74. For the wheel actuator mechanism shown in Figure P4.20, graphically create a displacement diagram of the angular position of the wheel assembly as the crank rotates a full revolution clockwise.

4–75. For the rock-crushing mechanism shown in Figure P4.34, graphically create a displacement diagram of the angular position of the ram as the crank rotates a full revolution clockwise.

Displacement Diagrams—Analytical

4–76. For the compressor mechanism shown in Figure P4.5, analytically create a displacement diagram for the position of the piston as the crank rotates a full revolution counterclockwise.

4–77. For the shearing mechanism shown in Figure P4.7, analytically create a displacement diagram for the position of the blade as the crank rotates a full revolution counterclockwise.

4–78. For the rock-crushing mechanism shown in Figure P4.13, analytically create a displacement diagram of the angular position of the ram as the crank rotates a full revolution counterclockwise.

4–79. For the windshield wiper mechanism shown in Figure P4.15, analytically create a displacement diagram of the angular position of the wiper as the crank rotates a full revolution counterclockwise.

4–80. For the wheel actuator mechanism shown in Figure P4.20, analytically create a displacement diagram of the angular position of the wheel assembly as the crank rotates a full revolution counterclockwise.

4–81. For the rock-crushing mechanism shown in Figure P4.34, analytically create a displacement diagram of the angular position of the ram as the crank rotates a full revolution counterclockwise.

Displacement Problems Using Working Model

4–82. For the compressor mechanism shown in Figure P4.5, use the Working Model software to create a simulation and plot a displacement diagram for the position of the piston as the crank rotates a full revolution counterclockwise.

4–83. For the shearing mechanism shown in Figure P4.7, use the Working Model software to create a simulation and plot a displacement diagram for the position of the blade as the crank rotates a full revolution counterclockwise.

4–84. For the rock-crushing mechanism shown in Figure P4.13, use the Working Model software to create a simulation and plot a displacement diagram of the angular position of the ram as the crank rotates a full revolution counterclockwise.

4–85. For the windshield wiper mechanism shown in Figure P4.15, use the Working Model software to create a simulation and plot a displacement diagram of the angular position of the wiper as the crank rotates a full revolution counterclockwise.

4–86. For the wheel actuator mechanism shown in Figure P4.20, use the Working Model software to create a simulation and plot a displacement diagram of the angular position of the wheel assembly as the crank rotates a full revolution counterclockwise.

4–87. For the rock-crushing mechanism shown in Figure P4.34, use the Working Model software to create a simulation and plot a displacement diagram of the angular position of the ram as the crank rotates a full revolution counterclockwise.

CASE STUDIES

4–1. Figure C4.1 shows a mechanism that was designed to impart motion on a machine slide. Carefully examine the configuration of the components in the mechanism, then answer the following leading questions to gain insight into the operation.

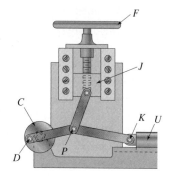

FIGURE C4.1 *(Courtesy, Industrial Press)*

1. As wheel *C* is rotated clockwise, and slide J remains stationary, what is the continual motion of pin *D*?
2. What is the continual motion of pin *P*?
3. What is the continual motion of pin *K*?
4. What effect does turning the handwheel *F* have on slide *J*?
5. What effect does turning the handwheel *F* have on the motion of the mechanism? Be sure to comment on all characteristics of the motion.
6. What is the purpose of this device?
7. Draw a kinematic diagram and calculate the mobility of the mechanism.

4–2. Figure C4.2 presents an interesting materials handling system for advancing small parts onto a feed track. Carefully examine the configuration of the components in the mechanism, then answer the following leading questions to gain insight into the operation.

1. The small, round-head screw blanks are fed into a threading machine through tracks *B* and *C*. How do the screws get from bowl *A* to track *B*?
2. Although not clearly shown, track *B* is of a parallel finger design. Why is a parallel finger arrangement used to carry the screws?
3. As a second mechanism intermittently raises link *D*, what is the motion of track *B*?
4. What is the purpose of link *E*?
5. As a second mechanism intermittently raises link *D*, what is the motion of the screws?
6. What determines the lowest position that link *D* can travel? Notice that the tips of track *B* do not contact the bottom of bowl *A*.
7. As screws are congested in the outlet track *C*, what happens to finger *F* as link *D* is forced lower?
8. As screws are congested in the outlet track *C*, what happens to the tips of track *B*?
9. What is the purpose of this device? Discuss its special features.
10. What type of mechanism could be operating link *D*?

4–3. Figure C4.3 depicts a production transfer machine that moves clutch housings from one station to another. Platform *A* supports the housings during the transfer. Carefully examine the configuration of the components in the mechanism, then answer the following leading questions to gain insight into the operation.

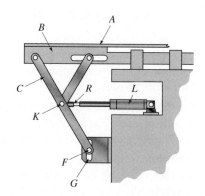

FIGURE C4.3 *(Courtesy, Industrial Press)*

1. To what type of motion is bar *B* restricted?
2. What motion does link *C* encounter as air cylinder *L* is shortened?
3. What is the motion of point *K* as air cylinder *L* is shortened?
4. Why does joint *F* need to ride in slot *G*?
5. What is the purpose of this mechanism?
6. What effect does turning the threaded rod end *R* have, thus lengthening the cylinder rod?
7. Draw a kinematic sketch of this mechanism.
8. Compute the mobility of this mechanism.

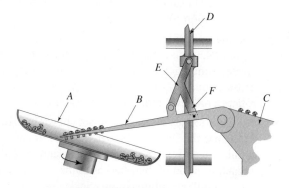

FIGURE C4.2 *(Courtesy, Industrial Press)*

MECHANISM DESIGN

5.1 INTRODUCTION

Up to this point in the text, an emphasis was placed on the analysis of existing mechanisms. The previous chapter explored methods to determine the displacement of a mechanism whose link lengths are given. Compared to this analysis, the design of a mechanism presents the opposite task: That is, given a desired motion, a mechanism form and dimensions must be determined. *Synthesis* is the term given to describe the process of designing a mechanism that produces a desired output motion for a given input motion. The selection of a particular mechanism capable of achieving the desired motion is termed *type synthesis*. A designer should attempt to use the simplest mechanism capable of performing the desired task. For this reason, slider-crank and four-bar mechanisms are the most widely used. This chapter focuses on these two mechanisms.

After selecting a mechanism type, appropriate link lengths must be determined in a process called *dimensional synthesis*. This chapter focuses on dimensional synthesis. To design a mechanism, intuition can be used along with

analysis methods described in the previous chapter. Often, this involves an iterate-and-analyze methodology, which can be an inefficient process, especially for inexperienced designers. However, this iteration process does have merit, especially in problems where synthesis procedures have not or cannot be developed. However, several methods for dimensional synthesis have been developed and can be quite helpful. This chapter serves as an introduction to these methods. Because analytical techniques can become quite complex, the focus is on graphical techniques. As stated throughout the text, employing graphical techniques on a CAD system produces accurate results.

5.2 TIME RATIO

Many mechanisms that produce reciprocating motion are designed to produce symmetrical motion. That is, the motion characteristics of the outward stroke are identical to those of the inward stroke. These mechanisms often accomplish work in both directions. An engine mechanism and windshield wiper linkages are examples of these kinematically balanced mechanisms.

However, other machine design applications require a difference in the average speed of the forward and return strokes. These machines typically work only on the forward stroke. The return stroke needs to be as fast as possible, so maximum time is available for the working stroke. Cutting machines and package-moving devices are examples of these quick-return mechanisms.

A measure of the quick return action of a mechanism is the *time ratio*, Q, which is defined as follows:

$$Q = \frac{\text{Time of slower stroke}}{\text{Time of quicker stroke}} \geq 1 \qquad (5.1)$$

An imbalance angle, β, is a property that relates the geometry of a specific linkage to the timing of the stroke. This angle can be related to the time ratio, Q:

$$Q = \frac{180° + \beta}{180° - \beta} \qquad (5.2)$$

Equation 5.2 can be rewritten to solve the imbalance angle as follows:

$$\beta = 180° \frac{(Q - 1)}{(Q + 1)} \qquad (5.3)$$

Thus, in the dimensional synthesis of a mechanism, the desired time ratio is converted to a necessary geometric constraint through the imbalance angle β.

The total cycle time for the mechanism is:

$$\Delta t_{cycle} = \text{Time of slower stroke} + \text{Time of quicker stroke} \qquad (5.4)$$

For mechanisms that are driven with a constant speed rotational actuator, the required crank speed, ω_{crank}, is related to the cycle time as follows:

$$\omega_{crank} = (\Delta t_{cycle})^{-1} \qquad (5.5)$$

EXAMPLE PROBLEM 5.1

A quick-return mechanism is to be designed, where the outward stroke must consume 1.2 s and the return stroke 0.8 s. Determine the time ratio, imbalance angle, cycle time, and speed at which the mechanism should be driven.

SOLUTION:

1. **Calculate the Time Ratio and Imbalance Angle**

   The time ratio can be determined from equation (5.1):

   $$Q = \left(\frac{1.2\text{ s}}{0.8\text{ s}}\right) = 1.5$$

   The resulting imbalance angle can be determined from equation (5.3):

   $$\beta = 180° \frac{(1.5 - 1)}{(1.5 + 1)} = 36°$$

2. **Calculate Cycle Time for the Mechanism**

   The total time for the forward and return stroke is as follows:

   $$\Delta t_{cycle} = 1.2 + 0.8 = 2.0 \text{ s/rev}$$

3. **Calculate the Required Speed of the Crank**

   Because one cycle of machine operation involves both the forward and return strokes the time for the crank to complete one revolution is also 2.0 s. The required crank speed, ω_{crank}, is determined as

   $$\omega_{crank} = \left(\Delta t_{cycle}\right)^{-1}$$

   $$= \frac{1}{2 \text{ s/rev}} = 0.5 \text{ rev/s} \left(\frac{60 \text{ s}}{1 \text{ min}}\right)$$

   $$= 30 \text{ rev/min}$$

   In Chapter 6, the concept of angular speed will be formally presented.

5.3 TIMING CHARTS

Timing charts are often used in the mechanism design process to assist in the synchronization of motion between mechanisms. For example, a pair of mechanisms may be used to transfer packages from one conveyor to another. One mechanism could lift a package from the lower conveyor and the other mechanism would push the package onto the upper conveyor while the first remains stationary. Both mechanisms would then return to the start position and set another cycle. A timing chart is used to graphically display this information. Additionally, timing charts can be used to estimate the magnitudes of the velocity and acceleration of the follower links. The velocity of a link is the time rate at which its position is changing. Acceleration is the time rate at which its velocity is changing and is directly related to the forces required to operate the mechanism. Chapter 6 provides comprehensive coverage of linkage velocity analysis and Chapter 7 focuses on linkage acceleration. Both velocity and acceleration are vector quantities, but only their magnitudes, v and a, are used in timing charts.

Timing charts that are used to synchronize the motion of multiple mechanisms typically assume constant acceleration. While the actual acceleration values produced in the mechanism can be considerably different (as will be presented in Chapter 7), the constant acceleration assumption produces polynomial equations for the velocity and position as a function of time. The timing chart involves plotting the magnitude of the output velocity versus time. Assuming constant acceleration, the velocity-time graph appears as straight lines. The displacement is related to the maximum velocity, acceleration, and time through the following equations.

$$\Delta R = \frac{1}{2} v_{peak} \Delta t \qquad (5.6)$$

$$\Delta R = \frac{1}{4} a (\Delta t)^2 \qquad (5.7)$$

For the package-moving scenario previously described, the lift mechanism is desired to raise 8.0 in. in 1.5 s, remain stationary for 1.0 s, and return in 1.0 s. The push mechanism should remain stationary for 1.5 s, push 6.0 in. in 1.0 s, and return in 1.0 s. The timing charts for both mechanisms are shown in Figure 5.1. The figures illustrate that as one mechanism is lifting (velocity appears as a triangle), the other remains stationary (no velocity). Also, while the second mechanism is pushing, the first remains stationary. Thus, synchronization is verified. Further, the maximum speed and acceleration are related to the displacement and the time for the motion by rewriting Equations (5.6) and (5.7), respectively. For the lifting mechanism,

Lift stroke: $v_{peak} = 2 \dfrac{\Delta R}{\Delta t} = 2 \dfrac{(8.0\,\text{in.})}{(1.5\,\text{s})} = 10.67\ \text{in./s}$

$a = 4 \dfrac{\Delta R}{\Delta t^2} = 4 \dfrac{(8.0\,\text{in.})}{(1.5\,\text{s})^2} = 14.22\ \text{in./s}^2$

In similar computations, the peak velocity of the return stroke is -16.00 in./s, and the acceleration is -32.00 in./s^2. For the pushing mechanism, the peak velocity of the push stroke 12.00 in./s, and the acceleration is 24.00 in./s^2. For the pushing mechanism, the peak velocity of the return stroke -12.00 in./s, and the acceleration is -24.00 in./s^2.

It is noted that because the velocity is the time rate of change of the position, principles of calculus dictate that the displacement of the mechanism is the area under the v-t chart. This is seen in Equation (5.6) and the displacement is the area of the velocity triangle, which is $1/2(v_{peak}\Delta t)$. The displacement for each motion is labeled in Figure 5.1. While it must be emphasized that the velocity and acceleration are estimates, they can be useful in the early design stage as seen in the following example problem.

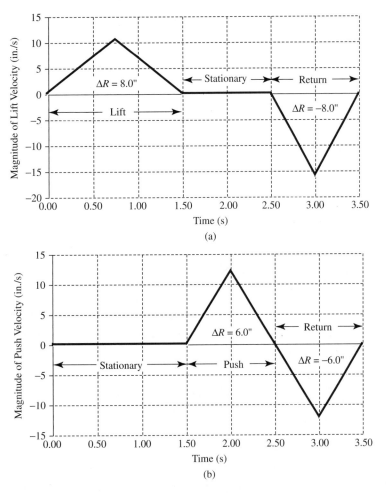

FIGURE 5.1 Timing charts.

EXAMPLE PROBLEM 5.2

A sleeve bearing insertion process requires a conveyor to index 8.0 in. in 0.4 s and dwell while a bearing is pressed into a housing on the conveyor. The bearing must travel 4.0 in. to meet the housing, then be pressed 2.0 in. into the housing. The entire press stroke should take 0.6 s, and return in 0.4 s while the conveyor is indexing.

a. Determine the time ratio, cycle time, and motor speed of the press mechanism.

b. Sketch the synchronized timing charts.

c. Estimate the peak velocity and acceleration of the housing on the conveyor.

d. Estimate the peak velocity and acceleration of the bearing press movement.

e. Estimate the peak velocity and acceleration of the bearing press return.

f. Optimize the motion so the maximum acceleration of any part is less than $1g$ ($1g = 386.4$ in./s^2).

SOLUTION: 1. *Calculate the Time Ratio, Cycle Time, and Crank Speed*

The time ratio can be determined from equation (5.1):

$$Q = \left(\frac{0.6 \text{ s}}{0.4 \text{ s}}\right) = 1.5$$

The total time for the forward and return stroke is as follows:

$$\Delta t_{\text{cycle}} = 0.6 + 0.4 \text{ s/rev}$$

Because one cycle of machine operation involves both the forward and return stroke, the time for the crank to complete one revolution is 1.0 s. The required crank speed is determined as

$$\omega_{\text{crank}} = \left(\Delta t_{\text{cycle}}\right)^{-1}$$

$$= \frac{1}{1.0 \text{ s/rev}} = 1.0 \text{ rev/s} \left(\frac{60 \text{ s}}{1 \text{ min}}\right)$$

$$= 60.0 \text{ rev/min}$$

2. *Sketch the Timing Charts*

The timing charts are constructed and shown in Figure 5.2.

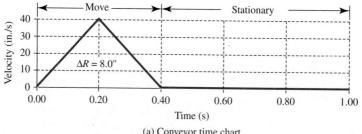

(a) Conveyor time chart

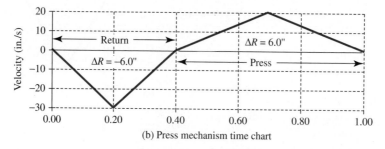

(b) Press mechanism time chart

FIGURE 5.2 Timing charts for Example Problem 5.2.

3. *Calculate Motion Parameters of the Housing on the Conveyor*

The estimated velocity and acceleration magnitudes for the housing on the conveyor is

$$v_{peak} = 2\frac{\Delta R}{\Delta t_1} = 2\frac{(8.0\,in.)}{(0.4\,s)} = 40.00\,in./s$$

$$a = 4\frac{\Delta R}{\Delta t_1^2} = 4\frac{(8.0\,in.)}{(0.4\,s)^2} = 200.00\,in./s^2$$

4. *Calculate Motion Parameters of the Bearing Press Return Stroke*

The estimated velocity and acceleration magnitudes for the bearing press return stroke is

$$v_{peak} = 2\frac{\Delta R}{\Delta t_1} = 2\frac{(-6.0\,in.)}{(0.4\,s)} = -30.00\,in./s$$

$$a = 4\frac{\Delta R}{\Delta t_1^2} = 4\frac{(-6.0\,in.)}{(0.4\,s)^2} = -150.00\,in./s^2$$

5. *Calculate Motion Parameters of the Bearing Press Working Stroke*

The estimated velocity and acceleration magnitudes for the bearing press working stroke is

$$v_{peak} = 2\frac{\Delta R}{\Delta t_2} = 2\frac{(6.0\,in.)}{(0.6\,s)} = 20.00\,in./s$$

$$a = 4\frac{\Delta R}{\Delta t_2^2} = 4\frac{(6.0\,in.)}{(0.6\,s)^2} = 66.67\,in./s^2$$

6. *Optimize Motion*

The largest acceleration magnitude is the housing on the conveyor at 200 in/s^2 = 200/386.4 = 0.517g. The motion can be optimized and production can be increased by substituting $a = 386.4$ in./s^2 (1g) into equation (5.7) and solving for a reduced conveyor movement time.

$$\Delta t_1 = \sqrt{4\frac{\Delta R}{a}} = \sqrt{4\frac{(8.0\,in.)}{(386.4\,in./s^2)}} = 0.288\,s$$

Maintaining the time ratio, the reduced bearing press stroke can be determined by rewriting equation (5.1).

$$\Delta t_2 = Q\,\Delta t_1 = 1.5(0.288\,s) = 0.432\,s$$

The increased crank speed can be determined from equation 5.5

$$\omega_{crank} = (0.288 + 0.432\,s)^{-1}$$

$$= \frac{1}{0.720\,s/rev}$$

$$= 1.389\,rev/s\left(\frac{60\,s}{1\,min}\right) = 83.3\,rev/min$$

Since production rate is related to line speed, the production is increased 39 percent by using time charts and optimizing the motion while keeping within acceptable acceleration limits.

5.4 DESIGN OF SLIDER-CRANK MECHANISMS

Many applications require a machine with reciprocating, linear sliding motion of a component. Engines and compressors require a piston to move through a precise distance, called the stroke, as a crank continuously rotates. Other applications such as sewing machines and power hacksaws require a similar, linear, reciprocating motion. A form of the slider-crank mechanism is used in virtually all these applications.

5.4.1 In-Line Slider-Crank Mechanism

An in-line slider-crank mechanism has the crank pivot coincident with the axis of the sliding motion of the piston pin. An in-line slider-crank mechanism is illustrated in Figure 5.3. The *stroke*, $|\Delta R_4|_{max}$, is defined as the linear distance that the sliding link exhibits between the extreme positions. Because the motion of the crank (L_2) and connecting arm (L_3) is symmetric about the sliding axis, the crank angle required to execute a forward stroke is the same as that for the return stroke. For this reason, the in-line

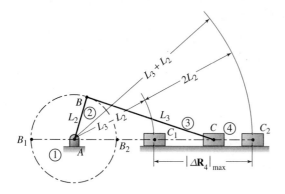

FIGURE 5.3 In-line slider-crank mechanism.

slider-crank mechanism produces balanced motion. Assuming that the crank is driven with a constant velocity source, as an electric motor, the time consumed during a forward stroke is equivalent to the time for the return stroke.

The design of an in-line slider-crank mechanism involves determining the appropriate length of the two links, L_2 and L_3, to achieve the desired stroke, $|\Delta R_4|_{max}$. As can be seen from Figure 5.3, the stroke of the in-line slider-crank mechanism is twice the length of the crank. That is, the distance between B_1 and B_2 is the same as the distance between C_1 and C_2. Therefore, the length of crank, L_2, for an in-line slider-crank can be determined as follows:

$$L_2 = \frac{|\Delta R_4|_{max}}{2} \qquad (5.8)$$

The length of the connecting arm, L_3, does not affect the stroke of an in-line slider-crank mechanism. However, a shorter connecting arm yields greater acceleration values. Figure 5.4 illustrates the effect of the connecting arm length

and offset distance (if any) on the maximum acceleration of the sliding link. These data clearly show that the connecting arm length should be made as large as possible. (Note that for an in-line slider-crank, the offset value, L_1, is zero.) As a general rule of thumb, the connecting arm should be at least three times greater than the length of the crank. A detailed analysis, as presented in Chapter 7, should be completed to determine the precise accelerations of the links and resulting inertial loads.

5.4.2 Offset Slider-Crank Mechanism

The mechanism illustrated in Figure 5.5a is an offset slider-crank mechanism. With an offset slider-crank mechanism, an offset distance is introduced. This offset distance, L_1, is the distance between the crank pivot and the sliding axis. With the presence of an offset, the motion of the crank and connecting arm is no longer symmetric about the sliding axis. Therefore, the crank angle required to execute the forward stroke is different from the crank angle required for the return stroke. An offset slider-crank mechanism provides a quick return when a slower working stroke is needed.

In Figure 5.5a, it should be noted that A, C_1, and C_2 are not collinear. Thus, the stroke of an offset slider-crank mechanism is always greater than twice the crank length. As the offset distance increases, the stroke also becomes larger. By inspecting Figure 5.5a, the feasible range for the offset distance can be written as:

$$L_1 < L_3 - L_2 \qquad (5.9)$$

Locating the limiting positions of the sliding link is shown in Figure 5.5a and was discussed in Chapter 4. The design of a slider-crank mechanism involves determining an appropriate offset distance, L_1, and the two links lengths, L_2 and L_3, to achieve the desired stroke, $|\Delta R_4|_{max}$, and imbalance

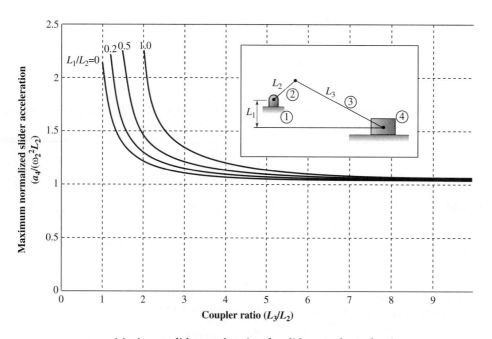

FIGURE 5.4 Maximum slider acceleration for slider-crank mechanisms.

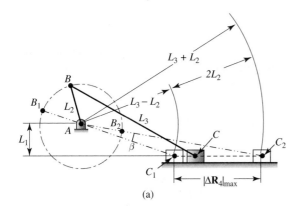

(a)

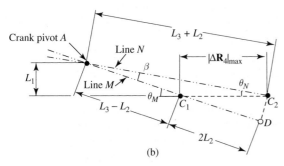

(b)

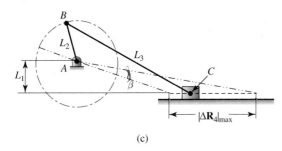

(c)

FIGURE 5.5 Offset slider-crank mechanism.

angle, β. The graphical procedure to synthesize a slider-crank mechanism is as follows:

1. Locate the axis of the pin joint on the sliding link. This joint is labeled as point C in Figure 5.5a.

2. Draw the extreme positions of the sliding link, separated by the stroke, $|\Delta R_4|_{max}$.

3. At one of the extreme positions, construct *any* line M through the sliding link pin joint, inclined at an angle θ_M. This point is labeled C_1 in Figure 5.5b.

4. At the other extreme position, draw a line N through the sliding link pin joint, labeled C_2 in Figure 5.5b, inclined at an angle β from line M. Note that $\theta_N = \theta_M - \beta$.

5. The intersection of lines M and N defines the pivot point for the crank, point A. The offset distance, L_1, can be scaled from the construction (Figure 5.5b).

6. From the construction of the limiting positions, it is observed that the length between C_1 and D is $2L_2$. Note that this arc, C_2D, is centered at point A. Because both lines are radii of the same arc, the radius AC_2 is equal to

the lengths $AC_1 + C_1D$. Rearranging this relationship gives

$$C_1D = AC_2 - AC_1$$

Substituting and rearranging, the length of the crank, L_2, for this offset slider-crank mechanism can be determined as

$$L_2 = \frac{1}{2}(AC_2 - AC_1) \qquad (5.10)$$

7. From the construction of the limiting positions, it is also observed that

$$AC_1 = L_3 - L_2$$

Rearranging, the length of the coupler, L_3, for this offset slider-crank mechanism is

$$L_3 = AC_1 + L_2 \qquad (5.11)$$

The complete mechanism is shown in Figure 5.5c. The design procedure, implemented with a CAD system, achieves accurate results.

Note that any line M can be drawn though point C_1 at an arbitrary inclination angle, θ_M. Therefore, an infinite number of suitable mechanisms can be designed. In general, the mechanisms that produce the longest connecting arm have lower accelerations, and subsequently lower inertial forces. Figure 5.4 can be used to determine the ramifications of using a short connecting arm. As a general rule of thumb, the connecting arm should be at least three times greater than the length of the crank. A detailed acceleration analysis, as presented in Chapter 7, should be completed to determine the inherent inertial loads.

Analytical methods can be incorporated by viewing the triangle in Figure 5.5b to generate expressions for the link lengths L_1, L_2, and L_3, as a function of the stroke $|\Delta R_4|_{max}$, the imbalance angle β, and the inclination of the arbitrary line M, θ_M.

$$L_1 = |\Delta R_4|_{max} \left[\frac{\sin(\theta_M)\sin(\theta_M - \beta)}{\sin(\beta)} \right] \qquad (5.12)$$

$$L_2 = |\Delta R_4|_{max} \left[\frac{\sin(\theta_M) - \sin(\theta_M - \beta)}{2\sin(\beta)} \right] \qquad (5.13)$$

$$L_3 = |\Delta R_4|_{max} \left[\frac{\sin(\theta_M) + \sin(\theta_M - \beta)}{2\sin(\beta)} \right] \qquad (5.14)$$

5.5 DESIGN OF CRANK-ROCKER MECHANISMS

A crank-rocker mechanism has also been discussed on several occasions. It is common for many applications where repeated oscillations are required. Figure 5.6a illustrates the geometry of a crank-rocker. Comparable to the stroke of a slider-crank mechanism, the crank-rocker mechanism exhibits a *throw angle*, $(\Delta\theta_4)_{max}$ (Figure 5.6a). This throw angle is defined as the angle between the extreme positions of the rocker link.

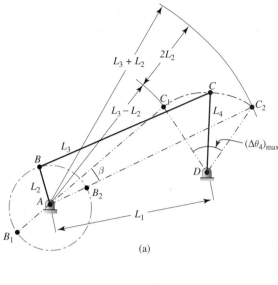

(a)

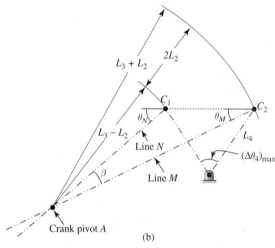

(b)

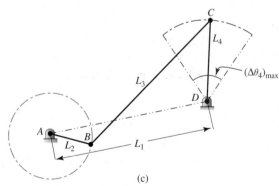

(c)

FIGURE 5.6 Crank-rocker mechanism.

Similar to the offset slider-crank mechanism, a crank-rocker can be used as a quick-return mechanism. The time ratio defined in equations (5.1) and (5.2) equally applies to a crank-rocker. The imbalance angle, β, of a crank-rocker mechanism is also shown in Figure 5.6a.

The limiting positions of a crank-rocker are shown in Figure 5.6a and were discussed extensively in Chapter 4. Note that the radial length between the two extreme positions is twice the crank length. This notion becomes important when designing a crank-rocker mechanism.

The design of a crank-rocker mechanism involves determining appropriate lengths of all four links to achieve the desired throw angle, $(\Delta\theta_4)_{max}$, and imbalance angle, β. The graphical procedure to synthesize a crank-rocker mechanism is as follows:

1. Locate the pivot of the rocker link, point D in Figure 5.6b.

2. Choose *any* feasible rocker length, L_4. This length is typically constrained by the spatial allowance for the mechanism.

3. Draw the two positions of the rocker, separated by the throw angle, $(\Delta\theta_4)_{max}$.

4. At one of the extreme positions, construct *any* line M through the end of the rocker link, inclined at an angle θ_M. This point is labeled C_2 in Figure 5.6b.

5. At the other extreme position, draw a line N through the end of the rocker link, which is inclined at angle β from line M. Note that $\theta_N = \theta_M - \beta$.

6. The intersection of lines M and N defines the pivot point for the crank, point A. The length between the two pivots, L_1, can be scaled from the construction (Figure 5.6c). For cases where a balanced timing is required ($Q = 1$), lines M and N are collinear. Thus, a pivot point for the crank, point A, can be selected anywhere along lines M and N.

7. From the construction of the limiting positions, it is observed that the length between C_1 and E is $2L_2$. Note that this arc, C_2E, is centered at point A. Because both lines are radii of the same arc, the radius AC_2 is equal to the lengths $AC_1 + C_1E$. Rearranging this relationship gives

$$C_1E = AC_2 - AC_1$$

Substituting and rearranging, the length of the crank, L_2, for this crank-rocker mechanism can be determined as

$$L_2 = \frac{1}{2}(AC_2 - AC_1) \tag{5.15}$$

8. From the construction of the limiting positions, it is also observed that

$$AC_1 = L_3 - L_2$$

Rearranging, the length of the coupler, L_3, for this crank-rocker mechanism is

$$L_3 = AC_1 + L_2 \tag{5.16}$$

The completed mechanism is shown in Figure 5.6c. In step 4, line M is drawn through point C_1, at an arbitrary inclination angle, θ_M. Therefore, an infinite number of suitable mechanisms can be designed to achieve the desired throw angle and time ratio As with slider-crank mechanisms, four-bar mechanisms that include a longer coupler will have lower accelerations and subsequently lower inertial forces.

An additional measure of the "quality" of a four-bar mechanism is the *transmission angle*, γ. This is the angle between the coupler and the rocker, as illustrated in

Figure 5.6c. A common function of a four-bar linkage is to transform rotary into oscillating motion. Frequently, in such applications, a large force must be transmitted. In these situations, the transmission angle becomes of extreme importance. When the transmission angle becomes small, large forces are required to drive the rocker arm. For best results, the transmission angle should be as close to 90° as possible during the entire rotation of the crank. This will reduce bending in the links and produce the most favorable force-transmission conditions. The extreme values of the transmission angle occur when the crank lies along the line of the frame. A common rule of thumb is that a four-bar linkage should not be used when the transmission angle is outside the limits of 45° and 135°. Force analysis, as presented in Chapters 13 and 14, can be used to determine the effect of the actual transmission angle encountered.

In some instances, the length of one of the links must be a specific dimension. Most commonly, a target length of the frame (L_1) is specified. However, only the rocker length (L_4) is directly specified in the procedure just outlined. Since the four-bar mechanism was designed to attain specific angular results, the length of all the links can be appropriately scaled to achieve the desired link dimension and maintain the design objective. All CAD systems have the ability to scale the constructed geometry of Figure 5.6b.

Analytical methods can be incorporated by analyzing the triangles in Figure 5.6b to generate expressions for the link lengths L_2, L_3, and L_4, as a function of the throw ($\Delta\theta_4$)$_{max}$, the frame length (L_1), the imbalance angle β, and the inclination of the arbitrary line M, θ_M.

$$L_4 = \frac{L_1 \sin \beta}{\sqrt{\kappa}} \qquad (5.17)$$

Where:

$$\kappa = \sin^2\beta + 4\sin^2\left((\Delta\theta_4)_{max}/2\right)\sin^2(\theta_M + \beta)$$
$$- 4\sin\beta\sin\left((\Delta\theta_4)_{max}/2\right)\sin(\theta_M + \beta)$$
$$\sin\left((\Delta\theta_4)_{max}/2 + \theta_M)\right)$$

$$L_3 = \frac{L_4 \sin\left((\Delta\theta_4)_{max}/2\right)}{\sin \beta}\left[\sin\theta_M + \sin(\theta_M + \beta)\right] \quad (5.18)$$

$$L_2 = L_3 - \frac{2L_4 \sin\left((\Delta\theta_4)_{max}/2\right)\sin\theta_M}{\sin\beta} \qquad (5.19)$$

5.6 DESIGN OF CRANK-SHAPER MECHANISMS

A crank-shaper mechanism that is capable of higher time ratios is shown in Figure 5.7. It is named for its use in metal shaper machines, where a slow cutting stroke is followed by a rapid return stroke when no work is being performed. The design of an crank-shaper mechanism involves determining the appropriate length of the three primary links—L_1, L_2, and L_3—to achieve the desired stroke, $|\Delta\mathbf{R}_E|_{max}$.

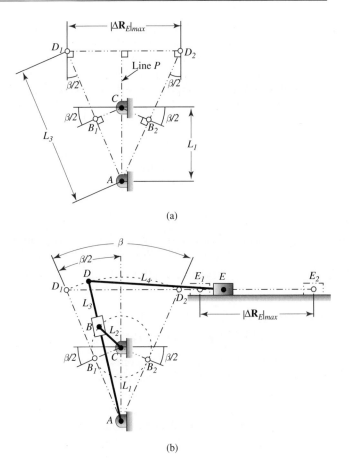

FIGURE 5.7 Crank-shaper mechanism.

The graphical procedure to synthesize the crank-shaper mechanism is as follows:

1. Construct a line whose length is equal to the desired stroke, $|\Delta\mathbf{R}_E|_{max}$. The endpoints are labeled at each D_1 and D_2 as shown in Figure 5.7a.

2. Construct an inclined line from D_1 and another from D_2 at an angle $\beta/2$ as shown in Figure 5.7a.

3. The intersection of the two inclined lines locates the rocker pivot, point A in Figure 5.7a. The line between points A and D_1 or between A and D_2 represents the rocker and will be designated L_3.

4. Draw a line perpendicular to line D_1D_2 through A. This line is labeled P in Figure 5.7a.

5. The crank pivot, point C, can be placed anywhere along line P. The distance between points A and C represents the frame and will be designated L_1.

6. Draw a line perpendicular to line AD_1, through point C. The intersection will be designated B_1 as shown in Figure 5.5a. Line B_1C represents the crank and will be designated L_2. Similarly, draw a line perpendicular to line AD_2, through point C. The intersection will be designated B_2.

7. The length of L_4, as shown in Figure 5.7b, can be made an appropriate value to fit the application. As with other slider-crank mechanisms, longer lengths will reduce maximum accelerations.

Note that the crank pivot, point C, can be placed along line P. Therefore, an infinite number of suitable mechanisms can be designed. A longer L_1 will dictate a longer crank, L_2, which will exhibit less force at joint B, but higher sliding speeds. It is common to compromise and select point C near the middle of line P.

Analytical methods can be incorporated by viewing the triangle in Figure 5.7a to generate expressions for the link lengths L_2 and L_3, as a function of the stroke $|\Delta\mathbf{R}_E|_{max}$, the imbalance angle β, and the frame length L_1 selected. As stated, L_4 should be made as long as the application allows.

$$L_3 = \frac{|\Delta\mathbf{R}_E|_{max}}{2\sin(\beta/2)} \qquad (5.20)$$

$$L_2 = L_1\sin(\beta/2) \qquad (5.21)$$

5.7 MECHANISM TO MOVE A LINK BETWEEN TWO POSITIONS

In material handling machines, it is common to have a link that moves from one general position to another. When two positions of a link are specified, this class of design problems is called *two-position synthesis*. This task can be accomplished by either rotating a link about a single pivot point or by using the coupler of a four-bar mechanism.

5.7.1 Two-Position Synthesis with a Pivoting Link

Figure 5.8a illustrates two points, A and B, that reside on a common link and move from A_1B_1 to A_2B_2. A single link can be designed to produce this displacement. The problem reduces to determining the pivot point for this link and the angle that the link must be rotated to achieve the desired displacement.

The graphical procedure to design a pivoting link for two-position synthesis is as follows:

1. Construct two lines that connect A_1A_2 and B_1B_2, respectively.

2. Construct a perpendicular bisector of A_1A_2.

3. Construct a perpendicular bisector of B_1B_2.

4. The intersection of these two perpendicular bisectors is the required location for the link pivot and shown as point C in Figure 5.8b. The center of rotation between two task positions is termed the *displacement pole*. Point C is the displacement pole of position 1 and 2.

5. The angle between the pivot point C, A_1, and A_2 is the required angle that the link must be rotated to produce the desired displacement. This angle is labeled $\Delta\theta$ in Figure 5.8c. A crank-rocker linkage can be subsequently designed to achieve this rotational motion, if it is desired to drive the mechanism with a continually rotating crank.

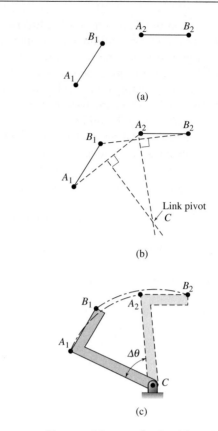

FIGURE 5.8 Two-position synthesis with a pivoting link.

5.7.2 Two-Position Synthesis of the Coupler of a Four-Bar Mechanism

In an identical problem to the one presented in the preceding section, Figure 5.9a illustrates two points, A and B, that must sit on a link and move from A_1B_1 to A_2B_2. Some applications may make a single pivoted link unfeasible, such as when the pivot point of the single link is inaccessible. In these cases, the coupler of a four-bar linkage can be designed to produce the required displacement. Appropriate lengths must be determined for all four links and the location of pivot points so that the coupler achieves the desired displacement.

The graphical procedure to design a four-bar mechanism for two-position synthesis is as follows:

1. Construct two lines that connect A_1A_2 and B_1B_2, respectively.

2. Construct a perpendicular bisector of A_1A_2.

3. Construct a perpendicular bisector of B_1B_2.

4. The pivot points of the input and output links can be placed anywhere on the perpendicular bisector. These pivot points are shown as C and D in Figure 5.9b.

5. The length of the two pivoting links are determined by scaling lengths A_1C and B_1D (Figure 5.9c).

The completed linkage is shown in Figure 5.9c. Because the pivot points C and D can be placed anywhere along the perpendicular bisectors, an infinite number of mechanisms

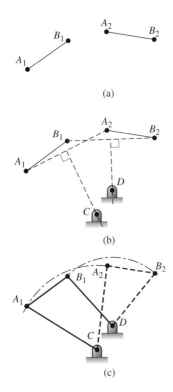

FIGURE 5.9 Two-position synthesis with a coupler link.

can be designed to accomplish the desired displacement. Note that longer pivoting links rotate at a smaller angle to move the coupler between the two desired positions. This produces larger transmission angles and reduces the force required to drive the linkage. The CAD system produces accurate results.

5.8 MECHANISM TO MOVE A LINK BETWEEN THREE POSITIONS

In some material handling machines, it is desired to have a link move between three positions. When three positions of a link are specified, this class of design problem is called *three-position synthesis*. For three-point synthesis, it generally is not possible to use a single pivoting link. This task is accomplished with the coupler of a four-bar mechanism.

Figure 5.10a illustrates two points, A and B, that must sit on a link and move from A_1B_1 to A_2B_2 to A_3B_3. Appropriate lengths must be determined for all four links and the location of pivot points so that the coupler achieves this desired displacement.

The graphical procedure to design a four-bar mechanism for three-point synthesis is as follows:

1. Construct four lines connecting A_1 to A_2, B_1 to B_2, A_2 to A_3, and B_2 to B_3.

2. Construct a perpendicular bisector of A_1A_2, a perpendicular bisector of B_1B_2, a perpendicular bisector of A_2A_3, and a perpendicular bisector of B_2B_3.

3. The intersection of the perpendicular bisector of A_1A_2 and the perpendicular bisector of A_2A_3

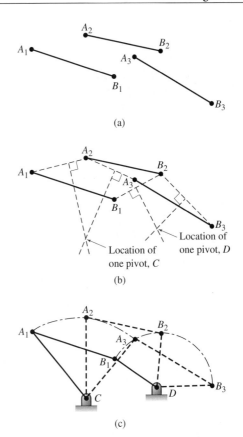

FIGURE 5.10 Three-position synthesis with a coupler link.

locates one pivot point. This is shown as point C in Figure 5.10b.

4. The intersection of the perpendicular bisector of B_1B_2 and the perpendicular bisector of B_2B_3 locates the other pivot point. This is shown as point D in Figure 5.10b.

5. The length of the two pivoting links is determined by scaling lengths A_1C and B_1D, as shown in Figure 5.7c.

The completed linkage is shown in Figure 5.10c. Again, the CAD system produces accurate results.

5.9 CIRCUIT AND BRANCH DEFECTS

As introduced in Chapter 4, an assembly circuit is all possible configurations of the mechanism links that can be realized without disconnecting the joints. Figure 4.24 illustrates two assembly circuits for the four-bar mechanism. As the procedure in Sections 5.5 and 5.6 is followed, it is possible that the one position will sit on a different assembly circuit as the other position(s). This is called a *circuit defect*, and is a fatal flaw in a linkage design. Once a four-bar mechanism has been synthesized, a position analysis should be performed to verify that the target positions can be reached from the original linkage configuration without disconnecting the joints.

A *branch defect* occurs if the mechanism reaches a toggle configuration between target positions. Unlike a circuit defect, a branch defect is dependent on the selection of driving link. In a toggle configuration, the mechanism becomes locked and the driving link is unable to actuate. The branch defect may not be a fatal flaw in the design, as an alternate link can be actuated to drive the mechanism between target positions.

PROBLEMS

Time Ratio Calculations

In Problems 5–1 through 5–3, a quick-return mechanism is to be designed where the outward stroke consumes t_1 and the return stroke t_2. Determine the time ratio, imbalance angle, and speed at which the mechanism should be driven.

5–1. $t_1 = 1.1\,\text{s}; t_2 = 0.8\,\text{s}.$

5–2. $t_1 = 0.35\,\text{s}; t_2 = 0.20\,\text{s}.$

5–3. $t_1 = 0.041\,\text{s}; t_2 = 0.027\,\text{s}.$

For Problems 5–4 through 5–6, a quick-return mechanism drives at ω rpm and has an imbalance angle of β. Determine the time ratio and the time to complete the outward and return strokes.

5–4. $\omega = 180\,\text{rpm}; \beta = 25°.$

5–5. $\omega = 75\,\text{rpm}; \beta = 37°.$

5–6. $\omega = 500\,\text{rpm}; \beta = 20°.$

Timing Charts

5-7. A reciprocating saw needs to move the blade downward 0.75 in. in 0.10 s and return in 0.08 s. Determine the time ratio and crank speed. Also, sketch the timing chart and estimate the peak velocity and acceleration of the motion.

5-8. A punch press needs to move a stamp downward 1.5 in. in 0.60 s and return in 0.35 s. Determine the time ratio and crank speed. Also, sketch the timing chart and estimate the peak velocity and acceleration of the motion.

5-9. A process requires a conveyor to index packages 6.0 in. in 0.6 s and dwell while a stamp is applied to the package. The stamp head must travel 8.0 in. to meet the package. The entire stamp stroke should take 0.8 s. Determine the time ratio and crank speed of the mechanism. Also, sketch the synchronized timing charts and estimate the peak velocity and acceleration of the different motion elements.

5-10. A process requires a conveyor to index cans 2.0 in. in 0.12 s and dwell while a cover is pressed onto a can. The cap head must travel 3.0 in. to approach the can. The entire cover pressing stroke should take 0.25 s. Determine the time ratio and crank speed of the mechanism. Also, sketch the synchronized timing charts and estimate the peak velocity and acceleration of the different motion elements.

Design of Slider-Crank Mechanisms

For Problems 5–11 through 5–18, design a slider-crank mechanism with a time ratio of Q, stroke of $|\Delta R_4|_{max}$ and time per cycle of t. Use either the graphical or analytical method. Specify the link lengths L_2, L_3, offset distance L_1 (if any), and the crank speed.

5–11. $Q = 1; |\Delta R_4|_{max} = 2\,\text{in.}; t = 1.2\,\text{s}.$

5–12. $Q = 1; |\Delta R_4|_{max} = 8\,\text{mm}; t = 0.08\,\text{s}.$

5–13. $Q = 1; |\Delta R_4|_{max} = 0.9\,\text{mm}; t = 0.4\,\text{s}.$

5–14. $Q = 1.25; |\Delta R_4|_{max} = 2.75\,\text{in.}; t = 0.6\,\text{s}.$

5–15. $Q = 1.37; |\Delta R_4|_{max} = 46\,\text{mm}; t = 3.4\,\text{s}.$

5–16. $Q = 1.15; |\Delta R_4|_{max} = 1.2\,\text{in.}; t = 0.014\,\text{s}.$

5–17. $Q = 1.20; |\Delta R_4|_{max} = 0.375\,\text{in.}; t = 0.025\,\text{s}.$

5–18. $Q = 1.10; |\Delta R_4|_{max} = 0.625\,\text{in.}; t = 0.033\,\text{s}.$

Design of Crank-Rocker Mechanisms

For Problems 5–19 through 5–28, design a crank-rocker mechanism with a time ratio of Q, throw angle of $(\Delta\theta_4)_{max}$, and time per cycle of t. Use either the graphical or analytical method. Specify the link lengths L_1, L_2, L_3, L_4, and the crank speed.

5–19. $Q = 1; (\Delta\theta_4)_{max} = 78°; t = 1.2\,\text{s}.$

5–20. $Q = 1; (\Delta\theta_4)_{max} = 100°; t = 3.5\,\text{s}.$

5–21. $Q = 1.15; (\Delta\theta_4)_{max} = 55°; t = 0.45\,\text{s}.$

5–22. $Q = 1.24; (\Delta\theta_4)_{max} = 85°; t = 1.8\,\text{s}.$

5–23. $Q = 1.36; (\Delta\theta_4)_{max} = 45°; t = 1.2\,\text{s}.$

5–24. $Q = 1.20; (\Delta\theta_4)_{max} = 96°; t = 0.3\,\text{s}.$

5–25. $Q = 1.18; (\Delta\theta_4)_{max} = 72°; t = 0.08\,\text{s}; L_1 = 8.0\,\text{in.}$

5–26. $Q = 1.10; (\Delta\theta_4)_{max} = 115°; t = 0.2\,\text{s}; L_1 = 6.5\,\text{in.}$

5–27. $Q = 1.22; (\Delta\theta_4)_{max} = 88°; t = 0.75\,\text{s}; L_1 = 8.0\,\text{in.}$

5–28. $Q = 1.08; (\Delta\theta_4)_{max} = 105°; t = 1.50\,\text{s};$ $L_1 = 100.0\,\text{mm}.$

Design of Crank-Shaper Mechanisms

For Problems 5–29 through 5–32, design a crank-shaper mechanism with a time ratio of Q, stroke of $|\Delta R_E|_{max}$ and time per cycle of t. Use either the graphical or analytical method. Specify the link lengths L_1, L_2, L_3, L_4, and the crank speed.

5–29. $Q = 1.50; |\Delta R_E|_{max} = 2.75\,\text{in.}; t = 0.6\,\text{s}.$

5–30. $Q = 1.75; |\Delta R_E|_{max} = 46\,\text{mm}; t = 3.4\,\text{s}.$

5–31. $Q = 2.00; |\Delta R_E|_{max} = 0.375\,\text{in.}; t = 0.014\,\text{s}.$

5–32. $Q = 1.80; |\Delta R_E|_{max} = 1.2\,\text{in.}; t = 0.25\,\text{s}.$

Two-Position Synthesis, Single Pivot

For Problems 5–33 through 5–36, a link containing points A and B must assume the positions listed in the table for each problem. Graphically determine the location of a fixed pivot for a single pivoting link that permits the motion listed. Also determine the degree that the link must be rotated to move from position 1 to position 2.

5–33.

Coordinates:	A_x (in.)	A_y (in.)	B_x (in.)	B_y (in.)
Position 1	0.0000	9.0000	5.0000	9.0000
Position 2	6.3600	6.3600	9.9005	2.8295

5–34.

Coordinates:	A_x (in.)	A_y (in.)	B_x (in.)	B_y (in.)
Position 1	2.2800	5.3400	6.8474	7.3744
Position 2	9.7400	8.5000	12.5042	4.336

5–35.

Coordinates:	A_x (mm)	A_y (mm)	B_x (mm)	B_y (mm)
Position 1	−53.000	41.000	75.205	19.469
Position 2	−36.000	40.000	87.770	−8.112

5–36.

Coordinates:	A_x (mm)	A_y (mm)	B_x (mm)	B_y (mm)
Position 1	25.507	47.312	83.000	11.000
Position 2	97.000	30.000	150.676	71.748

Two-Position Synthesis, Two Pivots

In Problems 5–37 through 5–40, a link containing points A and B is to assume the positions listed in the table for each problem. Graphically find the location of two fixed pivots and the lengths of all four links of a mechanism with a coupler that will exhibit the motion listed. Also, determine the amount that the pivot links must be rotated to move the coupler from position 1 to position 2.

5–37.

Coordinates:	A_x (in.)	A_y (in.)	B_x (in.)	B_y (in.)
Position 1	−0.3536	4.8501	4.4000	3.3000
Position 2	−3.1000	3.2000	1.5562	5.0220

5–38.

Coordinates:	A_x (in.)	A_y (in.)	B_x (in.)	B_y (in.)
Position 1	0.9000	4.5000	9.0380	7.7150
Position 2	−1.0000	5.6000	5.5727	11.3760

5–39.

Coordinates:	A_x (mm)	A_y (mm)	B_x (mm)	B_y (mm)
Position 1	−40.000	−60.000	28.936	−30.456
Position 2	−65.350	−26.352	8.000	−42.000

5–40.

Coordinates:	A_x (mm)	A_y (mm)	B_x (mm)	B_y (mm)
Position 1	−37.261	−2.041	−18.000	1.000
Position 2	−18.000	−3.000	0.858	−7.963

Three-Position Synthesis

For Problems 5–41 through 5–44, a link containing points A and B must assume the three positions listed in the table for each problem. Graphically find the location of two fixed pivots and the lengths of all four links of a mechanism with a coupler that will exhibit the motion listed. Also, determine the amount that the pivot links must be rotated to move the coupler from position 1 to position 2, then from position 2 to position 3.

5–41.

Coordinates:	A_x (in.)	A_y (in.)	B_x (in.)	B_y (in.)
Position 1	−1.0000	−0.9000	5.2862	−1.7980
Position 2	−2.7000	−1.3000	3.6428	−0.9980
Position 3	−4.4000	−2.0000	1.7719	−0.5068

5–42.

Coordinates:	A_x (in.)	A_y (in.)	B_x (in.)	B_y (in.)
Position 1	−5.5000	−0.1000	7.9836	5.2331
Position 2	−2.4000	0.5000	12.0831	1.1992
Position 3	−0.6000	1.6000	13.6483	−1.0902

5–43.

Coordinates:	A_x (mm)	A_y (mm)	B_x (mm)	B_y (mm)
Position 1	0.000	40.000	54.774	44.980
Position 2	21.000	51.000	72.204	30.920
Position 3	39.000	49.000	82.143	14.887

5–44.

Coordinates:	A_x (mm)	A_y (mm)	B_x (mm)	B_y (mm)
Position 1	43.000	−76.000	149.890	−50.027
Position 2	3.000	−52.000	111.127	−72.211
Position 3	−12.000	−33.000	91.840	−69.294

CASE STUDIES

5–1. Figure C5.1 shows a mechanism that drives a sliding block, I. Block I, in turn, moves the blade of a power hacksaw. Carefully examine the configuration of the components in the mechanism. Then, answer the following questions to gain insight into the operation of the mechanism.

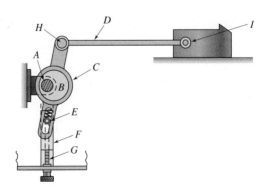

FIGURE C5.1 (Courtesy, Industrial Press.)

1. As shaft A rotates 90° cw, what is the motion of lobe B, which is attached to shaft A?
2. As shaft A rotates 90° cw, what is the motion of item C?

3. Is a slot necessary at roller *E*?
4. As shaft *A* rotates 90° cw, what is the motion of pin *H*?
5. As shaft *A* rotates 90° cw, what is the motion of pin *I*?
6. Determine the mobility of this mechanism.
7. As thread *G* rotates to pull roller *E* downward, how does that alter the motion of link *C*?
8. As thread *G* rotates to pull roller *E* downward, how does that alter the motion of pin *H*?
9. What is the purpose of this mechanism?

5–2. Figure C5.2 shows a mechanism that also drives a sliding block *B*. This sliding block, in turn, drives a cutting tool. Carefully examine the configuration of the components in the mechanism. Then, answer the following questions to gain insight into the operation of the mechanism.

1. As rod *A* drives to the right, what is the motion of sliding block *B*?
2. Describe the motion of sliding block *B* as roller *C* reaches groove *D*.

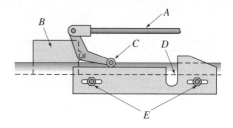

FIGURE C5.2 (Courtesy, Industrial Press.)

3. Describe the motion of sliding block *B* as rod *A* drives to the left, bringing *C* out of groove *D*.
4. Describe the continual motion of sliding block *B* as rod *A* oscillates horizontally.
5. What is the purpose of this mechanism?
6. Describe a device that could drive rod *A* to the left and right.
7. The adjustment slots at *E* provide what feature to the mechanism?

VELOCITY ANALYSIS

OBJECTIVES

Upon completion of this chapter, the student will be able to:

1. Define linear, rotational, and relative velocities.

2. Convert between linear and angular velocities.

3. Use the relative velocity method to graphically solve for the velocity of a point on a link, knowing the velocity of another point on that link.

4. Use the relative velocity method to graphically and analytically determine the velocity of a point of interest on a floating link.

5. Use the relative velocity method to analytically solve for the velocity of a point on a link, knowing the velocity of another point on that link.

6. Use the instantaneous center method to graphically and analytically determine the velocity of a point.

7. Construct a velocity curve to locate extreme velocity values.

6.1 INTRODUCTION

Velocity analysis involves determining "how fast" certain points on the links of a mechanism are traveling. Velocity is important because it associates the movement of a point on a mechanism with time. Often the timing in a machine is critical.

For instance, the mechanism that "pulls" video film through a movie projector must advance the film at a rate of 30 frames per second. A mechanism that feeds packing material into a crate must operate in sequence with the conveyor that indexes the crates. A windshield wiper mechanism operating on high speed must sweep the wiper across the glass at least 45 times every minute.

The determination of velocity in a linkage is the purpose of this chapter. Two common analysis procedures are examined: the relative velocity method and the instantaneous center method. Consistent with other chapters in this book, both graphical and analytical techniques are included.

6.2 LINEAR VELOCITY

Linear velocity, **V**, of a point is the linear displacement of that point per unit time. Recall that linear displacement of a point, $\Delta\mathbf{R}$, is a vector and defined as a change in position of that point. This concept was introduced in Section 4.3.

As described in Chapter 4, the displacement of a point is viewed as translation and is discussed in linear terms. By definition, a point can only have a linear displacement. When the time elapsed during a displacement is considered, the velocity can be determined.

As with displacement, velocity is a vector. Recall that vectors are denoted with the boldface type. The magnitude of velocity is often referred to as "speed" and designated as $v = |\mathbf{V}|$. Understanding the direction of linear velocity requires determining the direction in which a point is moving at a specific instant.

Mathematically, linear velocity of a point is expressed as

$$\mathbf{V} = \lim_{\Delta t \to 0} \frac{d\mathbf{R}}{dt} \qquad (6.1)$$

and for short time periods as

$$\mathbf{V} \cong \frac{\Delta\mathbf{R}}{\Delta t} \qquad (6.2)$$

Since displacement is a vector, equation (6.1) indicates that velocity is also a vector. As with all vectors, a direction is also required to completely define velocity. Linear velocity is expressed in the units of length divided by time. In the U.S. Customary System, the common units used are feet per second (ft/s or fps), feet per minute (ft/min or fpm), or inches per second (in./s or ips). In the International System, the common units used are meters per second (m/s) or millimeters per second (mm/s).

6.2.1 Linear Velocity of Rectilinear Points

A point can move along either a straight or curved path. As seen in the earlier chapters, many links are constrained to straight-line, or rectilinear, motion. For points that are attached to a link that is restricted to rectilinear motion, equations (6.1) and (6.2) can be used to calculate the magnitude of the velocity. The orientation of the linear velocity vector is simply in the direction of motion, which is usually obvious.

EXAMPLE PROBLEM 6.1

Crates on a conveyor shown in Figure 6.1 move toward the left at a constant rate. It takes 40 s to traverse the 25-ft conveyor. Determine the linear velocity of a crate.

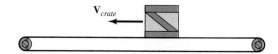

FIGURE 6.1 Translating crate for Example Problem 6.1.

SOLUTION: Because the crates travel at a constant rate, equation (6.2) can be used to determine the linear velocity of the crate.

$$\mathbf{V}_{crate} = \frac{\Delta \mathbf{R}}{\Delta t} = \frac{25 \text{ ft}}{40 \text{ s}} = .625 \text{ ft/s} \left(\frac{12 \text{ in.}}{1 \text{ ft}} \right) = 7.5 \text{ in./s} \ \leftarrow$$

6.2.2 Linear Velocity of a General Point

For points on a link undergoing general motion, equations (6.1) and (6.2) are still valid. The direction of the linear velocity of a point is the same as the direction of its instantaneous motion. Figure 6.2 illustrates the velocity of two points on a link. The velocities of points A and B are denoted as $\mathbf{V}_A$ and $\mathbf{V}_B$, respectively. Note that although they are on the same link, both these points can have different linear velocities. Points that are farther from the pivot travel faster. This can be "felt" when sitting on the outer seats of amusement rides that spin.

From Figure 6.2, the velocity of point A, $\mathbf{V}_A$, is directed along the path that point A is moving at this instant—that is, tangent to an arc centered at O, which is also perpendicular to link OA. In casual terms, if point A were to break away from link 2 at this instant, point A would travel in the direction of its linear velocity.

6.2.3 Velocity Profile for Linear Motion

Advances in technology have allowed precise motion control for many applications, such as automation, test, and measurement equipment. These systems implement servo drives, which are motors controlled by a microprocessor. The intended motion is specified to a controller. Sensors monitor the motion of the moving link and provide feedback to the controller. If a difference between the intended motion and the actual motion is measured, the controller will alter the signal to the motor and correct the deviation. Because of the precision, responsiveness, and lowering cost, the use of servo systems is growing rapidly.

For optimal motion control, smooth high-speed motion is desired, with a minimal effort required from the motor. The controller must direct the motor to change velocity wisely to

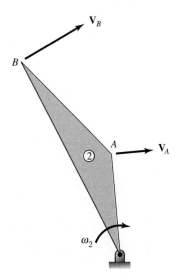

FIGURE 6.2 Linear velocities of points on a link.

achieve maximum results. For a linear servo system, the motion characteristics of a translating machine component are usually specified with a shaped velocity profile. The velocity profile prescribes the speed-up, steady-state, and slow-down motion segments for the translating link. The actual displacement can be calculated from the velocity profile. Rewriting equation (6.1),

$$d\mathbf{R} = \mathbf{V} \ dt$$

Solving for the displacement, $\Delta \mathbf{R}$, gives

$$\Delta \mathbf{R} = \int \mathbf{V} \ dt \qquad (6.3)$$

With knowledge from elementary calculus, equation (6.3) states that the displacement for a certain time interval is the area under the v–t curve for that time interval.

EXAMPLE PROBLEM 6.2

Servo-driven actuators are programmed to move according to a specified velocity profile. The linear actuator, shown in Figure 6.3a, is programmed to extend according to the velocity profile shown in Figure 6.3b. Determine the total displacement during this programmed move.

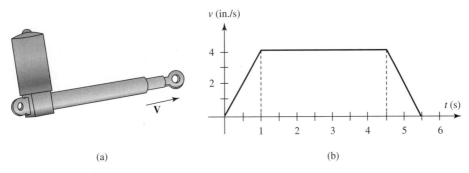

(a)   (b)

FIGURE 6.3 Velocity profile for Example Problem 6.2.

SOLUTION:

1. ***Displacement during the Speed-Up Portion of the Move***

   During the first one-second portion of the move, the actuator is speeding up to its steady-state velocity. The area under the v–t curve forms a triangle and is calculated as

   $$\Delta R_{\text{speed-up}} = \tfrac{1}{2} \left(V_{\text{steady-state}} \right) \left(\Delta t_{\text{speed-up}} \right) = \tfrac{1}{2} \left(4 \text{ in./s} \right) [(1-0) \text{ s}] = 2 \text{ in.} \longrightarrow$$

2. ***Displacement during the Steady-State Portion of the Move***

   During the time interval between 1 and 4.5, the actuator is moving at its steady-state velocity. The area under the v–t curve forms a rectangle and is calculated as

   $$\Delta R_{\text{steady-state}} = \left(V_{\text{steady-state}} \right) \left(\Delta t_{\text{steady-state}} \right) = \left(4 \text{ in./s} \right) [(4.5-1) \text{ s}] = 14 \text{ in.} \longrightarrow$$

3. ***Displacement during the Slow-Down Portion of the Move***

   During the time interval between 4.5 and 5.5, the actuator is slowing down from its steady-state velocity. The area under the v–t curve forms a triangle and is calculated as

   $$\Delta R_{\text{slow-down}} = \tfrac{1}{2} \left(V_{\text{steady-state}} \right) \left(\Delta t_{\text{slow-down}} \right) = \tfrac{1}{2} \left(4 \text{ in./s} \right) [(5.5-4.5) \text{ s}] = 2 \text{ in.} \longrightarrow$$

4. ***Total Displacement during the Programmed Move***

   The total displacement during the programmed move is the sum of the displacement during the speed-up, steady-state, and slow-down portions of the move.

   $$\Delta R_{\text{total}} = \Delta R_{\text{speed-up}} + \Delta R_{\text{steady-state}} + \Delta R_{\text{slow-down}} = 2 + 14 + 2 = 18 \text{ in.} \longrightarrow \text{(extension)}$$

6.3 VELOCITY OF A LINK

Several points on a link can have drastically different linear velocities. This is especially true as the link simply rotates about a fixed point, as in Figure 6.2. In general, the motion of a link can be rather complex as it moves (translates) and spins (rotates).

Any motion, however complex, can be viewed as a combination of a straight-line movement and a rotational movement. Fully describing the motion of a link can consist of identification of the linear motion of one point and the rotational motion of the link.

Although several points on a link can have different linear velocities, being a rigid body, the entire link has the same angular velocity. *Angular velocity, ω,* of a link is the angular displacement of that link per unit of time. Recall that rotational displacement of a link, $\Delta\theta$, is defined as the angular change in orientation of that link. This was introduced in Section 4.3.

Mathematically, angular velocity of a link is expressed as:

$$\omega = \lim_{\Delta t \to o} \frac{\Delta \theta}{\Delta t} = \frac{d\theta}{dt} \qquad \textbf{(6.4)}$$

and for short time periods, or when the velocity can be assumed linear,

$$\omega \cong \frac{\Delta \theta}{\Delta t} \qquad \textbf{(6.5)}$$

The direction of angular velocity is in the direction of the link's rotation. In planar analyses, it can be fully described by specifying either the term *clockwise* or *counterclockwise*. For example, the link shown in Figure 6.2 has an angular velocity that is consistent with the linear velocities of the points that are attached to the link. Thus, the link has a clockwise rotational velocity.

Angular velocity is expressed in the units of angle divided by time. In both the U.S. Customary System and the International System, the common units used are revolutions per minute (rpm), degrees per second (deg/s), or radians per second (rad/s or rps).

EXAMPLE PROBLEM 6.3

The gear shown in Figure 6.4 rotates counterclockwise at a constant rate. It moves 300° in. 0.5 s. Determine the angular velocity of the gear.

FIGURE 6.4 Rotating gear for Example Problem 6.3.

SOLUTION: Because the gear rotates at a constant rate, equation (6.4) can be used to determine the angular velocity of the gear.

$$\omega_{\text{gear}} = \frac{\Delta\theta_{\text{gear}}}{\Delta t} = \frac{300°}{0.5 \text{ s}} = 600 \text{ deg/s} \left(\frac{1 \text{ rev}}{360 \text{ deg}}\right)\left(\frac{60 \text{ s}}{1 \text{ min}}\right) = 50 \text{ rpm, counterclockwise}$$

6.4 RELATIONSHIP BETWEEN LINEAR AND ANGULAR VELOCITIES

For a link in pure rotation, the magnitude of the linear velocity of any point attached to the link is related to the angular velocity of the link. This relationship is expressed as

$$v = r\omega \qquad (6.6)$$

where:

$v = |\mathbf{V}|$ = magnitude of the linear velocity of the point of consideration

r = distance from the center of rotation to the point of consideration

ω = angular velocity of the rotating link that contains the point of consideration

Linear velocity is always perpendicular to a line that connects the center of the link rotation to the point of consideration. Thus, linear velocity of a point on a link in pure rotation is often called the *tangential velocity*. This is because the linear velocity is tangent to the circular path of that point, or perpendicular to the line that connects the point with the pivot.

It is extremely important to remember that the angular velocity, ω, in equation (6.6) must be expressed as units of radians per time. The radian is a dimensionless unit of angular measurement that can be omitted. Linear velocity is

expressed in units of length per time and not radians times length per time, as equation (6.6) would imply.

Often, the conversion must be made from the more common unit of revolutions per minute (rpm):

$$\omega(\text{rad/min}) = \left[\omega\,(\text{rad/min})\right]\left[\frac{2\pi \text{ rad}}{\text{rev}}\right]$$
$$= 2\pi\left[\omega(\text{rad/min})\right] \qquad (6.7)$$

and

$$\omega\,(\text{rad/s}) = \left[\omega\,(\text{rev/min})\right]\left[\left(\frac{2\pi \text{ rad}}{1 \text{ rev}}\right)\left(\frac{1 \text{ min}}{60 \text{ s}}\right)\right]$$
$$= \frac{\pi}{30}\left[\omega(\text{rev/min})\right] \qquad (6.8)$$

As mentioned, a radian is a dimensionless measure of an angle. To be exact, an angle expressed in radians is the ratio of the arc length swept by the angle to the radius. When an angle expressed in radians is multiplied by another value, the radian designation is omitted.

As stated in the previous section, the angular velocity of the link and the linear velocities of points on the link are consistent. That is, the velocities (rotational or linear) are in the direction that the object (link or point) is instantaneously moving. As mentioned, linear velocity is always perpendicular to a line that connects the center of link rotation to the point of consideration.

EXAMPLE PROBLEM 6.4

Figure 6.5 illustrates a cam mechanism used to drive the exhaust port of an internal combustion engine. Point *B* is a point of interest on the rocker plate. At this instant, the cam forces point *B* upward at 30 mm/s. Determine the angular velocity of the rocker plate and the velocity of point *C*.

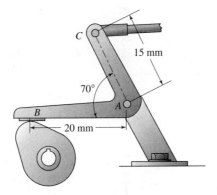

FIGURE 6.5 Mechanism for Example Problem 6.4.

SOLUTION: 1. *Draw a Kinematic Diagram and Calculate Degrees of Freedom*

The rocker plate is connected to the frame with a pin joint at point A. The velocity of point B is a vector directed upward with a magnitude of 30 mm/s. Figure 6.6 shows a kinematic diagram.

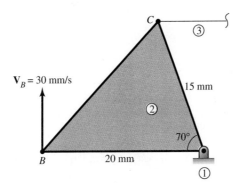

FIGURE 6.6 Kinematic diagram for Example Problem 6.4.

2. *Calculate the Angular Velocity of Link 2*

It should be apparent that as point B travels upward, the rocker plate, link 2, is forced to rotate clockwise. Therefore, as point B has upward linear velocity, the rocker plate must have a clockwise angular velocity. The magnitude of the angular velocity is found by rearranging equation (6.5):

$$\omega_2 = \frac{v_B}{r_{AB}} = \frac{30 \text{ mm/s}}{20 \text{ mm}} = 1.5 \text{ rad/s}$$

This can be converted to rpm by rearranging equation (6.6):

$$\omega_2 \, (\text{rev/min}) = \frac{30}{\pi}\left[\omega_2 \,(\text{rad/s})\right] = \frac{30}{\pi}[1.5 \text{ rad/s}] = 14.3 \text{ rpm}$$

Including the direction,

$$\omega_2 = 1.5 \text{ rad/s, cw}$$

3. *Calculate the Linear Velocity of Point C*

The linear velocity of point C can also be computed from equation (6.5):

$$v_C = r_{AC}\omega_2 = (15 \text{ mm})(1.5 \text{ rad/s}) = 22.5 \text{ mm/s}$$

The direction of the linear velocity of C must be consistent with the angular velocity of link 2. The velocity also is perpendicular to the line that connects the center of rotation of link 2, point A, to point C. Therefore, the velocity of point C is directed 20° (90°–70°) above the horizontal. Including the direction,

$$\mathbf{V}_C = 22.5 \text{ mm/s} \, \diagup\!\!\underline{20°}$$

6.5 RELATIVE VELOCITY

The difference between the motion of two points is termed *relative motion.* Consider a situation where two cars travel on the interstate highway. The car in the left lane travels at 65 miles per hour (mph), and the car in the right lane travels at 55 mph. These speeds are measured in relationship to a stationary radar unit. Thus, they are a measurement of *absolute motion.*

Although both are moving forward, it appears to the people in the faster car that the other car is actually moving backward. That is, the relative motion of the slower car to the faster car is in the opposite direction of the absolute motion. Conversely, it appears to the people in the slower car that the faster car is traveling at 10 mph. That is, the relative velocity of the faster car to the slower car is 10 mph.

Relative velocity is a term used when the velocity of one object is related to that of another reference object, which can also be moving. The following notation distinguishes between absolute and relative velocities:

V_A = absolute velocity of point A

V_B = absolute velocity of point B

$V_{B/A}$ = relative velocity of point B with respect to A
= velocity of point B "as observed" from point A

Relative motion, that is, the difference between the motion of two points, can be written mathematically as

$$V_{B/A} = V_B \rightarrow V_A \qquad (6.9)$$

or rearranged as $\quad \vec{V_B} = \vec{V_A} + \vec{V_{B/A}}$

$$V_B = V_A + > V_{B/A} \qquad (6.10)$$

Note that equations (6.9) and (6.10) are vector equations. Therefore, in order to use the equations, vector polygons must be prepared in accordance with the equations. The techniques discussed in Section 3.16 must be used in dealing with these equations.

EXAMPLE PROBLEM 6.5

Figure 6.7 shows a cargo lift mechanism for a delivery truck. At this instant, point A has a velocity of 12 in./s in the direction shown, and point B has a velocity of 10.4 in./s, also in the direction shown. Determine the angular velocity of the lower link and the relative velocity of point B relative to point A.

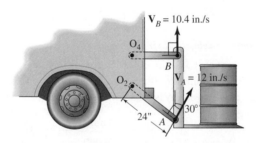

FIGURE 6.7 Mechanism for Example Problem 6.5.

SOLUTION: 1. ***Draw a Kinematic Diagram and Identify Mobility***

Figure 6.8a shows the kinematic diagram of this mechanism. Notice that it is the familiar four-bar mechanism, having a single degree of freedom.

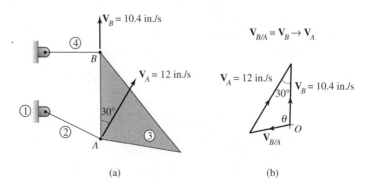

(a) (b)

FIGURE 6.8 Kinematic diagram for Example Problem 6.5.

2. ***Calculate the Angular Velocity of Link 2***

    From the kinematic diagram, it should be apparent that as point *A* travels up and to the right, link 2 rotates counterclockwise. Thus, link 2 has a counterclockwise angular velocity. The magnitude of the angular velocity is found by rearranging equation (6.6) as follows:

    $$\omega_2 = \frac{v_A}{r_{AO_2}} = \frac{(12 \text{ in./s})}{(24 \text{ in.})} = 0.5 \text{ rad/s}$$

    This can be converted to rpm by rearranging equation (6.7) as

    $$\omega_2(\text{rev/min}) = \frac{30}{\pi}\left[\omega_2(\text{rad/s})\right] = \frac{30}{\pi}[0.5 \text{ rad/s}] = 4.8 \text{ rpm}$$

    Including the direction,

    $$\omega_2 = 4.8 \text{ rpm, counterclockwise}$$

3. ***Calculate the Linear Velocity of Point*** B ***Relative to Point*** A

    The relative velocity of *B* with respect to *A* can be found from equation (6.9):

    $$\mathbf{V}_{B/A} = \mathbf{V}_B \mathbin{->} \mathbf{V}_A$$

    A vector polygon is formed from this equation and is given in Figure 6.8b. Notice that this is a general triangle. Either a graphical or analytical solution can be used to determine the vector $v_{B/A}$.

    Arbitrarily using an analytical method, the velocity magnitude $v_{B/A}$ can be found from the law of cosines.

    $$v_{B/A} = \sqrt{[v_A{}^2 + v_B{}^2 - 2(v_A)(v_B)(\cos 30°)]}$$
    $$= \sqrt{(12 \text{ in./s})^2 + (10.4 \text{ in./s})^2 - 2(12 \text{ in./s})(10.4 \text{ in./s})(\cos 30°)} = 6.0 \text{ in./s}$$

    The angle between the velocity magnitudes $v_{B/A}$ and v_B is shown as θ in Figure 6.8b. It can be found by using the law of sines:

    $$\theta = \sin^{-1}\left[\frac{(12 \text{ in./s})}{6 \text{ in./s}} \sin 30°\right] = 90°$$

    Thus, this vector polygon actually formed a right triangle. The relative velocity of *B* with respect to *A* is stated formally as follows:

    $$\mathbf{V}_{B/A} = 6.0 \text{ in./s} \leftarrow$$

Relative velocity between two points on a link is useful in determining velocity characteristics of the link. Specifically, the relative velocity of any two points on a link can be used to determine the angular velocity of that link. Assuming that points *A*, *B*, and *C* lay on a link, the angular velocity can be stated as

$$\omega = \frac{v_{A/B}}{r_{AB}} = \frac{v_{B/C}}{r_{BC}} = \frac{v_{A/C}}{r_{AC}} \qquad (6.11)$$

The direction of the angular velocity is consistent with the relative velocity of the two points. The relative velocity of *B* with respect to *A* implies that *B* is seen as rotating about *A*. Therefore, the direction of the relative velocity of *B* "as seen from" *A* suggests the direction of rotation of the link that is shared by points *A* and *B*. Referring to Figure 6.9, when $v_{B/A}$ is directed up and to the left, the angular velocity of the link is counterclockwise. Conversely, when $v_{B/A}$ is directed down and to the right, the angular velocity of the link is clockwise.

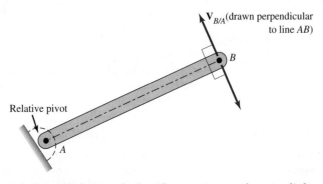

FIGURE 6.9 Relative velocity of two points on the same link.

6.6 GRAPHICAL VELOCITY ANALYSIS: RELATIVE VELOCITY METHOD

Graphical velocity analysis will determine the velocity of mechanism points in a single configuration. It must be emphasized that the results of this analysis correspond to the current position of the mechanism. As the mechanism moves, the configuration changes, and the velocities also change.

The basis of the relative velocity method of analysis is derived from the following fact:

Two points that reside on the same link can only have a relative velocity that is in a direction perpendicular to the line that connects the two points.

This fact is an extension of the definition of relative velocity. Figure 6.9 illustrates two points, A and B, that are on the same link. Recall that $v_{B/A}$ is the velocity of B "as observed" from A. For an observer at A, it appears that B is simply rotating around A, as long as both A and B are on the same link. Thus, the velocity of B with respect to A must be perpendicular to the line that connects B to A.

With this fact and vector analysis techniques, the velocity of points on a mechanism can be determined.

6.6.1 Points on Links Limited to Pure Rotation or Rectilinear Translation

The most basic analysis using the relative velocity method involves points that reside on links that are limited to pure rotation or rectilinear translation. The reason is that the direction of the motion of the point is known. Pin joints are convenient points of analysis because they reside on two links, where one is typically constrained to pure rotation or rectilinear translation.

Figure 6.10 shows a slider-crank mechanism. Point B resides on links 2 and 3. Notice that the direction of the velocity of point B is known because link 2 is constrained to pure rotation. Point C resides on links 3 and 4. Likewise, the direction of the velocity of point C is known because link 4 is constrained to rectilinear translation. If the velocity of point B is known, the velocity of point C can be quickly found because the direction of that velocity is

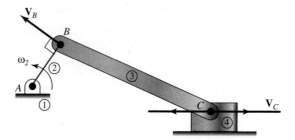

FIGURE 6.10 Links constrained to pure rotation and rectilinear translation.

also known. Only the magnitude and sense need to be determined.

The general solution procedure to problems of this type can be summarized as:

1. Determine the direction of the unknown velocity by using the constraints imposed by the joint, either pure rotation or pure translation.

2. Determine the direction of the relative velocity between the two joints. For two points on the same link, the relative velocity is always perpendicular to the line that connects the points.

3. Use the following relative velocity equation to draw a vector polygon:

$$\mathbf{V}_{\text{Unknown point}} = \mathbf{V}_{\text{Known point}}$$
$$+> \mathbf{V}_{\text{Unknown point/Known point}}$$

4. Using the methods outlined in Section 3.18, and the vector equation above, determine the magnitudes of

$$\mathbf{V}_{\text{Unknown point}} \text{ and } \mathbf{V}_{\text{Unknown point/Known point}}$$

This analysis procedure describes the logic behind graphical velocity analysis. The actual solution can be completed using manual drawing techniques (using a protractor and compass) or can be completed on a CAD system (using a rotate and copy command). The logic is identical; however, the CAD solution is not susceptible to limitations of drafting accuracy. Regardless of the method being practiced, the underlying concepts of graphical position analysis can be further illustrated and expanded through the following example problem.

EXAMPLE PROBLEM 6.6

Figure 6.11 shows a rock-crushing mechanism. It is used in a machine where large rock is placed in a vertical hopper and falls into this crushing chamber. Properly sized aggregate, which passes through a sieve, is discharged at the bottom. Rock not passing through the sieve is reintroduced into this crushing chamber.

Determine the angular velocity of the crushing ram, in the shown configuration, as the 60-mm crank rotates at 120 rpm, clockwise.

SOLUTION: 1. **Draw a Kinematic Diagram and Calculate Degrees of Freedom**

Figure 6.12a shows a kinematic diagram of this mechanism. Notice that this mechanism is the familiar four-bar linkage, having a single degree of freedom. With one degree of freedom, this mechanism is fully operated with the one input motion. Of course, this motion is the rotation of link 2, at a rate of 120 rpm.

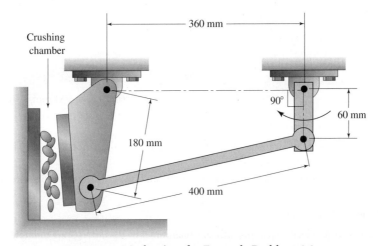

FIGURE 6.11 Mechanism for Example Problem 6.6.

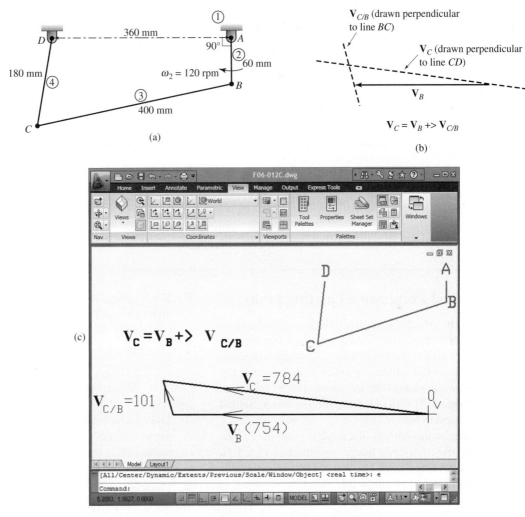

FIGURE 6.12 Diagrams for Example Problem 6.6.

2. **Decide on an Appropriate Relative Velocity Equation**

The objective of the analysis is to determine the angular velocity of link 4. Link 2 contains the input motion (velocity). Point B resides on both links 2 and 3. Point C resides on both links 3 and 4. Because points B and C reside on link 3, the relative velocity method can be used to relate the input velocity (link 2) to the desired velocity (link 4). The relative velocity equation for this analysis becomes

$$\mathbf{V}_C = \mathbf{V}_B + > \mathbf{V}_{C/B}$$

3. ***Determine the Velocity of the Input Point***

The velocity of point B is calculated as

$$\omega_2 \left(\text{rad/s} \right) = \frac{\pi}{30} (120 \text{ rpm}) = 12.56 \text{ rad/s, cw}$$

$$V_B = \omega_2 \, r_{AB} = \left(12.56 \text{ rad/s} \right) (60 \text{ mm}) = 754 \text{ mm/s} \;\; \leftarrow$$

4. ***Determine the Directions of the Desired Velocities***

Because link 4 is fixed to the frame at D, link 4 is limited to rotation about D. Therefore, the velocity of point C must be perpendicular to the line CD.

Also, as earlier stated, points B and C reside on link 3. Therefore, the relative velocity of C with respect to B must lie perpendicular to the line BC.

5. ***Draw a Velocity Polygon***

In the relative velocity equation, only the magnitudes of $\mathbf{V}_C$ and $\mathbf{V}_{C/B}$ are unknown. This is identical to the problems illustrated in Section 3.18. The vector polygon used to solve this problem is shown in Figure 6.12b. The magnitudes can be determined by observing the intersection of the directed lines of $\mathbf{V}_C$ and $\mathbf{V}_{C/B}$. The completed vector polygon is shown in Figure 6.12c.

6. ***Measure the Velocities from the Velocity Polygon***

The velocities are scaled from the velocity diagram to yield

$$\mathbf{V}_C = 784.0 \text{ mm/s} \;\underline{\;7.0°\;} \nwarrow$$

$$\mathbf{V}_{C/B} = 101.1 \text{ mm/s} \;\underline{\;72.7°\;} \searrow$$

7. ***Calculate Angular Velocities***

Ultimately, the angular velocities of link 4 is desired. The angular velocities of both links 3 and 4 can be determined from equation (6.6):

$$\omega_4 = \frac{v_C}{r_{CD}} = \frac{\left(789.4 \text{ mm/s} \right)}{(180 \text{ mm})} = 4.36 \text{ rad/s, cw}$$

$$\omega_3 = \frac{v_{C/B}}{r_{BC}} = \frac{101.1 \text{ mm/s}}{(400 \text{ mm})} = 0.25 \text{ rad/s, cw}$$

6.6.2 General Points on a Floating Link

Determining the velocity of general points on a floating link presents a slightly more complicated analysis. A floating link is simply a link that is not limited to pure rotation or rectilinear translation. The difficulty arises in that neither the direction nor magnitude of the unknown velocity is known. This is fundamentally different from the analysis presented in Example Problem 6.6.

To determine the velocity of a general point on a floating link, the velocity of two additional points on the link must be already determined. The two points are commonly pin joints constrained to either translation or rotation, as presented in Section 6.6.1. The velocities of these special points are readily obtained in an analysis similar to Example Problem 6.6.

Figure 6.13a illustrates a link in which the velocities of points A and B are already determined. To determine the velocity of point C, the following procedure can be followed:

1. Two equations can be written.

$$\mathbf{V}_C = \mathbf{V}_A +> \mathbf{V}_{C/A}$$

$$\mathbf{V}_C = \mathbf{V}_B +> \mathbf{V}_{C/B}$$

Since points A, B, and C are on the same link, the directions of $\mathbf{V}_{C/A}$ and $\mathbf{V}_{C/B}$ are perpendicular to lines CA and CB, respectively.

2. The individual relative velocity equations can be set equal to each other. In this case, this yields the following:

$$\mathbf{V}_C = \mathbf{V}_A +> \mathbf{V}_{C/A} = \mathbf{V}_B +> \mathbf{V}_{C/B}$$

3. The relative velocities can be solved by again using the techniques outlined in Section 3.18. This involves constructing the vector polygon as shown in Figure 6.13b.

4. The relative velocity magnitudes can be measured from the vector polygon.

5. Knowing the relative velocities, the velocity of the point of interest, point C, can be determined using one of the individual equations outlined in step 1. This can be readily found from the original vector polygon, as shown in Figure 6.13c.

Again, vector polygons can be constructed using identical logic with either manual drawing techniques or CAD. This logic behind the analysis is illustrated in the following example problem.

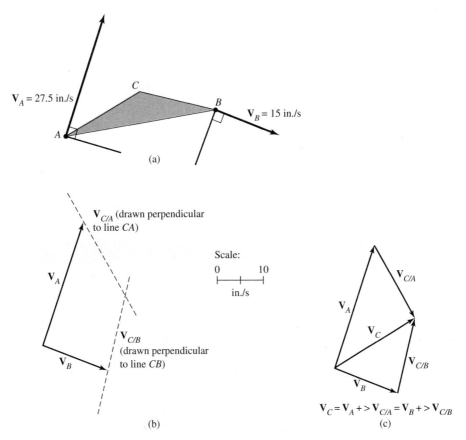

FIGURE 6.13 Velocity of a point of interest.

EXAMPLE PROBLEM 6.7

Figure 6.14 illustrates a mechanism that extends reels of cable from a delivery truck. It is operated by a hydraulic cylinder at *A*. At this instant, the cylinder retracts at a rate of 5 mm/s. Determine the velocity of the top joint, point *E*.

SOLUTION: 1. *Draw a Kinematic Diagram and Calculate Degrees of Freedom*

Figure 6.15a shows the kinematic diagram of this mechanism. To fully understand this mechanism, the mobility is computed.

$$n = 6 \qquad j_p = (5 \text{ pins } + 2 \text{ sliders}) = 7 \qquad j_h = 0$$

and

$$M = 3(n - 1) - 2j_p - j_h = 3(6 - 1) - 2(7) - 0 = 1$$

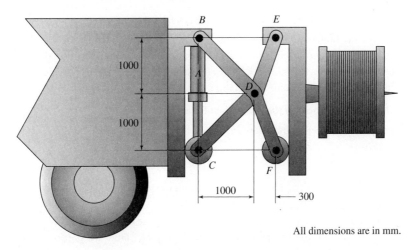

All dimensions are in mm.

FIGURE 6.14 Mechanism for Example Problem 6.7.

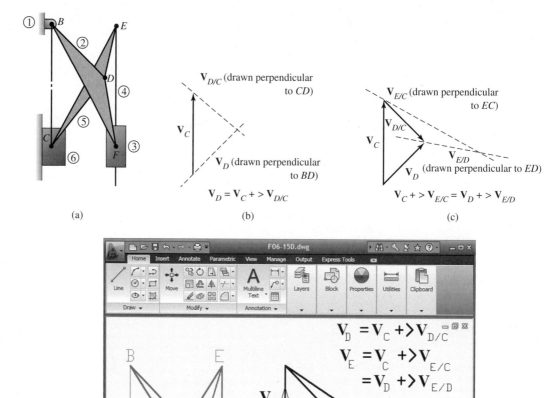

FIGURE 6.15 Diagrams for Example Problem 6.7.

With one degree of freedom, this mechanism is fully operated with the one input motion. Of course, this motion is the actuation of the hydraulic cylinder upward at a rate of 5 mm/s.

2. **Decide on a Method to Achieve the Desired Velocity**

Link 5 carries both point C (known velocity) and point E (unknown velocity). However, link 5 is a floating link, as it is not constrained to either pure rotation or pure translation. Therefore, prior to determining the velocity of point E, one other velocity on link 5 must be established. Point D is a convenient point because it resides on link 5 and a link that is constrained to rotation (link 2).

3. **Determine the Velocity of the Convenient Point (Point D)**

The equation that will allow determination of the velocity of point D can be written as

$$\mathbf{V}_D = \mathbf{V}_C +> \mathbf{V}_{D/C}$$

Because link 2 is fixed to the frame at B, point D is constrained to rotation about B. Therefore, the velocity of point D must be perpendicular to the line BD.

In addition, both points D and C reside on the same link, namely, link 5. Therefore, the relative velocity of D with respect to C must be perpendicular to the line DC. From the previous two statements, the directions of both velocities $\mathbf{V}_D$ and $\mathbf{V}_{D/C}$ are known.

The vector polygon used to solve this problem is shown in Figure 6.15b. The magnitudes can be determined by observing the intersection of the directed lines, $\mathbf{V}_D$ and $\mathbf{V}_{D/C}$. The magnitudes of the velocities can be scaled, yielding the following equations:

$$\mathbf{V}_{D/C} = 3.5 \text{ mm/s} \diagdown 45°$$

$$\mathbf{V}_D = 3.5 \text{ mm/s} \diagup 45°$$

4. ***Determine the Velocity of the Point on the Floating Link (Point 5)***

Now that the velocities of two points on link 5 are fully known, the velocity of point E can be determined. Using two forms of the relative velocity equation, the velocity of the points C, D, and E can be related:

$$\mathbf{V}_E = \mathbf{V}_C +> \mathbf{V}_{E/C} = \mathbf{V}_D +> \mathbf{V}_{E/D}$$

The velocities of C and D as well as the direction of the relative velocities are known. A vector polygon is constructed in Figure 6.15c.

Once the magnitudes of the relative velocities are determined, the polygon can be completed. The completed polygon is shown in Figure 6.15d. The velocity of E can be included in the polygon according to the vector equation above. Measuring the vectors from the completed polygon yields

$$\mathbf{V}_{E/D} = 2.65 \text{ mm/s} \diagdown 16.7°$$

$$\mathbf{V}_{E/C} = 5.95 \text{ mm/s} \diagdown 33.0°$$

$$\mathbf{V}_E = 5.29 \text{ mm/s} \diagup 19.4°$$

6.6.3 Coincident Points on Different Links

Calculating velocities of moving links that are connected through a sliding joint involves using coincident points that reside on the two bodies. Typically, the direction of the sliding motion is known. Therefore, the direction of the relative velocity of the coincident points is known. This is sufficient information to determine the motion of the driven links. The concept is best illustrated through an example problem.

EXAMPLE PROBLEM 6.8

Figure 6.16 shows a mechanism that tips the bed of a dump truck. Determine the required speed of the hydraulic cylinder in order to tip the truck at a rate of 5 rad/min.

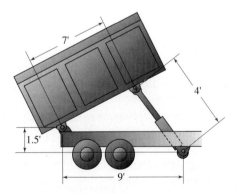

FIGURE 6.16 Dump truck mechanism for Example Problem 6.8.

SOLUTION: 1. ***Draw a Kinematic Diagram and Identify the Degrees of Freedom***

Kinematically, this mechanism is an inversion to the common slider-crank mechanism. The slider-crank has one degree of freedom, which in this case is the extension and contraction of the hydraulic cylinder. Figure 6.17a shows the kinematic diagram of this mechanism.

Link 1 represents the bed frame, link 4 is the cylinder, link 3 is the piston/rod, and link 2 is the bed. Notice that the pin joint that connects links 2 and 3 is labeled as point B. However, because links 2, 3, and 4 are located at point B, these coincident points are distinguished as B_2, B_3, and B_4.

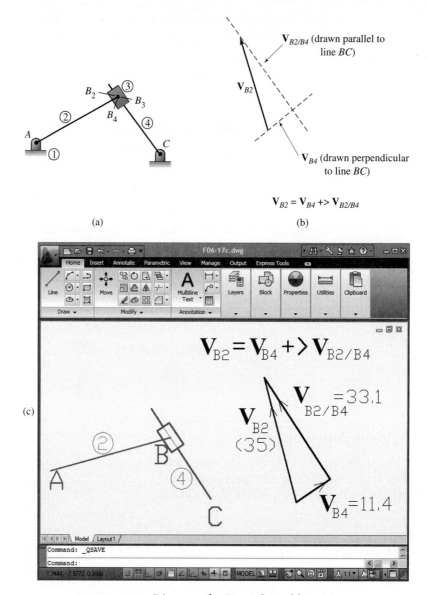

FIGURE 6.17 Diagrams for Example Problem 6.8.

2. ***Decide on a Method to Achieve the Desired Velocity***

   The problem is to determine the speed of the hydraulic cylinder that will cause link 2 to rotate at a rate of 5 rad/min, counterclockwise. In terms of the kinematic model, the velocity of B_2 relative to B_4 must be determined.

   The velocities of the coincident points are related through equation (6.9):

   $$\mathbf{V}_{B2} = \mathbf{V}_{B4} + > \mathbf{V}_{B2/B4}$$

   In this equation, the magnitude, $\mathbf{V}_{B2}$, can be calculated from the rotational speed of link 2. Additionally, because the links are constrained to pure rotation, the directions of $\mathbf{V}_{B2}$ and $\mathbf{V}_{B4}$ are perpendicular to links 2 and 4, respectively.

   Finally, because B_2 and B_4 are connected through a sliding joint and the direction of the sliding is known, the relative velocity, $v_{B2/B4}$, must be along this sliding direction. Therefore, enough information is known to construct a velocity polygon.

3. ***Determine the Velocity of the Input Point (Point $\mathbf{B_2}$)***

   The velocity of B_2 can be found with the following:

   $$v_{B2} = \omega_2 r_{AB2} = \left(5 \text{ rad/min}\right)(7 \text{ ft}) = 35 \text{ ft/min}$$

   The direction of the velocity of point B_2 is perpendicular to link 2, which is up and to the left.

4. *Determine the Velocity of the Point on the Follower Link (Point B₄)*

The vector polygon used to solve this problem is shown in Figure 6.17b. The magnitudes can be determined by observing the intersection of the directed lines of v_{B4} and $v_{B2/B4}$.

5. *Measure the Desired Velocities from the Polygon*

The magnitudes of the velocities can be scaled from the CAD layout in Figure 6.17c, yielding the following:

$$\mathbf{V}_{B2/B4} = 33.1 \text{ ft/min} \quad \underline{\text{56°}} \nwarrow$$

$$\mathbf{V}_{B4} = 11.4 \text{ ft/min} \quad \nearrow \underline{34°}$$

Therefore, at this instant, the cylinder must be extended at a rate of 33 ft/min to have the bed tip at a rate of 5 rad/min.

6.7 VELOCITY IMAGE

A useful property in velocity analysis is that each link in a mechanism has an image in the velocity polygon. To illustrate, a mechanism is shown in Figure 6.18a, with its associated velocity diagram in Figure 6.18b.

Examine the triangle drawn using the terminus of the three absolute velocity vectors. This triangle is shaped with proportional dimensions to the floating link itself and rotated 90°. The shape in the velocity polygon is termed a *velocity image* of the link. The velocity image of link 5 in Example Problem 6.7 can be seen in Figure 6.15d.

If this concept of velocity image is known initially, the solution process can be reduced considerably. Once the velocity of two points on a link is determined, the velocity of any other point that sits on the link can be readily found. The two points can be used as the base of the velocity image. The shape of that link can be scaled and constructed on the velocity polygon. Care must be taken, however, not to allow the shape of the link to be inverted between the kinematic diagram and the velocity polygon.

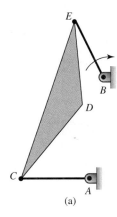

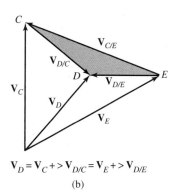

$$\mathbf{V}_D = \mathbf{V}_C +> \mathbf{V}_{D/C} = \mathbf{V}_E +> \mathbf{V}_{D/E}$$

(a)   (b)

FIGURE 6.18 Velocity image.

6.8 ANALYTICAL VELOCITY ANALYSIS: RELATIVE VELOCITY METHOD

Analytical velocity analysis involves exactly the same logic as employed in graphical analysis. The vector polygons are created according to the appropriate relative velocity equations. Because analytical techniques are used, the accuracy of the polygon is not a major concern, although a rough scale allows insight into the solutions. The vector equations can be solved using the analytical techniques presented in Chapter 3.

Analytical solutions are presented in the following example problems.

EXAMPLE PROBLEM 6.9

Figure 6.19 shows a primitive well pump that is common in undeveloped areas. To maximize water flow, the piston should travel upward at a rate of 50 mm/s. In the position shown, determine the angular velocity that must be imposed on the handle to achieve the desired piston speed.

SOLUTION: 1. *Draw a Kinematic Diagram and Identify the Degrees of Freedom*

Figure 6.20a shows the kinematic diagram of this mechanism. Notice that this is a variation of a slider-crank mechanism, which has one degree of freedom. Link 2 represents the handle. Therefore, the goal of this problem is to determine ω_2.

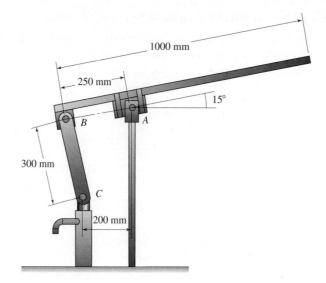

FIGURE 6.19 Well pump for Example Problem 6.9.

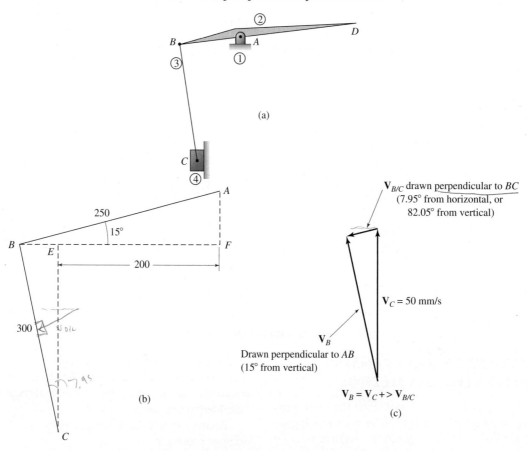

FIGURE 6.20 Diagrams for Example Problem 6.9.

2. *Analyze the Mechanism Geometry*

Figure 6.20b isolates the geometry of the core mechanism links. Notice that this geometry was used to form two right triangles. Focusing on the upper triangle, *ABF*, and using the trigonometric functions, the length of sides *BF* and *AF* can be determined.

$$BF = (250 \text{ mm}) \cos 15° = 241.48 \text{ mm}$$

$$AF = (250 \text{ mm}) \sin 15° = 64.70 \text{ mm}$$

The length of *BE* is calculated by

$$BE = BF - EF = 241.48 \text{ mm} - 200 \text{ mm} = 41.48 \text{ mm}$$

Focusing on the lower triangle, the interior angle at C can be found with the following:

$$\angle BCE = \sin^{-1}\left(\frac{41.48}{300}\right) = 7.95°$$

3. ***Assemble the Velocity Polygon***

    To solve for the angular velocity of link 2, the linear velocity of point B, which resides on link 2, must be determined. Link 3 is of special interest because it carries both point C (known velocity) and point B (unknown velocity).

    Because link 2 is fixed to the frame at A, point B is limited to rotation about A. Therefore, the velocity of point B must be perpendicular to line AB. In addition, since both points B and C reside on the same link (link 3), the relative velocity of B with respect to C must lie perpendicular to the line BC.

    From the previous two statements, the directions of both velocities $\mathbf{V}_B$ and $\mathbf{V}_{B/C}$ are known. Velocity $\mathbf{V}_B$ is perpendicular to AB, 15° from the vertical. Velocity $v_{B/C}$ is perpendicular to BC, 7.95° from the horizontal, or $90° - 7.95° = 82.05°$ from the vertical. These velocities can be related using equation (6.10):

    $$\mathbf{V}_B = \mathbf{V}_C +> \mathbf{V}_{B/C}$$

    In this equation, only the magnitudes of $\mathbf{V}_B$ and $\mathbf{V}_{B/C}$ are unknown. The vector polygon that is used to solve this problem is shown in Figure 6.20c. The magnitudes can be determined by solving for the length of the sides (vector magnitudes) of the general triangle.

    The remaining interior angle of this vector triangle is

    $$180° - 82.05° - 15° = 82.95°$$

4. ***Calculate the Velocity of Point*** **B**

    The law of sines is used to determine the vector magnitudes:

    $$\mathbf{V}_{B/C} = \mathbf{V}_C\left(\frac{\sin 15°}{\sin 82.95°}\right) = 13.04 \text{ mm/s } \overline{7.95°}$$

    $$\mathbf{V}_B = \mathbf{V}_C\left(\frac{\sin 82.05°}{\sin 82.95°}\right) = 49.90 \text{ mm/s } 15° = 49.9 \text{ mm/s } \underline{75°}$$

5. ***Determine the Angular Velocity of Link 2***

    Now that the velocity B is determined, the angular velocity of link 2 can be solved. Notice that consistent with the direction of $\mathbf{V}_B$, link 2 must rotate clockwise:

    $$\omega_2 = \frac{v_B}{r_{AB}} = \frac{49.9 \text{ mm/s}}{250 \text{ mm}} = 0.20 \text{ rad/s, cw}$$

    Convert this result to rpm with the following:

    $$\omega\,(\text{rev/min}) = \frac{30}{\pi}\left[\omega(\text{rad/s})\right] = \frac{30}{\pi}\,[0.20 \text{ rad/s}] = 1.9 \text{ rpm, cw}$$

EXAMPLE PROBLEM 6.10

Figure 6.21 illustrates a roofing material delivery truck conveyor. Heavy roofing materials can be transported on the conveyor to the roof. The conveyor is lifted into place by extending the hydraulic cylinder. At this instant, the cylinder is extending at a rate of 8 fpm (ft/min). Determine the rate that the conveyor is being lifted.

SOLUTION: 1.   ***Draw the Kinematic Diagram and Identify the Degrees of Freedom***

Figure 6.22a shows the kinematic diagram of this mechanism. Link 4 represents the conveyor, link 2 represents the cylinder, and link 3 represents the piston/rod. Because a sliding joint is used to connect two rotating links, defining coincident points will aid problem solution. Point B_2 is attached to link 2, and point B_4 is attached, as a point of reference, to link 4. The goal of this problem is to determine ω_4.

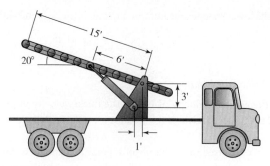

FIGURE 6.21   Conveyor for Example Problem 6.10.

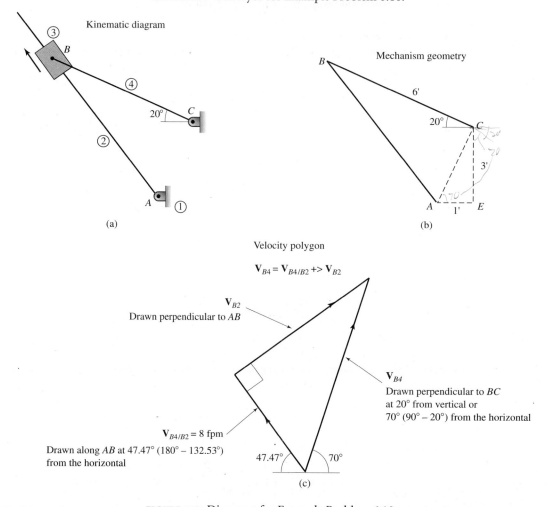

FIGURE 6.22   Diagrams for Example Problem 6.10.

2. ***Analyze the Mechanism Geometry***

Figure 6.22b isolates the geometry of the core mechanism links. Notice that this geometry was used to form two triangles. Focusing on the lower right, triangle *ACE* yields the following:

$$AC = \sqrt{[AE^2 + CE^2]}$$

$$= \sqrt{(1 \text{ ft})^2 + (3 \text{ ft})^2} = 3.16 \text{ ft}$$

$$\angle CAE = \tan^{-1}\left(\frac{CE}{AE}\right) = \tan^{-1}\left(\frac{3 \text{ ft}}{1 \text{ ft}}\right) = 71.57°$$

$$\angle ACE = \tan^{-1}\left(\frac{AE}{CE}\right) = \tan^{-1}\left(\frac{1 \text{ ft}}{3 \text{ ft}}\right) = 18.43°$$

Because link 4 is inclined at 20° above horizontal, the full angle at C is

$$\angle BCE = 90° + 20° = 110°$$

then the angle at C in the upper triangle is

$$\angle ACB = \angle BCE - \angle ACE = 110° - 18.43° = 91.57°$$

The geometry of the upper triangle can be fully determined by the law of cosines

$$AB = \sqrt{AC^2 + BC^2 - 2(AC)(BC)\cos\angle ACB}$$

$$= \sqrt{(3.16 \text{ ft})^2 + (6 \text{ ft})^2 - 2(3.16 \text{ ft})(6 \text{ ft})\cos 91.57°} = 6.86 \text{ ft}$$

and the law of sines

$$\angle BAC = \sin^{-1}\left\{\left(\frac{6 \text{ ft}}{6.86 \text{ ft}}\right)\sin 91.57°\right\} = 60.96°$$

$$\angle CBA = \sin^{-1}\left\{\left(\frac{3.16 \text{ ft}}{6.86 \text{ ft}}\right)\sin 91.57°\right\} = 27.42°$$

Finally, the total included angle at A is

$$\angle BAE = \angle CAE + \angle BAC = 71.57° + 60.96° = 132.53°$$

3. **Assemble a Velocity Polygon**

To solve for the angular velocity of link 2, the linear velocity of point B_2, which resides on link 2, must be determined. The extension of the hydraulic cylinder is given, which represents the velocity of point B on link 4, relative to point B on link 2 ($\mathbf{V}_{B4/B2}$). These velocities can be related using equation (6.10):

$$\mathbf{V}_{B4} = \mathbf{V}_{B4/B2} + > \mathbf{V}_{B2}$$

Because link 4 is fixed to the frame at C, point B_4 is limited to rotation about C. Therefore, the velocity of point B_4 must be perpendicular to the line BC.

In addition, link 2 is fixed to the frame at A, and point B_2 is limited to rotation about A. Therefore, the velocity of point B_2 must be perpendicular to the line AB.

From the previous two statements, the directions of both velocities $\mathbf{V}_{B4}$ and $\mathbf{V}_{B2}$ are known.

The vector polygon that is used to solve this problem is shown in Figure 6.22c. Notice that these vectors form a right triangle. The magnitudes can be determined by solving for the length of the sides (vector magnitudes) of the right triangle.

The bottom interior angle of this vector triangle is

$$180° - 70° - 47.47° = 62.53°$$

4. **Calculate the Velocity of Point B**

The velocity of B_2 is found from the following trigonometric relationships of a right triangle:

$$\mathbf{V}_{B4} = \left(\frac{\mathbf{V}_{B4/B2}}{\cos 62.53°}\right) = 17.43 \text{ ft/min} \nearrow 70°$$

5. **Determine the Angular Velocity of Link 2**

Now that velocity B_4 is known, the angular velocity of link 4 can be solved. Notice that consistent with the direction of v_{B4}, link 4 must rotate clockwise:

$$\omega_4 = \frac{v_{B4}}{r_{BC}} = \frac{17.43 \text{ ft/min}}{6 \text{ ft}} = 2.89 \text{ rad/min, cw}$$

Convert this result to rpm by

$$\omega_4 = \left(\frac{2.89 \text{ rad}}{\text{min}}\right)\left(\frac{1 \text{ rev}}{2\pi \text{ rad}}\right) = 0.46 \text{ rev/min, cw}$$

6.9 ALGEBRAIC SOLUTIONS FOR COMMON MECHANISMS

For the common slider-crank and four-bar mechanisms, closed-form algebraic solutions have been derived [Ref. 12]. They are given in the following sections.

6.9.1 Slider-Crank Mechanism

A general slider-crank mechanism was illustrated in Figure 4.19 and is uniquely defined with dimensions L_1, L_2, and L_3. With one degree of freedom, the motion of one link must be specified to drive the other links. Most often the crank is driven. Therefore, knowing θ_2, ω_2, and the position of all the links, from equations (4.6) and (4.7), the velocities of the other links can be determined. As presented in Chapter 4, the position equations are

$$\theta_3 = \sin^{-1}\left\{\frac{L_1 + L_2 \sin \theta_2}{L_3}\right\} \qquad (4.6)$$

$$L_4 = L_2 \cos(\theta_2) + L_3 \cos(\theta_3) \qquad (4.7)$$

The velocity equations are given as [Refs. 10, 11, 12, 14]

$$\omega_3 = -\omega_2\left(\frac{L_2 \cos \theta_2}{L_3 \cos \theta_3}\right) \qquad (6.12)$$

$$v_4 = -\omega_2 L_2 \sin \theta_2 + \omega_3 L_3 \sin \theta_3 \qquad (6.13)$$

6.9.2 Four-Bar Mechanism

A general four-bar mechanism was illustrated in Figure 4.23 and is uniquely defined with dimensions L_1, L_2, L_3, and L_4. With one degree of freedom, the motion of one link must be specified to drive the other links. Most often the crank is driven. Therefore, knowing θ_2, ω_2, and the position of all the links, from equations (4.9) through (4.12), the velocities of the other links can be determined. As presented in Chapter 4, the position equations are as follows:

$$BD = \sqrt{L_1^2 + L_2^2 - 2(L_1)(L_2)\cos \theta_2} \qquad (4.9)$$

$$\gamma = \cos^{-1}\left[\frac{(L_3)^2 + (L_4)^2 - (BD)^2}{2(L_3)(L_4)}\right] \qquad (4.10)$$

$$\theta_3 = 2\tan^{-1}\left[\frac{-L_2 \sin \theta_2 + L_4 \sin \gamma}{L_1 + L_3 - L_2 \cos \theta_2 - L_4 \cos \gamma}\right] \qquad (4.11)$$

$$\theta_4 = 2\tan^{-1}\left[\frac{L_2 \sin \theta_2 - L_3 \sin \gamma}{L_2 \cos \theta_2 + L_4 - L_1 - L_3 \cos \gamma}\right] \qquad (4.12)$$

The velocity equations are as follows [Refs. 10, 11, 12, 14]:

$$\omega_3 = -\omega_2\left[\frac{L_2 \sin(\theta_4 - \theta_2)}{L_3 \sin \gamma}\right] \qquad (6.14)$$

$$\omega_4 = -\omega_2\left[\frac{L_2 \sin(\theta_3 - \theta_2)}{L_4 \sin \gamma}\right] \qquad (6.15)$$

6.10 INSTANTANEOUS CENTER OF ROTATION

In determining the velocity of points on a mechanism, the concept of instant centers can be used as an alternative approach to the relative velocity method. This approach is based on the fact that any link, regardless of the complexity of its motion, instantaneously appears to be in pure rotation about a single point. This instantaneous pivot point is termed the *instant center* of rotation for the particular link. The instant center for a floating link, link 3, in relation to the frame is shown as (13) in Figure 6.23.

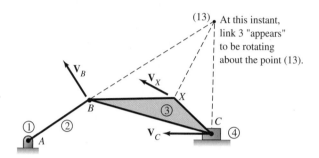

FIGURE 6.23 Instantaneous center.

Using this concept, each link can be analyzed as if it were undergoing pure rotation. An instant center may exist on or off the body, and its position is not fixed in time. As a link moves, its instant center also moves. However, the velocities of different points on a mechanism are also instantaneous. Therefore, this fact does not place a serious restriction on the analysis.

This concept also extends to relative motion. That is, the motion of any link, relative to any other link, instantaneously appears to be rotating only about a single point. Again, the imagined pivot point is termed the instant center between the two links. For example, if two links were designated as 1 and 3, the instant center would be the point at which link 3 instantaneously appears to be rotating relative to link 1. This instant center is designated as (13) and verbalized as "one three," not thirteen. Note that the instant center shown in Figure 6.23 is designated as (13). If link 1 were the frame, as is the typical designation, this instant center would describe the absolute motion of link 3. From kinematic inversion, this point is also the center of instantaneous motion of link 1 relative to link 3. Thus, the instant center (13) is the same as (31).

Because every link has an instant center with every other link, each mechanism has several instant centers. The total number of instant centers in a mechanism with n links is

$$\text{Total number of instant centers} = \frac{n(n-1)}{2} \qquad (6.16)$$

6.11 LOCATING INSTANT CENTERS

In a typical analysis, it is seldom that every instant center is used. However, the process of locating each center should be understood because every center could conceivably be employed.

6.11.1 Primary Centers

Some instant centers can be located by simply inspecting a mechanism. These centers are termed *primary centers.* In locating primary centers, the following rules are used:

1. When two links are connected by a pin joint, the instant center relating the two links is at this pivot point. This first rule is illustrated in Figure 6.24a.

2. The instant center for two links in rolling contact with no slipping is located at the point of contact. This second rule is illustrated in Figure 6.24b.

3. The instant center for two links with straight line sliding is at infinity, in a direction perpendicular to the direction of sliding. The velocity of all points on a link, which is constrained to straight sliding relative to another link, is identical and in the direction of sliding. Therefore, it can be imagined that this straight motion is rotation about a point at a great distance because a straight line can be modeled as a portion of a circle with an infinitely large radius. Because velocity is always perpendicular to a line drawn to the pivot, this instant center must be perpendicular to the sliding direction. This center could be considered to be on any line parallel to the sliding direction because the lines meet at infinity. This third rule is illustrated in Figure 6.24c.

4. The instant center for two links having general sliding contact must lie somewhere along the line normal to the direction of sliding contact. This fourth rule is illustrated in Figure 6.24d.

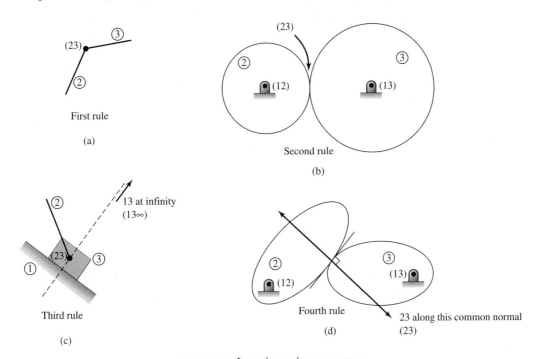

FIGURE 6.24 Locating primary centers.

EXAMPLE PROBLEM 6.11

Figure 6.25 illustrates an air compressor mechanism. For this mechanism, locate all the primary instant centers.

SOLUTION: 1.   ***Draw a Kinematic Diagram***

The kinematic diagram for the air compressor is illustrated in Figure 6.26.

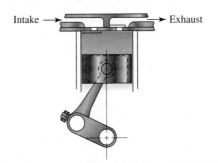

FIGURE 6.25 Air compressor for Example Problem 6.11.

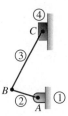

FIGURE 6.26 Kinematic diagram for Example Problem 6.11.

2. **Use the First Rule to Locate Primary Centers**

The four links are numbered on the kinematic diagram. The pin joints are also lettered. The first pin joint, A, connects link 1 and link 2. From the first rule for primary instant centers, this joint is the location of instant center (12). Similarly, pin joint B is instant center (23) and pin joint C is instant center (34).

It is clear from the kinematic diagram in Figure 6.26 that rolling contact does not join any links. Therefore, the second rule does not apply to this mechanism.

3. **Use the Third Rule to Locate Primary Centers**

Because a straight sliding joint occurs between links 4 and 1, this instant center is visualized at infinity, in a direction perpendicular to the sliding direction. Figure 6.27 illustrates the notation used to identify this, along with labeling all other primary instant centers. Recall that this instant center could be on a line parallel to line (14 ∞) because it can be considered that parallel lines intersect at infinity.

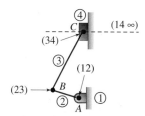

FIGURE 6.27 Primary instant centers for Example Problem 6.11.

It is clear from the kinematic diagram in Figure 6.26 that general sliding does not join any links. Therefore, the fourth rule does not apply to this mechanism.

6.11.2 Kennedy's Theorem

Instant centers that cannot be found from the four rules for primary centers are located with the use of *Kennedy's theorem*. It states that

"The three instant centers corresponding with any three bodies all lie on the same straight line"

For example, imagine three arbitrary links—links 3, 4, and 5. Kennedy's theorem states that instant centers (34), (45), and (35) all lie on a straight line. By applying this theorem, after locating all primary instant centers, all other instant centers can be found. Locating the precise position of the instant centers can be accomplished by using either graphical or analytical methods. Of course, graphical methods include both manual drawing techniques or CAD.

6.11.3 Instant Center Diagram

An instant center diagram is a graphical technique used to track the instant centers that have been located and those that still need to be found. In addition, it indicates the

combinations of instant centers that can be used in applying Kennedy's theorem. It is rare that all instant centers need to be located to perform a velocity analysis. The mechanism and the actuation link(s) and required output should be studied to determine the specific instant centers required. Then, the instant center diagram can be used to find those specific instant centers.

The instant center diagram is a circle divided into segments, one for each link in the mechanism being analyzed. The segment separators are labeled with the numbers corresponding to the links. An instant center diagram for a four-bar mechanism is shown in Figure 6.28a.

Any line that connects two points on the diagram represents an instant center, relating the two links identified by the endpoints. For example, the line that connects point 1 and point 4 represents the instant center (14). For instant centers that have been located, the corresponding line on the diagram is drawn heavy. Figure 6.28b indicates that instant centers (12), (23), (34), and (14) have been located. Instant centers needing to be located may then be represented by dashed lines. Figure 6.28c indicates that instant centers (13) and (24) have not yet been found. All instant

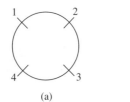 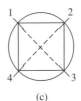

(a)   (b)   (c)

FIGURE 6.28 Instant center diagram.

centers are located when each point is connected to every other point.

Note that the lines in the diagram form triangles. Each triangle represents three instant centers, relating the three links at the vertices. From Kennedy's theorem, the three instant centers represented by the sides of a triangle must lie in a straight line. For example, refer to Figure 6.28c and isolate the triangle formed by lines (12), (23), and (13). Kennedy's theorem states that these three instant centers must be collinear.

If two sides of a triangle are drawn heavy, a line can be drawn on the mechanism diagram connecting the two known instant centers. This line contains the third instant center. If a second line can be drawn, the intersection of these two lines will locate the third center. Summarizing, to locate an instant center, two triangles must be found in the diagram that have two known sides and have as the unknown side the instant center being sought.

The following example problems illustrate the procedure for finding all instant centers.

EXAMPLE PROBLEM 6.12

Figure 6.29 illustrates a self-locking brace for a platform used on shipping docks. For this mechanism, locate all the instant centers.

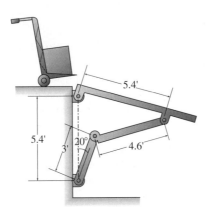

FIGURE 6.29 Locking brace for Example Problem 6.12.

SOLUTION: 1. ***Draw a Kinematic Diagram***

The kinematic diagram for the loading platform is illustrated in Figure 6.30a. The four links are numbered on the kinematic diagram. The pin joints are also lettered. Compute the total number of instant centers, with $n = 4$ links, as follows:

$$\text{Total number of instant centers} = \frac{n(n-1)}{2} = \frac{4(4-1)}{2} = 6$$

2. ***Sketch an Instant Center Diagram***

An instant center diagram is shown in Figure 6.30b. Table 6.1 can be used to systematically list all possible instant centers in a mechanism.

3. ***Locate the Primary Instant Centers***

The first pin joint, A, connects links 1 and 2. From the first rule for primary instant centers, this joint is the location of instant center (12). Similarly, pin joints B, C, and D are instant centers (23), (34), and (14), respectively. In Figure 6.30c, the instant center diagram is redrawn to reflect locating the primary instant centers (12), (23), (34), and (14). The instant centers (13) and (24) remain to be determined.

4. ***Use Kennedy's Theorem to Locate (13)***

The instant center diagram that is used to obtain (13) is shown in Figure 6.30d. Focus on the lower triangle formed by (13), (14), and (34). Applying Kennedy's theorem, (13) must lie on a straight line

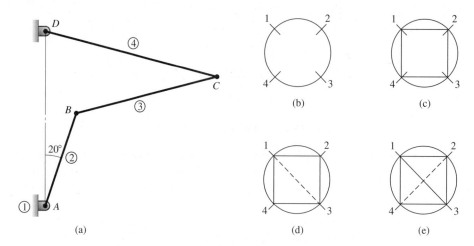

FIGURE 6.30 Kinematic and instant center diagram for Example Problem 6.12.

TABLE 6.1	Possible Instant Centers in a Mechanism ($n = 4$)		
1	**2**	**3**	**4**
12	23	34	
13	24		
14			

formed by (14) and (34), both of which have already been located, as indicated by the solid lines in Figure 6.30d.

Also notice the upper triangle created by (13), (12), and (23). Likewise, (13) must also lie on a straight line formed by (12) and (23), both of which have been previously located.

Thus, the intersection of these lines, (14)–(34) and (12)–(23), will determine the location of (13). Recall that at this instant, link 3 appears to be rotating around point (13).

5. ***Use Kennedy's Theorem to Locate (24)***

The instant center diagram that is used to obtain (24) is shown in Figure 6.30e. In an identical process, Kennedy's theorem states that instant center (24) must lie on the same line as (14) and (12), which have been located. Likewise, (24) must also lie on the same line as (23) and (34), also located. Thus, if a straight line is drawn through (14) and (12) and another straight line is drawn through (23) and (34), the intersection of these lines will determine the location of (24). At this instant, link 2 appears to be rotating, relative to link 4, around point (24).

Figure 6.31 illustrates the mechanism with all instant centers located.

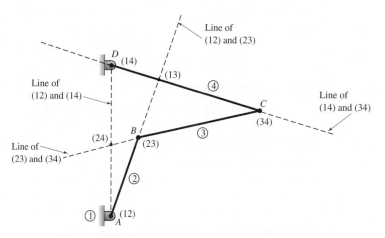

FIGURE 6.31 Instant centers for Example Problem 6.12.

EXAMPLE PROBLEM 6.13

Figure 6.32 illustrates a rock crusher. For this mechanism, locate all the instant centers.

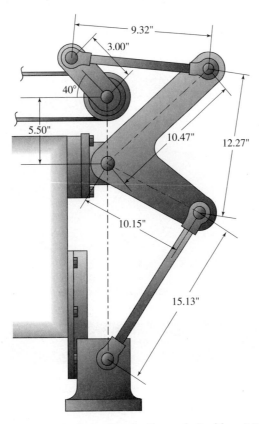

FIGURE 6.32 Rock crusher for Example Problem 6.13.

SOLUTION:

1. **Draw a Kinematic Diagram**

   The kinematic diagram for the rock crusher is illustrated in Figure 6.33a. The six links are numbered on the kinematic diagram. The pin joints are also lettered. Compute the total number of instant centers, with n = 6 links, as follows:

   $$\text{Total number of instant centers} = \frac{n(n-1)}{2} = \frac{6(6-1)}{2} = 15$$

2. **Sketch an Instant Center Diagram**

   An instant center diagram is shown in Figure 6.33b. Table 6.2 systematically lists all possible instant centers in a mechanism.

3. **Locate the Primary Instant Centers**

   The first pin joint, A, connects links 1 and 2. From the first rule for primary instant centers, this joint is the location of instant center (12). Similarly, pin joints B–F locate instant centers (23), (34), (14), (45), and (56), respectively.

   Because a straight sliding joint exists between links 6 and 1, this instant center (16) is located at infinity, in a direction perpendicular to the sliding direction. Recall that this instant center could be on a line parallel to this line because the lines meet at infinity. In Figure 6.33c, the instant center diagram is redrawn to locate (12), (23), (34), (45), (56), (14), and (16).

4. **Use Kennedy's Theorem to Locate the Other Instant Centers**

   The remaining combinations that need to be determined are instant centers (13), (24), (35), (46), (25), (36), (15), and (26).

   The instant center diagram that is used to obtain (13) is shown in Figure 6.33d. Focus on the triangle formed by (12), (23), and (13). Applying Kennedy's theorem, (13) must lie on a straight line formed by (12) and (23), which have already been located, as indicated by the solid lines in Figure 6.33d.

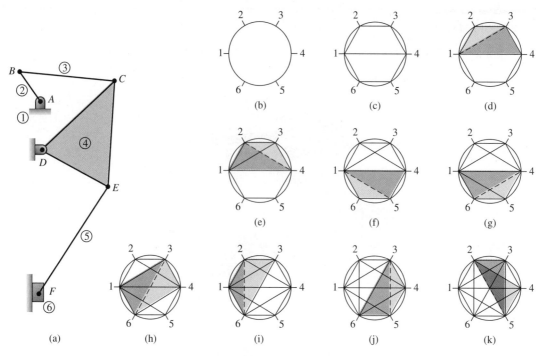

FIGURE 6.33 Kinematic diagram for Example Problem 6.13.

TABLE 6.2	Possible Instant Centers in a Mechanism ($n = 6$)				
1	**2**	**3**	**4**	**5**	**6**
12	23	34	45	56	
13	24	35	46		
14	25	36			
15	26				
16					

Also notice the triangle created by (13), (34), and (14). Likewise, (13) must also lie on a straight line formed by (13) and (34), which have been previously located. Thus, the intersection of these lines, (12)–(23) and (13)–(34), will determine the location of (13).

Table 6.3 is formulated to locate all remaining instant centers. Note that the order in which instant centers are found is extremely dependent on which instant centers are already located. This becomes quite an iterative process, but the instant center diagram becomes valuable in devising this approach. Figure 6.34 illustrates the mechanism with all instant centers located.

TABLE 6.3	Locating Instant Centers for Example Problem 6.13	
To Locate Instant Center	**Use Intersecting Lines**	**Instant Center Diagram**
13	(12)–(23) and (14)–(34)	Figure 6.33d
24	(12)–(14) and (23)–(34)	Figure 6.33e
15	(16)–(56) and (14)–(45)	Figure 6.33f
46	(14)–(16) and (45)–(56)	Figure 6.33g
36	(13)–(16) and (34)–(46)	Figure 6.33h
26	(12)–(16) and (23)–(36)	Figure 6.33i
35	(56)–(36) and (34)–(45)	Figure 6.33j
25	(24)–(45) and (23)–(35)	Figure 6.33k

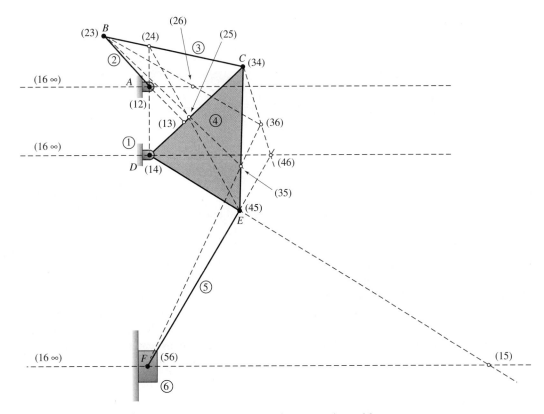

FIGURE 6.34 Instant centers for Example Problem 6.13.

6.12 GRAPHICAL VELOCITY ANALYSIS: INSTANT CENTER METHOD

The instant center method is based on the following three principles:

I. The velocity of a rotating body is proportional to the distance from the pivot point.

II. The instant center that is common to two links can be considered attached to either link.

III. The absolute velocity of the point, which serves as the common instant center, is the same, no matter which link is considered attached to that point.

Using these principles, the absolute velocity of any point on the mechanism can be readily obtained through a general method. This method is outlined in the following six steps:

1. Isolate the link with a known velocity (link *A*), the link containing the point for which the velocity is desired (link *B*), and the fixed link (link *C*).

2. Locate the instant center that is common to the link with the known velocity and the fixed link (instant center *AC*).

3. Locate the instant center that is common to the link with the known velocity and the link that contains the point where the velocity is desired (instant center *AB*).

4. Determine the velocity of the instant center (*AB*). This can be done by understanding that the velocity of a point on a link is proportional to the distance from the pivot. The instant center (*AC*) serves as the pivot. The known velocity on link *A* can be proportionally scaled to determine the velocity of the instant center (*AB*).

5. Locate the instant center that is common to the link with the point whose velocity is desired and the fixed link (instant center *BC*).

6. Determine the desired velocity. This can be done by understanding that the velocity on a link is proportional to the distance from the pivot. The instant center (*BC*) serves as this pivot. The velocity of the common instant center (*AB*) can be proportionally scaled to determine the desired velocity.

A graphical technique for proportionally scaling a vector uses a *line of centers, LC.* This is a line drawn from the pivot point of the link to the start of the known vector. A *line of proportion, LP,* must also be constructed. This is a line drawn from the pivot point to the end of the known vector. Figure 6.35a illustrates both the line of centers and the line of proportion. The distance from the pivot to the desired point can be transferred to the line of centers. The magnitude of the proportionally scaled vector is determined as parallel to the known vector and extending from *LC* to *LP* at the transferred distance. This is also illustrated in Figure 6.35a.

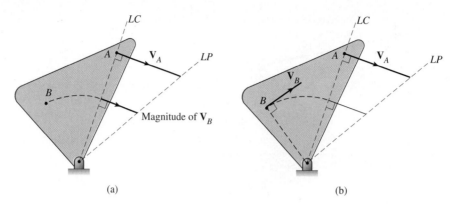

(a)   (b)

FIGURE 6.35 Using a line of centers and line of proportion.

Of course, the magnitude of the velocity is perpendicular to the line that connects the point with unknown velocity to the pivot point. Determining the magnitude and positioning of that vector in the proper direction fully defines the vector. Thus, the vector is graphically proportioned. The result is shown in Figure 6.35b.

We have described the logic behind the instantaneous center method of velocity analysis using graphical techniques. The actual solution can be completed with identical logic whether using manual drawing or CAD. Regardless of the process used, the underlying concepts of a graphical approach to the instantaneous center method of velocity analysis can be illustrated through the following example problems.

EXAMPLE PROBLEM 6.14

Figure 6.29 illustrated an automated, self-locking brace for a platform used on shipping docks. Example Problem 6.12 located all instant centers for the mechanism. Determine the angular velocity of link 4, knowing that link 2 is rising at a constant rate of 3 rad/s.

SOLUTION:

1. ***Draw a Kinematic Diagram with Instant Centers Located***

   The kinematic diagram, with the instant centers and scale information, is reproduced as Figure 6.36a.

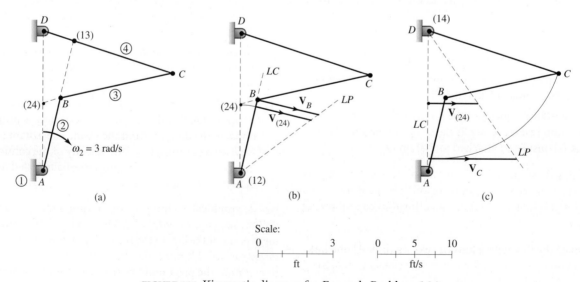

(a)   (b)   (c)

Scale:

0   3   0   5   10

ft   ft/s

FIGURE 6.36 Kinematic diagram for Example Problem 6.14.

2. ***Determine the Linear Velocity of a Convenient Point*** (**B**)

   The linear velocity of point *B* can be determined from the angular velocity of link 2. Point *B* has been measured to be 3 ft from the pivot of link 2 (point *A*).

   $$\mathbf{V}_B = r_{AB}\omega_2 = (3\ \text{ft})\left(3\ \text{rad/s}\right) = 9\ \text{ft/s}\ \underline{\big\backslash 20°}$$

3. *Incorporate the General Instant Center Velocity Procedure*

    a. Isolate the links.

        Link 2 contains the known velocity,
        Link 4 contains the point for which the velocity is desired, and
        Link 1 is the fixed link.

    b. The common instant center between the known and fixed link velocities is (12).

    c. The common instant center between the known and unknown link velocities is (24).

    d. The velocity of instant center (24) is graphically proportioned from the velocity of point B. Link 2 contains both point B and instant center (24); therefore, the velocity is proportionally scaled relative to instant center (12). This construction is shown in Figure 6.36b. The magnitude of this velocity, $v_{(24)}$, is scaled to 7.4 ft/s.

    e. The common instant center between the unknown and fixed link velocities is (14).

    f. The velocity of point C is graphically proportioned from the velocity of instant center (24). Link 4 contains both point C and instant center (24); therefore, the velocity is proportionally scaled relative to instant center (14). This construction is shown in Figure 6.36c. The magnitude of this velocity, v_C, is scaled to 13.8 ft/s.

4. *Determine the Angular Velocity of Link 4*

    Finally, the angular velocity of link 4 can be found from the velocity of point C. Point C has been scaled to be positioned at a distance of 5.4 ft from the pivot of link 4 (point D).

    $$\omega_4 = \frac{v_C}{r_{CD}} = \frac{13.8 \text{ ft/s}}{5.4 \text{ ft}} = 2.6 \text{ rad/s}$$

    Because the direction of the angular velocity is consistent with the velocity of point C, the link rotates counterclockwise. Therefore,

    $$\omega_4 = 2.6 \text{ rad/s, counterclockwise}$$

    Note that this rotational velocity could also be determined from the velocity of instant center (24) because this point is considered to consist of both links 2 and 4. However, as the first example problem on the topic, it can be difficult to visualize this point rotating with link 4.

EXAMPLE PROBLEM 6.15

Figure 6.32 illustrates a rock-crushing device. Example Problem 6.13 located all instant centers for the mechanism. In the position shown, determine the velocity of the crushing ram when the crank is rotating at a constant rate of 60 rpm clockwise.

SOLUTION: 1. *Draw a Kinematic Diagram with Instant Centers Located*

The kinematic diagram with the scale information is reproduced as Figure 6.37a.

2. *Determine the Linear Velocity of a Convenient Point* B

The linear velocity of point B can be determined from the angular velocity of link 2. Point B has been scaled to be positioned at a distance of 4.5 in. from the pivot of link 2 (point A):

$$\omega_2 = 60 \text{ rpm} \left(\frac{\pi}{30} \right) = 6.28 \text{ rad/s}$$

$$\mathbf{V}_B = r_{AB}\omega_2 = (4.5 \text{ in.})\left(6.28 \text{ rad/s}\right) = 28.3 \text{ in./s} \; \nearrow 40°$$

The purpose of this problem is to determine the linear velocity of point C.

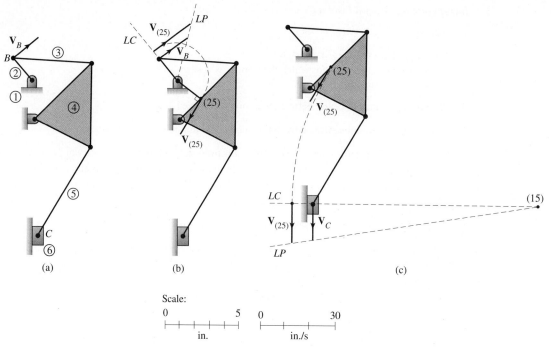

FIGURE 6.37 Diagrams for Example Problem 6.15.

3. **Incorporate the General Instant Center Velocity Procedure**

   a. Isolate the links.

   Link 2 contains the known velocity,
   Link 5 (or 6) contains the point for which the velocity is desired, and
   Link 1 is the fixed link.

   b. The common instant center between the known and fixed link velocities is (12).

   c. The common instant center between the known and unknown link velocities is (25).

   d. The velocity of the instant center (25) is graphically proportioned from the velocity of point B. Link 2 contains both point B and instant center (25); therefore, the velocity is proportionally scaled relative to instant center (12). This construction is shown in Figure 6.37b. The magnitude of this velocity, $v_{(25)}$, is scaled to 37.1 in./s.

   e. The common instant center between the unknown and fixed link velocities is (15).

   f. The velocity of point C is graphically proportioned from the velocity of instant center (25). Link 5 contains both point C and instant center (25); therefore, the velocity of instant center (25) is rotated to a line of centers created by point C and instant center (15). The velocity of instant center (25) is used to create a line of proportions. This line of proportions is then used to construct the velocity of C. This construction is shown in Figure 6.37c. The magnitude of this velocity, v_C, is scaled to 33.8 in./s.

   Formally stated,

$$\mathbf{V}_C = 33.8 \text{ in./s} \downarrow$$

6.13 ANALYTICAL VELOCITY ANALYSIS: INSTANT CENTER METHOD

The instant center method is virtually unaltered when an analytical approach is used in the solution. The only difference is that the positions of the instant centers must be determined through trigonometry, as opposed to constructing lines and locating the intersection points. This can be a burdensome task; thus, it is common to locate only the instant centers required for the velocity analysis. An analytical approach is illustrated through the following example problem.

EXAMPLE PROBLEM 6.16

Figure 6.38 shows a mechanism used in a production line to turn over cartons so that labels can be glued to the bottom of the carton. The driver arm is 15 in. long and, at the instant shown, it is inclined at a 60° angle with a clockwise angular velocity of 5 rad/s. The follower link is 16 in. long. The distance between the pins on the carriage is 7 in., and they are currently in vertical alignment. Determine the angular velocity of the carriage and the slave arm.

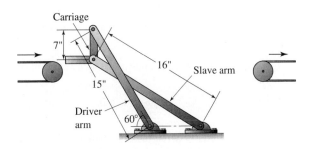

FIGURE 6.38 Turnover mechanism for Example Problem 6.16.

SOLUTION: 1.   ***Draw a Kinematic Diagram***

The kinematic diagram is shown in Figure 6.39a. A point of interest, X, was included at the edge of the carriage.

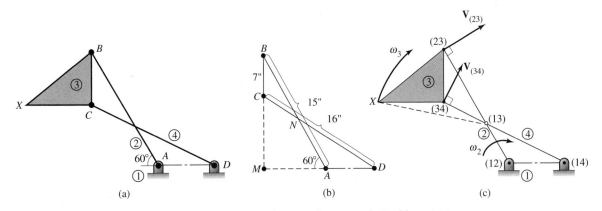

FIGURE 6.39 Kinematic diagram for Example Problem 6.16.

2.   ***Analyze the Mechanism Geometry***

Trigonometry is used to determine the distances and angles inherent in this mechanism's configuration. Triangles used to accomplish this are shown in Figure 6.39b. The distances BM and AM can be determined from triangle ABM.

$$BM = AB \, \sin \, (60°) = (15 \text{ in.}) \, \sin \, (60°) = 13.0 \text{ in.}$$

$$AM = AB \, \cos \, (60°) = (15 \text{ in.}) \, \cos \, (60°) = 7.5 \text{ in.}$$

Along the vertical BCM,

$$CM = BM - BC = 13.0 - 7.0 = 6.0 \text{ in.}$$

The angle ADC and the distance DN can be determined from triangle CDM.

$$\angle ADC = \sin^{-1} \left(\frac{CM}{CD} \right) = \sin^{-1} \left[\left(\frac{6 \text{ in.}}{16 \text{ in.}} \right) \right] = 22.0°$$

$$DM = CD \cos (22°) = (16 \text{ in.}) \cos (22°) = 14.8 \text{ in.}$$

3. ***Incorporate the General Instant Center Velocity Procedure***

    At this point, the general method for using the instant center method can be followed to solve the problem.

    a. Isolate the links.

        Link 2 contains the known velocity,
        Link 3 contains the point for which the velocity is desired, and
        Link 1 is the fixed link.

    b. The common instant center between the known and fixed link velocities is (12). By inspection, this instant center is located at point A.

    c. The common instant center between the known and unknown link velocities is (23). By inspection, this instant center is located at point B.

    d. The velocity of instant center (23) is simply the velocity of point B. This can be determined as

    $$\mathbf{V}_B = r_{AB}\omega_2 = (15 \text{ in.})\left(5 \text{ rad/s}\right) = 75 \text{ in./s } \nearrow 30°$$

    e. The common instant center between the unknown and fixed link velocities is (13). This instant center is located at the intersection of instant centers (12)–(23) and (14)–(34). By inspection, instant center (34) is located at point C and (14) is located at point D. Therefore, instant center (13) is located at the intersection of links 2 and 4. This point is labeled N in Figure 6.39b. The angles DAN, AND, and the distance AN can be determined from the general triangle AND.

    $$\angle DAN = 180° - 60° = 120°$$

    $$\angle AND = 180° - (120° + 22°) = 38°$$

    $$AN = \left[\sin(\angle ADN)\left(\frac{AD}{\sin(\angle AND)}\right)\right] = \left[\sin 22°\left(\frac{7.3 \text{ in.}}{\sin(38°)}\right)\right] = 5.5 \text{ in.}$$

    $$BN = BA - AN = 15 - 5.5 = 9.5 \text{ in.}$$

    Similarly,

    $$DN = \left[\sin \angle DAN\left(\frac{AN}{\sin \angle AND}\right)\right] = \left[\sin 120°\left(\frac{5.5 \text{ in.}}{\sin 38°}\right)\right] = 7.7 \text{ in.}$$

    $$CN = CD - DN = 16 - 7.7 = 8.3 \text{ in.}$$

    f. Link 3 instantaneously rotates around instant center (13). Thus, the angular velocity of link 3 can be calculated from the velocity of the common instant center (23) relative to instant center (13). This is illustrated in Figure 6.39c and is calculated as follows:

    $$\omega_3 = \frac{v_{23}}{r_{(13)-(23)}} = \frac{(75 \text{ in./s})}{(9.5 \text{ in.})} = 7.9 \text{ rad/s}$$

    Because the direction of the angular velocity is consistent with the velocity of point (23) relative to (13), the link rotates clockwise. Therefore,

    $$\omega_3 = 7.9 \text{ rad/s, cw}$$

    The velocity of point (34) can also be obtained using the angular velocity of link 3 because it is instantaneously rotating around instant center (13).

    $$\mathbf{V}_{(34)} = \omega_3\, r_{(13)-(34)} = \left(7.9 \text{ rad/s}\right)(8.3 \text{ in.}) = 65.6 \text{ in./s } \Big|\underset{\nearrow}{22°} = 65.6 \text{ in./s} \nearrow 68°$$

    Because link 4 is rotating relative to (14), the slave link velocity is

    $$\omega_4 = \frac{v_{(23)}}{r_{(14)-(23)}} = \frac{65.6 \text{ in./s}}{16 \text{ in.}} = 4.1 \text{ rad/s, cw}$$

6.14 VELOCITY CURVES

The analyses presented up to this point in the chapter are used to calculate the velocity of points on a mechanism at a specific instant. Although the results can be useful, they provide only a "snapshot" of the motion. The obvious short-coming of this analysis is that determination of the extreme conditions is difficult. It is necessary to investigate several positions of the mechanism to discover the critical phases.

It is convenient to trace the velocity magnitude of a certain point, or link, as the mechanism moves through its cycle. A *velocity curve* is such a trace. A velocity curve can be generated from a displacement diagram, as described in Section 4.11.

Recall that a displacement diagram plots the movement of a point or link as a function of the movement of an input point or link. The measure of input movement can be readily converted to time. This is particularly common when the driver operates at a constant velocity.

As discussed throughout the chapter, velocity is the time rate of change of displacement. Restating equations (6.1) and (6.2),

Linear velocity magnitude = v = Change in linear displacement per change in time

$$v = \frac{dR}{dt} \cong \frac{\Delta R}{\Delta t}$$

Restating equations (6.4) and (6.5),

Rotational velocity = ω = Change in angular displacement per change in time

$$\omega = \frac{d\theta}{dt} \cong \frac{\Delta \theta}{\Delta t}$$

Often, the driver of a mechanism operates at a constant velocity. For example, an input link driven by an electric motor, in steady state, operates at constant velocity. The motor shaft could cause the crank to rotate at 300 rpm, thus providing constant angular velocity. This constant velocity of the driver link converts the x-axis of a displacement diagram from rotational displacement to time. In linear terms, rearranging equation (6.2) yields:

$$\Delta t = \frac{\Delta R}{v} \tag{4.17}$$

In rotational terms, rearranging equation (6.5) yields:

$$\Delta t = \frac{\Delta \theta}{\omega} \tag{6.18}$$

Thus, equations (6.17) and (6.18) can be used to convert the displacement increment of the x-axis to a time increment. This is illustrated with Example Problem 6.17.

EXAMPLE PROBLEM 6.17

A displacement diagram of the piston operating in a compressor was plotted in Example Problem 4.11. This diagram was plotted relative to the crankshaft rotation. Use this data to plot the piston displacement relative to time when the crankshaft is driven by an electric motor at 1750 rpm.

SOLUTION: 1. ***Calculate the Time for 30° of Crank Rotation***

The main task of this problem is to convert the increment of crank angle in Figure 4.41 to time. The x-axis increment is 30° and the crankshaft rotates at 1750 rpm. To keep units consistent, the x-axis increment is converted to revolutions.

$$\Delta \theta = 30° \left(\frac{1 \text{ rev}}{360°} \right) = 0.08333 \text{ rev}$$

The time increment for the crank to rotate 0.08333 rev (30°) can be computed from equation (6.18).

$$\Delta t = \frac{\Delta \theta}{\omega} = \frac{(0.08333 \text{ rev})}{(1750 \text{ rev/min})}$$

$$= 0.0000476 \text{ min}$$

$$= (0.0000476 \text{ min}) \left(\frac{60 \text{ s}}{1 \text{ min}} \right) = 0.00286 \text{ s}$$

2. ***Add Time Column to Displacement Table***

The results of position analysis are reproduced with the time increment inserted into a spreadsheet. This is shown as Figure 6.40, which shows time tabulated in thousandths of a second. If not familiar with a spreadsheet, refer to Chapter 8.

3. ***Use Displacement and Time Data to Plot a Displacement Curve***

Using a spreadsheet, these values are plotted in Figure 6.41 to form a displacement diagram relative to time.

	Crank Angle (deg)	Time (0.001 sec)	Piston Displacement (in)
4	0	0.00	0.000
5	30	2.86	0.136
6	60	5.72	0.483
7	90	8.57	0.896
8	120	11.43	1.233
9	150	14.29	1.435
10	180	17.15	1.500
11	210	20.00	1.435
12	240	22.86	1.233
13	270	25.72	0.896
14	300	28.58	0.483
15	330	31.43	0.136
16	360	34.29	0.000

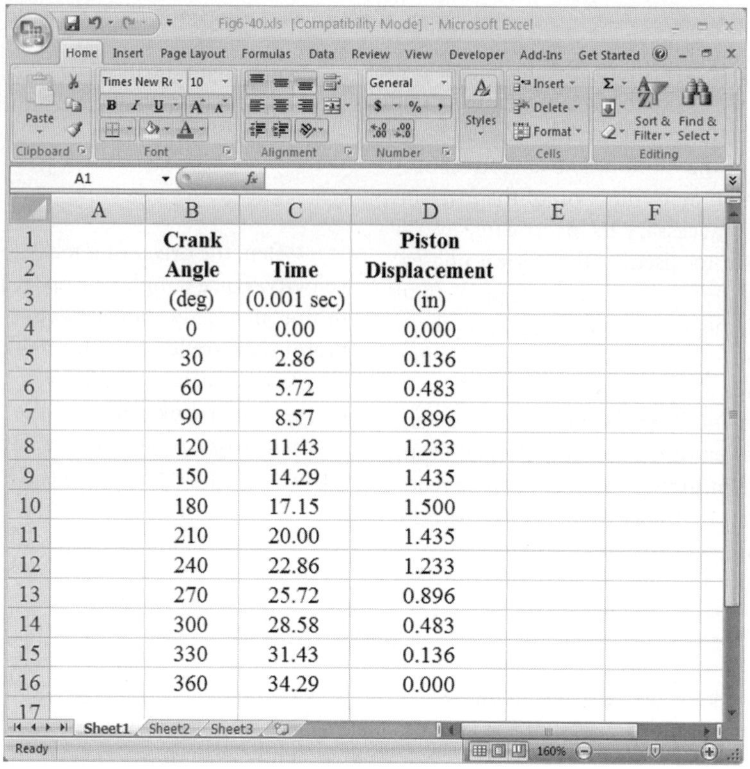

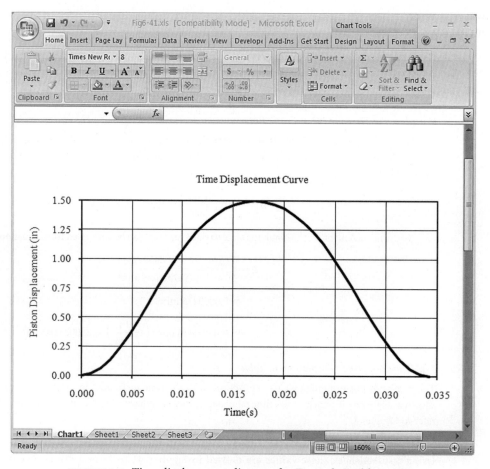

FIGURE 6.40 Time and displacement values for Example Problem 6.17.

FIGURE 6.41 Time displacement diagram for Example Problem 6.17.

These displacement diagrams relative to time can be used to generate a velocity curve because

$$\text{Velocity} = \frac{d\,(\text{displacement})}{d\,(\text{time})}$$

Differential calculus declares that the velocity at a particular instant is the slope of the displacement diagram at that instant. The task is to estimate the slope of the displacement diagram at several points.

6.14.1 Graphical Differentiation

The slope at a point can be estimated by sketching a line through the point of interest, tangent to the displacement curve. The slope of the line can be determined by calculating the change in *y*-value (displacement) divided by the change in *x*-value (time).

The procedure is illustrated in Figure 6.42. Notice that a line drawn tangent to the displacement diagram at t_1 is horizontal. The slope of this tangent line is zero. Therefore, the magnitude of the velocity at t_1 is zero.

A line drawn tangent to the displacement diagram at t_2 is slanted upward as shown. The slope of this line can be calculated as the change of displacement divided by the corresponding change in time. Notice that this $\Delta R, \Delta t$ triangle was drawn rather large to improve measurement accuracy. The velocity at t_2 is found as $\Delta R/\Delta t$ and is positive due to the upward slant of the tangent line. Also notice that this is the steepest section of the upward portion of the displacement curve. This translates to the greatest positive velocity magnitude.

This procedure can be repeated at several locations along the displacement diagram. However, only the velocity extremes and abrupt changes between them are usually desired. Using the notion of differential calculus and slopes, the positions of interest can be visually detected. In general, locations of interest include:

- The steepest portions of the displacement diagram, which correspond to the extreme velocities

- The locations on the displacement diagram with the greatest curvature, which correspond to the abrupt changes of velocities

As mentioned, the velocity at t_2 is greatest because t_2 is the steepest portion of the displacement diagram. The velocity at t_4 is the greatest velocity in the negative direction because t_4 is the steepest downward portion of the displacement diagram.

Identifying the positions of extreme velocities is invaluable. A complete velocity analysis, as presented in the previous sections of this chapter, can then be performed at these locations. Thus, comprehensive velocity analysis is performed only during important mechanism configurations.

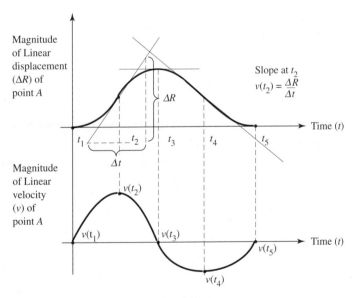

FIGURE 6.42 Velocity curves.

EXAMPLE PROBLEM 6.18

A displacement diagram relative to time was constructed for a compressor mechanism in Example Problem 6.17. Use this data to plot a velocity curve relative to time.

SOLUTION:

1. *Identify Horizontal Portions of the Displacement Diagram*

   The main task of constructing a velocity curve is to determine the slope of many points on the displacement curve. This curve is reprinted as Figure 6.43.

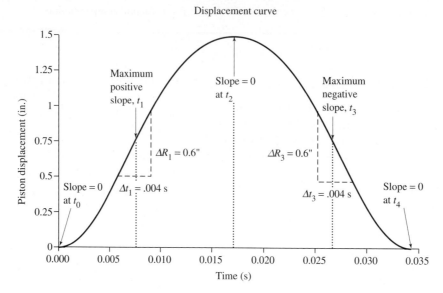

FIGURE 6.43 Displacement curve for Example Problem 6.18.

   From this curve, it is apparent that the curve has a horizontal tangent, or zero slope, at 0, 0.017, and 0.034 s. Therefore, the velocity of the piston is zero at 0, 0.017, and 0.034 s. These points are labeled t_0, t_2, and t_4, respectively.

2. *Calculate the Slope at the Noteworthy Portions of the Displacement Diagram*

   The maximum upward slope appears at 0.008 s. This point is labeled as t_1. An estimate of the velocity can be made from the values of ΔR_1 and Δt_1 read off the graph. The velocity at 0.008 s is estimated as

   $$v\,(t_1) = v_1 = \frac{0.60 \text{ in.}}{0.004 \text{ s}} = -150 \text{ in./s}$$

   Likewise, the maximum downward slope appears at 0.027 s. This point is labeled as t_3. Again, an estimate of the velocity can be made from the values of ΔR_3 and Δt_3 read off the graph. The velocity at 0.027 s is estimated as

   $$v\,(t_3) = v_3 = \frac{-0.60 \text{ in.}}{0.004 \text{ s}} = -150 \text{ in./s}$$

   The procedure of determining the slope of the displacement curve can be repeated at other points in time.

3. *Sketch the Velocity Curve*

   Compiling the slope and time information, a velocity curve can be constructed as shown in Figure 6.44.

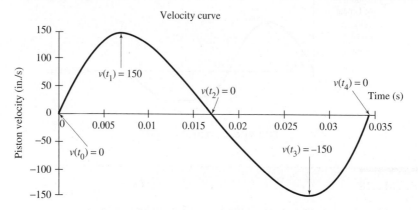

FIGURE 6.44 Velocity curve for Example Problem 6.18.

6.14.2 Numerical Differentiation

In creating a velocity curve using graphical differentiation, the theories of differential calculus are strictly followed. However, even with careful attention, inaccuracies are commonly encountered when generating tangent curves. Thus, other methods, namely numerical approaches, are often used to determine the derivative of a curve defined by a series of known points. The most popular method of numerically determining the derivative is The Richardson Method [Ref. 2]. It is valid for cases where the increments between the independent variables are equal. This limits the analysis to a constant time interval, which is not typically difficult. The derivative of the displacement–time curve can be numerically approximated from the following equation:

$$v_i = \left[\frac{\Delta R_{i+1} - \Delta R_{i-1}}{2\,\Delta t} \right]$$
$$- \left[\frac{\Delta R_{i+2} - 2\,\Delta R_{i+1} + 2\,\Delta R_{i-1} - \Delta R_{i-2}}{12\,\Delta t} \right] \quad \textbf{(6.19)}$$

where:

i = data point index
ΔR_i = displacement at data point i
$\Delta t = t_2 - t_1 = t_3 - t_2 = t_4 - t_3$
t_i = time at data point i

Although the general form may look confusing with the terms i, $i+1$, and so on, actual substitution is straightforward. To illustrate the use of this equation, the velocity at the fifth data point can be found by the following equation:

$$v_5 = \left[\frac{\Delta R_6 - \Delta R_4}{2\,\Delta t} \right]$$
$$- \left[\frac{\Delta R_7 - 2\,\Delta R_6 + 2\,\Delta R_4 - \Delta R_3}{12\,\Delta t} \right]$$

Some confusion may occur when calculating the derivative at the endpoints of the curves. For mechanism analysis, the displacement diagram repeats with every revolution of crank rotation. Therefore, as the curve is repeated, the data points prior to the beginning of the cycle are the same points at the end of the cycle. Thus, when 12 points are used to generate the displacement curve, the displacement at point 1 is identical to the displacement at point 13. Then the velocity at point 1 can be calculated as

$$v_1 = \left[\frac{\Delta R_2 - \Delta R_{12}}{2\,\Delta t} \right] - \left[\frac{\Delta R_3 - 2\,\Delta R_2 + 2\,\Delta R_{12} - \Delta R_{11}}{12\,\Delta t} \right]$$

Because this equation is a numerical approximation, the associated error decreases drastically as the increment of the crank angle and time are reduced.

EXAMPLE PROBLEM 6.19

A displacement diagram of the piston operating in a compressor was plotted in Example Problem 4.11. This diagram was converted to a displacement curve relative to time in Example Problem 6.17. Use this data to numerically generate a velocity curve.

SOLUTION: 1.   ***Determine the Time Increment between Position Data Points***

The spreadsheet from Example Problem 6.17 is expanded by inserting an additional column to include the piston velocity. The time increment is calculated as follows:

$$\Delta t = t_2 - t_1 = (0.00289 - 0.0) = 0.00286 \text{ s}$$

2.   ***Use Equation (6.19) to Calculate Velocity Data Points***

To illustrate the calculation of the velocities, a few sample calculations are shown:

$$v_2 = \left[\frac{(\Delta R_3 - \Delta R_1)}{2\Delta t} \right] - \left[\frac{\Delta R_4 - 2\Delta R_3 + 2\Delta R_1 - \Delta R_{12}}{12\Delta t} \right]$$

$$= \left[\frac{(0.483 - 0.0)}{2(0.00286)} \right] - \left[\frac{0.896 - 2(0.483) + 2(0.0) - 0.136}{12(0.00286)} \right] = 142.67 \text{ in./s}$$

$$v_9 = \left[\frac{(\Delta R_{10} - \Delta R_8)}{\Delta t} \right] - \left[\frac{\Delta R_{11} - 2\Delta R_{10} + 2\Delta R_8 - \Delta R_7}{12\Delta t} \right]$$

$$= \left[\frac{(0.896 - 1.435)}{2(0.00286)} \right] - \left[\frac{0.483 - 2(0.896) + 2(1.435) - 1.50}{12(0.00286)} \right] = -95.48 \text{ in./s}$$

$$v_{12} = \left[\frac{(\Delta R_{13} - \Delta R_{11})}{2\Delta t} \right] - \left[\frac{\Delta R_2 - 2\Delta R_{13} + 2\Delta R_{11} - \Delta R_{10}}{12\Delta t} \right],$$

$$= \left[\frac{(0.0 - 0.483)}{2(0.00286)} \right] - \left[\frac{0.136 - 2(0.0) + 2(0.483) - 0.896}{2(0.00286)} \right] = -91.47 \text{ in./s}$$

3. **Compute the Velocity Data and Plot the Velocity Curve**

The results can be computed and tabulated as shown in Figure 6.45. A spreadsheet was used efficiently to perform these redundant calculations. For those who are unfamiliar with spreadsheets, refer to Chapter 8.

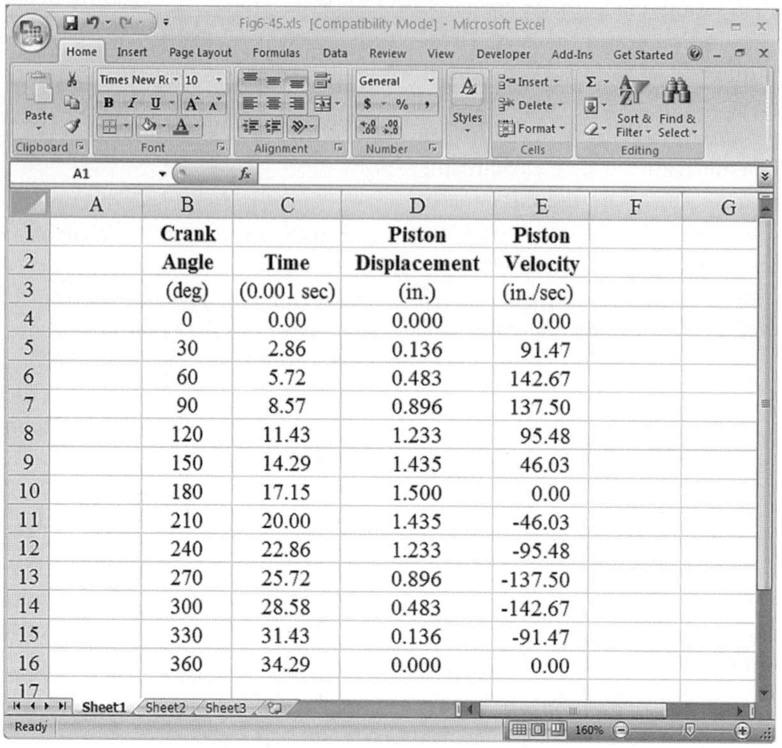

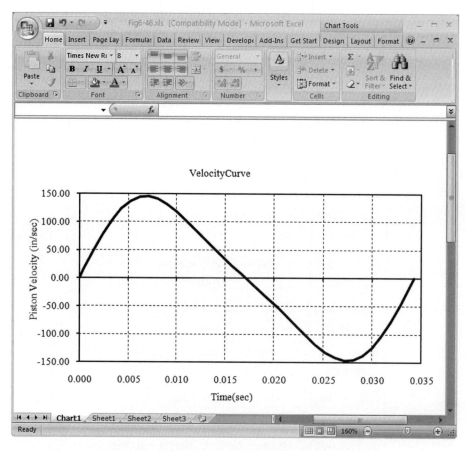

	Crank Angle (deg)	Time (0.001 sec)	Piston Displacement (in.)	Piston Velocity (in./sec)
4	0	0.00	0.000	0.00
5	30	2.86	0.136	91.47
6	60	5.72	0.483	142.67
7	90	8.57	0.896	137.50
8	120	11.43	1.233	95.48
9	150	14.29	1.435	46.03
10	180	17.15	1.500	0.00
11	210	20.00	1.435	-46.03
12	240	22.86	1.233	-95.48
13	270	25.72	0.896	-137.50
14	300	28.58	0.483	-142.67
15	330	31.43	0.136	-91.47
16	360	34.29	0.000	0.00

FIGURE 6.45 Velocity data for Example Problem 6.19

FIGURE 6.46 Velocity curve for Example Problem 6.19.

These values are plotted in Figure 6.46 to form a velocity diagram relative to time. Notice that this curve is still rather rough. For accuracy purposes, it is highly suggested that the crank angle increment be reduced to 10° or 15°. When a spreadsheet is used to generate the velocity data, even smaller increments are advisable to reduce the difficulty of the task.

PROBLEMS

General Velocity

6–1. A package is moved at a constant rate from one end of a 25-ft horizontal conveyor to the other end in 15 s. Determine the linear speed of the conveyor belt.

6–2. A hydraulic cylinder extends at a constant rate of 2 fpm (ft/min). Determine the time required to traverse the entire stroke of 15 in.

6–3. Determine the average speed (in mph) of an athlete who can run a 4-minute mile.

6–4. Determine the average speed (in mph) of an athlete who can run a 100-m dash in 10 s.

6–5. A gear uniformly rotates 270° clockwise in 2 s. Determine the angular velocity in rpm and rad/s.

6–6. Determine the angular velocity (in rpm) of the second, minute, and hour hand of a clock.

6–7. A servo-driven actuator is programmed to extend according to the velocity profile shown in Figure P6.7. Determine the total displacement during this programmed move.

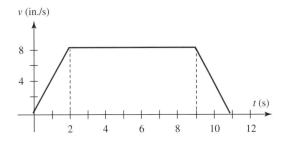

FIGURE P6.7 Problems 7 and 8.

6–8. A servo-driven actuator is programmed to extend according to the velocity profile shown in Figure P6.7. Use a spreadsheet to generate plots of velocity versus time and displacement versus time during this programmed move.

6–9. A linear motor is programmed to move according to the velocity profile shown in Figure P6.9. Determine the total displacement during this programmed move.

6–10. A linear motor is programmed to move according to the velocity profile shown in Figure P6.9. Use a spreadsheet to generate plots of velocity versus time and displacement versus time during this programmed move.

6–11. The drive roller for a conveyor belt is shown in Figure P6.11. Determine the angular velocity of the roller when the belt operates at 10 fpm (10 ft/min).

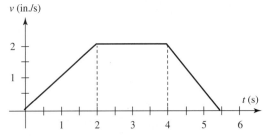

FIGURE P6.9 Problems 9 and 10.

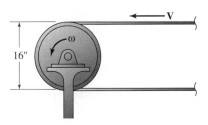

FIGURE P6.11 Problems 11 and 12.

6–12. The drive roller for a conveyor belt is shown in Figure P6.11. Determine the linear speed of the belt when the roller operates at 30 rpm counterclockwise.

6–13. Link 2 is isolated from a kinematic diagram and shown in Figure P6.13. The link is rotating counterclockwise at a rate of 300 rpm. Determine the velocity of points A and B. Use $\gamma = 50°$ and $\beta = 60°$.

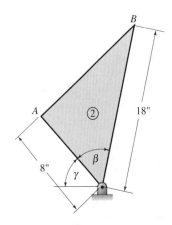

FIGURE P6.13 Problems 13 and 14.

6–14. Link 2 is isolated from a kinematic diagram and shown in Figure P6.13. The link is rotating clockwise, driving point A at a speed of 40 ft/s. Determine the velocity of points A and B and the angular velocity of link 2. Use $\gamma = 50°$ and $\beta = 60°$.

Relative Velocity

6–15. A kinematic diagram of a four-bar mechanism is shown in Figure P6.15. At the instant shown, $v_A = 800$ mm/s and $v_B = 888$ mm/s. Graphically determine the relative velocity of point B with respect to point A. Also determine the angular velocity of links 2 and 4.

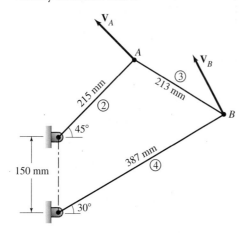

FIGURE P6.15 Problems 15 and 16.

6–16. A kinematic diagram of a four-bar mechanism is shown in Figure P6.15. At the instant shown, $v_A = 20$ mm/s and $v_B = 22.2$ mm/s. Graphically determine the relative velocity of point B with respect to point A. Also determine the angular velocity of links 2 and 4.

6–17. A kinematic diagram of a slider-crank mechanism is shown in Figure P6.17. At the instant shown, $v_A = 380$ ft/s and $v_B = 400$ ft/s. Graphically determine the relative velocity of point A with respect to point B. Also, determine the angular velocity of link 2.

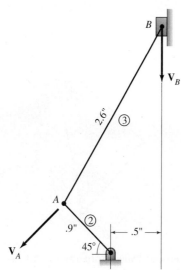

FIGURE P6.17 Problems 17 and 18.

6–18. A kinematic diagram of a slider-crank mechanism is shown in Figure P6.17. At the instant shown, $v_A = 20$ ft/s and $v_B = 21$ ft/s. Graphically determine the relative velocity of point A with respect to point B. Also, determine the angular velocity of link 2.

Relative Velocity Method—Graphical

6–19. For the compressor linkage shown in Figure P6.19, use the relative velocity method to graphically determine the linear velocity of the piston as the crank rotates clockwise at 1150 rpm.

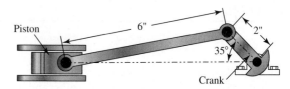

FIGURE P6.19 Problems 19, 20, 41, 52, 63, 74, 85, 96, 104, and 112.

6–20. For the compressor linkage shown in Figure P6.19, use the relative velocity method to graphically determine the linear velocity of the piston as the crank rotates counterclockwise at 1775 rpm.

6–21. For the reciprocating saw shown in Figure P6.21, use the relative velocity method to graphically determine the linear velocity of the blade as the crank wheel rotates counterclockwise at 1500 rpm.

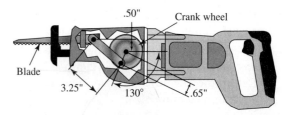

FIGURE P6.21 Problems 21, 22, 42, 53, 64, 75, 86, 97, 105, and 113.

6–22. For the reciprocating saw shown in Figure P6.21, use the relative velocity method to graphically determine the linear velocity of the blade as the crank wheel rotates clockwise at 900 rpm.

6–23. For the shearing mechanism in the configuration shown in Figure P6.23, use the relative velocity method to graphically determine the linear velocity of the blade as the crank rotates clockwise at 100 rpm.

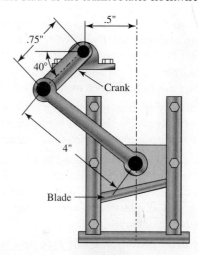

FIGURE P6.23 Problems 23, 24, 43, 54, 65, 76, 87, 98, 106, and 114.

6–24. For the shearing mechanism in the configuration shown in Figure P6.23, use the relative velocity method to graphically determine the linear velocity of the blade as the crank rotates counterclockwise at 80 rpm.

6–25. For the rear windshield wiper mechanism shown in Figure P6.25, use the relative velocity method to graphically determine the angular velocity of the wiper arm as the crank rotates counterclockwise at 40 rpm.

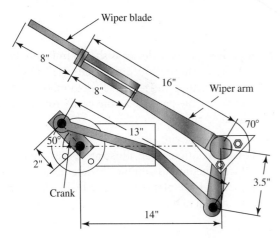

FIGURE P6.25 Problems 25, 26, 44, 55, 66, 77, 88, 99, 107, and 115.

6–26. For the rear windshield wiper mechanism shown in Figure P6.25, use the relative velocity method to graphically determine the angular velocity of the wiper arm as the crank rotates clockwise at 60 rpm.

6–27. The device in Figure P6.27 is a sloshing bath used to wash vegetable produce. For the configuration shown, use the relative velocity method to graphically determine the angular velocity of the water bath as the crank is driven counterclockwise at 100 rpm.

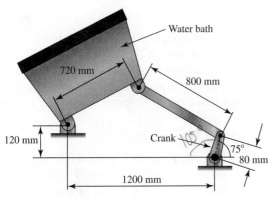

FIGURE P6.27 Problems 27, 28, 45, 56, 67, 78, 89, 100, 108, and 116.

6–28. The device in Figure P6.27 is a sloshing bath used to wash vegetable produce. For the configuration

shown, use the relative velocity method to graphically determine the angular velocity of the water bath as the crank is driven clockwise at 75 rpm.

6–29. The device in Figure P6.29 is a drive mechanism for the agitator on a washing machine. For the configuration shown, use the relative velocity method to graphically determine the angular velocity of the segment gear as the crank is driven clockwise at 50 rpm.

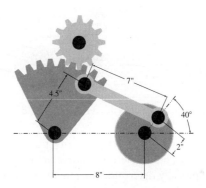

FIGURE P6.29 Problems 29, 30, 46, 57, 68, 79, 90, 101, 109, and 117.

6–30. The device in Figure P6.29 is a drive mechanism for the agitator on a washing machine. For the configuration shown, use the relative velocity method to graphically determine the angular velocity of the segment gear as the crank is driven counterclockwise at 35 rpm.

6–31. For the hand-operated shear shown in Figure P6.31, use the relative velocity method to graphically determine the angular velocity of the handle required to have the blade cut through the metal at a rate of 3 mm/s. Also determine the linear velocity of point X.

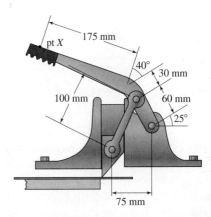

FIGURE P6.31 Problems 31, 32, 47, 58, 69, 80, and 91.

6–32. For the hand-operated shear shown in Figure P6.31, use the relative velocity method to graphically determine the linear velocity of the blade as the handle is rotated at a rate of 2 rad/s clockwise. Also determine the linear velocity of point X.

6–33. For the foot-operated air pump shown in Figure P6.33, use the relative velocity method to graphically determine the angular velocity of the foot pedal required to contract the cylinder at a rate of 5 in./s. Also determine the linear velocity of point X.

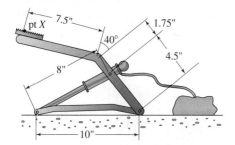

FIGURE P6.33 Problems 33, 34, 48, 59, 70, 81, and 92.

6–34. For the foot-operated air pump shown in Figure P6.33, use the relative velocity method to graphically determine the rate of cylinder compression when the angular velocity of the foot pedal assembly is 1 rad/s counterclockwise. Also determine the linear velocity of point X.

6–35. A two-cylinder compressor mechanism is shown in Figure P6.35. For the configuration shown, use the relative velocity method to graphically determine the linear velocity of both pistons as the 1.5-in. crank is driven clockwise at 1775 rpm. Also determine the instantaneous volumetric flow rate out of the right cylinder.

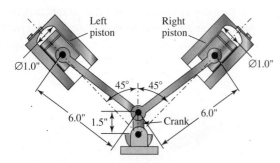

FIGURE P6.35 Problems 35, 36, 49, 60, 71, 82, 93, 102, 110, and 118.

6–36. A two-cylinder compressor mechanism is shown in Figure P6.35. For the configuration shown, use the relative velocity method to graphically determine the linear velocity of both pistons as the 1.5-in. crank is driven counterclockwise at 1150 rpm. Also determine the instantaneous volumetric flow rate out of the left cylinder.

6–37. A package-moving device is shown in Figure P6.37. For the configuration illustrated, use the relative velocity method to graphically determine the linear velocity of the package as the crank rotates clockwise at 40 rpm.

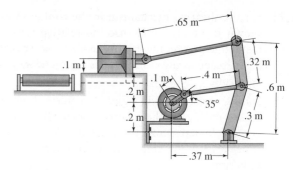

FIGURE P6.37 Problems 37, 38, 50, 61, 72, 83, 94, 103, 111, and 119.

6–38. A package-moving device is shown in Figure P6.37. For the configuration illustrated, use the relative velocity method to graphically determine the linear velocity of the package as the crank rotates clockwise at 65 rpm.

6–39. A package-moving device is shown in Figure P6.39. For the configuration illustrated, use the relative velocity method to graphically determine the linear velocity of the platform as the hydraulic cylinder extends at a rate of 16 fpm.

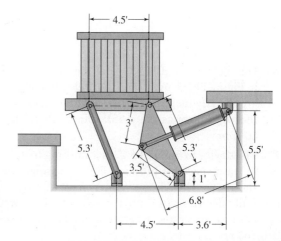

FIGURE P6.39 Problems 39, 40, 51, 62, 73, 84, and 95.

6–40. A package-moving device is shown in Figure P6.39. For the configuration illustrated, use the relative velocity method to graphically determine the linear velocity of the platform as the hydraulic cylinder retracts at a rate of 12 fpm.

Relative Velocity Method—Analytical

6–41. For the compressor linkage shown in Figure P6.19, use the relative velocity method to determine the linear velocity of the piston as the crank rotates clockwise at 950 rpm.

6–42. For the reciprocating saw shown in Figure P6.21, use the relative velocity method to analytically determine the linear velocity of the blade as the crank wheel rotates counterclockwise at 1700 rpm.

6–43. For the shearing mechanism in the configuration shown in Figure P6.23, use the relative velocity method to analytically determine the linear velocity of the blade as the crank rotates clockwise at 85 rpm.

6–44. For the rear windshield wiper mechanism shown in Figure P6.25, use the relative velocity method to analytically determine the angular velocity of the wiper arm as the crank rotates counterclockwise at 45 rpm.

6–45. The device in Figure P6.27 is a sloshing bath used to wash vegetable produce. For the configuration shown, use the relative velocity method to analytically determine the angular velocity of the water bath as the crank is driven counterclockwise at 90 rpm.

6–46. The device in Figure P6.29 is a drive mechanism for the agitator on a washing machine. For the configuration shown, use the relative velocity method to analytically determine the angular velocity of the segment gear as the crank is driven clockwise at 60 rpm.

6–47. For the links for the hand-operated shear shown in Figure P6.31, use the relative velocity method to analytically determine the angular velocity of the handle required to have the blade cut through the metal at a rate of 2 mm/s.

6–48. For the foot-operated air pump shown in Figure P6.33, use the relative velocity method to analytically determine the rate of cylinder compression as the foot pedal assembly rotates counterclockwise at a rate of 1 rad/s.

6–49. A two-cylinder compressor mechanism is shown in Figure P6.35. For the configuration shown, use the relative velocity method to analytically determine the linear velocity of both pistons as the 1.5-in. crank is driven clockwise at 2000 rpm. ~~Also determine the instantaneous volumetric flow rate out of the right cylinder.~~

6–50. A package-moving device is shown in Figure P6.37. For the configuration illustrated, use the relative velocity method to analytically determine the linear velocity of the package as the crank rotates clockwise at 80 rpm.

6–51. A package-moving device is shown in Figure P6.39. For the configuration illustrated, use the relative velocity method to analytically determine the linear velocity of the platform as the hydraulic cylinder retracts at a rate of 10 fpm.

Locating Instantaneous
Centers—Graphically

6–52. For the compressor linkage shown in Figure P6.19, graphically determine the location of all the instantaneous centers.

6–53. For the reciprocating saw shown in Figure P6.21, graphically determine the location of all the instantaneous centers.

6–54. For the shearing mechanism in the configuration shown in Figure P6.23, graphically determine the location of all the instantaneous centers.

6–55. For the rear windshield wiper mechanism shown in Figure P6.25, graphically determine the location of all the instantaneous centers.

6–56. For the produce-washing bath shown in Figure P6.27, graphically determine the location of all the instantaneous centers.

6–57. For the washing machine agitation mechanism shown in Figure P6.29, graphically determine the location of all the instantaneous centers.

6–58. For the hand-operated shear shown in Figure P6.31, graphically determine the location of all the instantaneous centers.

6–59. For the foot-operated air pump shown in Figure P6.33, graphically determine the location of all the instantaneous centers.

6–60. For the two-cylinder compressor mechanism shown in Figure P6.35, graphically determine the location of all the instantaneous centers.

6–61. For the package-moving device shown in Figure P6.37, graphically determine the location of all the instantaneous centers.

6–62. For the package-moving device shown in Figure P6.39, graphically determine the location of all the instantaneous centers.

Locating Instantaneous
Centers—Analytically

6–63. For the compressor linkage shown in Figure P6.19, analytically determine the location of all the instantaneous centers.

6–64. For the reciprocating saw shown in Figure P6.21, analytically determine the location of all the instantaneous centers.

6–65. For the shearing mechanism in the configuration shown in Figure P6.23, analytically determine the location of all the instantaneous centers.

6–66. For the rear windshield wiper mechanism shown in Figure P6.25, analytically determine the location of all the instantaneous centers.

6–67. For the produce-washing bath shown in Figure P6.27, analytically determine the location of all the instantaneous centers.

6–68. For the washing machine agitation mechanism shown in Figure P6.29, analytically determine the location of all the instantaneous centers.

6–69. For the hand-operated shear shown in Figure P6.31, analytically determine the location of all the instantaneous centers.

6–70. For the foot-operated air pump shown in Figure P6.33, analytically determine the location of all the instantaneous centers.

6–71. For the two-cylinder compressor mechanism shown in Figure P6.35, analytically determine the location of all the instantaneous centers.

6–72. For the package-moving device shown in Figure P6.37, analytically determine the location of all the instantaneous centers.

6–73. For the package-moving device shown in Figure P6.39, analytically determine the location of all the instantaneous centers.

Instantaneous Center Method—Graphical

6–74. For the compressor linkage shown in Figure P6.19, use the instantaneous center method to graphically determine the linear velocity of the piston as the crank rotates counterclockwise at 1500 rpm.

6–75. For the reciprocating saw shown in Figure P6.21, use the instantaneous center method to graphically determine the linear velocity of the blade as the crank wheel rotates clockwise at 1200 rpm.

6–76. For the shearing mechanism in the configuration shown in Figure P6.23, use the instantaneous center method to graphically determine the linear velocity of the blade as the crank rotates counterclockwise at 65 rpm.

6–77. For the rear windshield wiper mechanism shown in Figure P6.25, use the instantaneous center method to graphically determine the angular velocity of the wiper arm as the crank rotates clockwise at 55 rpm.

6–78. For the produce-sloshing bath shown in Figure P6.27, use the instantaneous method to graphically determine the angular velocity of the water bath as the crank is driven clockwise at 110 rpm.

6–79. For the washing machine agitator mechanism shown in Figure P6.29, use the instantaneous center method to graphically determine the angular velocity of the segment gear as the crank is driven counterclockwise at 70 rpm.

6–80. For the hand-operated shear in the configuration shown in Figure P6.31, use the instantaneous method to graphically determine the angular velocity of the handle required to have the blade cut through the metal at a rate of 4 mm/s.

6–81. For the foot-operated air pump shown in Figure P6.33, use the instantaneous center method to graphically determine the rate of cylinder compression as the foot pedal assembly rotates counterclockwise at a rate of 0.75 rad/s.

6–82. A two-cylinder compressor mechanism is shown in Figure P6.35. For the configuration shown, use the instantaneous center method to graphically determine the linear velocity of both pistons as the 1.5-in. crank is driven counterclockwise at 2200 rpm. Also determine the instantaneous volumetric flow rate out of the right cylinder.

6–83. A package-moving device is shown in Figure P6.37. For the configuration illustrated, use the graphical instantaneous center method to determine the linear velocity of the package as the crank rotates clockwise at 70 rpm.

6–84. A package-moving device is shown in Figure P6.39. For the configuration illustrated, use the instantaneous center method to graphically determine the linear velocity of the platform as the hydraulic cylinder extends at a rate of 8 fpm.

Instantaneous Center Method—Analytical

6–85. For the compressor linkage shown in Figure P6.19, use the instantaneous center method to analytically determine the linear velocity of the piston as the crank rotates clockwise at 1100 rpm.

6–86. For the reciprocating saw shown in Figure P6.21, use the instantaneous center method to analytically determine the linear velocity of the blade as the crank wheel rotates counterclockwise at 1375 rpm.

6–87. For the shearing mechanism in the configuration shown in Figure P6.23, use the instantaneous center method to analytically determine the linear velocity of the blade as the crank rotates clockwise at 55 rpm.

6–88. For the rear windshield wiper mechanism shown in Figure P6.25, use the instantaneous center method to analytically determine the angular velocity of the wiper arm as the crank rotates counterclockwise at 35 rpm.

6–89. For the produce-sloshing bath shown in Figure P6.27, use the instantaneous method to analytically determine the angular velocity of the water bath as the crank is driven counterclockwise at 95 rpm.

6–90. For the washing machine agitator mechanism shown in Figure P6.29, use the instantaneous center method to analytically determine the angular velocity of the segment gear as the crank is driven clockwise at 85 rpm.

6–91. For the hand-operated shear in the configuration shown in Figure P6.31, use the instantaneous method to analytically determine the angular velocity of the handle required to have the blade cut through the metal at a rate of 2.5 mm/s.

6–92. For the foot-operated air pump shown in Figure P6.33, use the instantaneous center method to analytically determine the rate of cylinder compression as the foot pedal assembly rotates counterclockwise at a rate of 0.6 rad/s.

6–93. A two-cylinder compressor mechanism is shown in Figure P6.35. For the configuration shown, use the instantaneous center method to analytically determine the linear velocity of both pistons as the 1.5-in. crank is driven clockwise at 1775 rpm. Also determine the instantaneous volumetric flow rate out of the right cylinder.

6–94. A package-moving device is shown in Figure P6.37. For the configuration illustrated, use the graphical instantaneous center method to determine the linear velocity of the package as the crank rotates clockwise at 30 rpm.

6–95. A package-moving device is shown in Figure P6.39. For the configuration illustrated, use the instantaneous center method to analytically determine the linear velocity of the platform as the hydraulic cylinder retracts at a rate of 7 fpm.

Velocity Curves—Graphical

6–96. The crank of the compressor linkage shown in Figure P6.19 is driven clockwise at a constant rate of 1750 rpm. Graphically create a curve for the linear displacement of the piston as a function of the crank angle. Convert the crank angle to time. Then graphically calculate the slope to obtain a velocity curve of the piston as a function of time.

6–97. The crank wheel of the reciprocating saw shown in Figure P6.21 is driven counterclockwise at a constant rate of 1500 rpm. Graphically create a curve for the linear displacement of the saw blade as a function of the crank angle. Convert the crank angle to time. Then graphically calculate the slope to obtain a velocity curve of the saw blade as a function of time.

6–98. The crank of the shearing mechanism shown in Figure P6.23 is driven clockwise at a constant rate of 80 rpm. Graphically create a curve for the linear displacement of the shear blade as a function of the crank angle. Convert the crank angle to time. Then graphically calculate the slope to obtain a velocity curve of the shear blade as a function of time.

6–99. The crank of the rear windshield wiper mechanism shown in Figure P6.25 is driven clockwise at a constant rate of 65 rpm. Graphically create a curve for the angular displacement of the wiper blade as a function of the crank angle. Convert the crank angle to time. Then graphically calculate the slope to obtain an angular velocity curve of the wiper blade as a function of time.

6–100. The crank of the sloshing bath shown in Figure P6.27 is driven counterclockwise at 90 rpm. Graphically create a curve for the angular displacement of the bath as a function of the crank angle. Convert the crank angle to time. Then graphically calculate the slope to obtain an angular velocity curve of the bath as a function of time.

6–101. The crank of the washing machine agitator mechanism shown in Figure P6.29 is driven clockwise at 80 rpm. Graphically create a curve for the angular displacement of the segment gear as a function of the crank angle. Convert the crank angle to time. Then graphically calculate the slope to obtain an angular velocity of the segment gear as a function of time.

6–102. The crank of the two-cylinder compressor mechanism shown in Figure P6.35 is driven clockwise at 1250 rpm. Graphically create a curve for the linear displacement of both pistons as a function of the crank angle. Convert the crank angle to time. Then graphically calculate the slope to obtain velocity curves of both pistons as a function of time.

6–103. The crank of the package-moving device shown in Figure. P6.37 is driven clockwise at 25 rpm. Graphically create a curve for the linear displacement of the ram as a function of the crank angle. Convert the crank angle to time. Then graphically calculate the slope to obtain a velocity curve of the ram as a function of time.

Velocity Curves—Analytical

6–104. The crank of the compressor linkage shown in Figure P6.19 is driven counterclockwise at a constant rate of 2150 rpm. Use a spreadsheet to analytically create a curve for the linear displacement of the piston as a function of the crank angle. Convert the crank angle axis to time. Then use numerical differentiation to obtain a velocity curve of the piston as a function of time.

6–105. The crank wheel of the reciprocating saw shown in Figure P6.21 is driven clockwise at a constant rate of 1900 rpm. Use a spreadsheet to analytically create a curve for the linear displacement of the saw blade as a function of the crank angle. Convert the crank angle axis to time. Then use numerical differentiation to obtain a velocity curve of the saw blade as a function of time.

6–106. The crank of the shearing mechanism shown in Figure P6.23 is driven clockwise at a constant rate of 80 rpm. Use a spreadsheet to analytically create a curve for the linear displacement of the shear blade as a function of the crank angle. Convert the crank angle axis to time. Then use numerical differentiation to obtain a velocity curve of the shear blade as a function of time.

6–107. The crank of the rear windshield wiper mechanism shown in Figure P6.25 is driven counterclockwise at a constant rate of 55 rpm. Use a spreadsheet to analytically create a curve for the angular displacement of the wiper blade as a function of the crank angle. Convert the crank angle axis to time. Then use numerical differentiation to obtain an angular velocity curve of the wiper blade as a function of time.

6–108. The crank of the sloshing bath shown in Figure P6.27 is driven clockwise at 65 rpm. Use a spreadsheet to analytically create a curve for the angular displacement of the bath as a function of the crank angle. Convert the crank angle axis to time. Then use numerical differentiation to obtain an angular velocity curve of the bath as a function of time.

6–109. The crank of the washing machine agitator mechanism shown in Figure P6.29 is driven counterclockwise at 65 rpm. Use a spreadsheet to analytically create a curve for the angular displacement of the segment gear as a function of the crank angle. Convert the crank angle axis to time. Then use numerical differentiation to obtain an angular velocity curve of the segment gear as a function of time.

6–110. The crank of the two-cylinder compressor mechanism shown in Figure P6.35 is driven counterclockwise at 1500 rpm. Use a spreadsheet to analytically create a curve for the linear displacement of both pistons as a function of the crank angle. Convert the crank angle axis to time. Then use numerical differentiation to obtain velocity curves of both pistons as a function of time.

6–111. The crank of the package-moving device shown in Figure P6.37 is driven counterclockwise at 30 rpm. Use a spreadsheet to analytically create a curve for the linear displacement of the ram as a function of the crank angle. Convert the crank angle axis to time. Then use numerical differentiation to obtain a velocity curve of the ram as a function of time.

Velocity Using Working Model

6–112. The crank of the compressor linkage shown in Figure P6.19 is driven clockwise at a constant rate of 1750 rpm. Use the Working Model software to create a simulation and plot the linear velocity of the piston as a function of the crank angle.

6–113. The crank wheel of the reciprocating saw shown in Figure P6.21 is driven counterclockwise at a constant rate of 1500 rpm. Use the Working Model software to create a simulation and plot the linear velocity of the saw blade as a function of the crank angle.

6–114. The crank of the shearing mechanism shown in Figure P6.23 is driven clockwise at a constant rate of 80 rpm. Use the Working Model software to create a simulation and plot the angular velocity of the wiper blade as a function of the crank angle.

6–115. The crank of the rear windshield wiper mechanism shown in Figure P6.25 is driven clockwise at a constant rate of 65 rpm. Use the Working Model software to create a simulation and plot the angular velocity of the wiper blade as a function of the crank angle.

6–116. The crank of the sloshing bath shown in Figure P6.27 is driven counterclockwise at 90 rpm. Use the Working Model software to create a simulation and plot the angular velocity of the bath as a function of the crank angle.

6–117. The crank of the washing machine agitator mechanism shown in Figure P6.29 is driven clockwise at 80 rpm. Use the Working Model software to create a simulation and plot the angular velocity of the segment gear as a function of the crank angle.

6–118. The crank of the two-cylinder compressor mechanism shown in Figure P6.35 is driven clockwise at 1250 rpm. Use the Working Model software to create a simulation and plot the angular velocity of both pistons as a function of the crank angle.

6–119. The crank of the package-moving device shown in Figure P6.37 is driven clockwise at 25 rpm. Use the Working Model software to create a simulation and plot the linear velocity of the ram as a function of the crank angle.

CASE STUDIES

6–1. Figure C6.1 illustrates a mechanism that is used to drive a power hacksaw. The mechanism is powered with an electric motor shaft, keyed to gear A. Carefully examine the configuration in question, then answer the following leading questions to gain insight into the operation of the mechanism.

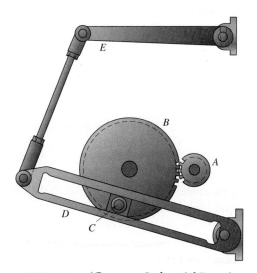

FIGURE C6.1 (Courtesy, Industrial Press.)

1. When gear A is forced to rotate counterclockwise, what is the motion of mating gear B?
2. When gear A is forced to rotate counterclockwise, what is the motion of stud pin C?
3. When gear A is forced to rotate counterclockwise, what is the motion of lever D?
4. How does the motion of lever D differ from the motion of lever E?
5. Determine the position of gear B that would place lever D at its lower extreme position.
6. Determine the position of gear B that would place lever D at its upper extreme position.
7. Examine the amount of rotation of gear B to raise lever D and the amount of rotation to lower the lever.
8. Approximately what is the difference between the time to raise and the time to lower lever D?
9. Comment on the continual motion of lever E.

6–2. Figure C6.2 illustrates the mechanism that drives a table for a special grinding operation. Carefully examine the configuration in question, then answer the following leading questions to gain insight into the operation of the mechanism.

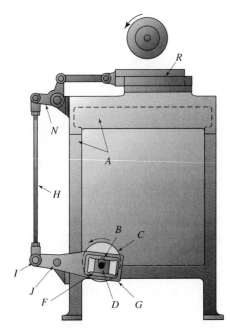

FIGURE C6.2 (Courtesy, Industrial Press.)

1. When wheel *C* is forced to rotate counterclockwise, what is the motion of pin *D*?
2. When wheel *C* is forced to rotate counterclockwise, what is the motion of link *G*?
3. Determine the position of wheel *C* that would place point *I* at its upper extreme position.
4. Determine the position of wheel *C* that would place point *I* at its lower extreme position.
5. Examine the amount of rotation of wheel *C* to raise point *I* and the amount of rotation to lower the point.
6. Approximately what is the difference between the time to raise and the time to lower point *I*?
7. Comment on the cyclical motion of lever *E*.
8. Describe the motion of table *R*.
9. What is the function of this mechanism?
10. Why are there screw threads on both ends of link *H*?
11. Compute the mobility of this mechanism.

6–3. Figure C6.3 illustrates the mechanism that drives a bellows for an artificial respiration machine. Carefully examine the configuration in question, then answer the following leading questions to gain insight into the operation of the mechanism.

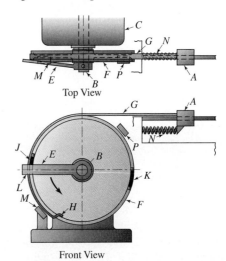

FIGURE C6.3 (Courtesy, Industrial Press.)

1. When link *E* drives continually counterclockwise and rides slot *J*, at the instant shown, what is the motion of disk *F*?
2. When link *E* drives continually counterclockwise and rides slot *J*, at the instant shown, what is the motion of strap *G*?
3. When link *E* drives continually counterclockwise and rides slot *J*, at the instant shown, what is the motion of slide *A*?
4. As link *E* approaches the ramped pad *M*, what happens to the spring *N*?
5. As link *E* contacts the ramped pad *M*, what happens to link *E*?
6. As link *E* contacts the ramped pad *M*, what is the motion of disk *F*?
7. As link *E* contacts the ramped pad *M*, what is the motion of slide *A*?
8. As link *E* continues to rotate beyond the ramped pad *M*, what is the motion of disk *F*?
9. As link *E* catches slot *K*, what is the motion of disk *F*?
10. Describe the continual motion of slide *A*, which drives one end of the bellows.

ACCELERATION ANALYSIS

7.1 INTRODUCTION

Acceleration analysis involves determining the manner in which certain points on the links of a mechanism are either "speeding up" or "slowing down." Acceleration is a critical property because of the inertial forces associated with it. In the study of forces, Sir Isaac Newton discovered that an inertial force is proportional to the acceleration imposed on a body. This phenomenon is witnessed anytime you lunge forward as the brakes are forcefully applied on your car.

Of course, an important part of mechanism design is to ensure that the strength of the links and joints is sufficient to withstand the forces imposed on them. Understanding all forces, especially inertia, is important. Force analysis is introduced in Chapters 13 and 14. However, as a preliminary step, acceleration analysis of a mechanism's links must be performed.

The determination of accelerations in a linkage is the purpose of this chapter. The primary procedure used in this

analysis is the relative acceleration method, which utilizes the results of the relative velocity method introduced in Chapter 6. Consistent with other chapters in this book, both graphical and analytical techniques are utilized.

7.2 LINEAR ACCELERATION

Linear acceleration, $\mathbf{A}$, of a point is the change of linear velocity of that point per unit of time. Chapter 6 was dedicated to velocity analysis. Velocity is a vector quantity, which is defined with both a magnitude and a direction. Therefore, a change in either the magnitude or direction of velocity produces an acceleration. The magnitude of the acceleration vector is designated $a = |\mathbf{A}|$.

7.2.1 Linear Acceleration of Rectilinear Points

Consider the case of a point having straight line, or rectilinear, motion. Such a point is most commonly found on a link that is attached to the frame with a sliding joint. For this case, only the magnitude of the velocity vector can change. The acceleration can be mathematically described as

$$\mathbf{A} = \lim_{\Delta t \to 0} \frac{\Delta \mathbf{V}}{\Delta t} = \frac{d\mathbf{v}}{dt} \tag{7.1}$$

However, because

$$\mathbf{V} = \frac{d\mathbf{R}}{dt}$$

then

$$\mathbf{A} = \frac{d^2 \mathbf{R}}{dt^2} \tag{7.2}$$

For short time periods, or when the acceleration can be assumed to be linear, the following relationship can be used:

$$\mathbf{A} \cong \frac{\Delta \mathbf{V}}{\Delta t} \tag{7.3}$$

Because velocity is a vector, equation (7.1) states that acceleration is also a vector. The direction of linear acceleration is in the direction of linear movement when the link accelerates. Conversely, when the link decelerates, the direction of linear acceleration is opposite to the direction of linear movement.

Linear acceleration is expressed in the units of velocity (length per time) divided by time, or length per squared time. In the U.S. Customary System, the common units used are feet per squared second (ft/s^2) or inches per squared second ($in./s^2$). In the International System, the common units used are meters per squared second (m/s^2) or millimeters per squared second (mm/s^2). For comparison purposes, linear acceleration is often stated relative to the acceleration due to gravity $g = 32.17 \, ft/s^2 = 386.4 \, in./s^2 = 9.81 \, m/s^2$. Thus, a $10g$ acceleration is equivalent to $3864 \, in./s^2$.

7.2.2 Constant Rectilinear Acceleration

Rewriting equation (7.3), the velocity change that occurs during a period of constant acceleration is expressed as

$$\Delta V = V_{final} - V_{initial} = A \, \Delta t \qquad (7.4)$$

Additionally, the corresponding displacement that occurs during a period of constant acceleration can be written as

$$\Delta R = \frac{1}{2} A \Delta t^2 + V_{initial} \, \Delta t \qquad (7.5)$$

Equations (7.4) and (7.5) can be combined to give

$$(V_{final})^2 = (V_{initial})^2 + 2A \, \Delta R \qquad (7.6)$$

Since rectilinear motion is along a straight line, the direction of the displacement, velocity, and acceleration (r, v, a) can be specified with an algebraic sign along a coordinate axis. Thus, equations (7.4), (7.5), and (7.6) can be written in terms of the vector magnitudes (r, v, a).

EXAMPLE PROBLEM 7.1

An express elevator used in tall buildings can reach a full speed of 15 mph in 3 s. Assuming that the elevator experiences constant acceleration, determine the acceleration and the displacement during the 3 s.

SOLUTION: 1. *Calculate Acceleration*

Assuming that the acceleration is constant, equation (7.3) can be accurately used. Because the elevator starts at rest, the velocity change is calculated as

$$\Delta V = (15 \, mph - 0) = 15 \, mph$$

$$= \left(\frac{15 \, miles}{hr}\right)\left(\frac{5280 \, ft}{1 \, mile}\right)\left(\frac{1 \, hr}{3600 \, s}\right) = 22 \, ft/s$$

Then, the acceleration is calculated as

$$A = \frac{\Delta V}{\Delta t} = \frac{(22 \, ft/s)}{3 \, s} = 7.3 \, ft/s^2 \uparrow$$

2. *Normalize the Acceleration with Respect to Gravity*

When people accelerate in an elevator, the acceleration is often "normalized" relative to the acceleration due to gravity. The standard acceleration due to gravity (g) on earth is $32.17 \, ft/s^2$ or $9.81 \, m/s^2$. Therefore, the acceleration of the elevator can be expressed as

$$A = 7.3 \, ft/s^2 \left(\frac{1 \, g}{32.2 \, ft/s^2}\right) = 0.22 \, g$$

3. *Calculate the Displacement during the 3-Second Interval*

The displacement can be determined from equation (7.5).

$$\Delta R = \frac{1}{2} a \Delta t^2 + v_{initial} \Delta t = \frac{1}{2} (7.3 \, ft/s^2)(3 \, s)^2 + (0)(3 \, s)$$

$$= 32.9 \, ft \uparrow \text{ (or roughly 3 floors)}$$

7.2.3 Acceleration and the Velocity Profile

As stated in equation (7.1), the instantaneous acceleration is the first derivative of the instantaneous velocity with respect to time. Occasionally, a closed-form equation for the instantaneous velocity of a point is available. In these cases, the derivative of the equation, evaluated at the specified time, will yield the instantaneous acceleration. More often, especially in programmable actuators used in automation, velocity profiles are specified as introduced in Chapter 6. Recall that the displacement for a certain time interval is the area under the v-t curve for that time interval. Conversely, the acceleration at a certain time is the slope of the v-t curve.

EXAMPLE PROBLEM 7.2

An automated assembly operation requires linear motion from a servo actuator. The total displacement must be 10 in. For design reasons, the maximum velocity must be limited to 2 in./s, and the maximum acceleration or deceleration should not exceed 4 in./s². Plot the velocity profile for this application.

SOLUTION:

1. **Determine the Motion Parameters during Speed-Up**

For the standard velocity profile for a servomotor, the speed-up portion of the motion is constant acceleration. Rewriting and substituting the magnitudes of the velocity v and acceleration a into equation (7.3) gives the time consumed during speed-up.

$$\Delta t = \frac{\Delta v}{a} = \frac{(2 - 0)\ \text{in./s}}{4\ \text{in./s}^2} = 0.5\ \text{s}$$

Equation (7.5) is used to calculate the magnitude of the displacement during speed-up.

$$\Delta R = \frac{1}{2}\,a\Delta t^2 + v_{\text{initial}}\,\Delta t$$

$$= \frac{1}{2}\left(4\ \text{in./s}^2\right)(.5\ \text{s})^2 + (0)(.5\ \text{s}) = 0.5\ \text{in.}$$

2. **Determine the Motion Parameters during Slow-Down**

For a standard velocity profile, the slow-down portion of the motion is constant acceleration. The time consumed during slow-down is

$$\Delta t = \frac{\Delta v}{a} = \frac{(0 - 2)\ \text{in./s}}{-4\ \text{in./s}^2} = 0.5\ \text{s}$$

The magnitude of the displacement during slow-down is

$$\Delta R = \frac{1}{2}\,a\Delta t^2 + v_{\text{initial}}\Delta t$$

$$= \frac{1}{2}\left(-4\ \text{in./s}^2\right)(.5\ \text{s})^2 + 22\ \text{in./s}(.5\ \text{s}) = 0.5\ \text{in.}$$

3. **Determine the Motion Parameters during Steady-State**

Because 0.5 in. of displacement occurs during speed-up and another 0.5 in. during shut-down, the remaining 9.0 in. of displacement is during constant velocity motion. Equation (6.2) is used to calculate the time consumed during constant velocity.

$$\Delta t = \frac{\Delta R}{v} = \frac{9\ \text{in.}}{2\ \text{in./s}} = 4.5\ \text{s}$$

4. **Determine the Motion Parameters during Steady-State**

Using the velocity and time information for this sequence, the velocity profile shown in Figure 7.1 is generated.

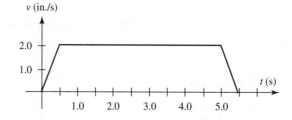

FIGURE 7.1 Velocity profile for Example Problem 7.2.

7.2.4 Linear Acceleration of a General Point

As earlier mentioned, the velocity of a point moving in a general fashion may change in two ways:

1. The magnitude of the velocity can change. This produces an acceleration acting along the path of motion, as presented in the previous section. This acceleration is termed *tangential acceleration*, $\mathbf{A}^t$.

2. The direction of the velocity vector can change over time. This occurs as the link, with which the point is associated, undergoes rotational motion. It produces a centrifugal acceleration that acts perpendicular to the direction of the path of motion. This acceleration is termed *normal acceleration*, $\mathbf{A}^n$.

Figure 7.2 illustrates point *A*, which is moving along a curved path. The tangential acceleration of point *A*, $\mathbf{A}_A^t$, is the linear acceleration along the direction of motion. Note that the vector points in the direction of motion because point *A* is accelerating. If point *A* were decelerating, the acceleration vector would point opposite to the direction of motion. Of course, the velocity vector always points in the direction of motion. Therefore, an accelerating point is associated with a tangential acceleration vector that is consistent with the velocity vector. Conversely, deceleration is associated with a tangential acceleration vector that opposes the velocity vector. The magnitude of tangential acceleration can be determined using equations (7.1), (7.2), or (7.3).

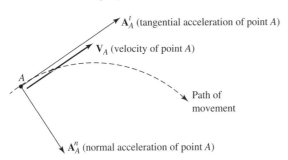

FIGURE 7.2 Acceleration of point *A*.

The normal acceleration of point *A*, $\mathbf{A}_A^n$, is a result of a change in the direction of the velocity vector. It acts along a line that is perpendicular to the direction of movement and toward the center of curvature of this path. Further details pertaining to tangential and normal accelerations are presented in Section 7.4.

7.3 ACCELERATION OF A LINK

Recall from Section 6.3 that any motion, however complex, can be viewed as a combination of a straight line movement and a rotational movement. Fully describing the motion of a link can consist of specifying the linear motion of one point and the rotational motion of the link about that point.

As with velocity, several points on a link can have different accelerations, yet the entire link has the same rotational acceleration.

7.3.1 Angular Acceleration

Angular acceleration, α, of a link is the angular velocity of that link per unit of time. Mathematically, angular acceleration of a link is described as

$$\alpha = \lim_{\Delta t \to 0} \frac{\Delta \omega}{\Delta t} = \frac{d\omega}{dt} \tag{7.7}$$

However, because

$$\omega = \frac{d\theta}{dt}$$

then

$$\alpha = \frac{d^2\theta}{dt^2} \tag{7.8}$$

For short time periods, or when the angular acceleration is assumed to be linear, the following relationship can be used:

$$\alpha \cong \frac{\Delta \omega}{\Delta t} \tag{7.9}$$

Similarly to the discussion in Section 7.2, the direction of angular acceleration is in the direction of motion when the angular velocity increases or the link accelerates. Conversely, the angular acceleration is in the opposite direction of motion when the angular velocity decreases, or the link is decreasing. In planar analyses, the direction should be described as either clockwise or counterclockwise.

Angular acceleration is expressed in the units of angular velocity (angle per time) divided by time, or angle per squared time. In both the U.S. Customary System and the International System, the common units used are degrees per squared second (deg/s^2), revolutions per squared second (rev/s^2), or the preferred unit of radians per squared second (rad/s^2).

7.3.2 Constant Angular Acceleration

Rewriting equation (7.7), the angular velocity change that occurs during a period of constant angular acceleration is expressed as

$$\Delta \omega = \omega_{\text{final}} - \omega_{\text{initial}} = \alpha \, \Delta t \tag{7.10}$$

Additionally, the corresponding angular displacement that occurs during a period of constant angular acceleration can be written as

$$\Delta \theta = \frac{1}{2} \alpha \Delta t^2 + \omega_{\text{initial}} \Delta t \tag{7.11}$$

Equations (7.10) and (7.11) can be combined to give

$$(\omega_{\text{final}})^2 = (\omega_{\text{initial}})^2 + 2\alpha \Delta \theta \tag{7.12}$$

EXAMPLE PROBLEM 7.3

An electric motor drives the grinding wheel clockwise, as shown in Figure 7.3. It will speed up to 1800 rpm in 2 s when the power is turned on. Assuming that this speed-up is at a constant rate, determine the angular acceleration of the grinding wheel. Also determine the number of revolutions that the wheel spins before it is at full speed.

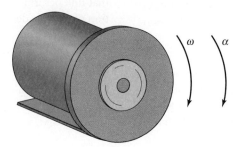

FIGURE 7.3 Grinding wheel for Example Problem 7.3.

SOLUTION: 1. *Calculate the Acceleration*

Since acceleration is typically specified in rad/s^2, convert the speed of the grinding wheel to rad/s with the following:

$$\Delta\omega = 1800 \text{ rpm}\left(\frac{2\pi \text{ rad}}{1 \text{ rev}}\right)\left(\frac{1 \text{ min}}{60 \text{ s}}\right) = 188.5 \text{ rad/s, cw}$$

With constant acceleration, equation (7.9) can be used, giving

$$\alpha = \frac{\Delta\omega}{\Delta t}$$

$$= \left(\frac{188.5 \text{ rad/s} - 0}{2 \text{ s}}\right) = 94.2 \text{ rad/s}^2, \text{ cw}$$

The direction of the acceleration is clockwise, which is in the direction of motion because the grinding wheel is speeding up.

2. *Calculate the Displacement during the 2-Second Interval*

The number of revolutions during this speed-up period can be determined through equation (7.11).

$$\Delta\theta = \frac{1}{2}\alpha\Delta t^2 + \omega_{\text{initial}}\Delta t = \frac{1}{2}\left(94.2 \text{ rad/s}^2\right)(2 \text{ s})^2 + (0)(2 \text{ s})$$

$$= 188.4 \text{ rad}\left(\frac{1 \text{ rev}}{2\pi \text{ rad}}\right) = 30.0 \text{ revolutions}$$

7.4 NORMAL AND TANGENTIAL ACCELERATION

As presented in Section 7.2.4, the velocity of a point moving in a general path can change in two independent ways. The magnitude or the direction of the velocity vector can change over time. Of course, acceleration is the time rate of velocity change. Thus, acceleration is commonly separated into two elements: normal and tangential components. The normal component is created as a result of a change in the direction of the velocity vector. The tangential component

is formed as a result of a change in the magnitude of the velocity vector.

7.4.1 Tangential Acceleration

For a point on a rotating link, little effort is required to determine the direction of these acceleration components. Recall that the instantaneous velocity of a point on a rotating link is perpendicular to a line that connects that point to the center of rotation. Any change in the magnitude of this velocity creates tangential acceleration, which is also perpendicular to the line that connects the point with the center of rotation. The magnitude of the tangential acceleration of point A on a rotating link 2 can be expressed as

$$a_A^t = \frac{dv_A}{dt} = \frac{d(\omega_2 r_{OA})}{dt} = r_{OA}\frac{d\omega_2}{dt} = r_{OA}\alpha_2 \quad (7.13)$$

It is extremely important to remember that the angular acceleration, α, in equation (7.13) must be expressed as units of radians per squared time. Radians per squared second is the most common unit. Similarly to the discussion in Section 7.2, tangential acceleration acts in the direction of motion when the velocity increases or the point accelerates. Conversely, tangential acceleration acts in the opposite direction of motion when the velocity decreases or the point decelerates.

7.4.2 Normal Acceleration

Any change in velocity direction creates normal acceleration, which is always directed toward the center of rotation. Figure 7.4a illustrates a link rotating at constant speed. The velocity of point A is shown slightly before and after the configuration under consideration, separated by a small angle $d\theta_2$. Because the link is rotating at constant speed, the magnitudes of $V_{A'}$ and $V_{A''}$ are equal. Thus, $V_{A'} = V_{A''}$.

Figure 7.4b shows a velocity polygon, vectorally solving for the change in velocity, dv. Notice that the change of the velocity vector, dv, is directed toward the center of link rotation. In fact, the normal acceleration will *always* be directed toward the center of link rotation. This is because, as the point rotates around a fixed pivot, the velocity vector will change along the curvature of motion. Thus, the normal

vector to this curvature will always be directed toward the fixed pivot.

In Figure 7.4a, because $\Delta\theta$ is small, the following relationship can be stated:

$$dv_A = v_A d\theta_2$$

Because acceleration is defined as the time rate of velocity change, both sides should be divided by time:

$$a_A^n = \frac{dv_A}{dt} = v_A\frac{d\theta_2}{dt} = v_A\omega_2$$

Using equation (6.6), the relationships between the magnitude of the linear velocity and angular velocity, the following equations for the magnitude of the normal acceleration of a point can be derived:

$$a_A^n = v_A\omega_2 = (\omega_2 r_{OA})\omega_2 = \omega_2^2 r_{OA} \quad (7.14)$$

$$a_A^n = v_A\omega_2 = v_A\left(\frac{v_A}{r_{OA}}\right) = \frac{v_A^2}{r_{OA}} \quad (7.15)$$

7.4.3 Total Acceleration

As previously mentioned, acceleration analysis is important because inertial forces result from accelerations. These loads must be determined to ensure that the machine is adequately designed to handle these dynamic loads. Inertial forces are proportional to the total acceleration of a body. The *total acceleration*, **A**, is the vector resultant of the tangential and normal components. Mathematically, it is expressed as

$$\mathbf{A}_A = \mathbf{A}_A^n +> \mathbf{A}_A^t \quad (7.16)$$

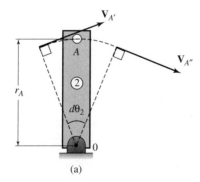

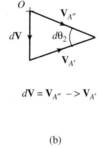

$$d\mathbf{V} = \mathbf{V}_{A''} -> \mathbf{V}_{A'}$$

(a) (b)

FIGURE 7.4 Normal acceleration.

EXAMPLE PROBLEM 7.4

The mechanism shown in Figure 7.5 is used in a distribution center to push boxes along a platform and to a loading area. The input link is driven by an electric motor, which, at the instant shown, has a velocity of 25 rad/s and accelerates at a rate of 500 rad/s². Knowing that the input link has a length of 250 mm, determine the instantaneous acceleration of the end of the input link in the position shown.

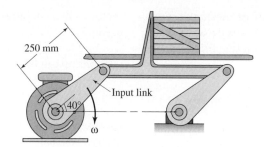

FIGURE 7.5 Transfer mechanism for Example Problem 7.4.

SOLUTION: 1.   ***Draw a Kinematic Diagram and Calculate Degrees of Freedom***

The kinematic diagram for the transfer mechanism is shown as Figure 7.6a. Notice that it is the familiar four-bar mechanism.

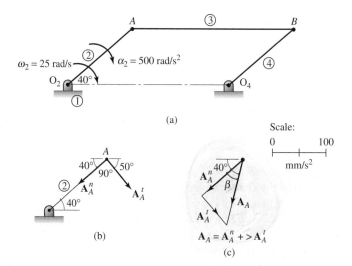

FIGURE 7.6 Diagrams for Example Problem 7.4.

2.   ***Determine the Tangential Acceleration of Point A***

Because the input link (link 2) is in pure rotation, the acceleration components of the end of the link can be readily obtained. Equation (7.13) can be used to determine the magnitude of the tangential acceleration.

$$a_A^t = r\alpha_2 = (250 \text{ mm})\left(500 \text{ rad/s}^2\right) = 125{,}000 \text{ mm/s}^2 = 125.0 \text{ m/s}^2$$

Because the link is accelerating, the direction of the vector is in the direction of the motion at the end of the link, which is perpendicular to the link itself. Thus, the tangential acceleration is

$$\mathbf{A}_A^t = 125.0 \text{ m/s}^2 \, \diagdown\!\!50°$$

3.   ***Determine the Normal Acceleration of Point A***

Equation (7.14) can be used to determine the magnitude of the normal acceleration.

$$a_A^n = r_{O_2A}\omega_2^2 = (250 \text{ mm})\left(25 \text{ rad/s}\right)^2 = 156{,}250 \text{ mm/s}^2 = 156.25 \text{ m/s}^2$$

Normal acceleration always occurs toward the center of rotation. Thus, normal acceleration is calculated as

$$\mathbf{A}_A^n = 156.25 \text{ m/s}^2 \, \diagdown\!\!40°$$

The components of the acceleration are shown in Figure 7.6b.

4. **Determine the Total Acceleration of Point A**

   The total acceleration can be found from analytical methods presented in Chapter 3. A sketch of the vector addition is shown in Figure7.6c. Because the normal and tangential components are orthogonal, the magnitude of the total acceleration is computed as

   $$a_A = \sqrt{(a_A^n)^2 + (a_A^t)^2}$$
   $$= \sqrt{(125.0 \text{ m/s}^2)^2 + (2156.25 \text{ m/s}^2)^2} = 200.10 \text{ m/s}^2$$

   The angle of the total acceleration vector from the normal component can be calculated as

   $$\beta = \tan^{-1}\left(\frac{a_A^t}{a_A^n}\right) = \tan^{-1}\left(\frac{125.0 \text{ m/s}^2}{156.25 \text{ m/s}^2}\right) = 38.7°$$

   The direction of the total acceleration vector from the horizontal axis is

   $$40.0° + 38.7° = 78.7°$$

   Formally, the total acceleration can then be written as

   $$\mathbf{A}_A = 200.10 \text{ m/s}^2 \quad \overline{78.7°}$$

   The total acceleration can also be determined through a graphical procedure using either CAD or traditional drawing techniques, as explained in Chapter 3.

7.5 RELATIVE MOTION

As discussed in detail in Chapter 6, the difference between the motion of two points is termed *relative motion. Relative velocity* was defined as the velocity of one object as observed from another reference object that is also moving. Likewise, *relative acceleration* is the acceleration of one object as observed from another reference object that is also moving.

7.5.1 Relative Acceleration

As with velocity, the following notation is used to distinguish between absolute and relative accelerations:

$\mathbf{A}_A$ = absolute acceleration (total) of point A

$\mathbf{A}_B$ = absolute acceleration (total) of point B

$\mathbf{A}_{B/A}$ = relative acceleration (total) of point B with respect to A

  = acceleration (total) of point B "as observed" from point A

From equation (6.10), the relationship between absolute velocity and relative velocity can be written as

$$\mathbf{V}_B = \mathbf{V}_A +> \mathbf{V}_{B/A}$$

Taking the time derivative of the relative velocity equation yields the relative acceleration equation. This can be written mathematically as

$$\mathbf{A}_B = \mathbf{A}_A +> \mathbf{A}_{B/A} \tag{7.17}$$

Typically, it is more convenient to separate the total accelerations in equation (7.17) into normal and tangential components. Thus, each acceleration is separated into its two components, yielding the following:

$$\mathbf{A}_B^n +> \mathbf{A}_B^t = \mathbf{A}_A^n +> \mathbf{A}_A^t +> \mathbf{A}_{B/A}^n +> \mathbf{A}_{B/A}^t \tag{7.18}$$

Note that equations (7.17) and (7.18) are vector equations and the techniques discussed in Chapter 3 must be used in dealing with these equations.

EXAMPLE PROBLEM 7.5

Figure 7.7 shows a power hacksaw. At this instant, the electric motor rotates counterclockwise and drives the free end of the motor crank (point B) at a velocity of 12 in./s. Additionally, the crank is accelerating at a rate of 37 rad/s². The top portion of the hacksaw is moving toward the left with a velocity of 9.8 in./s and is accelerating at a rate of 82 in./s². Determine the relative acceleration of point C with respect to point B.

SOLUTION: 1. **Draw a Kinematic Diagram and Identify the Degrees of Freedom**

   Figure 7.8a shows the kinematic diagram of the power hacksaw. Notice that it is the familiar slider-crank mechanism with one degree of freedom.

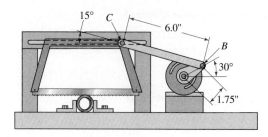

FIGURE 7.7 Power saw for Example Problem 7.5.

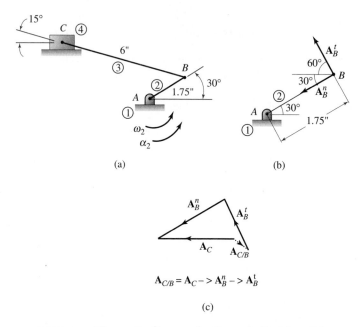

(a)   (b)

$$\mathbf{A}_{C/B} = \mathbf{A}_C \text{ --> } \mathbf{A}_B^n \text{ --> } \mathbf{A}_B^t$$

(c)

FIGURE 7.8 Kinematic diagram for Example Problem 7.5.

2. **Determine the Tangential Acceleration of Point B**

    From the kinematic diagram, it should be apparent that point *B* travels up and to the left as link 2 rotates counterclockwise. Because the motor crank (link 2) is in pure rotation, the components of the acceleration at the end of the link can be readily obtained. Equation (7.13) can be used to determine the magnitude of the tangential acceleration.

    $$a_B^t = r_{AB}\alpha_2 = (1.75 \text{ in.})\left(37 \text{ rad/s}^2\right) = 64.75 \text{ in./s}^2$$

    Because the link accelerates, the direction of the vector is in the direction of the motion at the end of the link. Thus, the tangential acceleration is calculated as

    $$\mathbf{A}_B^t = 64.75 \text{ in./s}^2 \quad \underline{60°}$$

3. **Determine the Normal Acceleration of Point B**

    Equation (7.15) can be used to determine the magnitude of the normal acceleration.

    $$a_B^n = \frac{v_B^2}{r_{AB}} = \frac{(12 \text{ in./s})^2}{1.75 \text{ in.}} = 82.29 \text{ in./s}^2$$

    Normal acceleration is always directed toward the center of rotation. Thus, normal acceleration is

    $$\mathbf{A}_B^n = 82.29 \text{ in./s}^2 \quad \underline{30°}$$

    Link 2 is isolated and the components of this acceleration are shown in Figure 7.8b.

4. **Specify the Acceleration of Point C**

    Point *C* is constrained to linear motion. Therefore, point *C* does not experience a normal acceleration. The total acceleration is given in the problem statement as

    $$\mathbf{A}_C = 82 \text{ in./s}^2 \ \leftarrow$$

5. **Construct the Velocity Polygon for the Acceleration of C Relative to B**

   To determine the relative acceleration, equation (7.16) can be written in terms of points B and C and rearranged as

   $$\mathbf{A}_{C/B} = \mathbf{A}_C {-}{>} \mathbf{A}_B$$

   Because two acceleration components of point B exist, the equation is written as

   $$\mathbf{A}_{C/B} = \mathbf{A}_C {-}{>} (\mathbf{A}_B^n {+}{>} \mathbf{A}_B^t) = \mathbf{A}_C {-}{>} \mathbf{A}_B^n {-}{>} \mathbf{A}_B^t$$

   A vector polygon is formed from this equation (Figure 7.8c). The unknown vector can be determined using the methods presented in Chapter 3. Either a graphical or analytical solution can be used to determine vector $\mathbf{A}_{C/B}$.

6. **Solve for the Unknown Vector Magnitudes**

   Arbitrarily using an analytical method, the acceleration $\mathbf{A}_{C/B}$ can be found by separating the vectors into horizontal and vertical components. See Table 7.1.

TABLE 7.1 Horizontal and Vertical Vector Components for Acceleration $\mathbf{A}_{C/B}$

Vector	Reference Angle (θ_x)	Horizontal Component $a_h = a \cos \theta_x$	Vertical Component $a_v = a \sin \theta_x$
$\mathbf{A}_c$	180°	-82.00	0
$\mathbf{A}_B^n$	210°	-71.26	-41.15
$\mathbf{A}_B^t$	120°	-32.83	56.08

Separate algebraic equations can be written for the horizontal and vertical components as follows:

$$\mathbf{A}_{C/B} = \mathbf{A}_C {-}{>} \mathbf{A}_B^n {-}{>} \mathbf{A}_B^t$$

<u>horizontal comp.:</u> $\mathbf{A}_{C/B}^h = (-82.0) - (-71.27) - (-32.38)$

$$= +21.35 = 21.35 \text{ in./s}^2$$

<u>vertical comp.:</u> $\mathbf{A}_{C/B}^v = (0) - (-41.15) - (+56.08) = -14.93 \text{ in./s}^2$

The magnitude of the acceleration can be found by

$$\mathbf{A}_{C/B} = \sqrt{\left(a_{C/B}^h\right)^2 + \left(a_{C/B}^v\right)^2}$$

$$= \sqrt{(21.35)^2 + (-14.93)^2} = 26.05 \text{ in./s}^2$$

The direction of the vector can be determined by

$$\theta_x = \tan^{-1}\left[\frac{a_{C/B}^h}{a_{C/B}^v}\right] = \tan^{-1}\left[\frac{-14.93 \text{ in./s}^2}{21.35 \text{ in./s}^2}\right] = -35° = \diagdown 35°$$

Finally, the relative acceleration of C with respect to B is

$$\mathbf{A}_{C/B} = 26.05 \text{ in./s}^2 \diagdown 35°$$

7.5.2 Components of Relative Acceleration

The acceleration of points on a mechanism can be much more easily analyzed when separated into normal and tangential components. For links that are attached directly to the frame, the direction of the acceleration components is obvious, as described in the previous section. The normal component is always directed to the center of rotation, and the tangential component is perpendicular to the normal component and in the direction that is consistent

with either the acceleration or deceleration of the point. Recall that tangential acceleration is in the direction of motion when the point accelerates. Conversely, tangential acceleration is opposite to the direction of motion when the point decelerates.

For points that are on the same link, a link that is not directly attached to the frame, the analysis focuses on the relative accelerations of these points. Figure 7.9 shows such a link that is not directly attached to the frame, typically called

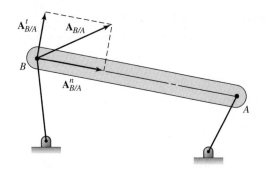

FIGURE 7.9 Relative normal and tangential accelerations.

a floating link. The relative acceleration between two points that reside on that link is shown. Notice that the normal and tangential components of this acceleration are also shown and are directed along the link (normal) and perpendicular to the link (tangential). Reiterating, the relative acceleration of two points is the acceleration of one point as seen from the other reference point.

As with velocity analysis, relative motion consists of pure relative rotation of the observed point about the reference point. In other terms, the relative motion of B with respect to A is visualized as if point B were rotating around point A. Thus, a normal component of relative acceleration is directed toward the center of relative rotation, or the reference point. The tangential relative acceleration is directed perpendicular to the normal relative acceleration. The magnitudes of these components are computed in a similar fashion to the absolute acceleration of points rotating around fixed points.

$$a_{B/A}^t = \frac{dv_{B/A}}{dt} = \frac{d(\omega_3 r_{BA})}{dt} = r_{BA}\alpha_3 \qquad (7.19)$$

$$a_{B/A}^n = r_{BA}\omega^2 = \frac{(v_{B/A})^2}{r_{BA}} \qquad (7.20)$$

The direction of the relative tangential acceleration is consistent with the angular acceleration of the floating link, and vice versa. Referring to Figure 7.9, the relative tangential acceleration shows the tangential acceleration of point B as it rotates around point A directed upward and toward the right, which infers a clockwise angular acceleration of link 3.

EXAMPLE PROBLEM 7.6

For the power hacksaw in Example Problem 7.5, determine the angular acceleration of the 6-in. connecting link (link 3).

SOLUTION: 1. ***Identify the Relevant Link Geometry***

The relative acceleration of C with respect to B was determined as

$$\mathbf{A}_{C/B} = 26.05 \text{ in./s}^2 \enspace \diagdown\!\underline{35°}$$

Also note from Figure 7.7 that the connecting link is inclined at a 15° angle. Using this data, the total relative acceleration can be resolved into normal and tangential components. These components are shown in Figure 7.10.

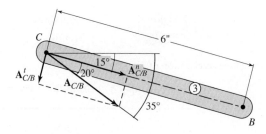

FIGURE 7.10 Relative accelerations for Example Problem 7.6.

2. ***Resolve the Total Relative Acceleration into Normal and Tangential Components***

Figure 7.10 illustrates that 20° (35°–15°) separates the total relative acceleration vector and the normal component. Thus, the magnitudes of the relative acceleration components can be analytically determined from the following trigonometric relationships:

$$\mathbf{A}_{C/B}^t = a_{C/B}(\sin 20°) = 26.05 \text{ in./s}^2 (\sin 20°) = 8.91 \text{ in./s}^2 \enspace \diagup\!\underline{15°}$$

$$\mathbf{A}_{C/B}^n = a_{C/B}(\cos 20°) = 26.05 \text{ in/.s}^2 (\cos 20°) = 24.48 \text{ in./s}^2 \enspace \underline{75°}\!\diagdown$$

3. *Calculate the Rotational Acceleration of Link 3*

From Figure 7.10, the tangential acceleration of point C with respect to B is downward and to the right. This implies that the angular acceleration of link 3 is counterclockwise. The magnitude can be determined as

$$\alpha_3 = \frac{a_{C/B}^t}{r_{CB}} = \frac{8.91 \text{ in./s}^2}{6 \text{ in.}} = 1.49 \text{ rad/s}^2$$

Therefore, the angular acceleration of the connecting link is determined by

$$\alpha_3 = 1.49 \text{ rad/s}^2, \text{ counterclockwise}$$

7.6 RELATIVE ACCELERATION ANALYSIS: GRAPHICAL METHOD

Acceleration analysis is usually employed to determine the acceleration of several points on a mechanism at a single configuration. It must be understood that the results of this analysis yield instantaneous motion characteristics. As the mechanism moves, even an infinitesimal amount, the motion characteristics change. Nonetheless, the instantaneous characteristics are needed, particularly the extreme values. It was emphasized earlier that accelerations impose inertial forces onto the links of a mechanism. The resulting stresses must be fully understood to ensure safe operation of a machine.

The strategy for determining the acceleration of a point involves knowing the acceleration of another point on that same link. In addition, the velocity of the desired point and the relative velocity between the two points must be known. This information can complete a relative velocity analysis as described in Chapter 6.

Analysis can proceed throughout a mechanism by using points that are common to two links. For example, a point that occurs on a joint is common to two links. Therefore, determining the acceleration of this point enables one to subsequently determine the acceleration of another point on either link. In this manner, the acceleration of any point on a mechanism can be determined by working outward from the input link.

Recall from equation (7.18) that the relative acceleration equation can be expanded to include the normal and tangential components.

$$\mathbf{A}_B^n +> \mathbf{A}_B^t = \mathbf{A}_A^n +> \mathbf{A}_A^t +> \mathbf{A}_{B/A}^n +> \mathbf{A}_{B/A}^t$$

Assume that the acceleration of point B needs to be determined and the acceleration of point A is already known. Also assume that a full velocity analysis, involving the two points, has already been conducted. In a typical situation, the directions of all six components are known. All normal components are directed toward the center of relative rotation. All tangential components are perpendicular to the normal components. In addition, the magnitudes of all the normal acceleration vectors can be found from equation (7.14) or (7.15). Of course, the magnitude of the tangential acceleration of the known point (point A) is also established. Therefore, the vector analysis only needs to determine the magnitude of the tangential component of the point desired and the magnitude of the relative tangential component.

Relative acceleration analysis forms a vector problem identical to the general problems presented in Sections 3.18 and 3.19. Both graphical and analytical solutions are feasible, as seen throughout Chapter 3. In many problems, the magnitude of certain terms may be zero, eliminating some of the six vector components in equation (7.18). For example, when the known point is at a joint that is common to a constant angular velocity link, the point has no tangential acceleration. Another example occurs when a point is common to a link that is restricted to linear motion. The velocity of the point does not change direction and the point has no normal acceleration.

As in velocity analysis, the graphical solution of acceleration polygons can be completed using manual drawing techniques or on a CAD system. The logic is identical; however, the CAD solution is not susceptible to limitations of drafting accuracy. Regardless of the method being practiced, the underlying concepts of graphical position analysis can be further illustrated and expanded through the following example problems.

EXAMPLE PROBLEM 7.7

The mechanism shown in Figure 7.11 is designed to move parts along a conveyor tray and then rotate and lower those parts to another conveyor. The driving wheel rotates with a constant angular velocity of 12 rpm. Determine the angular acceleration of the rocker arm that rotates and lowers the parts.

SOLUTION: 1. *Draw the Kinematic Diagram and Identify the Degrees of Freedom*

The portion of the mechanism that is under consideration includes the driving wheel, the follower arm, and the link that connects the two. Notice that, once again, this is the common four-bar mechanism having one degree of freedom. A scaled, kinematic diagram is shown in Figure 7.12a.

2. ***Decide on a Method to Achieve the Desired Acceleration***

The angular acceleration of the rocker (link 4) can be determined from the tangential component of the acceleration of point *C*. Thus, the crux of the problem is to determine the acceleration of point *C*. In turn, the acceleration of point *C*, which also resides on link 3, can be determined from knowing the acceleration of point *B*. Point *B* is positioned on both links 2 and 3. Therefore, the acceleration of point *B* can be determined from knowing the motion of the input link, link 2.

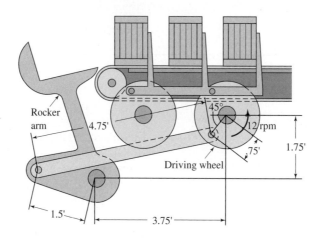

FIGURE 7.11 Mechanism for Example Problem 7.7.

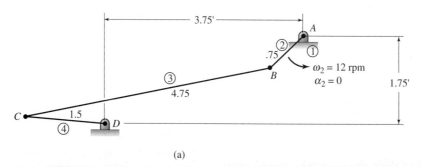

(a)

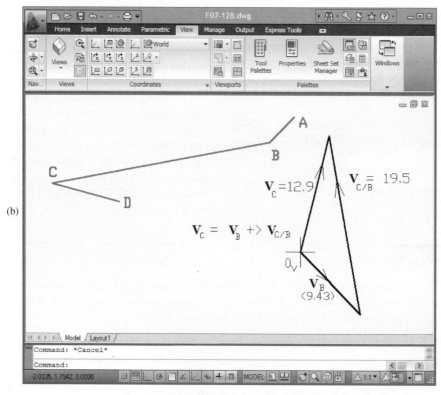

(b)

FIGURE 7.12 Diagrams for Example Problem 7.7.

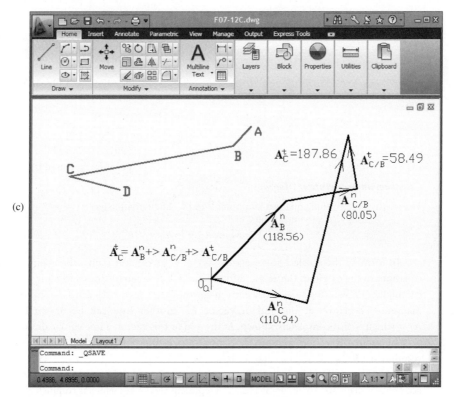

FIGURE 7.12 Continued

3. **Determine the Velocity of Points B and C**

The first step is to construct a velocity diagram that includes points B and C. Calculating the magnitude of the velocity of point B can be accomplished with the following:

$$\omega_2(\text{rad/s}) = \frac{\pi}{30}(\omega\text{ rpm}) = \frac{\pi}{30}(12\text{ rpm}) = 1.26\text{ rad/s, counterclockwise}$$

$$\mathbf{V}_B = \omega_2\mathbf{r}_{AB} = (1.26\text{ rad/s})(0.75\text{ ft}) = .943\text{ ft/s} \quad \angle 45°$$

The direction of $\mathbf{V}_B$ is perpendicular to link 2 and in the direction consistent with ω_2, down and to the right. Using CAD, a vector can be drawn to scale, from the velocity diagram origin, to represent this velocity.

The relative velocity equation for points B and C can be written as

$$\mathbf{V}_C = \mathbf{V}_B +> \mathbf{V}_{C/B}$$

Thus, at the origin of the velocity diagram, a line can be drawn to represent the direction of vector $\mathbf{V}_C$. This is perpendicular to link 4 because point C resides on a link that pivots about a fixed center. At the end of the vector $\mathbf{V}_B$, a line can also be drawn to represent the direction of $\mathbf{V}_{C/B}$. As with all relative velocity vectors, the direction is perpendicular to the line that connects points C and B. The intersection of the $\mathbf{V}_C$ and $\mathbf{V}_{C/B}$ direction lines determines the magnitudes of both vectors. The completed velocity diagram is shown in Figure 7.12b.

Scaling the vectors from the diagram yields the following:

$$\mathbf{V}_C = 1.290\text{ ft/s} \quad \angle 76°$$

$$\mathbf{V}_{C/B} = 1.950\text{ ft/s} \quad \angle 80°$$

4. **Calculate Acceleration Components**

The next step is to construct an acceleration diagram that includes points B and C. Calculating the magnitudes of the known accelerations is accomplished by

$$\mathbf{A}_B^n = \frac{(\mathbf{V}_B)^2}{r_{AB}} = \frac{(0.943\text{ ft/s})^2}{0.75\text{ ft}} = 1.186\text{ ft/s}^2 \quad \angle 45°$$

(directed toward the center of rotation, point A)

$$a_B^t = \alpha_2 r_{AB} = (0)\,(0.75\text{ ft}) = 0\text{ ft/s}^2$$

$$\mathbf{A}_{C/B}^n = \frac{(\mathbf{V}_{C/B})^2}{r_{CB}} = \frac{(1.950\text{ ft/s})^2}{4.75\text{ ft}} = .800\text{ ft/s}^2 \quad \underline{/10°}$$

(directed from C toward B, measured
from CAD)

$$\mathbf{A}_C^n = \frac{(\mathbf{V}_C)^2}{r_{CD}} = \frac{(1.290\text{ ft/s})^2}{1.5\text{ ft}} = 1.109\text{ ft/s}^2 \quad \underline{\searrow 14°}$$

(directed toward the center of rotation,
point D, measured from CAD)

5. **Construct an Acceleration Diagram**

The relative acceleration equation for points B and C can be written as

$$\mathbf{A}_C^n +> \mathbf{A}_C^t = \mathbf{A}_B^n +> \mathbf{A}_B^t +> \mathbf{A}_{C/B}^n +> \mathbf{A}_{C/B}^t$$

In forming the acceleration diagram, vector construction arbitrarily begins on the right side of the equation. At the origin of the acceleration diagram, a line can be drawn to represent the vector $\mathbf{A}_B^n$ which is completely known. Because it has zero magnitude, the vector $\mathbf{A}_B^t$ can be eliminated in the acceleration diagram. Therefore, at the end of vector $\mathbf{A}_B^n$ another line can be drawn to represent the vector $\mathbf{A}_{C/B}^n$, which is also completely known. At the end of this vector, a line can be drawn to represent the direction of vector $\mathbf{A}_{C/B}^t$. The magnitude is not known, but the direction is perpendicular to the normal component, $\mathbf{A}_{C/B}^n$.

Focusing on the left side of the equation, a new series of vectors will begin from the origin of the acceleration diagram. A line can be drawn to represent vector $\mathbf{A}_C^n$, which is completely known. At the end of this vector, a line can be drawn to represent the direction of vector $\mathbf{A}_C^t$; however, the vector magnitude is unknown. The line is directed perpendicular to the normal component, $\mathbf{A}_C^n$. Finally, the intersection of the $\mathbf{A}_C^t$ and $\mathbf{A}_{C/B}^t$ direction lines determines the magnitudes of both vectors. The completed acceleration diagram is shown in Figure 7.12c.

6. **Measure the Desired Acceleration Components**

Scaling the vector magnitudes from the diagram yields the following:

$$\mathbf{A}_C^t = 1.879\text{ ft/s}^2 \quad \underline{/76°}$$

$$\mathbf{A}_{C/B}^t = .585\text{ ft/s}^2 \quad \underline{80°\searrow}$$

Notice that the tangential acceleration of point C is in the same direction as the velocity. This indicates that point C is accelerating (speeding up), not decelerating.

7. **Calculate the Desired Angular Acceleration**

Finally, the angular acceleration of link 4 can be determined. By observing the direction of the tangential component of the acceleration of point C (up and to the right), it is obvious that link 4 accelerates in a clockwise direction. The magnitude of this angular acceleration is computed as

$$\alpha_4 = \frac{a_c^t}{r_{CD}} = \frac{(1.879\text{ ft/s}^2)}{1.5\text{ ft}} = 1.25\text{ rad/s}^2$$

Therefore, the angular acceleration of the rocker arm is

$$\alpha_4 = 1.25\text{ rad/s}^2,\ cw$$

EXAMPLE PROBLEM 7.8

The mechanism shown in Figure 7.13 is a common punch press designed to perform successive stamping operations. The machine has just been powered and at the instant shown is coming up to full speed. The driveshaft rotates clockwise with an angular velocity of 72 rad/s and accelerates at a rate of 250 rad/s^2. At the instant shown, determine the acceleration of the stamping die, which will strike the workpiece.

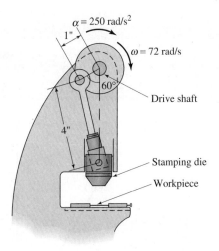

FIGURE 7.13 Mechanism for Example Problem 7.8.

SOLUTION: 1. **Draw the Kinematic Diagram and Identify the Degrees of Freedom**

The portion of the mechanism that is under consideration includes the driving wheel, the stamping die, and the link that connects the two. Notice that this is the common slider-crank mechanism, having a single degree of freedom. A scaled kinematic diagram is shown in Figure 7.14a.

2. **Decide on a Method to Achieve the Desired Acceleration**

The acceleration of the die (link 4) is strictly translational motion and is identical to the motion of point A. The acceleration of point A, which also resides on link 3, can be determined from knowing the acceleration of point B. Point B is positioned on both links 2 and 3. Therefore, the acceleration of point B can be determined from knowing the motion of the input link, link 2.

3. **Determine the Velocity of Points A and B**

Calculating the magnitude of the velocity of point B is as follows:

$$\mathbf{V}_B = \omega_2 r_{AB} = (72 \text{ rad/s})(1.0 \text{ in.}) = 72 \text{ in./s} \quad \underline{60°} \nwarrow$$

The direction of $\mathbf{V}_B$ is perpendicular to link 2 and consistent with the direction of ω_2, up and to the left. Using CAD, a vector can be drawn to scale, from the velocity diagram origin, to represent this velocity.

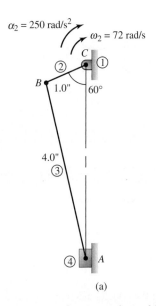

(a)

FIGURE 7.14 Diagrams for Example Problem 7.8.

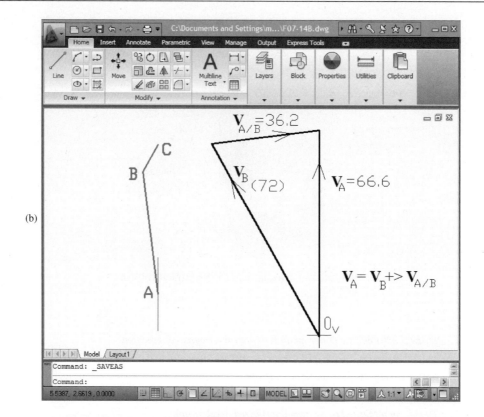

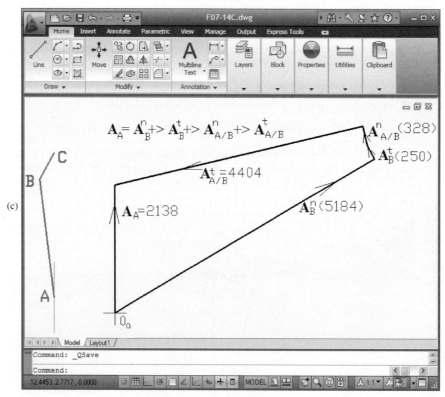

FIGURE 7.14 Continued

The next step is to construct a velocity diagram that includes points A and B. The relative velocity equation for points A and B can be written as

$$\mathbf{V}_A = \mathbf{V}_B +> \mathbf{V}_{A/B}.$$

Thus, at the origin of the velocity diagram, a line can be drawn to represent the direction of vector $\mathbf{V}_A$. This is parallel to the sliding surface because link 4 is constrained to vertical sliding motion. At the end of the vector

V_B, a line is drawn to represent the direction of $V_{A/B}$. As with all relative velocity vectors between two points on the same line, the direction is perpendicular to the line that connects points A and B. The intersection of the V_A and $V_{A/B}$ direction lines determines the magnitudes of both vectors. The completed velocity diagram is shown in Figure 7.14b.

Scaling the vector magnitudes from the diagram is determined as follows:

$$V_A = 70.3 \text{ in./s} \uparrow$$

$$V_{A/B} = 36.8 \text{ in./s} \quad \diagup 13°$$

4. ***Calculate the Acceleration Components***

The next step is to construct an acceleration diagram that includes points A and B. Calculating the magnitudes of the known accelerations is accomplished by the equations:

$$A_B^n = \frac{(v_B)^2}{r_{BC}} = \frac{(72 \text{ in./s})^2}{1.0 \text{ in.}} = 5184 \text{ in./s}^2 \quad \diagup 30°$$

(directed toward the center of rotation,
point C)

$$A_B^t = \alpha_2 r_{AB} = (250 \text{ rad/s}^2)(1.0 \text{ in.}) = 250 \text{ in./s}^2 \quad \underline{60°} \diagdown$$

(directed perpendicular to BC,
in the direction of rotational
acceleration)

$$A_{A/B}^n = \frac{\left(v_{A/B}\right)^2}{r_{AB}} = \frac{\left(36.8 \text{ in./s}\right)^2}{4.0 \text{ in.}} = 338 \text{ in./s}^2 \quad \underline{77°} \diagdown$$

(directed from A toward B,
measured from CAD)

Note that point A does not have a normal acceleration because the motion is strictly translational.

5. ***Construct an Acceleration Diagram***

The relative acceleration equation for points A and B can be written as

$$A_A^n +> A_A^t = A_B^n +> A_B^t +> A_{A/B}^n +> A_{A/B}^t$$

In forming the acceleration diagram, vector construction will arbitrarily start on the right side of the equation. At the origin of the acceleration diagram, a line can be drawn to represent the vector A_B^n which is known. At the end of A_B^n, a line can be drawn to represent vector A_B^t which is also known. At the end of vector A_B^t another line can be drawn to represent vector $A_{A/B}^n$ which is also known. At the end of this vector, a line can be drawn to represent the direction of vector $A_{A/B}^t$. This is perpendicular to the normal component, $A_{A/B}^n$, but has an unknown magnitude.

Focusing on the left side of the equation, a new series of vectors will begin from the origin of the acceleration diagram. The vector A_A^n has zero magnitude and is ignored. A line can be drawn to represent the direction of vector A_A^t; however, the magnitude is unknown. The line is directed parallel to the sliding motion of link 4. Finally, the intersection of the A_A^t and $A_{A/B}^t$ direction lines determines the magnitudes of both vectors. The completed acceleration diagram is shown in Figure 7.14c.

6. ***Measure the Desired Acceleration Components***

Scaling the vector magnitudes from the diagram is done with the following:

$$A_{A/B}^t = 4404 \text{ in./s}^2 \quad \overline{13°} \diagdown$$

$$A_A^t = 2138 \text{ in./s}^2 \uparrow$$

Thus the total acceleration of point A is

$$A_A = A_A^t = 2138 \text{ in./s}^2 = 178 \text{ ft/s}^2 = 5.53 \text{ g} \uparrow$$

Notice that the tangential acceleration of point A is in the same direction as the velocity. This indicates that point A is accelerating (speeding up), not decelerating.

7.7 RELATIVE ACCELERATION ANALYSIS: ANALYTICAL METHOD

The strategy for analytically determining the acceleration of various points on a mechanism is identical to the method outlined in the previous section. The difference is that vector polygons only need to be roughly sketched.

The magnitude and angles can be solved using the analytical methods introduced in Chapter 3 and incorporated in Chapter 6 and earlier sections in this chapter. The most effective manner of presenting the analytical method of acceleration analysis is through an example problem.

EXAMPLE PROBLEM 7.9

The mechanism shown in Figure 7.15 is used to feed cartons to a labeling machine and, at the same time, to prevent the stored cartons from moving down. At full speed, the driveshaft rotates clockwise with an angular velocity of 200 rpm. At the instant shown, determine the acceleration of the ram and the angular acceleration of the connecting rod.

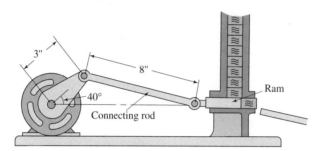

FIGURE 7.15 Mechanism for Example Problem 7.9.

SOLUTION:

1. **Draw a Kinematic Diagram**

   The portion of the mechanism that is under consideration includes the drive crank, the pusher ram, and the link that connects the two. Once again, notice that this is the common in-line, slider-crank mechanism. A kinematic diagram is shown in Figure 7.16a.

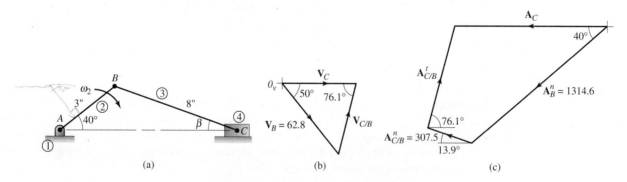

FIGURE 7.16 Diagrams for Example Problem 7.9.

2. **Decide on a Method to Achieve the Desired Acceleration**

   As in Example Problem 7.8, the acceleration of the ram (link 4) is strictly translational motion and is identical to the motion of point C. The acceleration of point C, which also resides on link 3, can be determined from knowing the acceleration of point B. Point B is positioned on both links 2 and 3. Therefore, the acceleration of point B can be determined from knowing the motion of the input link, link 2.

3. **Analyze the Mechanism Geometry**

   The angle between link 3 and the horizontal sliding surface of link 4, β in Figure 7.16a, can be determined from the law of sines.

$$\beta = \sin^{-1}\left(\frac{r_{AB}\sin 40°}{r_{BC}}\right) = \sin^{-1}\left(\frac{(3 \text{ in.})\sin 40°}{(8 \text{ in.})}\right) = 13.9°$$

4. **Determine the Velocity of Points B *and* C**

   Calculate the magnitude of the velocity of point B using the following equation:

   $$\omega_2 = \frac{\pi}{30}(200 \text{ rpm}) = 20.9 \text{ rad/s}$$

   $$\mathbf{V}_B = \omega_2 r_{AB} = (20.9 \text{ rad/s})(3.0 \text{ in.}) = 62.8 \text{ in./s} \quad \diagdown 50°$$

   The direction of $\mathbf{V}_B$ is perpendicular to link 2 and consistent with the direction of ω_2, down and to the right. The velocity of point C is parallel to the horizontal sliding surface, and the velocity of C with respect to B is perpendicular to the link that connects points B and C. Calculating this angle,

   $$90° + (-\beta) = 90° + (-13.9°) = 76.1°$$

   By understanding the directions of the vectors of interest, a velocity polygon can be assembled (Figure 7.16b). The magnitude of the third angle in the velocity polygon can be determined because the sum of all angles in a triangle is 180°.

   $$180° - (50° + 76.1°) = 53.9°$$

   The magnitudes of the velocities can be found from the law of sines.

   $$\mathbf{V}_C = \mathbf{V}_B\left(\frac{\sin 53.9°}{\sin 76.1°}\right) = 62.8 \text{ in./s}\left(\frac{\sin 53.9°}{\sin 76.1°}\right) = 52.3 \text{ in./s} \rightarrow$$

   Solve for the unknown velocities with the following:

   $$\mathbf{V}_{C/B} = \mathbf{V}_B\left(\frac{\sin 50°}{\sin 76.1°}\right) = 62.8 \text{ in./s}\left(\frac{\sin 50°}{\sin 76.1°}\right) = 49.6 \text{ in./s} \quad \diagup 76.1°$$

5. **Calculate Acceleration Components**

   The next step is to construct an acceleration diagram that includes points B and C. Calculate the magnitudes of the known accelerations using the following equations:

   $$A_B^n = \frac{(V_B)^2}{r_{AB}} = \frac{(62.8 \text{ in./s})^2}{3.0 \text{ in.}} = 1314.6 \text{ in./s}^2 \quad \diagdown 40°$$

   (directed toward the center of rotation, point A)

   $$A_B^t = \alpha_2 r_{AB} = (0 \text{ rad/s}^2)(3.0 \text{ in.}) = 0$$

   (because the driving link is rotating at constant velocity)

   $$A_{C/B}^n = \frac{(V_{C/B})^2}{r_{BC}} = \frac{(49.6 \text{ in./s})^2}{8.0 \text{ in.}} = 307.5 \text{ in./s}^2 \quad \diagdown 13.9°$$

   (directed from C toward B)

   Note that point C does not have a normal acceleration because the motion is strictly translational.

6. **Using Vector Methods, Solve the Relative Acceleration Equation**

   The relative acceleration equation for points B and C can be written as

   $$\mathbf{A}_C^n +> \mathbf{A}_C^t = \mathbf{A}_B^n +> \mathbf{A}_B^t +> \mathbf{A}_{C/B}^n +> \mathbf{A}_{C/B}^t$$

   In forming an acceleration diagram, vector placement arbitrarily starts on the right side of the equation. At the origin of the acceleration diagram, vector $\mathbf{A}_B^n$, which is completely known, is placed. Because no tangential component of the acceleration of point B exists, $\mathbf{A}_B^t$ is ignored. Vector $\mathbf{A}_{C/B}^n$, which is also completely known, is placed at the end of $\mathbf{A}_B^n$. At the end of $\mathbf{A}_{C/B}^n$, the vector $\mathbf{A}_{C/B}^t$ is placed; however, only the direction of this vector is known. It is directed perpendicular to the normal component, $\mathbf{A}_{C/B}^n$, and thus, perpendicular to the line that connects B and C. The angle has been calculated as

   $$90° + (-\beta) = 90° + (-13.4°) = 76.1°$$

   The first term on the left side of the equation can be ignored because there is no normal component of the acceleration of point C. Therefore, the vector representing the tangential acceleration of point C is placed at the origin. However, only the direction of this vector is known: It is parallel to the horizontal surface that link 4 is constrained to slide upon. The vector polygon is illustrated in Figure 7.16c. The unknown vector magnitudes, $\mathbf{A}^t_{C/B}$ and $\mathbf{A}^t_C$, can be determined using the methods presented in Chapter 3. First, each vector can be separated into horizontal and vertical components, as shown in Table 7.2.

	TABLE 7.2 Acceleration Components for Example Problem 7.9		
Vector	**Reference Angle (θ_x)**	**Horizontal Component $a_h = a \cos \theta_x$**	**Vertical Component $a_v = a \sin \theta_x$**
$a^n_B \mathbf{A}^n_B$	220°	1007.0	845.0
$\mathbf{A}^n_{C/B}$	166.1°	−298.5	73.9
$\mathbf{A}^t_{C/B}$	76.1°	$.240\,a^t_{C/B}$	$.971\,a^t_{C/B}$
$\mathbf{A}_C$	180°	$-a_C$	0

Separate algebraic equations can be written for the horizontal and vertical components.

$$\mathbf{A}_C = \mathbf{A}^n_B +> \mathbf{A}^n_{C/B} +> \mathbf{A}^t_{C/B}$$

horizontal comp.: $+ a_C = (-1007.0) + (-298.5) + \left(+0.240\,a^t_{C/B}\right)$

vertical comp.: $0 = (-845.0) + (+73.9) + \left(+0.971\,a^t_{C/B}\right)$

The vertical component equation can be solved algebraically to give the magnitude

$$a^t_{C/B} = 794.1 \text{ in./s}^2$$

This result can then be substituted into the horizontal equation to give the magnitude

$$a_C = 1496.1 \text{ in./s}^2$$

7. ***Clearly Specify the Desired Results***

Formally stated, the motion of the ram is

$$\mathbf{V}_c = 52.3 \text{ in./s} \quad \rightarrow$$

$$\mathbf{A}_c = 1496.1 \text{ in./s} \leftarrow$$

Notice that because the acceleration is in the opposite direction of the ram movement and velocity, the ram is decelerating.

8. ***Calculate the Angular Acceleration***

Finally, the motion of the connecting arm is calculated.

$$\omega_3 = \frac{v_{C/B}}{r_{CB}} = \frac{49.6 \text{ in./s}}{8 \text{ in.}} = 6.2 \text{ rad/s, counterclockwise}$$

where the direction is consistent with the velocity of C relative to B, counterclockwise. Also

$$\alpha_3 = \frac{a^t_{C/B}}{r_{CB}} = \frac{794.1 \text{ in./s}}{8.0 \text{ in.}} = 99.3 \text{ rad/s}^2, \text{ counterclockwise}$$

where the direction is consistent with the tangential acceleration of C relative to B, counterclockwise.

7.8 ALGEBRAIC SOLUTIONS FOR COMMON MECHANISMS

For the common slider-crank and four-bar mechanisms, closed-form algebraic solutions have been derived [Ref. 12]. They are given in the following sections.

7.8.1 Slider-Crank Mechanism

A general slider-crank mechanism was illustrated in Figure 4.20 and is uniquely defined with dimensions L_1, L_2, and L_3. With one degree of freedom, the motion of one link must be specified to drive the other links. Most often the crank is driven and θ_2, ω_2, and α_2 are specified. To readily address the slider-crank mechanism, position, velocity, and acceleration equations (as a function of θ_2, ω_2, and α_2) are available. As presented in Chapter 4, the position equations include

$$\theta_3 = \sin^{-1}\left(\frac{L_1 + L_2 \sin \theta_2}{L_3}\right) \tag{4.6}$$

$$L_4 = L_2 \cos(\theta_2) + L_3 \cos(\theta_3) \tag{4.7}$$

As presented in Chapter 6, the velocity equations are

$$\omega_3 = -\omega_2\left(\frac{L_2 \cos \theta_2}{L_3 \cos \theta_3}\right) \tag{6.12}$$

$$v_4 = -\omega_2 L_2 \sin \theta_2 + \omega_3 L_3 \sin \theta_3 \tag{6.13}$$

The acceleration equations are then given as [Ref. 12]

$$\alpha_3 = \frac{\omega_2^2 L_2 \sin \theta_2 + \omega_3^2 L_3 \sin \theta_3 - \alpha_2 L_2 \cos \theta_2}{L_3 \cos \theta_3} \tag{7.21}$$

$$\alpha_4 = -\alpha_2 L_2 \sin \theta_2 - \alpha_3 L_3 \sin \theta_3 \\ - \omega_2^2 L_2 \cos \theta_2 - \omega_3^2 L_3 \cos \theta_3 \tag{7.22}$$

Note that an in-line slider-crank is analyzed by substituting zero for L_1 in equation (4.6).

7.8.2 Four-Bar Mechanism

A general four-bar mechanism was illustrated in Figure 4.22 and is uniquely defined with dimensions L_1, L_2, L_3, and L_4. With one degree of freedom, the motion of one link must be specified to drive the other links. Most often the crank is driven and θ_2, ω_2, and α_2 are specified. To readily address the four-bar mechanism, position, velocity, and acceleration equations (as a function of θ_2, ω_2, and α_2) are available. As presented in Chapter 4, the position equations are

$$BD = \sqrt{L_1^2 + L_2^2 - 2(L_1)(L_2)\cos(\theta_2)} \tag{4.9}$$

$$\gamma = \cos^{-1}\left(\frac{L_3^2 + L_4^2 - BD^2}{2(L_3)(L_4)}\right) \tag{4.10}$$

$$\theta_3 = 2\tan^{-1}\left[\frac{-L_2 \sin\theta_2 + L_4 \sin\gamma}{L_1 + L_3 - L_2 \cos\theta_2 - L_4 \cos\gamma}\right] \tag{4.11}$$

$$\theta_4 = 2\tan^{-1}\left[\frac{L_2 \sin\theta_2 - L_3 \sin\gamma}{L_2 \cos\theta_2 + L_4 - L_1 - L_3 \cos\gamma}\right] \tag{4.12}$$

As presented in Chapter 6, the velocity equations are

$$\omega_3 = -\omega_2\left[\frac{L_2 \sin(\theta_4 - \theta_2)}{L_3 \sin\gamma}\right] \tag{6.14}$$

$$\omega_4 = -\omega_2\left[\frac{L_2 \sin(\theta_3 - \theta_2)}{L_4 \sin\gamma}\right] \tag{6.15}$$

The acceleration equations can be presented as

$$\alpha_3 = \frac{\alpha_2 L_2 \sin(\theta_2 - \theta_4) + \omega_2^2 L_2 \cos(\theta_2 - \theta_4) - \omega_4^2 L_4 + \omega_3^2 L_3 \cos(\theta_4 - \theta_3)}{L_3 \sin(\theta_4 - \theta_3)} \tag{7.23}$$

$$\alpha_4 = \frac{\alpha_2 L_2 \sin(\theta_2 - \theta_3) + \omega_2^2 L_2 \cos(\theta_2 - \theta_3) - \omega_3^2 L_4 \cos(\theta_4 - \theta_3) + \omega_3^2 L_3}{L_4 \sin(\theta_4 - \theta_3)} \tag{7.24}$$

7.9 ACCELERATION OF A GENERAL POINT ON A FLOATING LINK

Recall that a floating link is not directly connected to the fixed link. Therefore, the motion of a floating link is not limited to only rotation or translation, but a combination of both. In turn, the direction of the motion of points that reside on the floating link is not generally known. Contrast this with the motion of a point on a link that is pinned to the fixed link. The motion of that point must pivot at a fixed distance from the pin connection. Thus, the direction of motion is known.

During the acceleration analyses presented in the preceding sections, the underlying premise of the solution is that the direction of the motion is known. For a general point on a floating link, this is not true. For these cases, two relative acceleration equations must be used and solved simultaneously.

To illustrate the strategy of determining the acceleration of a general point on a floating link, consider the kinematic sketch of the four-bar linkage shown in Figure 7.17.

Link 3 is a floating link because it is not directly attached to link 1, the fixed link. Because points A and B both reside on links attached to the fixed link, the acceleration of these points can be readily determined. That is, using the methods in the previous two sections, both the direction and magnitude of $\mathbf{A}_A^n$, $\mathbf{A}_A^t$, $\mathbf{A}_B^n$, and $\mathbf{A}_B^t$ can be established.

However, point C does not reside on a link that is directly attached to the fixed link. Therefore, the exact path of motion of point C is not obvious. However, two relative acceleration equations can be written as

$$\mathbf{A}_C = \mathbf{A}_B^n +> \mathbf{A}_B^t +> \mathbf{A}_{C/B}^n +> \mathbf{A}_{C/B}^t \tag{7.25}$$

$$\mathbf{A}_C = \mathbf{A}_A^n +> \mathbf{A}_A^t +> \mathbf{A}_{C/A}^n +> \mathbf{A}_{C/A}^t \tag{7.26}$$

In equation (7.25), both the magnitude and direction of a_C is unknown along with the magnitude of $a_{C/B}^t$. Equation (7.26) introduces an additional unknown, namely the magnitude of $\mathbf{A}_{C/A}^t$. Overall, two vector equations can be written, each with the capability of determining two

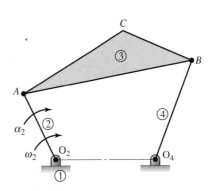

FIGURE 7.17 Point on a floating link.

unknowns. In the typical analysis, these two equations present four unknown quantities. Therefore, using the two equations simultaneously, the acceleration of point C can be determined either through a graphical or analytical procedure. The following example problem illustrates this method.

EXAMPLE PROBLEM 7.10

The mechanism shown in Figure 7.18 is used to pull movie film through a projector. The mechanism is driven by the drive wheel rotating at a constant 560 rpm. At the instant shown, graphically determine the acceleration of the claw, which engages with the film.

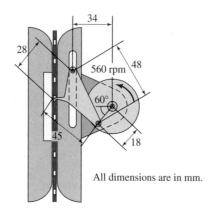

All dimensions are in mm.

FIGURE 7.18 Film advance mechanism for Example Problem 7.10.

SOLUTION:

1. ***Draw a Kinematic Diagram***

A scaled kinematic diagram of this mechanism is shown in Figure 7.19a. Notice that this is the basic slider-crank mechanism with a point of interest, point X, located at the claw.

The first step is to construct a velocity diagram that includes points B, C, and X. Calculate the magnitude of the velocity of point B with the following:

$$\omega = \frac{\pi}{30}(560 \text{ rpm}) = 58.6 \text{ rad/s, counterclockwise}$$

$$\mathbf{V}_B = \omega_2 r_{AB} = 258.6 \text{ rad/s}(18 \text{ mm}) = 1055 \text{ mm/s} = 1.055 \text{ mm/s} \quad \diagdown 30°$$

The direction of $\mathbf{V}_{B2}$ is perpendicular to link 2 and consistent with the direction of ω_2, down and to the right. Therefore, a vector can be drawn to scale from the velocity diagram origin to represent this velocity.

The relative velocity equation for points B and C can be written as

$$\mathbf{V}_C = \mathbf{V}_B +> \mathbf{V}_{C/B}$$

The velocity of C is constrained to translation in the vertical direction. Of course, the relative velocity of C with respect to B is perpendicular to the line that connects C and B. The velocity diagram shown in Figure 7.19b was drawn and the vector magnitudes were measured as

$$\mathbf{V}_C = 1.087 \text{ m/s} \downarrow$$

$$\mathbf{V}_{C/B} = 1.072 \text{ m/s} \quad \overline{31.5°\diagup}$$

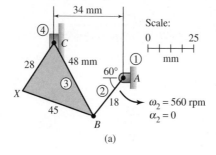

(a)

FIGURE 7.19 Diagrams for Example Problem 7.10.

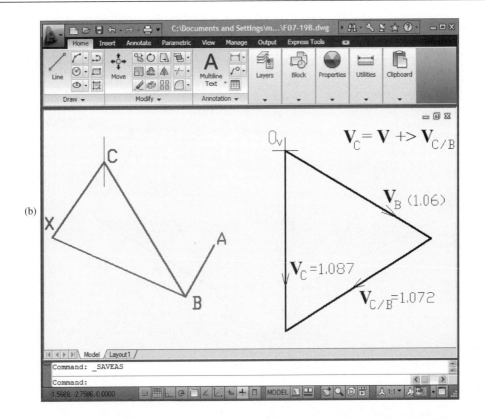

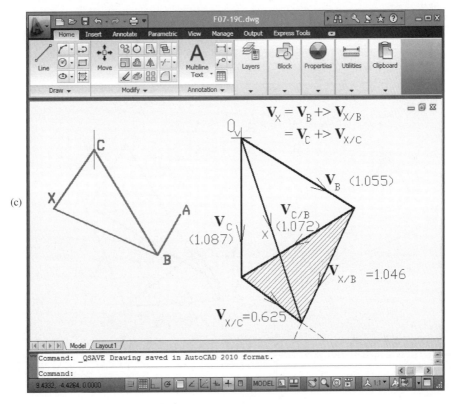

FIGURE 7.19 Continued

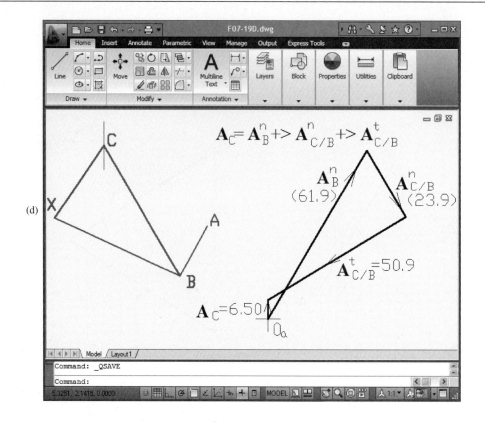

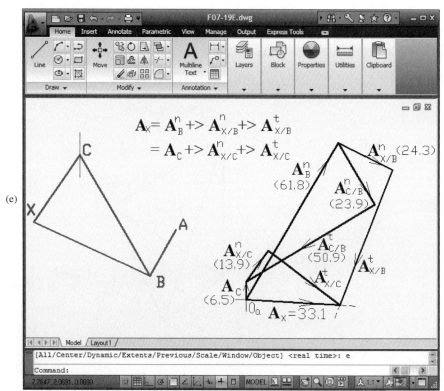

FIGURE 7.19 Continued

Because it is a general point on a floating link, the velocity of point X must be determined from solving the simultaneous vector equations.

$$\mathbf{V}_X = \mathbf{V}_B +> \mathbf{V}_{X/B}$$

$$\mathbf{V}_X = \mathbf{V}_C +> \mathbf{V}_{X/C}$$

The velocities of points B and C are already known and the directions of $V_{X/B}$ and $V_{X/C}$ are perpendicular to the lines that connect points X and B and X and C, respectively. These velocities were drawn to scale and added to the velocity polygon. The completed velocity diagram is shown in Figure 7.19c. The magnitudes of the unknown velocities were found as

$$\mathbf{V}_{X/C} = 0.625 \text{ m/s } \diagdown 35.2°$$

$$\mathbf{V}_{X/B} = 1.046 \text{ m/s } 66.4°\diagdown$$

2. **Calculate the Acceleration Components**

The next step is to construct an acceleration diagram that includes points A and B and, eventually, X. Calculate the magnitudes of the known accelerations with the following:

$$\mathbf{A}_B^n = \frac{(v_B)^2}{r_{AB}} = \frac{(1055 \text{ mm/s})^2}{18 \text{ mm}} = 61{,}834 \text{ mm/s}^2 = 61.8 \text{ m/s}^2 \diagup 60°$$

(directed toward the center of rotation, point A)

$$\mathbf{A}_B^t = 0 \text{ (because } \alpha_2 = 0 \text{)}$$

$$\mathbf{A}_{C/B}^n = \frac{(v_{C/B})^2}{r_{CB}} = \frac{(1072 \text{ mm/s})^2}{48 \text{ mm}} = 23{,}941 \text{ mm/s}^2 = 23.9 \text{ m/s}^2 \diagdown 58.5°$$

(directed from C toward B measured from CAD)

Note that point C does not have normal acceleration because the motion is strictly translational.

$$\mathbf{A}_{X/B}^n = \frac{(v_{X/B})^2}{r_{BX}} = \frac{(1046 \text{ mm/s})^2}{45 \text{ mm}} = 24{,}313 \text{ mm/s}^2 = 24.3 \text{ m/s}^2 \diagdown 23.6°$$

(directed from X toward B measured from CAD)

$$\mathbf{A}_{X/C}^n = \frac{(v_{X/C})^2}{r_{CX}} = \frac{(62.5 \text{ mm/s})^2}{28 \text{ mm}} = 13{,}950 \text{ mm/s}^2 = 13.9 \text{ m/s}^2 \diagup 54.8°$$

(directed from X toward C measured from CAD)

3. **Construct an Acceleration Diagram**

Understanding that there are no $\mathbf{A}_B^t$ and $\mathbf{A}_C^n$ components of acceleration, the relative acceleration equation for points B and C can be written as

$$\mathbf{A}_C = \mathbf{A}_C^t = \mathbf{A}_B^n +> \mathbf{A}_B^t +> \mathbf{A}_{C/B}^n +> \mathbf{A}_{C/B}^t$$

An acceleration diagram drawn to scale is shown in Figure 7.19d.

4. **Measure the Unknown Components**

Scaling the vector magnitudes from the diagram gives the following results.

$$\mathbf{A}_{C/B}^t = 50.9 \text{ m/s}^2 31.5°\diagup$$

$$\mathbf{A}_C = \mathbf{A}_C^t = 6.5 \text{ m/s}^2 \uparrow$$

5. **Continue the Acceleration Diagram**

As with velocities, because point X is a general point on a floating link, its acceleration must be determined from solving the simultaneous vector equations.

$$\mathbf{A}_X = \mathbf{A}_B^n +> \mathbf{A}_B^t +> \mathbf{A}_{X/B}^n +> \mathbf{A}_{X/B}^t$$

$$\mathbf{A}_X = \mathbf{A}_C^n +> \mathbf{A}_C^t +> \mathbf{A}_{X/C}^n +> \mathbf{A}_{X/C}^t$$

As determined, the accelerations $\mathbf{A}_B^t$ and $\mathbf{A}_C^n$ are zero. Also, $\mathbf{A}_B^n, \mathbf{A}_C^t, \mathbf{A}_{X/B}^n$, and $\mathbf{A}_{X/C}^n$ have been determined.

Again, in a similar fashion to the velocity analysis, the two acceleration equations are superimposed onto the original acceleration polygon. The accelerations were drawn to scale, and the completed acceleration diagram is shown in Figure 7.19e.

6. ***Measure the Desired Components***

The magnitudes of the unknown accelerations were measured as

$$\mathbf{A}^{t}_{X/C} = 31.6 \text{ m/s}^2 \quad \boxed{35.2°}$$

$$\mathbf{A}^{t}_{X/B} = 48.1 \text{ m/s}^2 \quad \boxed{66.4°}$$

and finally,

$$\mathbf{A}_{X} = 33.8 \text{ m/s}^2 \quad \boxed{0.3°}$$

7.10 ACCELERATION IMAGE

As with a velocity polygon, each link in a mechanism has an image in the acceleration polygon [Ref. 10]. To illustrate, a mechanism is shown in Figure 7.20a, with its associated velocity diagram in Figure 7.20b and acceleration diagrams in Figure 7.20c and 7.20d.

In Figure 7.20c, a triangle was drawn using the total acceleration vectors of points B and X. Notice that this triangle is a proportional image of the link that contains

points B and X. Similarly, Figure 7.20d shows a triangle that was constructed from the total acceleration vectors of points B, C, and X. Again, this triangle is a proportional image of the link that contains points B, C, and X. These shapes in the acceleration polygons are wisely termed the *acceleration images* of the links.

This concept provides a convenient means of constructing the acceleration polygon for a mechanism with complex links. The magnitudes of the relative acceleration

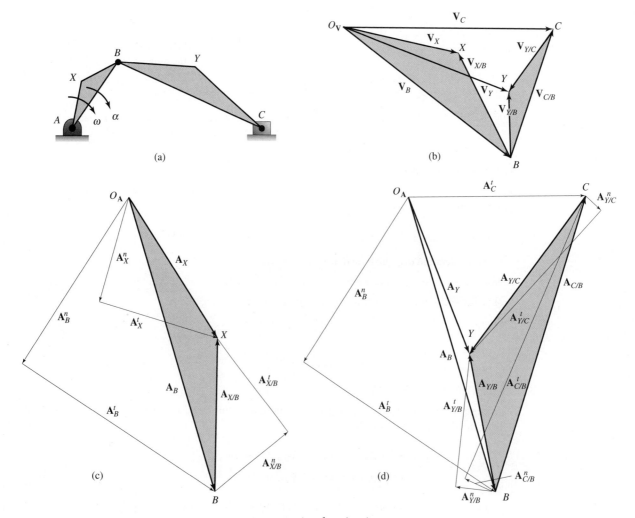

FIGURE 7.20 Acceleration image.

vectors for all points on a link will be proportional to the distance between the points. It means that the points on the acceleration diagram will form an image of the corresponding points on the kinematic diagram. Once the acceleration of two points on a link is determined, the acceleration of any other point can be readily found. The two points can be used as the base of the acceleration image. As with the velocity image, care must be taken, however, not to allow the shape of the link to be mirrored between the kinematic diagram and the velocity polygon.

7.11 CORIOLIS ACCELERATION

Throughout the preceding analyses, two components of an acceleration vector (i.e., normal and tangential) were thoroughly examined. In certain conditions, a third component of acceleration is encountered. This additional component is known as the *Coriolis component of acceleration* and is present in cases where sliding contact occurs between two rotating links.

Mechanisms used in machines have been known to fail due to the mistaken omission of this Coriolis component. Omitting the Coriolis component understates the acceleration of a link and the associated inertial forces. The actual stresses in the machine components can be greater than the design allows, and failure may occur. Therefore, every situation must be studied to determine whether a Coriolis acceleration component exists.

Specifically, the Coriolis component is encountered in the relative acceleration of two points when all of the following three conditions are simultaneously present:

1. The two points are coincident, but on different links;
2. The point on one link traces a path on the other link; and
3. The link that contains the path rotates.

Figure 7.21 illustrates a rear hatch of a minivan and the related kinematic diagram. Notice that point B can be associated with link 2, 3, or 4. To clarify the association to a link, point B is referred to as B_2, B_3, and B_4. Up to this point in the chapter, a coincident point on different links had the same acceleration because only pin joints were used to connect two rotating links. In Figure 7.21, both pin and sliding joints are used to connect the two rotating links, links 2 and 4. In this case, the velocities and accelerations of the coincident points B_2 and B_4 are not the same.

Relative motion equations can be used to relate the velocities and accelerations as follows:

$$\mathbf{V}_{B2} = \mathbf{V}_{B4} +> \mathbf{V}_{B2/B4}$$

$$\mathbf{A}_{B2} = \mathbf{A}_{B4} +> \mathbf{A}_{B2/B4}$$

This situation represents a mechanism analysis case where the Coriolis component must be included in the relative acceleration term, $\mathbf{A}_{B2/B4}$. Notice that

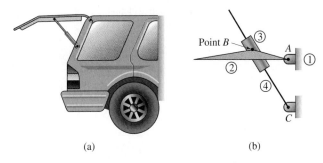

(a)   (b)

FIGURE 7.21 Case where Coriolis acceleration is encountered.

- The points are coincident, but not on the same link (condition 1);
- Point B_2 slides along and traces a path on link 4 (condition 2); and
- The link that contains the path, link 4, rotates (condition 3).

Separating the relative acceleration term into its components yields

$$\mathbf{A}_{B2/B4} = \mathbf{A}^n_{B2/B4} +> \mathbf{A}^t_{B2/B4} +> \mathbf{A}^c_{B2/B4} \qquad (7.27)$$

where

$\mathbf{A}^c_{B2/B4} =$ the Coriolis component of acceleration

The magnitude of the Coriolis component has been derived [Ref. 4] as

$$a^c_{B2/B4} = 2v_{B2/B4}\,\omega_4 \qquad (7.28)$$

Both the relative linear velocity and the absolute angular velocity can be determined from a thorough velocity analysis of the mechanism. The angular velocity, ω, must be of the link that contains the path of the sliding point. Care must be taken because a common error in calculating the Coriolis component is selecting the wrong angular velocity.

The direction of the Coriolis component is perpendicular to the relative velocity vector, $v_{B4/B2}$. The sense is obtained by rotating the relative velocity vector such that the head of the vector is oriented in the direction of the angular velocity of the path. Thus, when the angular velocity of the path, ω_4, rotates clockwise, the Coriolis direction is obtained by rotating the relative velocity vector 90° clockwise. Conversely, when the angular velocity of the path, ω_4, rotates counterclockwise, the Coriolis direction is obtained by rotating the relative velocity vector 90° counterclockwise. Figure 7.22 illustrates the four cases where the direction of the Coriolis component is determined.

Because both the magnitude and direction of the Coriolis component can be readily calculated from the velocity data, no additional unknown quantities are added to the acceleration equation. However, in solving problems, it is more convenient to write the acceleration equation with the point tracing on the left side. The technique for such acceleration analyses is best illustrated through the following example problem.

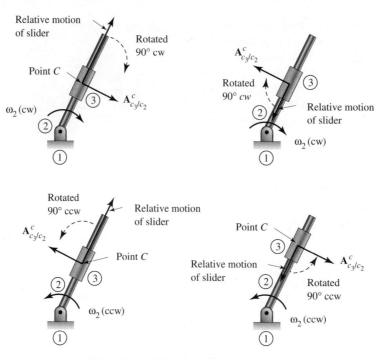

FIGURE 7.22 Directions of the Coriolis acceleration component.

EXAMPLE PROBLEM 7.11

Figure 7.23 illustrates handheld grass shears, used for trimming areas that are hard to reach with mowers or weed whackers. The drive wheel rotates counterclockwise at 400 rpm. Determine the angular acceleration of the oscillating blades at the instant shown.

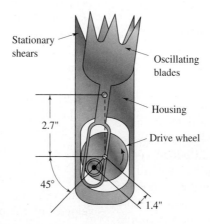

FIGURE 7.23 Grass shears for Example Problem 7.11.

SOLUTION: 1. **Draw a Kinematic Diagram**

A scaled kinematic diagram of this mechanism is shown in Figure 7.24a.

2. **Decide on a Method to Achieve the Desired Acceleration**

The acceleration of B_2 can be readily determined from the input information of link 2. The acceleration of B_4 must be found to determine the angular acceleration of link 4. Notice that sliding occurs between rotating links (2 and 4); thus, all three of the Coriolis conditions are met. The acceleration of link 4 will be obtained by incorporating equations (7.27) and (7.28).

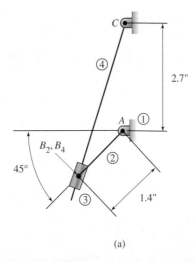

(a)

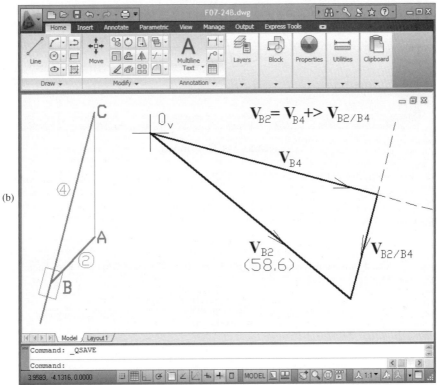

(b)

FIGURE 7.24 Diagrams for Example Problem 7.11.

3. **Complete a Full Velocity Analysis**

The first step is to construct a velocity diagram that includes points B_2 and B_4. Calculate the magnitude of the velocity of point B_2 with the following:

$$\omega_2 = \frac{\pi}{30}(400 \text{ rpm}) = 41.9 \text{ rad/s, counterclockwise}$$

$$V_{B2} = \omega_2 r_{AB} = 241.9 \text{ rad/s}(1.4 \text{ in.}) = 58.6 \text{ in./s} \quad \diagdown 45°$$

The direction of V_{B2} is perpendicular to link 2 and consistent with the direction of ω_2, down and to the right. Therefore, a vector can be drawn to scale from the velocity diagram origin to represent this velocity.

The relative velocity equation for points B_2 and B_4 can be written as

$$V_{B2} = V_{B4} +> V_{B2/B4}$$

Because link 4 is pinned to the fixed link, the velocity of B_4 is perpendicular to the line that connects B_4 with the center of rotation (point C). For this case, the relative velocity of B_2 with respect to B_4 is parallel to link 4

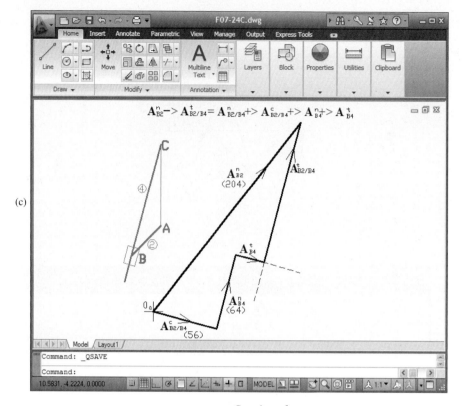

FIGURE 7.24 Continued

because B_2 slides along link 4. The velocity diagram shown in Figure 7.24b was drawn to scale to find the velocity magnitudes of

$$\mathbf{V}_{B4} = 50.7 \text{ in./s} \quad \angle 15°$$

$$\mathbf{V}_{B2/B4} = 29.3 \text{ in./s} \quad \angle 75°$$

The distance between points C and B_4 was measured from CAD as 3.8 in. Therefore, the angular velocity of link 4 can be calculated as

$$\omega_4 = \frac{V_{B4}}{r_{CB4}} = \frac{50.7 \text{ in/.s}}{3.8 \text{ in.}} = 13.3 \text{ rad/s, counterclockwise}$$

Because the velocity of B_4 has been found to be directed down and to the right, the angular velocity of link 4 must be counterclockwise.

4. *Calculate Acceleration Components*

   Calculate the magnitudes of the known accelerations with the following:

$$\mathbf{A}_{B2}^n = \frac{(V_{B2})^2}{r_{CB2}} = \frac{(58.6 \text{ in./s})^2}{1.4 \text{ in.}} = 2453 \text{ in./s}^2 = 204 \text{ ft/s}^2 \quad \angle 45°$$

(directed toward the center of rotation, point A)

$$\mathbf{A}_{B2}^t = 0 \ (\alpha_2 = 0)$$

$$\mathbf{A}_{B4}^n = \frac{(V_{B4})^2}{r_{CB4}} = \frac{(50.7 \text{ in./s})^2}{3.8 \text{ in.}} = 676 \text{ in./s}^2 = 56 \text{ ft/s}^2 \quad \angle 75°$$

(directed toward the center of rotation, point C)

$$\mathbf{A}_{B2/B4}^n = 0$$

(because B_2 is sliding on B_4 and the relative motion is purely translational)

$$\mathbf{A}_{B2/B4}^c = 2(v_{B2/B4})(\omega_4) = 2(29.3 \text{ in./s}) (13.3 \text{ rad/s})$$

$$= 779 \text{ in./s}^2 = 65 \text{ ft/s}^2 \quad \angle 15°$$

The direction of the Coriolis component is that of $v_{B2/B4}$, which is parallel to the path of B_2 relative to B_4, ($\overline{75°}$), rotated 90° in the direction of ω_4 (counterclockwise). Therefore, the Coriolis component is directed perpendicular to link 4, down and toward the right ($\underset{\downarrow}{\diagdown 15°}$).

5. ***Construct an Acceleration Diagram***

   The next step is to construct an acceleration diagram that includes points B_2 and B_4. As mentioned, it is typically more convenient to write the acceleration equation with the point doing the tracing, B_2, on the left side. Using this guideline, the acceleration equation is written as

   $$\mathbf{A}_{B\ \ 2}^{n} +> \mathbf{A}_{B2}^{t} = \mathbf{A}_{B4}^{n} +> \mathbf{A}_{B4}^{t} +> \mathbf{A}_{B2/B4}^{n} +> \mathbf{A}_{B2/B4}^{t} +> \mathbf{A}_{B2/B4}^{c}$$

   The unknown quantities in the acceleration equation are A_{B4}^{t} and $A_{B2/B4}^{t}$. Rewrite the acceleration equation so that each unknown is the last term on both sides of the equation:

   $$\mathbf{A}_{B2}^{n} +> \mathbf{A}_{B2}^{t} +> \mathbf{A}_{B2/B4}^{t} = \mathbf{A}_{B2/B4}^{n} +> \mathbf{A}_{B2/B4}^{c} +> \mathbf{A}_{B4}^{n} +> \mathbf{A}_{B4}^{t}$$

   An acceleration diagram drawn to scale is shown in Figure 7.24c.

6. ***Measure the Desired Acceleration Components***

   Scale the vector magnitudes from the diagram using the following equations:

   $$A_{B2/B4}^{t} = 112 \text{ ft/s}^2 \ \diagup\!\!\!75°$$
   $$A_{B4}^{t} = 37 \text{ ft/s}^2 = 444 \text{ in./s}^2 \ \diagdown\!\!\!\!15°$$

   and finally,

   $$\alpha_4 = \frac{a_{B4}^{t}}{r_{CB4}} = \frac{444 \text{ in./s}^2}{3.8 \text{ in.}} = 117 \text{ rad/s}^2$$

   Because the tangential acceleration of B_4 was determined to be down and to the right, the corresponding rotational acceleration of link 4 must be counterclockwise; therefore,

   $$\alpha_4 = 177 \text{ rad/s}^2, \text{ counterclockwise}$$

7.12 EQUIVALENT LINKAGES

Up to this point in the text, the examples of motion analysis involved only mechanisms with primary joints; that is, pin and sliding joints. Recall from Chapter 1 that a higher-order joint, such as a cam or gear joint, involves rolling and sliding motion. Both cams and gears are the focus in later chapters. However, the motion analysis of mechanisms with higher-order joints can be performed using the concepts already presented.

Velocity and acceleration analysis of mechanisms that utilize higher-order joints is greatly simplified by constructing an equivalent linkage. This method converts the instantaneous configuration of a mechanism to an *equivalent linkage*, where the links are connected by primary joints. Figure 7.25 illustrates two cam mechanisms that contain rolling and sliding joints. The dotted lines represent the equivalent linkages.

Notice that the coupler of these equivalent linkages is drawn from the respective centers of curvature of the two mating links. For a finite length of time, the two centers of curvature for the two mating links will remain a constant distance apart. Notice in Figure 7.25 that a coupler is used to replace the higher-order joint. This coupler extends between the center of curvature for the contacting surfaces of the two mating links. For a finite length of time, the centers of curvature for the two mating surfaces will remain a constant distance apart. The rationale behind this stems from the concept of instantaneous center, introduced in Section 6.10. Therefore, a coupler link, with two pin joints, can be used to replace the higher-order joint. It is important to note that the location of the center of curvature will change as the mechanism moves. However, once the equivalent linkage has been constructed, the method of analysis is identical to the problems previously encountered in this text.

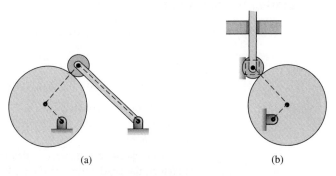

(a)   (b)

FIGURE 7.25 Equivalent linkages.

7.13 ACCELERATION CURVES

The analyses presented up to this point in the chapter are used to calculate the acceleration of points on a mechanism at a specific instant. Although they are important, the results provide only a snapshot of the motion. The obvious shortcoming of this analysis is that determination of the extreme conditions throughout a cycle is difficult. It is necessary to investigate several positions of the mechanism to discover the critical phases.

As shown with velocity, it is also convenient to trace the acceleration magnitude of a certain point, or link, as the mechanism moves through its cycle. Such a trace provides information about critical phases in the cycle. An acceleration curve provides this trace. An acceleration curve plots the acceleration of a point, or link, as a function of time. It can be generated from a velocity curve, which was introduced in Section 6.14.

Recall that a velocity curve plots the velocity magnitude of a point or link as a function of time. A velocity curve is generated from a displacement curve, which was introduced in Section 4.11. Thus, a displacement curve can be used to generate a velocity curve, which, in turn, can be used to generate an acceleration curve. This is because acceleration can be expressed as

$$\text{Acceleration} = \frac{d(\text{velocity})}{dt}$$

Differential calculus suggests that the acceleration at a particular instant is the slope of the velocity curve at that instant. Because velocity is the time derivative of displacement, acceleration can also be expressed as

$$\text{Acceleration} = \frac{d^2(\text{displacement})}{dt^2}$$

This equation suggests that acceleration at a particular instant is the curvature of the displacement curve. Admittedly, curvature may not be so convenient to determine as the slope. However, it is easy to visualize the locations of extreme accelerations by locating the regions of sharp curves on the displacement diagram. Although values may be difficult to calculate, the mechanism can be configured to the desired position, then a thorough acceleration analysis can be performed, as presented in the preceding sections.

To determine values for the acceleration curves, it is best to determine the slope at several regions of the velocity curve (see Section 6.14).

7.13.1 GRAPHICAL DIFFERENTIATION

The task is to estimate the slope of the velocity curve at several points. The slope of a curve, at a point, can be graphically estimated by sketching a line tangent to the curve at the point of interest. The slope of the line can be determined by calculating the measured change in "rise" (velocity) divided by the measured change in "run" (time).

This procedure can be repeated at several points along the velocity diagram. However, only the acceleration extremes and abrupt changes are usually desired. Using the notion of differential calculus and slopes, the positions of interest can be visually detected. They include:

- The steepest portions of the velocity diagram, which correspond to the extreme accelerations; and
- The locations on the velocity diagram with the greatest curvature, which correspond to the abrupt changes of accelerations.

It must be noted that errors can easily occur when determining the slope of a curve. These errors are magnified as the slope is measured from a derived curve. This is the case as an acceleration curve stems from a velocity curve, which stems from a displacement curve. Therefore, the values obtained for the acceleration diagram should be used cautiously.

Nevertheless, identifying the positions of extreme accelerations is invaluable. A complete acceleration analysis, as presented in the previous sections of this chapter, should then be performed at these mechanism orientations to obtain accurate acceleration values. The benefit of the acceleration curve is locating the important mechanism configurations; therefore, a comprehensive acceleration analysis can be performed.

EXAMPLE PROBLEM 7.12

A velocity curve was constructed for a compressor mechanism in Example Problem 6.18. Use these data to plot an acceleration curve.

SOLUTION: 1.   *Identify the Horizontal Portions of the Velocity Diagram*

The main task of constructing an acceleration curve is to determine the slope of many points on the velocity curve. This velocity curve was constructed in Example Problem 6.18 and is reprinted as Figure 7.26.

From this curve, it is apparent that the curve has a horizontal tangent, or zero slope, at 0.007 and 0.027 s. Therefore, the acceleration of the piston is zero at 0.007 and 0.027 s. These points are labeled t_1 and t_3, respectively.

Velocity curve

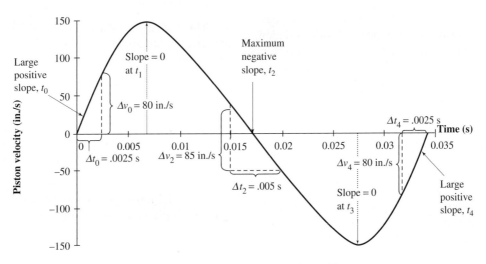

FIGURE 7.26 Velocity curve for Example Problem 7.12.

2. **Calculate the Slope at the Noteworthy Portions of the Velocity Curve**

The maximum upward slope appears at 0 s. This point was labeled as t_0. An estimate of the velocity can be made from the values of Δv_0 and Δt_0 read from the graph. Acceleration at 0 s is estimated as

$$a_0 = \frac{\Delta v_0}{\Delta t_0} = \frac{80 \text{ in./s}}{0.0025 \text{ s}} = 32,000 \text{ in./s}^2$$

Likewise, the maximum downward slope appears at 0.017 s. This point was labeled as t_2. Again, an estimate of the acceleration can be made from the values of Δv_2 and Δt_2 read from the graph. The velocity at 0.017 s is estimated as

$$a_2 = \frac{\Delta v_2}{\Delta t_2} = \frac{-85 \text{ in./s}}{0.005 \text{ s}} = -17,000 \text{ in./s}^2$$

3. **Sketch the Acceleration Curve**

The procedure for determining the slope of the velocity curve can be repeated at other points in time. By compiling the slope and time information, an acceleration curve can be constructed (Figure 7.27).

Acceleration curve

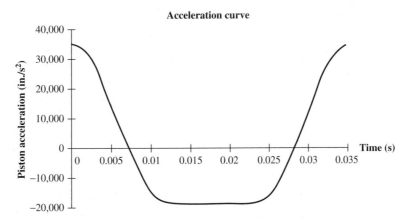

FIGURE 7.27 Acceleration curve for Example Problem 7.12.

7.13.2 Numerical Differentiation

Similar to Section 6.14.2, the acceleration curve can be determined from the velocity data by numerical differentiation. Again, the Richardson method [Ref. 3] is used for determining the derivative of a series of data points with an equally spaced, independent variable. Thus, the derivative of the velocity-time curve can be numerically approximated by using the following equation:

$$a_i = \left[\frac{v_{i+1} - v_{i-1}}{2\,\Delta t}\right] - \left[\frac{v_{i+2} - 2v_{i+1} + 2v_{i-1} - v_{i-2}}{12\,\Delta t}\right] \quad (7.29)$$

where:

i = data point index
v_i = velocity at data point i

$$\Delta t = t_2 - t_1 = t_3 - t_2 = t_4 - t_3$$
$$t_i = \text{time at data point } i$$

The second derivative can also be determined through numerical approximations. Although not as accurate, this allows the acceleration curve to be derived directly from the displacement–time curve. Again, the Richardson method is used to numerically determine the second derivative with the following equation:

$$a_i = \left[\frac{\Delta R_{i+1} - 2\Delta R_i + \Delta R_{i-1}}{\Delta t^2}\right] \quad (7.30)$$

where, in addition to the notation above,
ΔR_i = displacement at data point i

EXAMPLE PROBLEM 7.13

A displacement diagram of the piston operating in a compressor was plotted in Example Problem 4.11. From this diagram, a velocity curve was derived in Example Problem 6.18. Use this data to numerically generate an acceleration curve.

SOLUTION:

1. ***Determine the Time Increment Between Data Points***

   The spreadsheet data from Example Problem 6.17 (Figure 6.40) was expanded by inserting an additional column to include the magnitude of piston acceleration. In addition, in Example Problem 6.18 the time increment was calculated as

   $$\Delta t = t_2 - t_1 = (0.00286 - 0.0) = 0.00286 \text{ s}$$

2. ***Use Equation (7.29) to Calculate Acceleration Data Points***

   To illustrate the calculation of the accelerations, a few sample calculations using equation (7.29) follow:

   $$a_2 = \left[\frac{v_3 - v_1}{2\Delta t}\right] - \left[\frac{v_4 - 2v_3 + 2v_1 - v_{12}}{12\,\Delta t}\right]$$

   $$= \left[\frac{142.67 - 0.0}{2(.00286)}\right] - \left[\frac{137.50 - 2(142.67) + 2(0.0) - (-91.47)}{12(.00286)}\right]$$

   $$= 26{,}898 \text{ in./s}^2$$

   $$a_9 = \left[\frac{v_{10} - v_8}{2\Delta t}\right] - \left[\frac{v_{11} - 2v_{10} + 2v_8 - v_7}{12\,\Delta t}\right]$$

   $$= \left[\frac{(-142.67) - (-95.48)}{2(.00286)}\right] - \left[\frac{(-91.47) - 2(-142.67) + 2(-95.48) - (46.03)}{12(.00286)}\right]$$

   $$= -17{,}305 \text{ in./s}^2$$

   $$a_{12} = \left[\frac{v_{13} - v_{11}}{2\Delta t}\right] - \left[\frac{v_2 - 2v_{13} + 2v_{11} - v_{10}}{12\,\Delta t}\right]$$

   $$= \left[\frac{(0.0) - (-142.67)}{2(.00286)}\right] - \left[\frac{(91.47) - 2(0.0) + 2(-91.47) - (142.67)}{12(.00286)}\right]$$

   $$= 26{,}898 \text{ in./s}^2$$

3. ***Compile the Acceleration Results and Plot the Curve***

The resulting information, with all acceleration magnitudes calculated, is given in Figure 7.28. These values are plotted in Figure 7.29 to form an acceleration diagram, relative to time.

Notice that this curve is still rather rough. For accuracy purposes, it is highly suggested that the crank angle increment be reduced to 10° or 15°. When a spreadsheet is used to generate the acceleration data, even smaller increments are advisable and do not make the task any more difficult.

Data Point Index	Crank Angle (deg)	Time (0.001 sec)	Piston Displacement (in.)	Piston Velocity (in./sec)	Piston Acceleration (in./sec^2)
1	0	0.00	2.000	0.00	34655
2	30	2.86	0.136	91.47	26898
3	60	5.72	0.483	142.67	7901
4	90	8.57	0.896	137.50	-10181
5	120	11.43	1.233	95.48	-17305
6	150	14.29	1.435	46.03	-16761
7	180	17.15	1.500	0.00	-15759
8	210	20.00	1.435	-46.03	-16761
9	240	22.86	1.233	-95.48	-17305
10	270	25.72	0.896	-137.50	-10181
11	300	28.58	0.483	-142.67	7901
12	330	31.43	0.136	-91.47	26898
13	360	34.29	0.000	0.00	34655

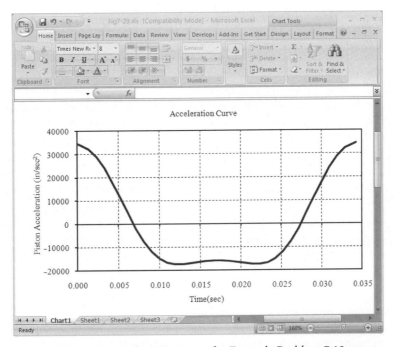

FIGURE 7.28 Acceleration data for Example Problem 7.13.

FIGURE 7.29 Acceleration curve for Example Problem 7.13.

PROBLEMS

Manual drawing techniques can be instructive for problems requiring graphical solution, but using a CAD system is highly recommended.

General Acceleration

7–1. Boxes are sitting on a conveyor belt as the conveyor is turned on, moving the boxes toward the right. The belt reaches full speed of 45 fpm (ft/min) in 0.5 s. Determine the linear acceleration of the boxes assuming that this acceleration is constant. Also determine the linear displacement of the boxes during this speed-up period.

7–2. A high-performance vehicle has a 0-to-60 (mph) time of 8.3 s. Determine the linear acceleration of the vehicle and the distance traveled in reaching 60 mph.

7–3. An elevator moves upward at a velocity of 12 ft/s. Determine the distance required to stop if the constant deceleration is not to exceed 10 ft/s^2.

7–4. Point A is on a slider that is accelerating uniformly upward along a vertical straight path. The slider has a velocity of 100 mm/s as it passes one point and 300 mm/s as it passes a second point, 0.2 s later. Determine the linear acceleration and the linear displacement of point A during this interval of time.

7–5. A linear actuator is used to push a load leftward. Starting at rest, it requires 1.5 s to reach a full speed of 0.75 m/s. Determine the linear acceleration of the load. Also determine the linear displacement of the load during this acceleration phase of the motion.

7–6. Starting from rest, a cam accelerates uniformly to 750 rpm clockwise in 8 s. Determine the angular acceleration of the cam.

7–7. The rotor of a jet engine rotates clockwise and idles at 10,000 rpm. When the fuel is shut off, the engine slows to a stop in 2 min. Assuming that the speed reduces uniformly, determine the angular acceleration of the engine. Also determine the rotational displacement of the rotor during this shutdown period.

7–8. The angular velocity of a shaft is increased with constant acceleration from 1000 rpm to 2500 rpm clockwise in 20 s. Determine the angular acceleration of the shaft.

7–9. A wheel rotates 400 revolutions counterclockwise while decelerating from 1100 rpm to 800 rpm. Determine the angular acceleration of the wheel.

Velocity Profiles

A servo-driven actuator is programmed to extend according to the velocity profile shown in Figure P7.10. Determine the maximum acceleration, maximum deceleration, and linear displacement during this programmed move.

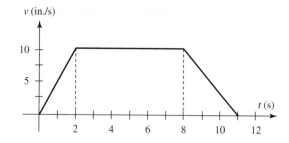

FIGURE P7.10 Problems 10 and 11.

7–11. A servo-driven actuator is programmed to extend according to the velocity profile shown in Figure P7.10. Use a spreadsheet to generate plots of displacement versus time, velocity versus time, and acceleration versus time during this programmed move.

7–12. A linear motor is programmed to move rightward according to the velocity profile shown in Figure P7.12. Determine the maximum acceleration, maximum deceleration, and linear displacement during this programmed move.

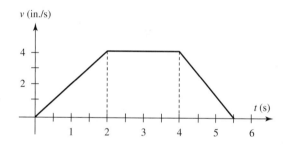

FIGURE P7.12 Problems 12 and 13.

7–13. A linear motor is programmed to move rightward according to the velocity profile shown in Figure P7.12. Use a spreadsheet to generate plots of displacement versus time, velocity versus time, and acceleration versus time during this programmed move.

7–14. A linear actuator is programmed to move 10 in. The maximum velocity is 4 in./s, and the constant acceleration and deceleration is limited to 6 in./s^2. Use a spreadsheet to generate plots of displacement versus time, velocity versus time, and acceleration versus time during this programmed move.

7–15. A linear actuator is programmed to move 75 mm. The maximum velocity is 50 mm/s, and the constant acceleration and deceleration is limited to 100 mm/s^2. Use a spreadsheet to generate plots of displacement versus time, velocity versus time, and acceleration versus time during this programmed move.

Normal and Tangential Acceleration

7–16. Link 2 is isolated from a kinematic diagram and shown in Figure P7.16. The link is rotating counterclockwise at a constant rate of 300 rpm. Determine

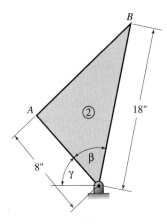

FIGURE P7.16 Problems 16, 17, and 18.

the total linear acceleration of points A and B. Use $\gamma = 50°$ and $\beta = 60°$.

7–17. Link 2 is isolated from a kinematic diagram and shown in Figure P7.16. The link is rotating counter-clockwise at a rate of 200 rpm, and accelerating at 400 rad/s^2. Determine the total linear acceleration of points A and B. Use $\gamma = 50°$ and $\beta = 60°$.

7–18. Link 2 is isolated from a kinematic diagram and shown in Figure P7.16. The link is rotating counter-clockwise at a rate of 300 rpm, and decelerating at 800 rad/s^2. Determine the total linear acceleration of points A and B. Use $\gamma = 50°$ and $\beta = 60°$.

Figure P7.19 shows a centrifugal clutch that engages two shafts at a threshold rotational velocity.

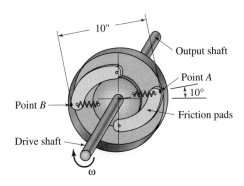

FIGURE P7.19 Problems 19, 20, and 21.

7–19. Determine the total acceleration of point A on the friction pad of the centrifugal clutch shown in Figure P7.19. At the instant shown, the driveshaft rotates at a constant 300 rpm clockwise.

7–20. Determine the total acceleration of point A on the friction pad of the centrifugal clutch shown in Figure P7.19. At the instant shown, the driveshaft is rotating at 300 rpm clockwise and is speeding up at a rate of 300 rad/s^2.

7–21. Determine the total acceleration of point A on the friction pad of the centrifugal clutch shown in

Figure P7.19. At the instant shown, the driveshaft is rotating at 300 rpm clockwise and is slowing down at a rate of 300 rad/s^2.

Relative Acceleration

7–22. For the kinematic diagram shown in Figure P7.22, the length of link AB is 100 mm and $\theta = 35°$. Box A moves upward at a velocity of 10 mm/s and accelerates at 5 mm/s^2. At the same time, the velocity of box B is 7 mm/s and accelerates at a rate of 25 mm/s^2. Graphically determine the linear velocity of A with respect to B and the linear acceleration of A with respect to B.

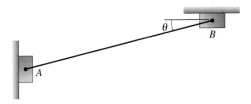

FIGURE P7.22 Problems 22 and 23.

7–23. For the kinematic diagram shown in Figure P7.22, the length of link AB is 15 in. and $\theta = 40°$. Box A moves upward at a velocity of 50 in./s and decelerates at 125 in./s^2. At the same time, the velocity of box B is 42 in./s and accelerates at a rate of 48.6 in./s^2. Analytically determine the linear velocity of A with respect to B and the linear acceleration of A with respect to B.

7–24. Figure P7.24 shows a device to open windows commonly found in elevated locations of gymnasiums and factories. At the instant when $\theta = 25°$, the drive nut moves to the right at a velocity of 1 ft/s and accelerates at 1 ft/s^2. At the same time, the velocity of the shoe is 0.47 ft/s, and it accelerates at a rate of 0.91 ft/s^2. Graphically determine the linear velocity of C with respect to B and the linear acceleration of C with respect to B.

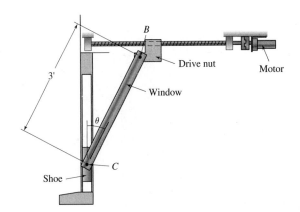

FIGURE P7.24 Problems 24 and 25.

7–25. For the window-opening mechanism shown in Figure P7.24, at the instant when $\theta = 55°$, the drive nut moves to the right at a velocity of 2 ft/s and accelerates at 1 ft/s^2. At the same time, the velocity of the shoe is 2.85 ft/s, and it accelerates at a rate of 8.51 ft/s². Graphically determine the linear velocity of *C* with respect to *B* and the linear acceleration of *C* with respect to *B*.

Relative Acceleration Method—Graphical

7–26. For the compressor linkage shown in Figure P7.26, use the relative acceleration method to graphically determine the linear velocity and linear acceleration of the piston as the crank rotates clockwise at a constant rate of 1150 rpm.

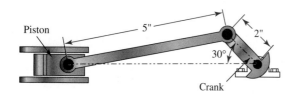

FIGURE P7.26 Problems 26, 27, 28, 44, 75, 81, and 87.

7–27. For the compressor linkage in Figure P7.26, at the instant shown, the crank is rotating counterclockwise at 2000 rpm and accelerating at 10,000 rad/s². Use the relative acceleration method to graphically determine the linear velocity and linear acceleration of the piston.

7–28. For the compressor linkage in Figure P7.26, at the instant shown, the crank is rotating clockwise at 1500 rpm and decelerating at 12,000 rad/s². Use the relative acceleration method to graphically determine the linear velocity and linear acceleration of the piston.

7–29. For the sewing machine linkage in Figure P7.29, at the instant when $\theta = 30°$, the drive wheel rotates counterclockwise at a constant 200 rpm. Use the relative acceleration method to graphically determine the linear velocity and linear acceleration of the needle.

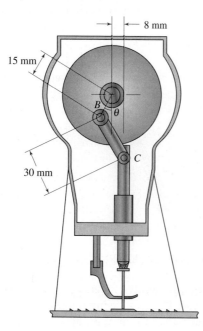

FIGURE P7.29 Problems 29, 30, 31, 45, 76, 82, and 88.

7–30. For the sewing machine linkage in Figure P7.29, at the instant when $\theta = 30°$, the drive wheel is rotating clockwise at 300 rpm and accelerating at 800 rad/s². Use the relative acceleration method to graphically determine the linear velocity and linear acceleration of the needle.

7–31. For the sewing machine linkage in Figure P7.29, at the instant when $\theta = 120°$, the drive wheel is rotating clockwise at 200 rpm and decelerating at 400 rad/s². Use the relative acceleration method to graphically determine the linear velocity and linear acceleration of the needle.

7–32. For the power hacksaw in Figure P7.32, at the instant shown, the 1.75-in. crank rotates clockwise at a constant 80 rpm. Use the relative acceleration method to graphically determine the linear velocity and linear acceleration of the saw blade.

7–33. For the power hacksaw in Figure P7.32, at the instant shown, the 1.75-in. crank rotates clockwise at 60 rpm

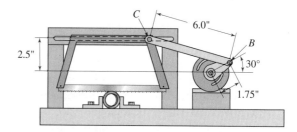

FIGURE P7.32 Problems 32, 33, 34, 46, 77, 83, and 89.

and accelerates at 40 rad/s². Use the relative acceleration method to graphically determine the linear velocity and linear acceleration of the saw blade.

7–34. For the power hacksaw in Figure P7.32, at the instant shown, the 1.75-in. crank rotates clockwise at 70 rpm and decelerates at 45 rad/s². Use the relative acceleration method to graphically determine the linear velocity and linear acceleration of the saw blade.

7–35. The motor on the coin-operated horse in Figure P7.35 rotates clockwise at a constant rate of 90 rpm. At the

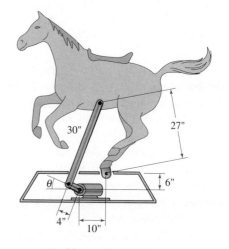

FIGURE P7.35 Problems 35, 36, 37, 47, 78, 84, and 90.

instant when $\theta = 30°$, use the relative acceleration method to graphically determine the angular velocity and angular acceleration of the horse.

7–36. At the instant when $\theta = 45°$, the motor on the coin-operated horse in Figure P7.35 rotates clockwise at 60 rpm and accelerates at 30 rad/s². Use the relative acceleration method to graphically determine the angular velocity and angular acceleration of the horse.

7–37. At the instant when $\theta = 120°$, the motor on the coin-operated horse in Figure P7.35 rotates clockwise at 40 rpm and decelerates at 20 rad/s². Use the relative acceleration method to graphically determine the angular velocity and angular acceleration of the horse.

7–38. The motor on the car wash sprayer in Figure P7.38 rotates counterclockwise at a constant rate of 120 rpm. At the instant when $\theta = 40°$, use the relative acceleration method to graphically determine the angular velocity and angular acceleration of the nozzle arm.

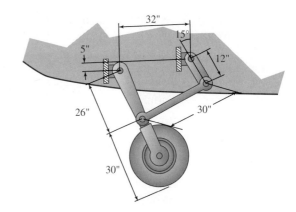

FIGURE P7.41 Problems 41, 42, 43, 49, 80, 86, and 92.

at a constant rate of 20 rpm. At the instant shown, use the relative acceleration method to graphically determine the angular velocity and angular acceleration of the wheel assembly.

7–42. At the instant shown, the 12-in. crank on the small aircraft landing gear actuation in Figure P7.41 rotates counterclockwise at 15 rpm and accelerates at 4 rad/s². Use the relative acceleration method to graphically determine the angular velocity and angular acceleration of the wheel assembly.

7–43. At the instant shown, the 12-in. crank on the small aircraft landing gear actuation in Figure P7.41 rotates counterclockwise at 18 rpm and decelerates at 3.5 rad/s². Use the relative acceleration method to graphically determine the angular velocity and angular acceleration of the wheel assembly.

Relative Acceleration Method—Analytical

7–44. For the compressor linkage in Figure P7.26, at the instant shown, the crank is rotating clockwise at 1800 rpm and accelerating at 12,000 rad/s². Use the relative acceleration method to analytically determine the linear velocity and linear acceleration of the piston.

7–45. For the sewing machine linkage in Figure P7.29, at the instant when $\theta = 30°$, the drive wheel is rotating clockwise at 250 rpm and accelerating at 6000 rad/s². Use the relative acceleration method to graphically determine the linear velocity and linear acceleration of the needle.

7–46. For the power hacksaw in Figure P7.32, at the instant shown, the 1.75-in. crank rotates clockwise at 55 rpm and decelerates at 35 rad/s². Use the relative acceleration method to graphically determine the linear velocity and linear acceleration of the saw blade.

7–47. At the instant when $\theta = 45°$, the motor on the coin-operated horse in Figure P7.35 rotates clockwise at 45 rpm and accelerates at 25 rad/s². Use the relative acceleration method to graphically determine the angular velocity and angular acceleration of the horse.

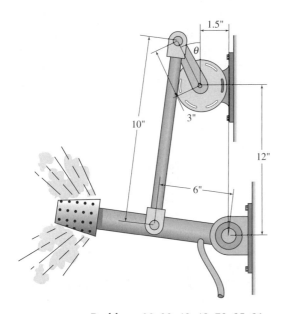

FIGURE P7.38 Problems 38, 39, 40, 48, 79, 85, 91.

7–39. At the instant when $\theta = 90°$, the motor on the car wash sprayer in Figure P7.38 rotates counterclockwise at 150 rpm and accelerates at 200 rad/s². Use the relative acceleration method to graphically determine the angular velocity and angular acceleration of the nozzle arm.

7–40. At the instant when $\theta = 120°$, the motor on the car wash sprayer in Figure P7.38 rotates counterclockwise at 100 rpm and decelerates at 100 rad/s². Use the relative acceleration method to graphically determine the angular velocity and angular acceleration of the nozzle arm.

7–41. The 12-in. crank on the small aircraft landing gear actuation in Figure P7.41 rotates counterclockwise

7–48. At the instant when $\theta = 90°$, the motor on the car wash sprayer in Figure P7.38 rotates counterclockwise at 130 rpm and decelerates at 180 rad/s². Use the relative acceleration method to graphically determine the angular velocity and angular acceleration of the nozzle arm.

7–49. At the instant shown, the 12-in. crank on the small aircraft landing gear actuation in Figure P7.41 rotates counterclockwise at 12 rpm and accelerates at 3 rad/s². Use the relative acceleration method to graphically determine the angular velocity and angular acceleration of the wheel assembly.

Acceleration of Points on a Floating Link—Graphical

7–50. The 3.25-in. link on the stamp mechanism in Figure P7.50 rotates clockwise at a constant rate of 20 rpm. At the instant when $\theta = 60°$, graphically determine the linear acceleration of point X on the stamp.

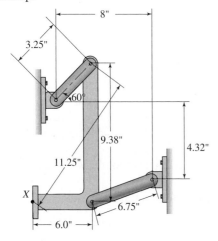

FIGURE P7.50 Problems 50–53, 58.

7–51. The 3.25-in. link on the stamp mechanism in Figure P7.50 rotates clockwise at a constant rate of 20 rpm. At the instant when $\theta = 120°$, graphically determine the linear acceleration of point X on the stamp.

7–52. The 3.25-in. link on the stamp mechanism in Figure P7.50 rotates clockwise at 30 rpm and is accelerating at 6 rad/s². At the instant when $\theta = 90°$, graphically determine the linear acceleration of point X on the stamp.

7–53. The 3.25-in. link on the stamp mechanism in Figure P7.50 rotates clockwise at 30 rpm and is decelerating at 6 rad/s². At the instant when $\theta = 90°$, graphically determine the linear acceleration of point X on the stamp.

7–54. The 0.5-m link on the lift mechanism in Figure P7.54 rotates counterclockwise at a constant rate of 12 rpm. At the instant when $\theta = 20°$, graphically determine the linear acceleration of point X.

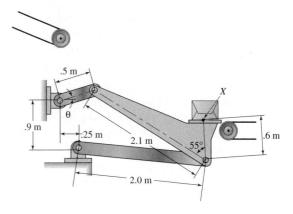

FIGURE P7.54 Problems 54–57, 59.

7–55. The 0.5-m link on the lift mechanism in Figure P7.54 rotates clockwise at a constant rate of 20 rpm. At the instant when $\theta = 30°$, graphically determine the linear acceleration of point X.

7–56. The 0.5-m link on the lift mechanism in Figure P7.54 rotates clockwise at 30 rpm and accelerates at 5 rad/s². At the instant when $\theta = 20°$, graphically determine the linear acceleration of point X.

7–57. The 0.5-m link on the lift mechanism in Figure P7.54 rotates counterclockwise at 18 rpm and decelerates at 5 rad/s². At the instant when $\theta = 0°$, graphically determine the linear acceleration of point X.

Acceleration of Points on a Floating Link—Analytical

7–58. The 3.25-in. link on the stamp mechanism in Figure P7.50 rotates clockwise at a constant rate of 20 rpm. At the instant when $\theta = 60°$, analytically determine the linear acceleration of point X on the stamp.

7–59. The 0.5-m link on the lift mechanism in Figure P7.54 rotates counterclockwise at a constant rate of 12 rpm. At the instant when $\theta = 20°$, graphically determine the linear acceleration of point X.

Coriolis Acceleration

7–60. For the kinematic diagram shown in Figure P7.60, at the instant when $\theta = 60°$, the angular velocity of link 2 is 30 rad/s clockwise. Slide 3 also moves outward on link 2 at a rate of 15 mm/s. Determine the Coriolis acceleration of point B on link 3 relative to link 2.

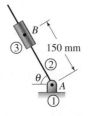

FIGURE P7.60 Problems 60–62.

7–61. For the kinematic diagram shown in Figure P7.60, at the instant when $\theta = 45°$, the angular velocity of link 2 is 30 rad/s counterclockwise. Slide 3 is also moving outward on link 2 at a rate of 15 mm/s. Determine the Coriolis acceleration of point B on link 3 relative to link 2.

7–62. For the kinematic diagram shown in Figure P7.60, at the instant when $\theta = 30°$, the angular velocity of link 2 is 30 rad/s clockwise. Slide 3 is also moving inward on link 2 at a rate of 15 mm/s. Determine the Coriolis acceleration of point B on link 3 relative to link 2.

Linkage Acceleration with

Coriolis—Graphical

7–63. For the kinematic diagram shown in Figure P7.63, the angular velocity of link 2 is 20 rad/s counterclockwise. Graphically determine the angular velocity of link 4, the sliding velocity of link 3 on link 4, and the angular acceleration of link 4.

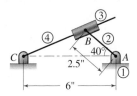

FIGURE P7.63 Problems 63, 64, and 71.

7–64. For the kinematic diagram shown in Figure P7.63, the angular velocity of link 2 is 20 rad/s counterclockwise, and it is accelerating at a rate of 5 rad/s². Graphically determine the angular velocity of link 4, the sliding velocity of link 3 on link 4, and the angular acceleration of link 4.

7–65. Figure P7.65 illustrates the driving mechanism in a saber saw. At the instant shown, the crank is rotating at a constant rate of 300 rpm clockwise. Graphically determine the linear acceleration of the saw blade.

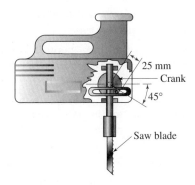

FIGURE P7.65 Problems 65, 66, and 72.

7–66. For the saber saw mechanism in Figure P7.65, the crank is rotating at a rate of 200 rpm clockwise and

accelerating at a rate of 45 rad/s². Graphically determine the linear acceleration of the saw blade.

7–67. Figure P7.67 illustrates a bicycle pump mechanism. At the instant shown, the cylinder is being retracted at a constant rate of 2 in./s. Graphically determine the rotational acceleration of the pedal assembly and the linear acceleration of point X.

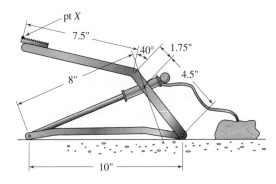

FIGURE P7.67 Problems 67, 68, and 73.

7–68. For the bicycle pump in Figure P7.67, the cylinder is retracting at a rate of 2 in./s and accelerating at a rate of 3 in./s². Graphically determine the rotational acceleration of the pedal assembly and the linear acceleration of point X.

7–69. Figure P7.69 illustrates a rudder mechanism used to steer ships. At the instant shown, the actuator is being extended at a constant rate of 0.1 m/s. Graphically determine the rotational velocity and acceleration of the rudder assembly.

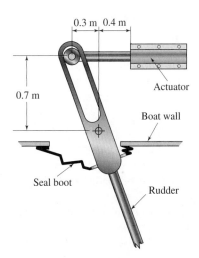

FIGURE P7.69 Problems 69, 70, and 74.

7–70. For the rudder mechanism in Figure P7.69, the actuator is extending at a rate of 0.1 m/s and decelerating at a rate of 0.3 m/s². Graphically determine the rotational velocity and acceleration of the rudder assembly.

Linkage Acceleration with
Coriolis—Analytical

7–71. For the kinematic diagram shown in Figure P7.63, the angular velocity of link 2 is 20 rad/s counterclockwise. Analytically determine the angular velocity of link 4, the sliding velocity of link 3 on link 4, and the angular acceleration of link 4.

7–72. Figure P7.65 illustrates the driving mechanism in a saber saw. At the instant shown, the crank is rotating at a constant rate of 300 rpm clockwise. Analytically determine the linear acceleration of the saw blade.

7–73. Figure P7.67 illustrates a bicycle pump mechanism. At the instant shown, the cylinder is being retracted at a constant rate of 2 in/s. Analytically determine the angular acceleration of the pedal assembly and the linear acceleration of point X.

7–74. Figure P7.69 illustrates a rudder mechanism used to steer ships. At the instant shown, the actuator is being extended at a constant rate of 0.1 m/s. Analytically determine the angular velocity and acceleration of the rudder assembly.

Acceleration Curves—Graphical

7–75. The crank of the compressor linkage shown in Figure P7.26 is driven clockwise at a constant rate of 1750 rpm. Graphically create a curve for the linear displacement of the piston as a function of the crank angle. Convert the crank angle axis to time. Then graphically calculate the slope to obtain velocity and acceleration curves of the piston as a function of time.

7–76. The crank of the sewing machine linkage shown in Figure P7.29 is driven counterclockwise at a constant rate of 175 rpm. Graphically create a curve for the linear displacement of the needle as a function of the crank angle. Convert the crank angle axis to time. Then graphically calculate the slope to obtain velocity and acceleration curves of the needle as a function of time.

7–77. The crank of the power hacksaw shown in Figure P7.32 is driven clockwise at a constant rate of 90 rpm. Graphically create a curve for the linear displacement of the blade as a function of the crank angle. Convert the crank angle axis to time. Then graphically calculate the slope to obtain velocity and acceleration curves of the blade as a function of time.

7–78. The motor on the coin-operated horse shown in Figure P7.35 is driven clockwise at a constant rate of 70 rpm. Graphically create a curve for the angular displacement of the horse as a function of the crank angle. Convert the crank angle axis to time. Then graphically calculate the slope to obtain angular velocity and angular acceleration curves of the horse as a function of time.

7–79. The motor on the car wash sprayer mechanism shown in Figure P7.38 is driven counterclockwise at a constant rate of 100 rpm. Graphically create a curve for the angular displacement of the nozzle arm as a function of the crank angle. Convert the crank angle axis to time. Then graphically calculate the slope to obtain angular velocity and angular acceleration curves of the nozzle arm as a function of time.

7–80. The crank on the landing gear mechanism shown in Figure P7.41 is driven counterclockwise at a constant rate of 18 rpm. Graphically create a curve for the angular displacement of the wheel assembly as a function of the crank angle. Convert the crank angle axis to time. Then graphically calculate the slope to obtain angular velocity and angular acceleration curves of the wheel assembly as a function of time.

Acceleration Curves—Graphical

7–81. The crank of the compressor linkage shown in Figure P7.26 is driven clockwise at a constant rate of 1450 rpm. Use a spreadsheet to analytically create a curve for the linear displacement of the piston as a function of the crank angle. Convert the crank angle to time. Then use numerical differentiation to obtain velocity and acceleration curves of the piston as a function of time.

7–82. The crank of the sewing machine linkage shown in Figure P7.29 is driven counterclockwise at a constant rate of 160 rpm. Use a spreadsheet to analytically create a curve for the linear displacement of the needle as a function of the crank angle. Convert the crank angle axis to time. Then use numerical differentiation to obtain velocity and acceleration curves of the needle as a function of time.

7–83. The crank of the power hacksaw shown in Figure P7.32 is driven clockwise at a constant rate of 85 rpm. Use a spreadsheet to analytically create a curve for the linear displacement of the blade as a function of the crank angle. Convert the crank angle axis to time. Then use numerical differentiation to obtain velocity and acceleration curves of the piston as a function of time.

7–84. The motor on the coin-operated horse shown in Figure P7.35 is driven clockwise at a constant rate of 80 rpm. Use a spreadsheet to analytically create a curve for the angular displacement of the horse as a function of the crank angle. Convert the crank angle axis to time. Then use numerical differentiation to obtain angular velocity and angular acceleration curves of the horse as a function of time.

7–85. The motor on the car wash sprayer mechanism shown in Figure P7.38 is driven counterclockwise at a constant rate of 110 rpm. Use a spreadsheet to analytically create a curve for the angular displacement of the nozzle as a function of the crank angle.

Convert the crank angle axis to time. Then use numerical differentiation to obtain angular velocity and angular acceleration curves of the nozzle as a function of time.

7–86. The crank on the landing gear mechanism shown in Figure P7.41 is driven counterclockwise at a constant rate of 16 rpm. Use a spreadsheet to analytically create a curve for the angular displacement of the wheel assembly as a function of the crank angle. Convert the crank angle axis to time. Then use numerical differentiation to obtain angular velocity and angular acceleration curves of the wheel assembly as a function of time.

Acceleration Using Working Model

7–87. The crank of the compressor linkage shown in Figure P7.26 is driven clockwise at a constant rate of 1750 rpm. Use the Working Model software to create a simulation and plot a linear acceleration curve of the piston as a function of time.

7–88. The crank of the sewing machine linkage shown in Figure P7.29 is driven counterclockwise at a constant rate of 175 rpm. Use the Working Model software to create a simulation and plot a linear acceleration curve of the needle as a function of time.

7–89. The crank of the power hacksaw shown in Figure P7.32 is driven clockwise at a constant rate of 90 rpm. Use the Working Model software to create a simulation and plot a linear acceleration curve of the blade as a function of time.

7–90. The motor on the coin-operated horse shown in Figure P7.35 is driven clockwise at a constant rate of 70 rpm. Use the Working Model software to create a simulation and plot an angular acceleration curve of the horse as a function of time.

7–91. The motor on the car wash sprayer mechanism shown in Figure P7.38 is driven counterclockwise at a constant rate of 100 rpm. Use the Working Model software to create a simulation and plot an angular acceleration curve of the nozzle arm as a function of time.

7–92. The crank on the landing gear mechanism shown in Figure P7.41 is driven counterclockwise at a constant rate of 18 rpm. Use the Working Model software to create a simulation and plot an angular acceleration curve of the wheel assembly as a function of time.

CASE STUDIES

7–1. Figure C7.1 shows a specialty machine that is driven by crankshaft I. The top cap H on the machine drives another mechanism, which is not shown. Carefully examine the components of the mechanism, then answer the following leading questions to gain insight into its operation.

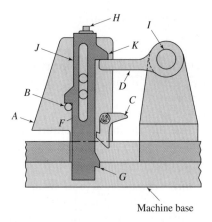

Machine base

FIGURE C7.1 (Courtesy, Industrial Press.)

1. As crankshaft I rotates clockwise 30° from the position shown, what is the motion of slide J?
2. As crankshaft I rotates a few more degrees clockwise, what happens to the mechanism?
3. What purpose does item C serve?
4. As crankshaft I continues to rotate, describe the motion of the slide.
5. What purpose does item B serve?
6. Describe the purpose of this mechanism.

7–2. Figure C7.2 shows a machine that feeds rivets to an automated assembly machine. Carefully examine the components of the mechanism, then answer the following leading questions to gain insight into its operation.

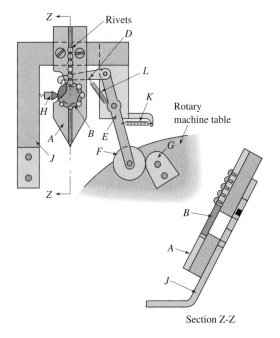

Section Z-Z

FIGURE C7.2 (Courtesy, Industrial Press.)

1. As the rotating machine table turns counterclockwise, what happens to lever E?
2. What purpose does spring K serve?

3. As the rotating table turns, what is the motion of item *D*?

4. What purpose does spring *L* serve?

5. What is the general name of the type of connection between items *B* and *D*? Describe the details of its function.

6. What is the purpose of the components at item *H*?

7. Describe the motion and actions that take place during the operation of this machine.

7–3. Figure C7.3 shows a specialty machine that accepts partially wrapped cartons from slot *B*. The machine folds the top and bottom wrappers down and moves the carton to another operation. In the position illustrated, a carton is shown at *A* and is being ejected from the machine. Carefully examine the components of the mechanism, then answer the following leading questions to gain insight into its operation.

1. As link *J* rotates clockwise 90° from the position shown, what is the motion of bellcrank *H*?

2. As link *J* rotates clockwise 90° from the position shown, what is the motion of pusher *E* and plate *C*?

3. As link *J* rotates clockwise 90° from the position shown, what is the motion of pin *S*? (Note that pin *S* is attached to slide *D* and is not constrained to ride in the groove.)

4. As link *J* rotates clockwise 90° from the position shown, what is the motion of guide pin *R*? (Note that pin *R* is constrained to ride in the groove.)

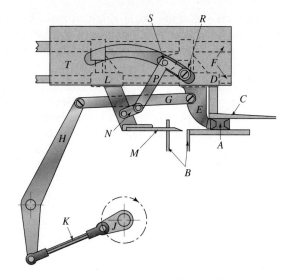

FIGURE C7.3 (Courtesy, Industrial Press.)

5. As link *J* rotates clockwise 90° from the position shown, what is the motion of bellcrank *P*?

6. As link *J* rotates clockwise 90° from the position shown, what is the motion of slide *L* and plate *M*?

7. Why is there a need for a short link *N*? Can link *P* be directly connected to slide *L*?

8. Comment on the relative spacing between plates *C* and *M* after link *J* rotates 90° clockwise.

9. Discuss the continual motion of plates *C* and *M* along with pusher *E*.

COMPUTER-AIDED MECHANISM ANALYSIS

OBJECTIVES

Upon completion of this chapter, the student will be able to:

1. Understand the basics of a general spreadsheet.

2. Understand the strategy for using a general spreadsheet for mechanism analysis.

3. Create computer routines for determining kinematic properties of either four-bar or slider-crank mechanisms.

8.1 INTRODUCTION

Throughout the text, both graphical and analytical techniques of mechanism analysis are introduced. As the more accurate, analytical solutions are desired for several positions of a mechanism, the number of calculations can become unwieldy. In these situations, the use of computer solutions is appropriate. Computer solutions are also valuable when several design iterations must be analyzed. In Section 2.2, "Computer Simulation of Mechanisms," the use of dedicated dynamic analysis software was introduced. This chapter focuses on other forms of computer approaches to mechanism analysis. These other forms include using spreadsheets and creating routines using programming languages.

8.2 SPREADSHEETS

Spreadsheets, such as Microsoft® Excel, are very popular in the professional environment for a variety of tasks. Spreadsheets have numerous built-in functions, ease of plotting results, and the ability to recognize formulas. These analytical features prompted widespread use of spreadsheets for more routine mechanism problems. Spreadsheets have been used in various problem solutions in this text. This section outlines the basics of using spreadsheets. Of course, the specific software manuals should be consulted for further details.

A spreadsheet is arranged in a large array of columns and rows. The number of columns and rows varies among the different software products. Column headings are lettered from A to Z, then from AA to AZ, then BA to BZ, and so on. Row headings are numbered 1, 2, 3, and so on. The top corner of a general spreadsheet is shown in Figure 8.1.

FIGURE 8.1 General spreadsheet.

The intersection of a column and a row is called a cell. Each cell is referred to by a cell address, which consists of the column and row that define the cell. Cell D3 is defined by the fourth (D) column and the third row. The cursor can be moved among cells with either the keyboard (arrow keys) or a mouse.

The value of a spreadsheet lies in storing, manipulating, and displaying data contained in a cell. This data commonly consists of either text, numbers, or formulas. The spreadsheet shown in Figure 8.2 has text entered into cells A1, F1, and F2 and numbers entered into cells A2 through A24, G1, and G2.

Although subtle differences may exist in the syntax among the spreadsheet programs, the logic behind creating formulas is identical. The syntax given here is applicable to Microsoft Excel. The user's manual of another product should be consulted for the details on any differences in syntax.

Entering a formula into a cell begins with an equal sign (=). The actual formula is then constructed using values, operators (+, −, *, /), cell references (e.g., G2), and functions (e.g., SIN, AVERAGE, ATAN, and RADIANS). Formulas for kinematic analysis can get rather complex. As an example, a simple formula can be placed in cell A8:

$$= A7 + 10 \tag{8.1}$$

Although the actual cell contents would contain this formula, the spreadsheet would visually show the number 60 in cell A8. The calculation would be automatically performed. For another example, the following expression can be inserted into cell B2:

$$= \text{ASIN}(G1 * \text{SIN}(A2 * \text{PI}()/180)/G2) * 180/\text{PI}() \tag{8.2}$$

This expression represents the angle between the connecting rod and the sliding plane for an in-line slider-crank mechanism. It was presented as equation (4.3) in Chapter 4:

$$\theta_3 = \sin^{-1}\left(\frac{L_2}{L_3} \sin\theta_2 \right) \tag{4.3}$$

The spreadsheet formula assumes that the following values have been entered:

- θ_2 in cell A2
- L_2 in cell G1
- L_3 in cell G2

It should be noted that as with most computer functions, any reference to angular values must be specified in radians. Notice that A2, an angle in degrees, is multiplied by $\pi/180$ to convert it to radians. After using the inverse sine function, ASIN, the resulting value also is an angle in radians. Therefore, it is converted back to degrees by multiplying by $180/\pi$. Excel has predefined RADIANS and DEGREES functions that can be convenient in conversions. Equation (4.3) can alternatively be inserted into a cell B2 of a spreadsheet with:

$$= \text{DEGREES}(\text{ASIN}(G1 * \text{SIN}(\text{RADIANS}(A2))/G2)) \tag{8.3}$$

If expression (8.1) were typed into A8 and expression (8.2) or (8.3) were typed into B2, the resulting spreadsheet would appear as depicted in Figure 8.3. It is important to remember that as a cell containing input data is changed, all results are updated. This allows design iterations to be completed with ease.

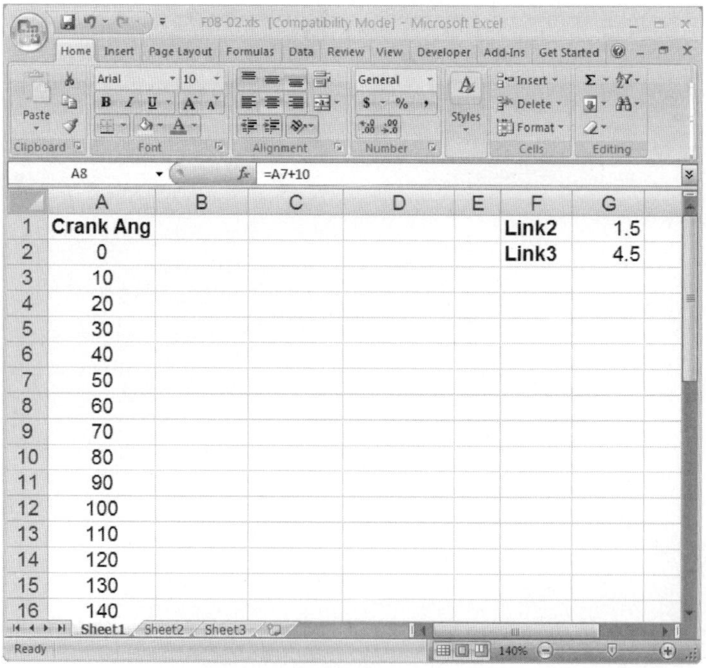

FIGURE 8.2 Spreadsheet with text and numbers entered into cells.

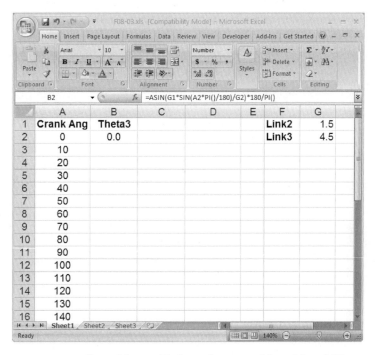

FIGURE 8.3 Spreadsheet with formulas entered into A8 and B2.

Another important feature of a spreadsheet is the copy and paste feature. The contents of a cell can be duplicated and placed into a new cell. The copy and paste feature eliminates redundant input of equations into cells.

Cell references in a formula can be either relative or absolute. Relative references are automatically adjusted when a copy of the cell is placed into a new cell. Consider the following formula entered in cell A8:

$$= A7 + 10$$

The cell reference A7 is a relative reference to the cell directly above the cell that contains the formula, A8. If this equation were copied and placed into cell A9, the new formula would become

$$= A8 + 10$$

Again, the cell reference A8 is a relative one; therefore, the spreadsheet would automatically adjust the formula.

An absolute address does not automatically adjust the cell reference after using the copy and paste feature. However, to specify an absolute reference, a dollar symbol must be placed prior to the row and column. For example, an absolute reference to cell G1 must appear as G1.

Consider expression (8.2) being placed into cell B2. To be most efficient, this formula should be slightly modified to read:

$$= ASIN(\$G\$1 * (SIN (A2 * PI()/180/\$G\$2)) * 180/PI()$$

In this manner, only the angle in cell A2 is a relative address. If the formula were copied to cell B3, the new formula would become

$$= ASIN(\$G\$1 * (SIN(A3 * PI()/180)/\$G\$2)) * 180PI()$$

Notice that the address of cell A2 has been automatically adjusted to read "A3." The connecting rod angle is calculated for the crank angle specified in cell A3.

To continue with an analysis of a mechanism, the following formula can be typed into cell C2:

$$= 180 - (A2 + B2)$$

This formula, shown in Figure 8.4, calculates the interior angle between the crank and connecting rod (equation 4.4):

$$\gamma = 180° - (\theta_2 + \theta_3) \qquad (4.4)$$

Because the angles are simply added, and a function is not called, a radian equivalent is not required.

Also, the following formula can be typed into cell D2:

$$= SQRT((\$G\$1)^2 + (\$G\$2)^2 - (2 * \$G\$1 * \$G\$2 * COS(C2 * PI()/180)))$$

This formula calculates the distance from the crank pivot to the slider pin joint (equation 4.5):

$$L_4 = \sqrt{L_2^2 + L_3^2 - 2(L_2)(L_3)\cos\gamma} \qquad (4.5)$$

If these two formulas were typed into C2 and D2, and text descriptions were typed into cells B1, C1, and D1, the resulting spreadsheet would appear as depicted in Figure 8.4.

Finally, because much care was taken with using absolute and relative cell addresses in creating the formulas in B2, C2, and D2, they can be copied into the cells down their respective columns. The user's manual should be consulted for the actual steps needed to copy the data into the remaining cells, which is usually a simple two- or three-step procedure. The resulting spreadsheet is shown in Figure 8.5.

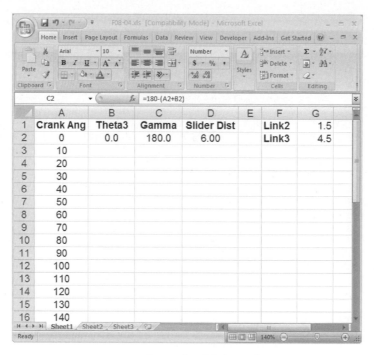

FIGURE 8.4 Formula added to cell C2.

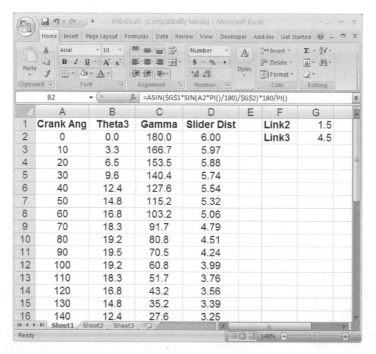

FIGURE 8.5 Final spreadsheet.

EXAMPLE PROBLEM 8.1

Figure 8.6 illustrates a linkage that operates a water nozzle at an automatic car wash. Using a spreadsheet, analytically determine the angular motion of the nozzle throughout the cycle of crank rotation.

SOLUTION:

The nozzle mechanism is a familiar four-bar linkage. Figure 8.7 shows the kinematic representation of this mechanism. A spreadsheet for this analysis has been set up and the upper portion is shown in Figure 8.8.

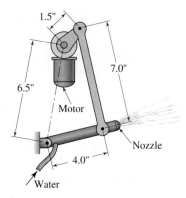

FIGURE 8.6 Water nozzle linkage for Example Problem 8.1.

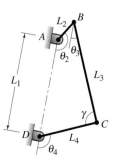

FIGURE 8.7 Kinematic sketch for Example Problem 8.1.

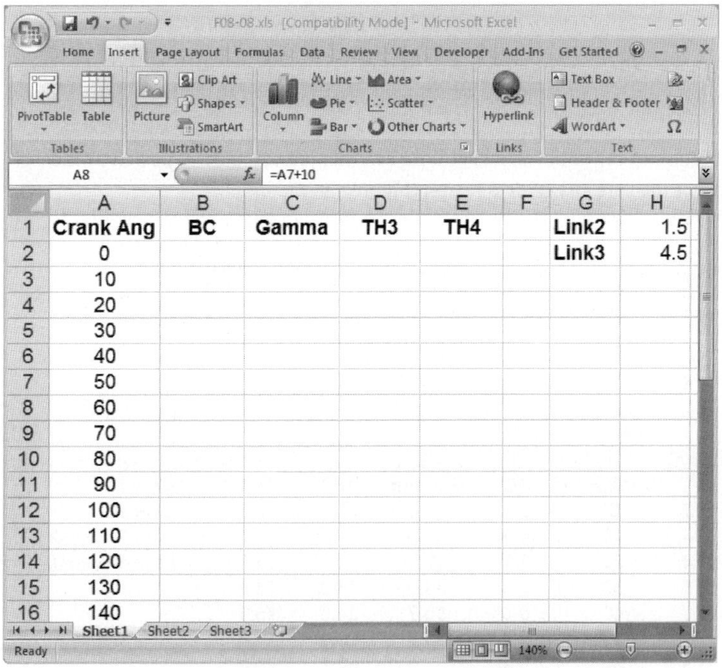

	A	B	C	D	E	F	G	H
1	**Crank Ang**	**BC**	**Gamma**	**TH3**	**TH4**		**Link2**	1.5
2	0						**Link3**	4.5
3	10							
4	20							
5	30							
6	40							
7	50							
8	60							
9	70							
10	80							
11	90							
12	100							
13	110							
14	120							
15	130							
16	140							

FIGURE 8.8 Spreadsheet for solution to Example Problem 8.1.

General equations, which govern the motion of the links of a four-bar mechanism, were given in Chapter 4. Equation 4.9 gave the general equation for the diagonal from point *B* to point *D*, as shown in Figure 8.7:

$$BD = \sqrt{L_1^2 + L_2^2 - 2(L_1)(L_2)\cos(\theta_2)}$$

A spreadsheet version of this equation can be placed and copied down column B. In cell B2, the following formula is inserted:

$$= \text{SQRT}(\$H\$1\char`^2 + \$H\$2\char`^2 - 2 * \$H\$1 * \$H\$2 * \text{COS}(\text{RADIANS}(A2)))$$

To facilitate copying the formula, note the use of absolute and relative addresses. Equation (4.10) gave the general equation for the transmission angle, γ, as shown in Figure 8.7:

$$\gamma = \cos^{-1} \frac{L_3^2 + L_4^2 + BD^2}{2L_3L_4}$$

A spreadsheet version of this equation can be placed and copied down column C. In cell C2, the following formula is inserted:

$$= \text{DEGREES(ACOS((H3$^\wedge$2 + H4$^\wedge$2 − B2$^\wedge$2)/(2 * H3 * H4)))}$$

Rewriting equation (4.11) will give the general equation for the angle of link 4, θ_4, as shown in Figure 8.7:

$$\theta_4 = 2 \tan^{-1} \left[\frac{L_2 \sin \theta_2 - L_3 \sin \gamma}{L_2 \cos \theta_2 + L_4 - L_1 - L_3 \cos \gamma} \right]$$

A spreadsheet version of this equation can be placed and copied down column E. In cell E2, the following formula is inserted:

$$= \text{DEGREES(2 * ATAN((H2 * SIN(RADIANS(A2)) − H3 * SIN(RADIANS(C2)))/}$$
$$\text{(H2 * COS(RADIANS(A2)) + H4 − H1 − H3 * COS(RADIANS(C2)))))}$$

Finally, equation (4.12) gave the general equation for the angle of link 3, θ_3, as shown in Figure 8.7:

$$\theta_3 = 2 \tan^{-1} \left[\frac{-L_2 \sin \theta_2 + L_4 \sin \gamma}{L_1 + L_3 - L_2 \cos \theta_2 - L_4 \cos \gamma} \right]$$

A spreadsheet version of this equation can be placed and copied down column D. In cell D2, the following formula is inserted:

$$= \text{DEGREES(2 * ATAN((−H2 * SIN(RADIANS(A2)) + H4 * SIN(RADIANS(C2)))/}$$
$$\text{(H1 + H3 − H2 * COS(RADIANS(A2)) − H4 * COS(RADIANS(C2)))))}$$

These formulas in cells B2, C2, D2, and E2 can then be copied and pasted in their respective columns. The resulting upper part of the spreadsheet is shown in Figure 8.9.

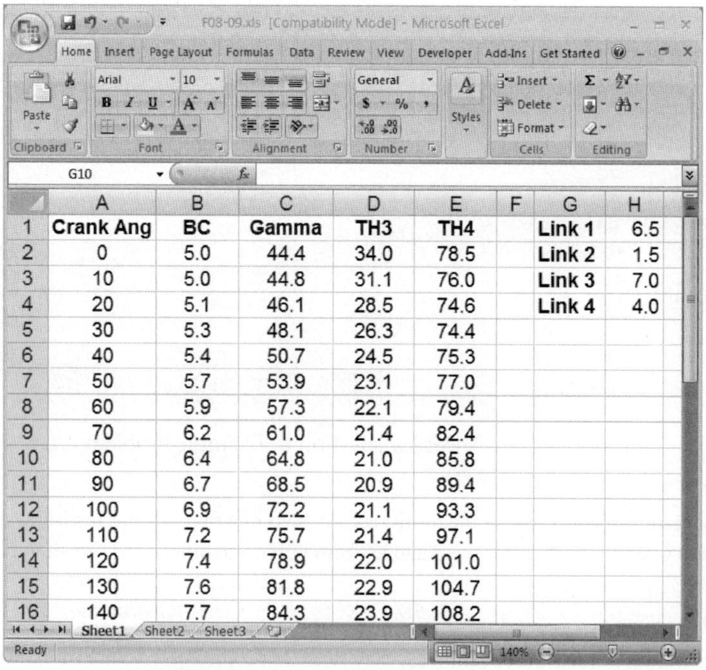

	A	B	C	D	E	F	G	H
1	Crank Ang	BC	Gamma	TH3	TH4		Link 1	6.5
2	0	5.0	44.4	34.0	78.5		Link 2	1.5
3	10	5.0	44.8	31.1	76.0		Link 3	7.0
4	20	5.1	46.1	28.5	74.6		Link 4	4.0
5	30	5.3	48.1	26.3	74.4			
6	40	5.4	50.7	24.5	75.3			
7	50	5.7	53.9	23.1	77.0			
8	60	5.9	57.3	22.1	79.4			
9	70	6.2	61.0	21.4	82.4			
10	80	6.4	64.8	21.0	85.8			
11	90	6.7	68.5	20.9	89.4			
12	100	6.9	72.2	21.1	93.3			
13	110	7.2	75.7	21.4	97.1			
14	120	7.4	78.9	22.0	101.0			
15	130	7.6	81.8	22.9	104.7			
16	140	7.7	84.3	23.9	108.2			

FIGURE 8.9 Completed spreadsheet for Example Problem 8.1.

8.3 USER-WRITTEN COMPUTER PROGRAMS

To solve mechanism problems, user-written computer routines can be written by using application software, such as MATHCAD or MATLAB, or a high-level language, such as VisualBasic or VisualC++. The programming language selected must have direct availability to trigonometric and inverse trigonometric functions. Due to the time and effort required to write a special program, they are most effective when a complex, yet commonly encountered, problem needs to be solved.

The logic behind writing computer programs to perform kinematic analysis is virtually identical to that for using a spreadsheet. The structure and syntax of the different high-level programming languages vary greatly. The following sections offer a strategy for writing computer programs to solve the kinematic properties of the two most common mechanisms, the slider-crank and the four-bar.

8.3.1 Offset Slider-Crank Mechanism

The following algorithm computes the position, velocity, and acceleration of all links of an offset slider-crank mechanism as the crank rotates at constant velocity. A kinematic sketch of a general offset slider-crank mechanism is shown in Figure 8.10. The general kinematic relationships used in the algorithm have been presented in various sections of this text [Ref. 12].

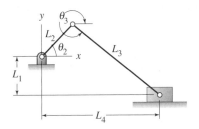

FIGURE 8.10 Offset slider-crank mechanism.

The dimensions of the mechanism are accepted as data, and the algorithm performs the calculations for one full cycle of crank rotation. The output can be either printed or written to a file. This file could then be converted to a spreadsheet if desired.

Step 1: Accept numeric data for L_1, L_2, L_3, and ω_2 and store
Step 2: Compute $\pi = 4 \tan^{-1}(1.0)$
Step 3: Enter a loop that indexes i from 0 to 360
Step 4: Compute $a = i (\pi/180)$
Step 5: Compute $b = L_2 \sin a$
Step 6: Compute $c = L_2 \cos a$
Step 7: Compute $d = -\sin^{-1}\{(L_1 + a)/L_3\}$
Step 8: Compute $\theta_3 = d(180/\pi)t$
Step 9: Compute $e = L_2 \sin d$
Step 10: Compute $f = L_2 \cos d$
Step 11: Compute $g = L_3 \sin d$
Step 12: Compute $h = L_3 \cos d$
Step 13: Compute $L_4 = c + f$
Step 14: Compute $\omega_3 = -\omega_2 (c/f)$

Step 15: Compute $v_4 = -\omega_2(b) - \omega_3(g)$
Step 16: Compute $\alpha_3 = \{b(\omega_2)^2 + g(\omega_3)^2\}/h$
Step 17: Compute $a_4 = -\{g(\alpha_3) + c(\omega_2)^2 + h(\omega_3)^2\}$
Step 18: Print (or write to file) i, θ_3, ω_3, α_3, L_4, v_4, a_4
Step 19: Increment i and return back to step 3

Recall that computer functions assume that angles are given in radians. Therefore, it is necessary to convert angular input and output as has been done in steps 4 and 8. This algorithm also works for an in-line slider-crank mechanism, by specifying $L_1 = 0$ as input.

8.3.2 Four-Bar Mechanism

The following algorithm computes the position, velocity, and acceleration of all links of a four-bar mechanism as the crank rotates at constant velocity. A kinematic sketch of a general four-bar mechanism is shown in Figure 8.11. Again, the general kinematic relationships used in this algorithm have been presented in various sections of this text [Ref. 12].

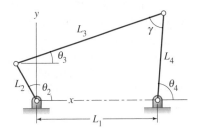

FIGURE 8.11 Four-bar mechanism.

As in the previous algorithm, the dimensions of the mechanism are accepted as data, and the algorithm performs the calculations for one full cycle of crank rotation. The output can be either printed or written to a file. This file could then be converted to a spreadsheet if desired.

Step 1: Accept numeric data for L_1, L_2, L_3, L_4, and ω_2 and store
Step 2: Compute $\pi = 4 \tan^{-1}(1.0)$
Step 3: Enter a loop that indexes i from 0 to 360
Step 4: Compute $a = i(\pi/180)$
Step 5: Compute $b = (L_3^2 + L_4^2 - L_1^2 - L_2^2)/(2L_3L_4)$
Step 6: Compute $c = L_1/L_3L_4$
Step 7: Compute $d = L_2 \sin a$
Step 8: Compute $e = L_2 \cos a$
Step 9: Compute $f = \cos^{-1}(b + ce)$
Step 10: Compute $\gamma = f(180/\pi)$
Step 11: Compute $g = \sin f$
Step 12: Compute $h = \cos f$
Step 13: Compute $p = 2 \tan^{-1}\{(-d + L_4g)/(-e + L_3 + L_1 - L_4h)\}$
Step 14: Compute $\theta_3 = p(180/\pi)$
Step 15: Compute $q = 2 \tan^{-1}\{(d - L_3g)/(e + L_4 - L_1 - L_3h)\}$
Step 16: Compute $\theta_4 = q(180/\pi)$
Step 17: Compute $\omega_3 = \omega_2L_2 \sin(q - a)/(L_3g)$
Step 18: Compute $\omega_4 = \omega_2L_2 \sin(p - a)/(L_4g)$
Step 19: Print (or write to file) i, γ, θ_3, ω_3, θ_4, ω_4
Step 20: Increment i and return back to step 3

Recall that computer functions will assume that angles are given in radians. Therefore, it is necessary to convert angular input and output as has been done in steps 4, 10, 14, and 16. This algorithm will give the solution for a four-bar mechanism in the first circuit. If the mechanism was assembled in the second circuit, this routine could be quickly modified to reflect that configuration. That can be accomplished by changing the plus and minus signs in the numerators of steps 13 and 15.

PROBLEMS

For Problems 8–1 and 8–2, develop a spreadsheet that can analyze the position of all links in an offset slider-crank mechanism for crank angles that range from 0 to 360. Keep it flexible so that the length of any link can be quickly altered. Using the listed values, produce a plot of the slider distance versus crank angle.

8–1 offset = 0.5 in.; crank = 1.25 in.; coupler = 7.0 in.

8–2 offset = 10 mm; crank = 25 mm; coupler = 140 mm.

For Problems 8–3 and 8–4, develop a spreadsheet that can analyze the position of all links in a four-bar mechanism for crank angles that range from 0 to 360. Keep it flexible so that the length of any link can be quickly altered. Using the listed values, produce a plot of the follower angle versus crank angle.

8–3 frame = 750 mm; crank = 50 mm; coupler = 750 mm; follower = 75 mm.

8–4 frame = 14 in.; crank = 1 in.; coupler = 16 in.; follower = 4.0 in.

For Problems 8–5 and 8–6, develop a spreadsheet that can determine the slider position, velocity, and acceleration for crank angles that range from 0 to 360. Keep it flexible so that the length of any link can be quickly altered. Using the listed values, produce a plot of the slider velocity versus crank angle.

8–5 offset = 1.25 in.; crank = 3.25 in.; coupler = 17.5 in.; crank speed = 20 rad/s; crank acceleration = 0 rad/s^2.

8–6 offset = 30 mm; crank = 75 mm; coupler = 420 mm; crank speed = 35 rad/s; crank acceleration = 100 rad/s^2.

For Problems 8–7 and 8–8, develop a spreadsheet that can determine the follower position and velocity for crank angles that range from 0 to 360. Keep it flexible so that the length of any link can be quickly altered. Using the following values, produce a plot of the follower velocity versus crank angle.

8–7 frame = 9 in.; crank = 1 in.; coupler = 10 in.; follower = 3.5 in.; crank speed = 200 rad/s; crank acceleration = 0 rad/s^2.

8–8 frame = 360 mm; crank = 40 mm; coupler = 400 mm; follower = 140 mm; crank speed = 6 rad/s; crank acceleration = 20 rad/s^2.

For Problems 8–9 and 8–10, develop a computer program that can determine the position, velocity, and acceleration of all links in a slider-crank mechanism for crank angles that range from 0 to 360. Keep it flexible so that the length of any link can be quickly altered. Using the listed values, determine the crank angle that produces the maximum slider acceleration.

8–9 offset = 3 in.; crank = 7.5 in.; coupler = 52.5 in.; crank speed = 4 rad/s; crank acceleration = 0 rad/s^2.

8–10 offset = 40 mm; crank = 94 mm; coupler = 525 mm; crank speed = 10 rad/s; crank acceleration = 10 rad/s^2.

For Problems 8–11 and 8–12, develop a computer program that can determine the position and velocity of all links in a four-bar mechanism for crank angles that range from 0 to 360. Using the listed values, determine the crank angle that produces the maximum slider acceleration.

8–11 frame = 18 in.; crank = 2 in.; coupler = 20 in.; follower = 7 in.; crank speed = 150 rad/s; crank acceleration = 0 rad/s^2.

8–12 frame = 60 mm; crank = 18 mm; coupler = 70 mm; follower = 32 mm; crank speed = 360 rad/s; crank acceleration = 20 rad/s^2.

CASE STUDY

8–1. The mechanism shown in Figure C8.1 is an elaborate crankshaft and crank for a slider-crank mechanism that is not shown, for which link K serves as the connecting rod. Carefully examine the components of the mechanism, then answer the following leading questions to gain insight into the operation.

1. In the position shown, as slide bar E pulls to the left, what is the motion of link D?

2. In the position shown, as slide bar E pulls to the left, what is the motion of slide block I?

3. Pulley J is keyed to shaft A. As pulley J rotates, what is the motion of crank pin C?

4. As pulley J rotates, what is the motion of slide bar E?

5. What effect does moving slide bar E to the left have on crank pin C and on the motion of the slider-crank mechanism it drives?

6. Sleeve F is keyed to the housing H. As pulley J drives shaft A, what is the motion of sleeve F?

7. Sleeve J is integrally molded with item G. What is item G?

8. Sleeve J has internal threads at its right end, and sleeve F has external threads at its right end. As item G rotates, what happens to sleeve F?

9. As item G rotates, what happens to slide bar E?

10. What is the purpose of this mechanism and how does it operate?

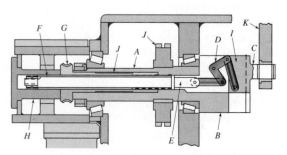

FIGURE C8.1 (Courtesy, Industrial Press.)

CAMS: DESIGN AND KINEMATIC ANALYSIS

OBJECTIVES

Upon completion of this chapter, the student will be able to:

1. Identify the different types of cams and cam followers.

2. Create a follower displacement diagram from prescribed follower motion criteria.

3. Understand the benefits of different follower motion schemes.

4. Use equations to construct cam follower displacement diagrams.

5. Geometrically construct cam follower displacement diagrams.

6. Graphically and analytically construct disk cam profiles with several types of followers.

7. Graphically and analytically construct cylindrical cam profiles.

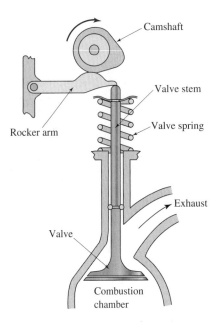

FIGURE 9.1 Engine valve train.

9.1 INTRODUCTION

A cam is a common mechanism element that drives a mating component known as a follower. From a functional viewpoint, a cam-and-follower arrangement is very similar to the linkages discussed throughout this book. The cam accepts an input motion similar to a crank and imparts a resultant motion to a follower.

Figure 9.1 illustrates one of the most common cam applications—namely, the valve train of an automotive engine. In this application, an oblong-shaped cam is machined on a shaft. This camshaft is driven by the engine. As the cam rotates, a rocker arm drags on its oblong surface. The rocker arm, in turn, imparts a linear, reciprocating motion to a valve stem. The motion of the valve must be such that the exhaust pathway is closed during a distinct portion of the combustion cycle and open during another distinct portion. Thus, the application is perfect for a cam because timing and motion must be precisely sequenced.

Notice that a spring is used around the valve stem. The rocker arm follower needs to maintain contact with the cam surface to achieve the desired motion. Thus, in most cam applications, the follower is forced against the cam surface through some mechanical means. Springs are very common

for this purpose. In cases where the follower is in the vertical plane, the weight of the follower may be sufficient to maintain contact. Some cam designs capture the follower in a groove to maintain contact. The important point is that contact between the cam and the follower must be sustained.

The unique feature of a cam is that it can impart a very distinct motion to its follower. In fact, cams can be used to obtain unusual or irregular motion that would be difficult, or impossible, to obtain from other linkages. Because the motion of cams can be prescribed, they are well suited for applications where distinct displacements and timing are paramount. Cams are often used in factory automation equipment because they can sequence displacements in a cost-efficient manner. Cams are precision machine components that generally cost more than conventional linkages. Figure 9.2 shows a selection of custom cams designed for special motion requirements. Note their precision machined outer profile. This chapter introduces the fundamentals of cam design.

9.2 TYPES OF CAMS

A great variety of cams are available from companies that specialize in design and manufacture. The manufacturers may classify cams into subcategories and market the cams

223

FIGURE 9.2 Various custom cams. (Courtesy of DE-STA-Co CAMCO Products.)

for different applications or different configurations. However, the great majority of cams can be separated into the following three general types:

> *Plate or disk cams* are the simplest and most common type of cam. A plate cam is illustrated in Figure 9.3a. This type of cam is formed on a disk or plate. The radial distance from the center of the disk is varied throughout the circumference of the cam. Allowing a follower to ride on this outer edge gives the follower a radial motion.
>
> A *cylindrical or drum cam* is illustrated in Figure 9.3b. This type of cam is formed on a cylinder. A groove is cut into the cylinder, with a varying location along the axis of rotation. Attaching a follower that rides in the groove gives the follower motion along the axis of rotation.
>
> A *linear cam* is illustrated in Figure 9.3c. This type of cam is formed on a translated block. A groove is cut into the block with a distance that varies from the plane of translation. Attaching a follower that rides in the groove gives the follower motion perpendicular to the plane of translation.

As mentioned, plate cams are the most common type of cam. Once the underlying theory is understood, it is also equally applicable to other types of cams.

9.3 TYPES OF FOLLOWERS

Followers are classified by their motion, shape, and position. The details of these classifications are shown in Figure 9.4 and discussed next.

9.3.1 Follower Motion

Follower motion can be separated into the following two categories:

> *Translating followers* are constrained to motion in a straight line and are shown in Figure 9.4a and c.
>
> *Swinging arm or pivoted followers* are constrained to rotational motion and are shown in Figure 9.4b and d.

9.3.2 Follower Position

The follower position, relative to the center of rotation of the cam, is typically influenced by any spacing requirements of

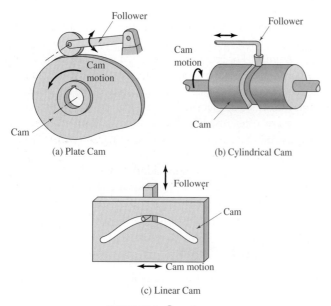

(a) Plate Cam   (b) Cylindrical Cam

(c) Linear Cam

FIGURE 9.3 Cam types.

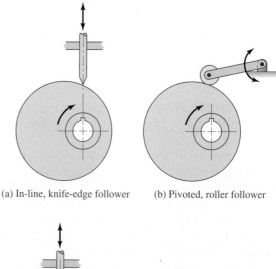

(a) In-line, knife-edge follower   (b) Pivoted, roller follower

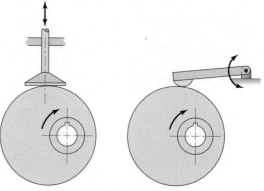

(c) Offset, flat-face follower   (d) Pivoted, spherical-face follower

FIGURE 9.4 Follower types.

the machine. The position of translating followers can be separated into two categories:

An *in-line follower* exhibits straight-line motion, such that the line of translation extends through the center of rotation of the cam and is shown in Figure 9.4a.

An *offset follower* exhibits straight-line motion, such that the line of the motion is offset from the center of rotation of the cam and is shown in Figure 9.4c.

In the case of pivoted followers, there is no need to distinguish between in-line and offset followers because they exhibit identical kinematics.

9.3.3 Follower Shape

Finally, the follower shape can be separated into the following four categories:

A *knife-edge follower* consists of a follower that is formed to a point and drags on the edge of the cam. The follower shown in Figure 9.4a is a knife-edge follower. It is the simplest form, but the sharp edge produces high contact stresses and wears rapidly. Consequently, this type of follower is rarely used.

A *roller follower* consists of a follower that has a separate part, the roller that is pinned to the follower stem. The follower shown in Figure 9.4b is a roller follower. As the cam rotates, the roller maintains contact with the cam and rolls on the cam surface. This is the most commonly used follower, as the friction and contact stresses are lower than those for the knife-edge follower. However, a roller follower can possibly jam during steep cam displacements. A more thorough discussion of the tendency for a follower to jam is presented later.

A *flat-faced follower* consists of a follower that is formed with a large, flat surface available to contact the cam. The follower shown in Figure 9.4c is a flat-faced follower. This type of follower can be used with a steep cam motion and does not jam. Consequently, this type of follower is used when quick motions are required. However, any follower deflection or misalignment causes high surface stresses. In addition, the frictional forces are greater than those of the roller follower because of the intense sliding contact between the cam and follower.

A *spherical-faced follower* consists of a follower formed with a radius face that contacts the cam. The follower shown in Figure 9.4d is a spherical-face follower. As with the flat-faced follower, the spherical-face can be used with a steep cam motion without jamming. The radius face compensates for deflection or misalignment. Yet, like the flat-faced follower, the frictional forces are greater than those of the roller follower.

Notice that these follower features are interchangeable. That is, any follower shape can be combined with either follower motion or position.

9.4 PRESCRIBED FOLLOWER MOTION

As mentioned, the unique feature of a cam is that it can impart a very distinct motion to its follower. Of course, the motion of the follower depends on the task required and can be prescribed to exacting detail.

For example, suppose a follower is used to drive pickup fingers on a paper-handling machine. Prescribing the desired follower involves separating the motion into segments and defining the action that must take place during the segments. To illustrate this process, assume that the pickup fingers must:

1. Remain closed for 0.03 s.
2. Open to a distance of 0.25 in., from the closed position, in 0.01 s.
3. Remain in this open position for 0.02 s.
4. Move to the closed position in 0.01 s.

Thus, by listing the exact requirements of the fingers, the motion of the follower has been prescribed.

Actually, the follower motion can be expressed in terms of angular cam displacement rather than time. This is more convenient in applications where the motion must be synchronized, such as the valve train shown in Figure 9.1.

For the pickup fingers just described, prescribed motion, stated in terms of cam rotation, could be listed as follows:

1. Remain closed for 154.3° of cam rotation.
2. Open to a distance of 0.25 in., from the closed position, in 51.4° of cam rotation.
3. Remain in this open position for 102.9° of cam rotation.
4. Move to the closed position in 51.4° of cam rotation.

Once the follower motion is prescribed, it is convenient to record it in a graphical form.

A plot of follower displacement versus time, or cam angular displacement, is termed a *follower displacement diagram*. This diagram is indispensable in that the follower motion and kinematics can be explored without regard to the shape of the cam itself. The vertical axis of this diagram displays the linear follower displacement, expressed in inches or millimeters. The horizontal axis displays time, measured in seconds or minutes, or angular cam displacements, measured in degrees or fractions of a revolution. This diagram is usually constructed to scale, and, along with follower kinematic analysis, it is extremely useful in determining cam shape.

For kinematic analysis, the follower displacement versus time curve is preferred. To assist in the task of designing a cam shape, the follower displacement versus cam angle curve is desired. Relating the cam rotation and time is a straightforward process using the theory presented in Chapter 6. Equation (6.4) gave the following:

$$\omega_{\text{cam}} = \frac{\Delta\theta}{\Delta t} \qquad (6.4)$$

When the cam is rotating at a constant velocity, which incorporates the overwhelming majority of applications, time can be related to angular displacement and vice versa. Cam rotation during an interval of follower motion is typically expressed by the symbol β. Likewise, the time consumed during an interval is designated T. The amount of follower rise, or fall, during an interval is designated H. Rewriting equation (6.4), using cam nomenclature, gives the relationship between cam rotation and time for an arbitrary interval, i:

$$\beta_i = (\omega_{cam})(T_i) \tag{9.1}$$

Equation (9.1) can also be used to determine the required speed of the cam, by observing the time consumed during one cycle.

$$\omega_{cam} = \frac{1\,rev}{\Sigma T_i} \tag{9.2}$$

where
ΣT_i = the total time for all motion intervals that comprise one cycle.

The period of cam rotation where there is no follower motion is termed a *dwell*. The details of motion during the follower raising and lowering intervals are primarily dictated by the task that needs to be accomplished and dynamic considerations. Because large forces are associated with large accelerations, there is a benefit to minimizing acceleration.

EXAMPLE PROBLEM 9.1

A cam is to be used for a platform that will repeatedly lift boxes from a lower conveyor to an upper conveyor. This machine is shown in Figure 9.5. Plot a displacement diagram and determine the required speed of the cam when the follower motion sequence is as follows:

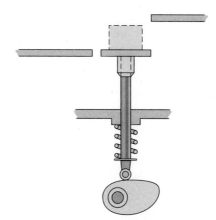

FIGURE 9.5 Cam system for Example Problem 9.1.

1. Rise 2 in. in 1.2 s.
2. Dwell for 0.3 s.
3. Fall 1 in. in 0.9 s.
4. Dwell 0.6 s.
5. Fall 1 in. in 0.9 s.

SOLUTION:

1. **Calculate the Time for a Full Cycle**

   The total time to complete the full cycle is needed to determine the required speed of the cam.

   $$\Sigma T_i = T_1 + T_2 + T_3 + T_4 + T_5 = (1.2 + 0.3 + 0.9 + 0.6 + 0.9)\,s = 3.9\,s$$

2. **Calculate the Required Rotational Speed of the Cam**

   Then from equation (9.2),

   $$\omega_{cam} = \frac{1\,rev}{\Sigma T_i} = \frac{1\,rev}{3.9\,s} = 0.256\,rev/s \left(\frac{60\,s}{1\,min}\right) = 15.38\,rpm$$

3. **Determine the Cam Rotation for Each Follower Motion Interval**

   The angular increment of the cam consumed by each follower motion sequence is determined by equation (9.1).

   $$\beta_1 = (\omega_{cam})(T_i) = (0.256\,rev/s)(1.2\,s) = 0.307\,rev$$
   $$= (0.307\,rev)(360°/1\,rev) = 110.5°$$

$$\beta_2 = (0.256\,\text{rev/s})(0.3\,\text{s}) = 0.077\,\text{rev} = 27.6°$$
$$\beta_3 = (0.256\,\text{rev/s})(0.9\,\text{s}) = 0.230\,\text{rev} = 82.9°$$
$$\beta_4 = (0.256\,\text{rev/s})(0.6\,\text{s}) = 0.154\,\text{rev} = 55.3°$$
$$\beta_5 = (0.256\,\text{rev/s})(0.9\,\text{s}) = 0.230\,\text{rev} = 82.9°$$

4. ***Plot the Displacement Diagram***

The resulting displacement diagram with both cam angle and time displayed on the horizontal axis is shown in Figure 9.6. Notice that a curved displacement profile was constructed during the rise and fall sequences. Dynamic considerations dictate the actual shape of the rise and fall sections.

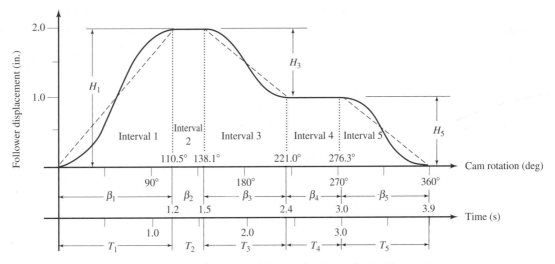

FIGURE 9.6 Displacement diagram for Example Problem 9.1.

9.5 FOLLOWER MOTION SCHEMES

In designing a cam, the objective is to identify a suitable shape for the cam. The primary interest is to ensure that the follower will achieve the desired displacements. Of course, these displacements are outlined in the displacement diagram. The shape of the cam is merely a means to obtain this motion.

In the previous section, the follower motion during rise and fall sequences was not fully identified. It was mentioned that the dynamic characteristics of the follower are important. Large accelerations cause large forces and, consequently, high stresses. Rapidly changing accelerations cause vibration and, consequently, noise. Due to these fundamental dynamic principles, the rise and fall portions of a cam displacement diagram are of vital importance.

For slow-moving cams, high accelerations are not a factor. Therefore, the cam is designed to merely yield the given displacements at the specified instant. The manner in which the follower arrives at the given point is trivial. In these cases, the cam is manufactured in the most convenient manner, as long as the given displacement is achieved. A plate cam can be simply composed of a combination of circular arcs and straight lines, which can be readily manufactured.

For high-speed applications, it is not enough to provide a given displacement. The dynamic characteristics of the follower during the rise and fall sequences must be specified in considerable detail in order to minimize the forces and vibrations.

A wide variety of motion schemes are available for moving the follower. The objective of these schemes is to produce the movement with smooth accelerations. In studying the dynamic characteristics of the follower for the different motion schemes, the following nomenclature is used:

H = Total follower displacement during the rise or fall interval under consideration. In the case of a pivoted follower, this is the total angular displacement of the follower link, $\Delta\theta_L$, during the particular interval.

T = Total time period for the rise or fall interval under consideration.

t = Time into rise or fall interval that defines the instantaneous follower properties.

β = Rotation angle of cam during the rise or fall interval under consideration (deg).

ϕ = Angle into rise or fall interval that defines the instantaneous follower properties (deg).

ω_{cam} = Speed of the cam (degrees per time).

ΔR = Magnitude of the instantaneous follower displacement at time t or cam angle β. In the case of a pivoted follower, this is analogous to the instantaneous angular displacement of the follower link, $\Delta\theta_L$.

v = Magnitude of the instantaneous follower velocity = dR/dt. In the case of a pivoted follower, this is analogous to the rotation of the follower link, ω_L.

a = Magnitude of the instantaneous follower acceleration = dv/dt.

9.5.1 Constant Velocity

The simplest follower motion during a rise or fall scheme is constant velocity. Constant velocity motion is characterized with a straight-line displacement diagram because velocity is uniform. The dynamic characteristics of a constant velocity rise are listed in Table 9.1.

Although the notion of zero acceleration is appealing, the ends of this motion scheme cause problems. Theoretically, the instantaneous jump from any constant value of velocity to another constant value of velocity results in an infinite acceleration. Because the machines driven by the follower will always have mass, this theoretically results in an infinite force. In practice, an instantaneous change in velocity is impossible due to the flexibility in machine members. Nevertheless, any shock is serious and must be kept to a minimum. Therefore, this motion in its pure form is impractical except for low-speed applications.

A constant velocity displacement diagram, along with velocity and acceleration curves, is shown in Figure 9.7.

9.5.2 Constant Acceleration

Constant acceleration motion during a rise or fall sequence produces the smallest possible values of acceleration for a given rise and time interval. The displacement diagram for a rise or fall interval is divided into two halves, one of constant acceleration and the other of constant deceleration. The shapes of each half of the displacement diagram are mirror-image parabolas. The dynamic characteristics of a constant acceleration rise are listed in Table 9.2.

This motion scheme, also known as parabolic or gravity motion, has constant positive and negative accelerations. However, it has an abrupt change of acceleration at the end of the motion and at the transition point between the acceleration and deceleration halves. These abrupt changes

result in abrupt changes in inertial forces, which typically cause undesirable vibrations. Therefore, this motion in its pure form is uncommon except for low-speed applications. A constant acceleration displacement diagram, along with velocity and acceleration curves, is shown in Figure 9.8.

A scaled displacement diagram is required to construct the actual cam profile. The equations presented in Table 9.2 can be used in conjunction with a spreadsheet or other equation-plotting package to complete this diagram. Although this analytical method is precise, care must be taken to plot the diagram to scale.

Graphical construction of a displacement diagram is an alternative method to generating a displacement diagram to scale. Such a construction using the constant acceleration motion scheme can be accomplished by referring to Figure 9.9 and using the following procedure:

1. Divide the follower rise (or fall) sequence into two halves. From Figure 9.9, AE represents the time period and EF the magnitude of rise for the first half of this motion scheme.

2. Divide both the horizontal and vertical axes of the quadrant AEFH into equal parts.

3. Construct vertical lines from the horizontal divisions.

4. Construct straight lines from corner A to the vertical divisions.

5. Draw a smooth curve through the points of intersection of the vertical lines and the lines drawn from corner A.

6. Repeat steps 2 through 5 for the remaining half of the curve as shown in quadrant FICG in Figure 9.9.

A constant acceleration fall is constructed as a mirror image to Figure 9.9.

9.5.3 Harmonic Motion

As seen with the polynomial follower schemes just described, inertial problems arise with discontinuities in the motion curves. To address that shortcoming, harmonic motion has been studied. Harmonic motion is derived from trigonometric functions, thus exhibiting very smooth motion curves. In a physical sense, it is the projection motion of a point on a rotating disk projected to a straight line. The dynamic characteristics of a harmonic rise are listed in Table 9.3.

This motion scheme is a definite improvement on the previous curves. It has a smooth, continuous acceleration.

TABLE 9.1	Cam Follower Kinematics for Constant Velocity Motion	
	Rise	**Fall**
Displacement:	$\Delta R_i = H_0 + \dfrac{H_i t_i}{T_i} = H_0 + \dfrac{H_i \phi_i}{\beta_i}$	$\Delta R_j = H_F + H_j\left(1 - \dfrac{t_j}{T_j}\right) = H_F + H_j\left(1 - \dfrac{\phi_j}{\beta_j}\right)$
Velocity:	$v_i = \dfrac{H_i}{T_i} = \dfrac{H_i \omega}{\beta_i}$	$v_j = \dfrac{-H_j}{T_j} = \dfrac{-H_j \omega}{\beta_j}$
Acceleration:	$a = 0$ (∞ at transitions)	$a = 0$ (∞ at transitions)

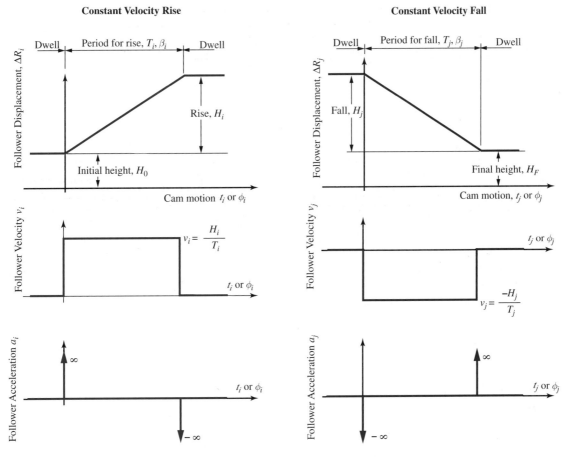

FIGURE 9.7 Constant velocity motion curves.

TABLE 9.2	Cam Follower Kinematics for Constant Acceleration Motion	
	Rise	**Fall**
For $0 < t < 0.5\,T\ (0 < \phi < 0.5\,\beta)$:		
Displacement:	$\Delta R_i = H_0 + 2H_i\left(\dfrac{t_i}{T_i}\right)^2$ $= H_0 + 2H_i\left(\dfrac{\phi_i}{\beta_i}\right)^2$	$\Delta R_i = H_F + H_j - 2H_j\left(\dfrac{t_j}{T_j}\right)^2$ $= H_F + H_j - 2H_j\left(\dfrac{\phi_j}{\beta_j}\right)^2$
Velocity:	$v_i = \dfrac{4H_i t_i}{T_i^2} = \dfrac{4H_i \omega \phi_i}{\beta_i^2}$	$v_j = \dfrac{-4H_j t_j}{T_j^2} = \dfrac{-4H_j \omega \phi_j}{\beta_j^2}$
Acceleration:	$a_i = \dfrac{4H_i}{T_i^2} = \dfrac{4H_i \omega^2}{\beta_i^2}$	$a_j = \dfrac{-4H_j}{T_j^2} = \dfrac{-4H_j \omega^2}{\beta_j^2}$
For $0.5\,T < t < T\ (0.5\,\beta < \phi < \beta)$:		
Displacement:	$\Delta R_i = H_0 + H_i - 2H_i\left(1 - \dfrac{t_i}{T_i}\right)^2$ $= H_0 + H_i + 2H_i\left(1 - \dfrac{\phi_i}{\beta_i}\right)^2$	$\Delta R_j = H_F + 2H_j\left(1 - \dfrac{t_j}{T_j}\right)^2$ $= H_F + 2H_j\left(1 - \dfrac{\phi_j}{\beta_j}\right)^2$
Velocity:	$v_i = \dfrac{4H_i}{T_i}\left(1 - \dfrac{t_i}{T_i}\right) = \dfrac{4H_i \omega}{\beta_i}\left(1 - \dfrac{\phi_i}{\beta_i}\right)$	$v_i = \dfrac{-4H_j}{T_j}\left(1 - \dfrac{t_j}{T_j}\right) = \dfrac{-4H_j \omega}{\beta_j}\left(1 - \dfrac{\phi_j}{\beta_j}\right)$
Acceleration:	$a_i = \dfrac{-4H_i}{T_i^2} = \dfrac{-4H_i \omega^2}{\beta_i^2}$	$a_j = \dfrac{4H_j}{T_j^2} = \dfrac{4H_j \omega^2}{\beta_j^2}$

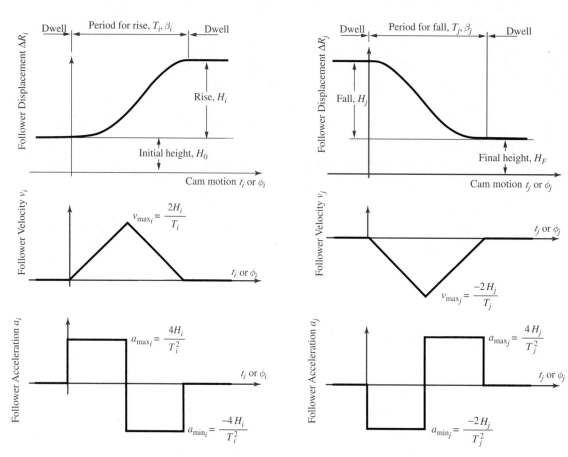

FIGURE 9.8 Constant acceleration motion curves.

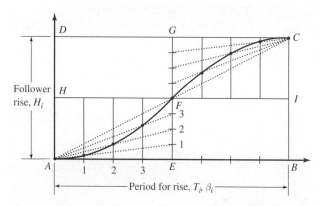

FIGURE 9.9 Construction of a constant acceleration displacement diagram.

However, it has a sudden change of acceleration at the ends of the motion. Again, this sudden change can be objectionable at higher speeds.

A harmonic displacement diagram, along with velocity and acceleration curves, is shown in Figure 9.10.

As with other schemes, a scaled displacement diagram is required to construct the actual cam profile. The equations in Table 9.3 can be used in conjunction with a spreadsheet or other equation-plotting package to complete this diagram. Although this analytical method is precise, care must be taken to plot the diagram accurately.

Graphical construction of a displacement diagram is an alternative method to generate a displacement diagram to scale. Such a construction using the harmonic motion scheme can be accomplished by referring to Figure 9.11 and using the following procedure:

1. Construct a semicircle having a diameter equal to the amount of rise (or fall) desired.

2. Divide the rise time period into incremental divisions.

3. Divide the semicircle into the same number of equal divisions of the follower rise period.

4. Draw vertical lines from the divisions on the time axis.

5. Draw horizontal lines from the division points on the semicircle to the corresponding division lines on the time axis.

6. Draw a smooth curve through the points of intersection found in the previous step.

A harmonic fall is constructed as a mirror image to Figure 9.11.

9.5.4 Cycloidal Motion

Cycloidal motion is another motion scheme derived from trigonometric functions. This scheme also exhibits very smooth motion curves and does not have the sudden change

TABLE 9.3   Cam Follower Kinematics for Harmonic Motion

	Rise	Fall
Displacement:	$\Delta R_i = H_0 + \dfrac{H_i}{2}\left[1 - \cos\left(\dfrac{\pi t_i}{T_i}\right)\right]$	$\Delta R_j = H_F + \dfrac{H_j}{2}\left[1 + \cos\left(\dfrac{\pi t_j}{T_j}\right)\right]$
	$= H_0 + \dfrac{H_i}{2}\left[1 - \cos\left(\dfrac{\pi \phi_i}{\beta_i}\right)\right]$	$= H_F + \dfrac{H_j}{2}\left[1 - \cos\left(\dfrac{\pi \phi_j}{\beta_j}\right)\right]$
Velocity:	$v_i = \dfrac{\pi H_i}{2T_i}\left[\sin\left(\dfrac{\pi t_i}{T_i}\right)\right]$	$v_j = \dfrac{-\pi H_j}{2T_j}\left[\sin\left(\dfrac{\pi t_j}{T_j}\right)\right]$
	$= \dfrac{\pi H_i \omega}{2\beta_i}\left[\sin\left(\dfrac{\pi \phi_i}{\beta_i}\right)\right]$	$= \dfrac{-\pi H_j \omega}{2\beta_j}\left[\sin\left(\dfrac{\pi \phi_j}{\beta_j}\right)\right]$
Acceleration:	$a_i = \dfrac{\pi^2 H_i}{2T_i^2}\left[\cos\left(\dfrac{\pi t_i}{T_i}\right)\right]$	$a_j = \dfrac{-\pi^2 H_i}{2T_i^2}\left[\cos\left(\dfrac{\pi t_j}{T_j}\right)\right]$
	$= \dfrac{\pi^2 H_i \omega^2}{2\beta_i^2}\left[\cos\left(\dfrac{\pi \phi_i}{\beta_i}\right)\right]$	$= \dfrac{-\pi^2 H_j \omega^2}{2\beta_j^2}\left[\cos\left(\dfrac{\pi \phi_j}{\beta_j}\right)\right]$

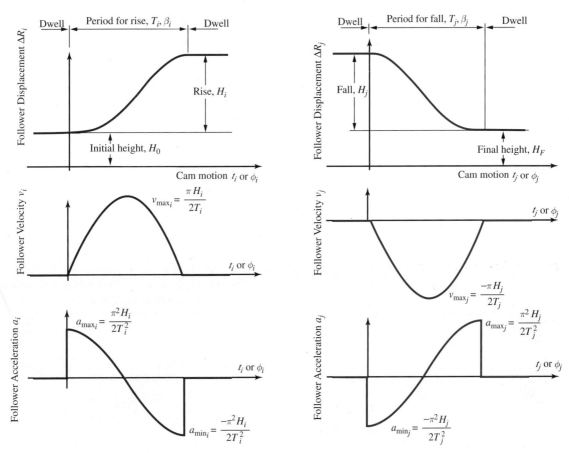

FIGURE 9.10 Harmonic motion curves.

in acceleration at the ends of the motion, which makes it popular for high-speed applications. It has low vibration wear and stress characteristics of all the basic curves described. In a physical sense, it is the motion of a point on a disk rolling on a straight line. The dynamic characteristics of

a cycloidal rise are listed in Table 9.4. A cycloidal displacement diagram, along with velocity and acceleration curves, is shown in Figure 9.12.

As before, a scaled displacement diagram is required to construct the actual cam profile. The equations presented in

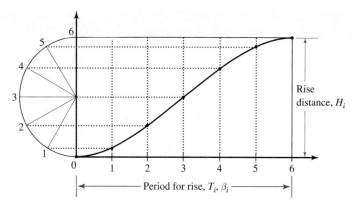

FIGURE 9.11 Construction of a harmonic displacement diagram.

Table 9.4 can be used in conjunction with a spreadsheet or other equation-plotting package to complete this diagram. Although this analytical method is precise, care must be taken to plot the diagram to full scale if graphical construction techniques will be used to design the cam.

Graphical construction of a displacement diagram is an alternative method to generating a displacement diagram to scale. Such a construction using the cycloidal motion scheme can be accomplished by referring to Figure 9.13 and using the following procedure:

1. On a displacement diagram grid, draw a line from the beginning point of the rise (or fall) to the final point. This line is drawn from A to C on Figure 9.13.

2. Extend the line drawn in the previous step and draw a circle, with radius $r = H/2\pi$ centered anywhere on that line.

3. Construct a vertical line through the center of the circle.

4. Divide the circle into an even number of parts.

5. Connect the circle division lines as shown in Figure 9.13 (1 to 4, 2 to 5, etc.).

6. Mark the intersection points of the lines drawn in step 5 with the vertical line drawn in step 3.

7. Divide the time period into the same number of equal parts as the circle. Construct vertical lines from these division points.

8. Project the points identified in step 6 along a line parallel with the line constructed in step 1.

9. Mark intersection points of the lines constructed in step 8 with the vertical lines drawn in step 7, as shown in Figure 9.13.

10. Construct a smooth curve through the points identified in step 9.

A cycloidal fall is constructed as a mirror image of Figure 9.13.

TABLE 9.4	Cam Follower Kinematics for Cycloidal Motion	
	Rise	**Fall**
Displacement:	$\Delta R_i = H_0 + H_i\left[\dfrac{t_i}{T_i} - \dfrac{1}{2\pi}\sin\left(\dfrac{2\pi t_i}{T_i}\right)\right]$ $= H_0 + H_i\left[\dfrac{\phi_i}{\beta_i} - \dfrac{1}{2\pi}\sin\left(\dfrac{2\pi \phi_i}{\beta_i}\right)\right]$	$\Delta R_j = H_F + H_j\left[1 - \dfrac{t_j}{T_j} + \dfrac{1}{2\pi}\sin\left(\dfrac{2\pi t_j}{T_j}\right)\right]$ $= H_F + H_j\left[\dfrac{\phi_j}{\beta_j} - \dfrac{1}{2\pi}\sin\left(\dfrac{2\pi \phi_j}{\beta_j}\right)\right]$
Velocity:	$v_i = \dfrac{H_i}{T_i}\left[1 - \cos\left(\dfrac{2\pi t_i}{T_i}\right)\right]$ $= \dfrac{H_i\omega}{\beta_i}\left[1 - \cos\left(\dfrac{2\pi \phi_i}{\beta_i}\right)\right]$	$v_j = \dfrac{-H_j}{T_j}\left[1 - \cos\left(\dfrac{2\pi t_i}{T_i}\right)\right]$ $= \dfrac{-H_j\omega}{\beta_j}\left[1 - \cos\left(\dfrac{2\pi \phi_j}{\beta_j}\right)\right]$
Acceleration:	$a_i = \dfrac{2\pi H_i}{T_i^2}\left[\sin\left(\dfrac{2\pi t_i}{T_i}\right)\right]$ $= \dfrac{2\pi H_i\omega^2}{\beta_i^2}\left[\sin\left(\dfrac{2\pi \phi_i}{\beta_i}\right)\right]$	$a_j = \dfrac{-2\pi H_j}{T_j^2}\left[\sin\left(\dfrac{2\pi t_i}{T_i}\right)\right]$ $= \dfrac{-2\pi H_j\omega^2}{\beta_j^2}\left[\sin\left(\dfrac{2\pi \phi_j}{\beta_j}\right)\right]$

Cycloidal Rise   **Cycloidal Fall**

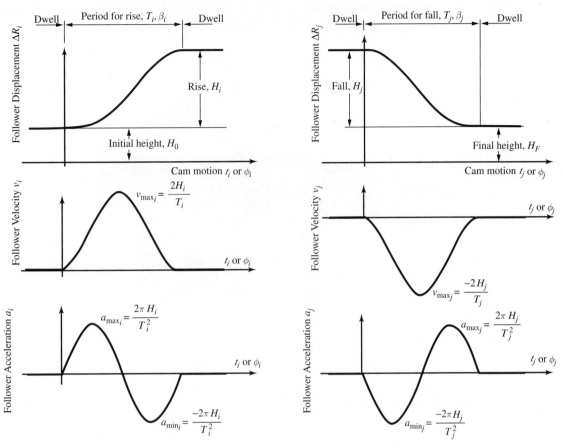

FIGURE 9.12 Cycloidal motion curves.

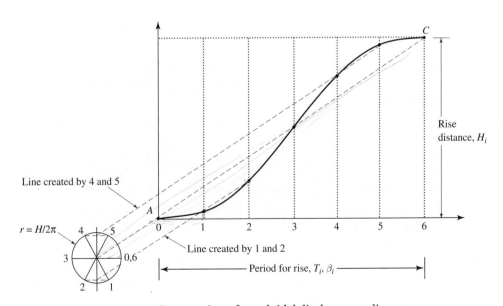

FIGURE 9.13 Construction of a cycloidal displacement diagram.

EXAMPLE PROBLEM 9.2

A cam is to be designed for an automated part loader as shown in Figure 9.14. Using the motion equations, construct a chart that tabulates follower displacement versus time and cam rotation. Also plot this data when the prescribed motion for this application is as follows:

1. Rise 50 mm in 1.5 s using the constant velocity motion scheme.
2. Return in 2.0 s using the cycloidal motion scheme.
3. Dwell for 0.75 s.
4. Repeat the sequence.

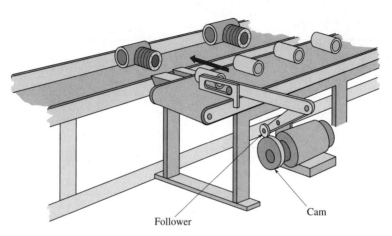

Cam

Follower

FIGURE 9.14 Part loader for Example Problem 9.2.

SOLUTION:

1. *Calculate Time to Complete a Full Cycle*

   The total time to complete the full cycle is needed to determine the required speed of the cam.

   $$\Sigma T_i = T_1 + T_2 + T_3$$
   $$= 1.5 + 2.0 + 0.75 = 4.25\,\text{s}$$

2. *Calculate the Required Rotational Speed of the Cam*

   From equation (9.2),

   $$\omega_{\text{cam}} = \frac{1\,\text{rev}}{\Sigma T_i} = \frac{1\,\text{rev}}{4.25\,\text{s}} = 0.235\,\text{rev/s}\left(\frac{60\,\text{s}}{1\,\text{min}}\right) = 14.12\,\text{rpm}$$

3. *Determine the Cam Rotation for Each Follower Motion Interval*

   The angular increment of the cam consumed by each follower motion sequence is determined by equation (9.1).

   $$\beta_1 = (\omega_{\text{cam}})(T_1) = (0.235\,\text{rev/s})(1.5\,\text{s}) = 0.353\,\text{rev} = 127.0°$$
   $$\beta_2 = (0.235\,\text{rev/s})(2.0\,\text{s}) = 0.470\,\text{rev} = 169.3°$$
   $$\beta_3 = (0.235\,\text{rev/s})(0.75\,\text{s}) = 0.177\,\text{rev} = 63.7°$$

4. *Calculate the Displacement during Each Follower Motion Interval*

   The first motion interval has $H_1 = 50$ mm and $T_1 = 1.5$ s. For a constant velocity rise, the displacement equation is given as

   $$\Delta R_1 = \frac{H_1 t_1}{T_1}$$

   The second motion interval has $H_2 = 50$ mm and $T_2 = 2.0$ s. For a cycloidal fall, the displacement equation is given as

   $$\Delta R_2 = H_2\left[1 - \left(\frac{t_2}{T_2}\right) + \frac{1}{2\pi}\sin\left(\frac{2\pi t_2}{T_2}\right)\right]$$

   The last motion interval is a dwell, where $\Delta R = $ constant. This dwell occurs at the retracted follower position; thus, $\Delta R_3 = 0$.

These equations were substituted into a spreadsheet (Figure 9.15). This data is used to produce the plot in Figure 9.16.

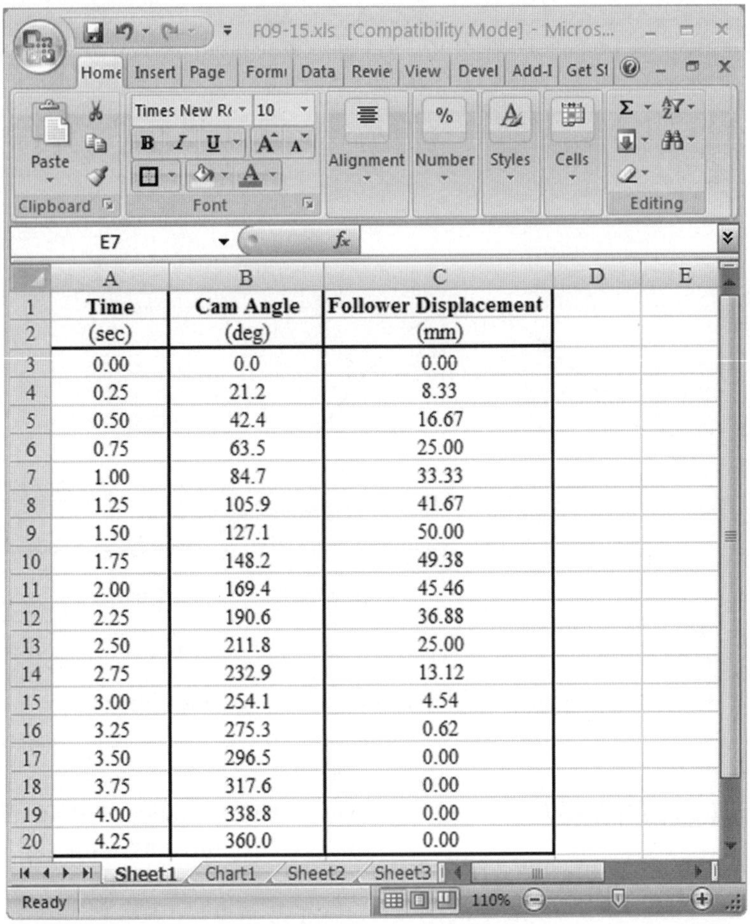

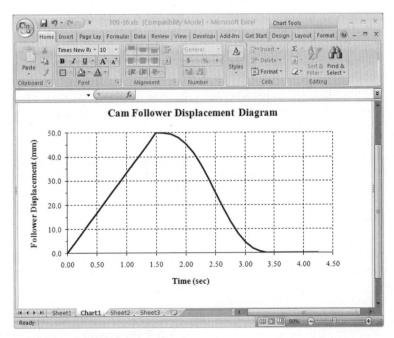

	A	B	C	D	E
1	**Time**	**Cam Angle**	**Follower Displacement**		
2	(sec)	(deg)	(mm)		
3	0.00	0.0	0.00		
4	0.25	21.2	8.33		
5	0.50	42.4	16.67		
6	0.75	63.5	25.00		
7	1.00	84.7	33.33		
8	1.25	105.9	41.67		
9	1.50	127.1	50.00		
10	1.75	148.2	49.38		
11	2.00	169.4	45.46		
12	2.25	190.6	36.88		
13	2.50	211.8	25.00		
14	2.75	232.9	13.12		
15	3.00	254.1	4.54		
16	3.25	275.3	0.62		
17	3.50	296.5	0.00		
18	3.75	317.6	0.00		
19	4.00	338.8	0.00		
20	4.25	360.0	0.00		

FIGURE 9.15 Follower displacement plot for Example Problem 9.2.

FIGURE 9.16 Spreadsheet for Example Problem 9.2.

EXAMPLE PROBLEM 9.3

For the application presented in Example Problem 9.2, graphically construct a follower displacement diagram.

SOLUTION: Using the data from Example Problem 9.2, the displacement diagram shown in Figure 9.17 can be constructed. Note that the circle used to construct the cycloidal fall has a radius of

$$r = \frac{H_1}{2\pi} = \frac{(50\,\text{mm})}{2\pi} = 7.96\,\text{mm}$$

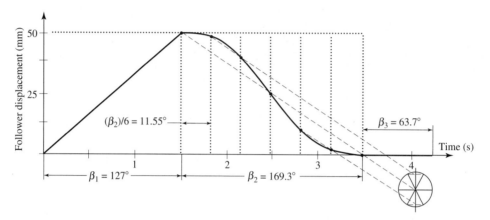

FIGURE 9.17   Follower displacement plot for Example Problem 9.3.

9.5.5 Combined Motion Schemes

In selecting a particular motion scheme, one goal is to minimize the dynamic forces induced during the rise or fall interval. This is done by minimizing the magnitude of the follower acceleration and keeping it continuous. Additionally, the kinetic energy stored in the follower is proportional to the square of the velocity. Thus, minimizing the maximum velocity is another goal that should be considered when specifying a motion scheme.

In addition to these goals, for high-speed applications it is wise to keep the acceleration smooth to avoid abrupt changes in dynamic loads. The time derivative of the acceleration is referred to as *jerk*. Sudden changes in acceleration are quantified by high magnitudes of jerk. Thus, reducing the magnitude and maintaining a continuous jerk versus time curve has advantages on machine loading.

Negative aspects of the constant velocity, constant acceleration, harmonic and cycloidal schemes are often adjusted to improve motion characteristics. The resulting motion is called a combined scheme. Descriptions of some of the more common combined schemes are given below. More comprehensive cam design sources should be consulted to obtain the details of the motion equations [Refs. 5, 11, 14]. Software, such as Dynacam, Analytix/Cams, and CamTrax, are available to construct motion follower displacement diagrams of these and other schemes.

Trapezoidal Acceleration   is a scheme that improves on the constant acceleration scheme shown in Figure 9.10,

where the acceleration versus time curve appears as a square wave. The difficulty with the square wave is that the acceleration, and consequently inertial force, changes abruptly. Thus, a jerk is induced onto the machine. The trapezoidal acceleration scheme softens the transitions where the acceleration versus time curve appears as a trapezoid. However, the area lost by knocking off the corners must be replaced by increasing the maximum acceleration.

Modified Trapezoidal Acceleration   improves on the trapezoidal scheme by replacing the sloped sides on the acceleration versus time curve with portions of a sine wave. By eliminating the corners, a smooth acceleration curve is created. The continuous slope (jerk) ensures that the change in dynamic forces is smooth.

3-4-5 Polynomial Displacement   is another scheme that improves on the constant acceleration scheme. Being second order polynomial, the constant acceleration scheme is hindered with a discontinuous acceleration curve. As with the trapezoidal scheme, another method of eliminating the discontinuity is to use a higher-order polynomial. Thus, a scheme has been formulated that incorporates third, fourth, and fifth order terms. Having fifth order term, this scheme provides continuous slope on the acceleration versus time curve. However, the jerk versus time curve will have discontinuities.

4-5-6-7 Polynomial Displacement   expands the 3-4-5 polyno-mial scheme and includes a seventh order term to provide a continuous and smooth jerk.

Modified Sinusoidal Acceleration improves on the cycloidal scheme by incorporating a second sinusoidal term with a different frequency. Smoothness of the cycloidal motion is retained and the maximum is reduced.

A summary of the peak velocity, peak acceleration, and peak jerk for the different motion schemes, as a function of the rise H and period of the interval T, is shown in Table 9.5.

TABLE 9.5 Motion Scheme Comparisons			
Motion Scheme	Peak Velocity	Peak Acceleration	Peak Jerk
Constant Velocity	1.000 H/T	∞	∞
Constant Acceleration	2.000 H/T	4.000 H/T^2	∞
Harmonic	1.571 H/T	4.945 H/T^2	∞
Cycloidal	2.000 H/T	6.283 H/T^2	40 H/T^3
Trapezoidal	2.000 H/T	5.300 H/T^2	44 H/T^3
Modified Trapezoidal	2.000 H/T	4.888 H/T^2	61 H/T^3
3-4-5 Polynomial	1.875 H/T	5.777 H/T^2	60 H/T^3
4-5-6-7 Polynomial	2.188 H/T	7.526 H/T^2	52 H/T^3
Modified Sine	1.760 H/T	5.528 H/T^2	69 H/T^3

9.6 GRAPHICAL DISK CAM PROFILE DESIGN

Once the desired motion of a cam and follower has been defined through a displacement diagram, the actual shape of the cam can be designed. The shape of the cam depends on the size of the cam along with the configuration of the follower. Prior to designing the profile of a disk cam, some geometric features must be defined. The following features are illustrated in Figure 9.18.

The *base circle* is the smallest circle centered on the cam rotation axis and tangent to the cam surface. The size of the base circle is typically dictated by the spatial restrictions of the application. In general, a large base circle causes fewer problems with force transmission. However, a large base circle and, hence, a large cam is contradictory to the common design goal of smaller products.

The *trace point* serves as a reference to determine the effective location of the follower. For a knife-edge follower, it is the point of cam and follower contact. For a roller follower, the trace point is chosen at the center of the roller. For a flat- or spherical-face follower, the trace point is chosen on the contact surface of the follower.

The *home position* of the cam is the orientation that corresponds to the 0° reference position on a displacement diagram.

The *prime circle* is a circle drawn through the trace point of the follower while the cam is at its home position.

The *pitch curve* is the path of the center of the follower.

For ease in cam profile construction, kinematic inversion will be used. The cam will be imagined as being stationary. The follower then should be rotated opposite to the direction of cam rotation. The desired location for the follower, at several positions, is constructed from the base circle. Conceptually, this is comparable to wrapping the displacement diagram around the base circle, creating the cam shape.

The specific procedures for different follower arrangements are illustrated in the following sections. The general displacement diagram shown in Figure 9.19 is used for all constructions. Notice that follower displacements have been identified at specific cam angles from the rise and fall portions of the diagram. These prescribed displacements are translated to the cam profile.

9.6.1 In-Line Knife-Edge Follower

The most efficient manner to describe the construction of a cam with a knife-edge follower is through an actual construction. Using the displacement diagram from Figure 9.19, a cam profile to be used with a knife-edge follower has been constructed and shown in Figure 9.20.

The following general procedure is used to graphically construct such a profile:

1. Draw the base circle of radius R_b. The size is typically a function of the spatial constraints of the application.

2. Draw the follower in the home position.

3. Draw radial lines from the center of the cam, corresponding to the cam angles identified on the displacement diagram. For construction purposes, the cam will remain stationary and the follower will be rotated in a direction opposite to the actual cam rotation.

4. Transfer the displacements from the displacement diagram to the radial lines. Measure these displacements from the base circle.

5. Draw a smooth curve through these prescribed displacements.

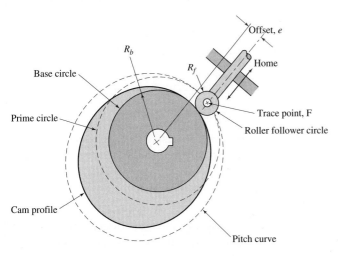

FIGURE 9.18 Cam nomenclature.

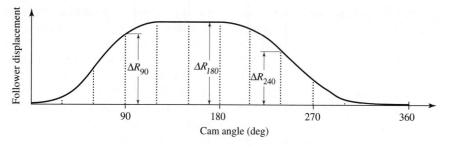

FIGURE 9.19 General follower displacement diagram.

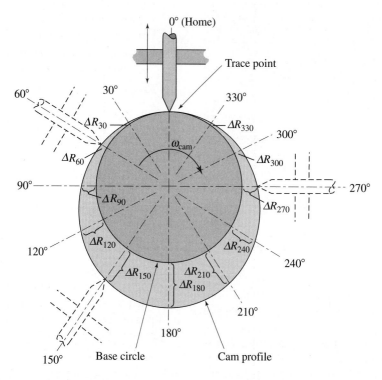

FIGURE 9.20 Cam profile design—in-line, knife-edge follower.

6. To accurately construct a profile consistent with the displacement diagram, it may be necessary to transfer additional intermediate points from the rise and fall intervals.

9.6.2 In-Line Roller Follower

Again, the most efficient manner of describing the construction of a cam with an in-line roller follower is through an actual construction. Using the displacement diagram from Figure 9.19, a cam profile to be used with an in-line roller follower has been constructed and shown in Figure 9.21. The following general procedure is used to construct such a profile:

1. Draw the base circle of radius R_b. The size is typically a function of the spatial constraints of the application.

2. Draw the follower of radius R_f in the home position, tangent to the base circle.

3. Draw radial lines from the center of the cam, corresponding to the cam angles identified on the displacement

diagram. For construction purposes, the cam will remain stationary and the follower will be rotated in a direction opposite to the actual cam rotation.

4. Identify the trace point at the home position. For a roller follower, this is the point at the center of the roller.

5. Draw the prime circle through the trace point at its home position.

6. Transfer the displacements from the displacement diagram to the radial lines. Measure these displacements from the prime circle.

7. Draw the roller outline of radius R_f, centered at the prescribed displacements identified in the previous step.

8. Draw a smooth curve tangent to the roller at these prescribed displacements.

9. To accurately construct a profile consistent with the displacement diagram, it may be necessary to transfer additional intermediate points from the rise and fall intervals.

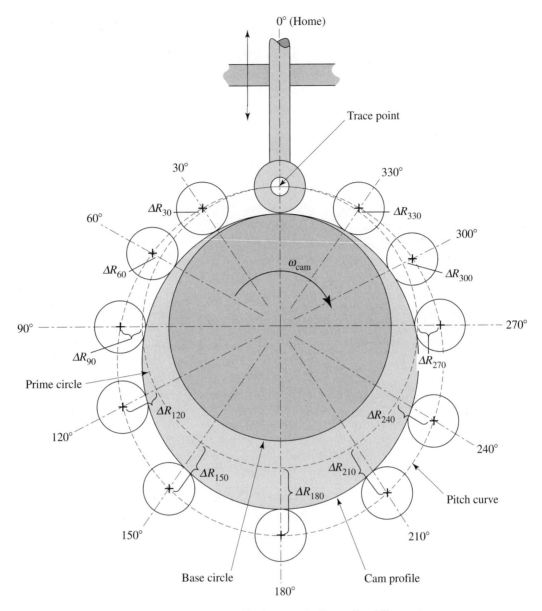

FIGURE 9.21 Cam profile design—in-line roller follower.

9.6.3 Offset Roller Follower

The most efficient manner of describing the construction of a cam with an offset roller follower is through an actual construction. Using the displacement diagram from Figure 9.19, a cam profile to be used with an offset roller follower has been constructed and shown in Figure 9.22. The following general procedure is used to construct such a profile.

1. Draw the base circle of radius R_b. The size is typically a function of the spatial constraints of the application.

2. Draw the follower centerline in the home position.

3. Draw the prime circle, whose radius is equal to the sum of the base and roller follower radii ($R_b < R_f$).

4. Draw the follower in the home position of radius R_f centered where the follower centerline intersects the prime circle.

5. Identify the trace point at the home position. For a roller follower, this is the point that is at the center of the roller.

6. Draw an offset circle of radius e, centered at the cam rotation axis. It will be tangent to the follower centerline.

7. Draw lines tangent to the offset circle, corresponding to the reference cam angles on the displacement diagram. For construction purposes, the cam will remain stationary and the follower will be rotated in a direction opposite to the actual cam rotation.

8. Transfer the displacements from the displacement diagram to the offset lines. Measure these displacements from the prime circle.

9. Draw the roller outline of radius R_f centered at the prescribed displacements identified in the previous step.

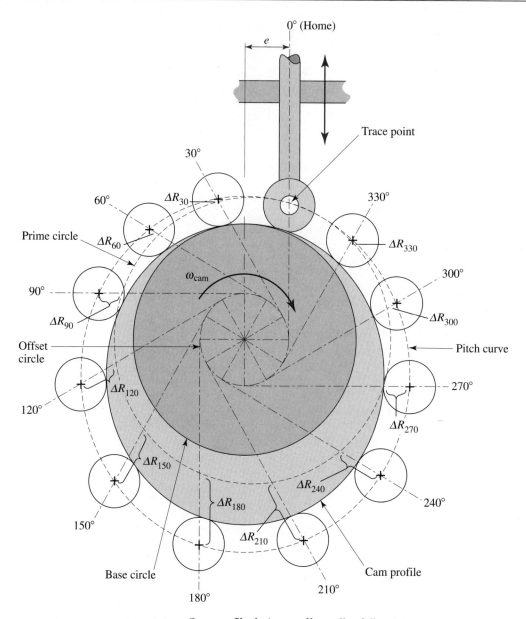

FIGURE 9.22 Cam profile design—offset roller follower.

10. Draw a smooth curve tangent to the roller at these prescribed displacements.

11. To accurately construct a profile consistent with the displacement diagram, it may be necessary to transfer additional intermediate points from the rise and fall intervals.

9.6.4 Translating Flat-Faced Follower

The most efficient manner for describing the construction of a cam with a flat-faced follower is through an actual construction. Using the displacement diagram from Figure 9.19, a cam profile to be used with a translating flat-faced follower has been constructed and shown in Figure 9.23.

The following general procedure is used to graphically construct such a profile.

1. Draw the base circle of radius R_b. The size is typically a function of the spatial constraints of the application.

Recall that for this type of follower, the base circle also serves as the prime circle.

2. Draw the follower in the home position, tangent to the base circle.

3. Draw radial lines from the center of the cam, corresponding to the cam angles on the displacement diagram. For construction purposes, the cam will remain stationary and the follower will be rotated in a direction opposite to the actual cam rotation.

4. Transfer the displacements from the displacement diagram to the radial lines, measured from the base circle.

5. Draw the flat-faced outline by constructing a line perpendicular to the radial lines at the prescribed displacements.

6. Draw a smooth curve tangent to the flat-faced outlines.

7. To accurately construct a profile consistent with the displacement diagram, it may be necessary to transfer additional intermediate points from the rise and fall motions.

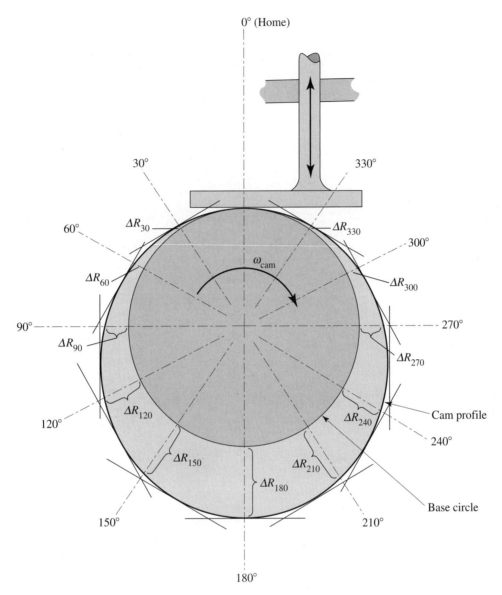

FIGURE 9.23 Cam profile design—flat-faced follower.

9.6.5 Pivoted Roller Follower

The pivoted follower provides rotational motion as the output of the cam and follower system. With translating followers, the equations presented in Section 9.5 are used to calculate the magnitude of the instantaneous linear displacement, ΔR_F, velocity, v_F, and acceleration, a_F, of the follower center, point F. For the pivoted follower, the equations presented in Section 9.5 can be used to calculate the instantaneous rotational displacement, $\Delta\theta_L$, velocity, ω_L, and acceleration, α_L, of the follower link. To use the equations in Section 9.5 for rotational motion analysis, the prescribed follower displacement must be angular, $\Delta\theta_L$, instead of linear, H.

Again, the most efficient manner of describing the construction of a cam with a pivoted roller follower is through an actual construction. Using the displacement diagram from Figure 9.19, a cam profile to be used with a pivoted roller follower has been constructed and shown in Figure 9.24.

The following general procedure is used to graphically construct such a profile.

1. Draw the base circle of radius R_b, where the size is a function of the spatial constraints of the application.

2. Draw the prime circle, whose radius is equal to the sum of the base and roller follower radii.

3. Draw the pivot circle of radius R_p. The distance between the pivot and the cam axis is also a function of the spatial constraints of the application.

4. Locate the home position of the pivot.

5. Draw an arc, centered at the home pivot, with a radius equal to the length of the pivoted follower link, R_L.

6. Draw the follower in the home position of radius R_f, centered where the arc drawn in step 5 intersects the prime circle.

7. Draw radial lines from the center of the cam to the pivot circle, corresponding to the cam angles on the displacement diagram. Recall that the follower is being rotated in a direction opposite to the cam rotation.

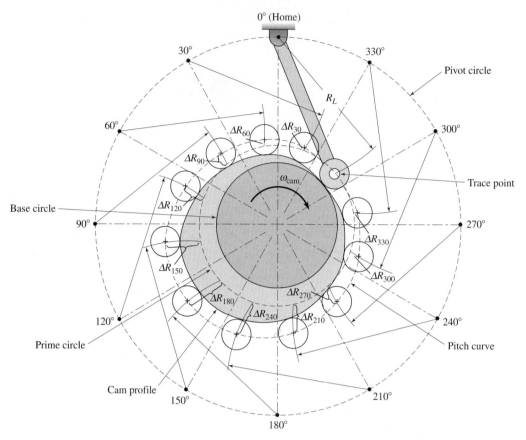

FIGURE 9.24 Cam profile design—pivoted roller follower.

8. From each pivot point, draw an arc with a radius equal to the length of the follower arm, R_L, outward from the prime circle.

9. Transfer the displacements from the displacement diagram to the pivot arcs drawn in step 8. As mentioned, the prescribed displacements for a pivoted follower can be angular. Equation (9.3) can be used to convert from angular displacement of the follower link, $\Delta\theta_L$, to linear displacement of the roller center, ΔR_F.

$$\Delta R_F = R_L \sqrt{2(1 - \cos \Delta\theta_L)}. \qquad (9.3)$$

10. Draw the roller outline, centered at the prescribed displacements identified in the previous step.

11. Draw a smooth curve tangent to the roller at these prescribed displacements.

12. To accurately construct a profile consistent with the displacement diagram, it may be necessary to transfer additional intermediate points from the rise and fall motions.

9.7 PRESSURE ANGLE

Because a force is always transmitted perpendicular to surfaces in contact, the cam does not always push the follower in the direction of its motion. As discussed in the previous section, the curvature of the cam affects the position between the follower centerline and the actual contact point.

The force required to push the follower depends on the application of the cam system. However, the contact force between the cam and follower can be much greater, depending on the location of the contact point. Actually, only one component of the contact force generates the follower motion. The other force component is undesirable as it generates a side load and must be absorbed by the follower guide bearings.

The *pressure angle*, δ, correlates these two components of the contact force. The pressure angle, at any point on the profile of a cam, is the angle between the follower movement and direction that the cam is pushing it. More precisely, it is the angle between the path of the follower motion and the line perpendicular to the cam profile at the point of follower contact. Each point on the cam surface has a pressure angle. The pressure angle is illustrated in Figure 9.25.

After graphically constructing a cam profile, the magnitude of the pressure angle can be visualized by observing the location of the contact point in relation to the follower centerline. The regions where the cam profile exhibits the greatest curvature should be identified. Measurements of the pressure angles in this region should be obtained. In general, the pressure angle should be kept as small as possible and should not exceed 30°. The magnitude of the pressure angle can be decreased by

1. Increasing the size of the base circle,

2. Decreasing the magnitude of follower displacement,

3. Increasing the angle of cam rotation prescribed for the follower rise or fall,

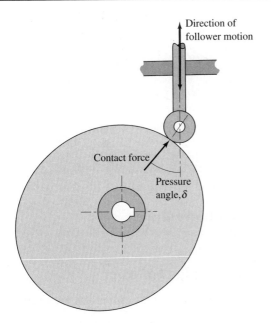

FIGURE 9.25 Pressure angle.

cam, the construction angle increments were refined for increased accuracy. The cam profile was constructed by locating the follower circles and drawing the cam profile tangent to the follower circles. Notice that the cam does not contact the follower at all locations. At a cam angle of 135°, the cam will not push the roller to the desired position.

This situation can be corrected by using a larger base circle or reducing the diameter of the roller follower. However, the contact stresses in the cam and follower are increased if the roller diameter is reduced. Thus, a roller of excessively small diameter should be avoided.

A similar situation can occur with a flat-faced follower. Figure 9.27a illustrates a cam segment whose follower also requires a rapid rise. Notice that once the flat-faced follower positions are located, a smooth curve cannot be constructed to represent the cam profile. One follower construction line (90°) falls outside the intersection of the adjacent follower construction lines. Thus, at a cam angle of 90°, the cam will not push the flat-faced follower to its desired position.

Figure 9.27b illustrates another cam segment, with a larger base circle. This cam has the exact same displacement requirements as the one in Figure 9.27a. In this case, a smooth cam profile can be constructed tangent to all follower construction lines. Again, a functional design was achieved by increasing the diameter of the base circle.

4. Decreasing the amount of follower offset, or
5. Modifying the follower motion scheme.

9.8 DESIGN LIMITATIONS

As seen in Section 9.6, the design of a cam profile cannot begin until first deciding on a follower type and the location and size of a base circle. These decisions are usually dependent on the magnitude of the transmitted forces and the size requirements of the cam-driven machinery. It must be understood that these decisions may not always be practical.

Figure 9.26 illustrates a cam with an in-line roller follower. Notice that there is a rapid rise and fall at a cam angle of 135°. Also note that during this portion of the

9.9 ANALYTICAL DISK CAM PROFILE DESIGN

The previous sections illustrated graphical methods used to design a cam profile. Depending on the precision required for the application, these methods can produce sufficiently accurate profiles. Of course, the accuracy is increased when the construction is accomplished on a CAD system. With CAD, splines are typically used to construct the smooth curve of the cam profile. Often,

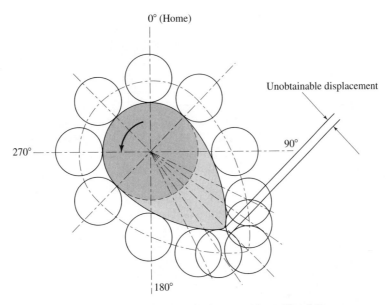

FIGURE 9.26 Impractical cam with a roller follower.

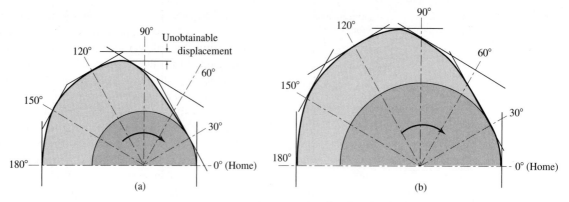

FIGURE 9.27 Impractical cam with a flat-faced follower.

splines have precision errors that may violate tangency constraints. To increase the accuracy, smaller cam angle increments can be employed.

In some situations, highly accurate cams are required. It is desirable to be able to analytically determine the coordinates of points on the cam surface as well as coordinates of a milling cutter that will be used to manufacture the cam. Equations have been developed for the coordinates of the different types of followers. This section merely presents these equations, and the reader is referred to more detailed sources for the derivations [Ref. 4]. Incorporating the equations into a spreadsheet or some other programmable device can quickly generate the profile coordinates.

In general, a Cartesian coordinate system is used so that the origin is at the cam center. The positive y-axis is along the direction of the follower motion in its home position. The positive x-axis is 90° clockwise from the y-axis, consistent with a right-hand coordinate system. Figure 9.28 illustrates this coordinate system.

9.9.1 Knife-Edge Follower

The x and y coordinates of the cam profile are given as

$$R_x = (R_b + \Delta R)\sin\phi \qquad (9.3)$$

$$R_y = (R_b + \Delta R)\cos\phi \qquad (9.4)$$

where the following notation is used:
R_x = x coordinate of cam surface profile
R_y = y coordinate of cam surface profile
R_b = Base circle radius
ϕ = Cam rotation angle measured against the direction of cam rotation from the home position
ΔR = Follower displacement at cam angle ϕ

Most cams are produced through a cutting operation, using computer numerical control milling machines. These machines are capable of rotating the cam by a fraction of a degree, while advancing the cutter by thousandths of a millimeter. Through this method, the cam profile can be precisely manufactured.

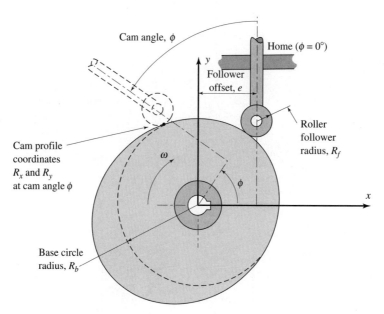

FIGURE 9.28 Cam profile coordinate system.

The x and y coordinates of the center of the milling cutter, or grinding wheel, are given as

$$C_x = (R_c + R_b + \Delta R)\sin\phi \qquad (9.6)$$

$$C_y = (R_c + R_b + \Delta R)\cos\phi \qquad (9.7)$$

where the following additional notation is used:

C_x = x coordinate of cutter center
C_y = y coordinate of cutter center
R_c = Mill cutter radius

EXAMPLE PROBLEM 9.4

For the application stated in Example Problem 9.2, analytically determine the cam profile coordinates when a knife-edge follower is incorporated. Because of the size constraints of the machine, a cam with a base circle diameter of 200 mm must be used. The cam is to rotate counterclockwise.

SOLUTION:

1. **Calculate Coordinates of the Cam Profile**

   The base circle radius is half of the base circle diameter; thus

   $$R_b = 100\,\text{mm}$$

   Substitution into equations (9.4) and (9.5) gives

   $$R_x = (R_b + \Delta R)\sin\phi = [(100\,\text{mm}) + \Delta R]\sin\phi$$

   $$R_y = (R_b + \Delta R)\cos\phi = [(100\,\text{mm}) + \Delta R]\cos\phi$$

2. **Summarize the Profile Coordinates for Several Cam Angles**

   Inserting these equations into a spreadsheet gives the results listed in Figure 9.29.

3. **Plot the Profile Coordinates**

   A spreadsheet can be used to easily create a plot of the profile coordinates. This plot is shown as Figure 9.30 and illustrates the cam profile.

t	ϕ	ΔR	Rx	Ry
(s)	(deg)	(mm)	(mm)	(mm)
0.00	0.0	0.00	0.0	100.0
0.25	21.2	8.33	39.1	101.0
0.50	42.4	16.67	78.6	86.2
0.75	63.5	25.00	111.9	55.7
1.00	84.7	33.33	132.8	12.3
1.25	105.9	41.67	136.3	-38.8
1.50	127.1	50.00	119.7	-90.4
1.75	148.2	49.38	78.6	-127.0
2.00	169.4	45.46	26.7	-143.0
2.25	190.6	36.88	-25.2	-134.5
2.50	211.8	25.00	-65.8	-106.3
2.75	232.9	13.12	-90.3	-68.2
3.00	254.1	4.54	-100.6	-28.6
3.25	275.3	0.62	-100.2	9.3
3.50	296.5	0.00	-89.5	44.6
3.75	317.6	0.00	-67.4	73.9
4.00	338.8	0.00	-36.1	93.2
4.25	360.0	0.00	0.0	100.0

FIGURE 9.29 Coordinates for Example Problem 9.4.

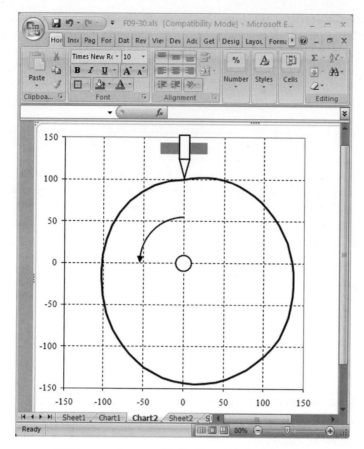

FIGURE 9.30 Cam profile for Example Problem 9.4.

9.9.2 In-Line Roller Follower

In general, a roller follower is complicated by the fact that the cam contact point is not in line with the roller center. The angle between the follower centerline and the cam contact point varies with curvature of the cam profile. For an in-line roller follower, this angle is the pressure angle. The instantaneous angle can be computed as

$$\alpha = \tan^{-1}\left[\frac{v}{\omega_{cam}} \frac{(R_f + R_b + \Delta R)}{(R_f + R_b + \Delta R)^2}\right] = \delta \quad \textbf{(9.8)}$$

For an in-line roller follower, this angle is also the pressure angle. In addition to the notation used in Section 9.9.1, the following terms are defined as

R_f = Radius of the roller follower

v = Magnitude of the instantaneous velocity of the cam follower at the cam angle ϕ

ω_{cam} = Rotational speed of the cam in radians per time

The term (v/ω_{cam}) is a measure of the rate of change of the follower displacement with respect to the cam angle. In

situations where the instantaneous follower velocity is not readily available, the slope of the displacement diagram can be estimated using equation (9.7).

$$\frac{v}{\omega_{cam}} = \frac{dR}{d\phi} \cong \frac{\Delta R}{\Delta \phi} \quad \textbf{(9.9)}$$

Then the x and y coordinates of the cam profile can be given as

$$R_x = -[R_f + R_b + \Delta R]\sin\phi + R_f\sin(\phi - \alpha) \quad \textbf{(9.10)}$$

$$R_y = [R_f + R_b + \Delta R]\cos\phi + R_f\cos(\phi - \alpha) \quad \textbf{(9.11)}$$

The x and y coordinates of the milling cutter are given as

$$C_x = -[R_f + R_b + \Delta R]\sin\phi + [R_c - R_f]\sin(\phi - \alpha) \quad \textbf{(9.12)}$$

$$C_y = [R_f + R_b + \Delta R]\cos\phi - [R_c - R_f]\cos(\phi - \alpha) \quad \textbf{(9.13)}$$

EXAMPLE PROBLEM 9.5

Two cams are used to drive a gripper of a mechanical part handler. The two cams can generate independent horizontal and vertical motions to the gripper. Such machines can relocate parts in a similar fashion to a robot at a fraction of the cost. The part handler is shown in Figure 9.31.

The prescribed motion for one of the cam followers is as follows:

1. Rise 1.5 in. in 1.5 s using the harmonic motion scheme.
2. Dwell for 2 s.
3. Return in 1.5 s using the harmonic motion scheme.
4. Dwell for 2 s.
5. Repeat the sequence.

An in-line roller follower with a radius of 0.5 in. is used on a cam with a base circle radius of 3.5 in. Tabulate the follower motion and specify the coordinates of the cam profile.

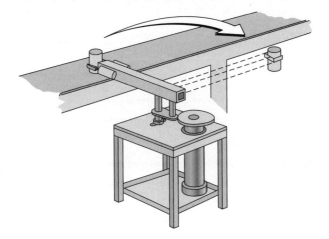

FIGURE 9.31 Part-handling machine for Example Problem 9.5.

SOLUTION:

1. ***Calculate the Time for a Full Cycle***

   The total time to complete the full cycle is needed to determine the required speed of the cam.

   $$\Sigma T_i = T_1 + T_2 + T_3 + T_4$$
   $$= 1.5 + 2.0 + 1.5 + 2.0 = 7.0\,s$$

2. ***Calculate the Required Rotational Speed of the Cam***

   From equation (9.2),

   $$\omega_{cam} = \frac{1\,rev}{\Sigma T_i} = \frac{1\,rev}{7\,s} = 0.143\,rev/s = 0.899\,rad/s = 8.57\,rpm$$

3. ***Determine the Cam Rotation for Each Follower Motion Interval***

   The angular increment of the cam consumed by each follower motion sequence is determined by equation (9.1).

   $$\beta_1 = (\omega_{cam})(T_1) = (0.143\,rev/s)(1.5\,s) = 0.214\,rev = 77.2°$$

   $$\beta_2 = (0.143\,rev/s)(2.0\,s) = 0.286\,rev = 102.8°$$

   $$\beta_3 = (0.143\,rev/s)(1.5\,s) = 0.214\,rev = 77.2°$$

   $$\beta_4 = (0.143\,rev/s)(2.0\,s) = 0.286\,rev = 102.8°$$

4. ***Calculate the Displacements during Each Follower Motion Interval***

   The equations for the harmonic rise and fall were given in Table 9.3. Substituting into the harmonic rise equations gives

   $$\Delta R_1 = \frac{H_1}{2}\left[1 - \cos\left(\frac{\pi t_1}{T_1}\right)\right] = \frac{(1.5\,in.)}{2}\left[1 - \cos\left(\frac{\pi t_1}{1.5\,s}\right)\right]$$

$$v_1 = \frac{\pi H_1}{2T_1}\left[\sin\left(\frac{\pi t_1}{T_1}\right)\right] = \frac{\pi(1.5\,\text{in.})}{2(1.5\,\text{s})}\left[\sin\left(\frac{\pi t_1}{1.5\,\text{s}}\right)\right]$$

Substituting into the harmonic fall equations gives

$$\Delta R_2 = \frac{H_2}{2}\left[1 + \cos\left(\frac{\pi t_2}{T_2}\right)\right] = \frac{(1.5\,\text{in.})}{2}\left[1 + \cos\left(\frac{\pi t_2}{1.5\,\text{s}}\right)\right]$$

$$v_2 = \frac{-\pi H_2}{2T_2}\left[\sin\left(\frac{\pi t_2}{T_2}\right)\right] = \frac{-\pi(1.5\,\text{in.})}{2(1.5\,\text{s})}\left[\sin\left(\frac{\pi t_2}{1.5\,\text{s}}\right)\right]$$

5. ***Calculate the Coordinates of the Cam Profile***

Substitution into equations (9.8), (9.10), and (9.11) gives

$$\alpha = \tan^{-1}\left[\frac{v}{\omega_{cam}}\frac{[R_f + R_b + s]}{[R_f + R_b + s]^2}\right] = \tan^{-1}\left[\frac{v}{(0.899\,\text{rad/s})}\frac{[(0.5\,\text{in.}) + (3.5\,\text{in.}) + s]}{[(0.5\,\text{in.}) + (3.5\,\text{in.}) + s]^2}\right]$$

$$R_x = -[R_f + R_b + \Delta R]\sin\phi + R_f\sin(\phi - \alpha) = -[0.5 + 3.5 + \Delta R]\sin\phi + 0.5\sin(\phi - \alpha)$$

$$R_y = -[R_f + R_b + \Delta R]\cos\phi + R_f\cos(\phi - \alpha) = -[0.5 + 3.5 + \Delta R]\cos\phi - 0.5\cos(\phi - \alpha)$$

6. ***Summarize the Profile Coordinates for Several Cam Angles***

Inserting these equations into a spreadsheet gives the results listed in Figure 9.32.

7. ***Plot the Profile Coordinates***

A spreadsheet can be used to easily create a plot of the profile coordinates. This plot is shown as Figure 9.33 and illustrates the cam profile.

t (s)	ϕ (deg)	ΔR (in.)	v (in./s)	δ (deg)	Rx (in.)	Ry (in.)
0.00	0	0.00	0.00	0.0	0.000	3.500
0.35	18	0.19	1.05	15.6	-1.274	3.488
0.70	36	0.67	1.56	20.4	-2.611	3.298
1.05	54	1.19	1.27	15.2	-3.886	2.661
1.40	72	1.48	0.33	3.8	-4.751	1.509
1.75	90	1.40	-0.79	-9.2	-4.906	0.080
2.10	108	1.50	0.00	0.0	-4.755	-1.545
2.45	126	1.50	0.00	0.0	-4.045	-2.939
2.80	144	1.50	0.00	0.0	-2.939	-4.045
3.15	162	1.50	0.00	0.0	-1.545	-4.755
3.50	180	1.50	0.00	0.0	0.000	-5.000
3.85	198	1.31	-1.05	-12.4	1.387	-4.616
4.20	216	0.83	-1.56	-19.8	2.425	-3.625
4.55	234	0.31	-1.27	-18.2	3.010	-2.380
4.90	252	0.02	-0.33	-5.2	3.332	-1.130
5.25	270	0.00	0.00	0.0	3.500	0.000
5.60	288	0.00	0.00	0.0	3.329	1.082
5.95	306	0.00	0.00	0.0	2.832	2.057
6.30	324	0.00	0.00	0.0	2.057	2.832
6.65	342	0.00	0.00	0.0	1.082	3.329
7.00	360	0.00	0.00	0.0	0.000	3.500

FIGURE 9.32 Cam profile coordinates for Example Problem 9.5.

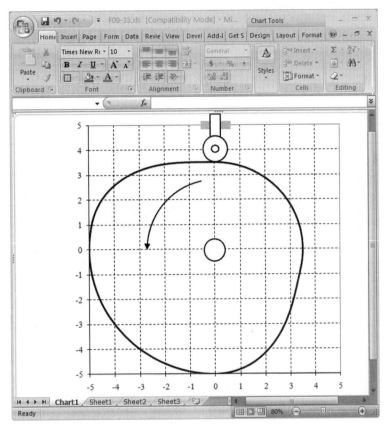

FIGURE 9.33 Cam profile for Example Problem 9.5.

9.9.3 Offset Roller Follower

An offset roller follower is further complicated by the fact that the follower motion is not in line with the cam contact point, which, in turn, is not in line with the roller center. Thus, the profile equations become a bit more complex. The angle between the lines connecting the follower center with the cam contact point, and the follower center with the cam center is computed as

$$\alpha = \tan^{-1}\left[\left(\frac{v}{\omega_{cam}}\right)\right.$$

$$\left.\left(\frac{R_f + R_b + \Delta R}{e^2 + (R_f + R_b + \Delta R)^2 - e(v/\omega_{cam})}\right)\right] \quad (9.14)$$

As in equation (9.8), the term (v/ω_{cam}) is a measure of the rate of change of the follower displacement with respect to the cam angle. In situations where the instantaneous follower velocity is not readily available, the slope of the displacement diagram can be estimated using equation (9.9). The pressure angle, δ, can be calculated as

$$\delta = \alpha - \tan^{-1}\left(\frac{e}{R_f + R_b + \Delta R}\right) \quad (9.15)$$

As before, the offset distance, e, is defined as the distance between the follower centerline and the cam center. A positive offset distance is defined in the positive x direction. Conversely, a negative offset distance is defined in the

negative x direction. The offset distance shown in Figure 9.28 is a positive value. Then the x and y coordinates of the cam profile can be given as

$$R_x = (e)\cos\phi - [R_f + R_b + \Delta R]\sin\phi + R_f\sin(\phi - \alpha) \quad (9.16)$$

$$R_y = (e)\sin\phi - [R_f + R_b + \Delta R]\cos\phi + R_f\cos(\phi - \alpha) \quad (9.17)$$

The x and y coordinates of the milling cutter are given as

$$C_x = (e)\cos\phi - [R_f + R_b + \Delta R]\sin\phi \quad (9.18)$$
$$+ [R_c - R_f]\sin(\phi - \alpha)$$

$$C_y = (e)\sin\phi - [R_f + R_b + \Delta R]\cos\phi \quad (9.19)$$
$$- [R_c - R_f]\cos(\phi - \alpha)$$

9.9.4 Translating Flat-Faced Follower

The analytical construction of a translating flat-faced follower can also exhibit a contact point that is not in line with the cam centerline. The angle between the follower centerline and the line connecting the cam contact point with the cam center varies with the curvature of the cam profile and can be computed as

$$\alpha = \tan^{-1}\left\{\left(\frac{v}{\omega_{cam}}\right)\frac{1}{(R_b + \Delta R)}\right\} \quad (9.20)$$

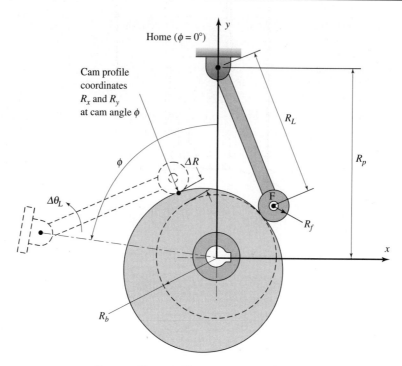

FIGURE 9.34 Cam profile coordinate system, with a pivoted follower.

As in equations (9.8) and (9.14), the term (v/ω_{cam}) is a measure of the rate of change of the follower displacement with respect to the cam angle. In situations where the instantaneous follower velocity is not readily available, the slope of the displacement diagram can be estimated using equation (9.9).

Then the x and y coordinates of the cam profile can be given as

$$R_x = \left(\frac{R_b + \Delta R}{\cos \alpha} \right) \cos(\phi + \alpha) \qquad (9.21)$$

$$R_y = \left(\frac{R_b + \Delta R}{\cos \alpha} \right) \sin(\phi + \alpha) \qquad (9.22)$$

The x and y coordinates of the milling cutter are given as

$$C_x = \left[\frac{R_b + \Delta R + R_c}{\cos \gamma} \right] \cos(\phi + \gamma) \qquad (9.23)$$

$$C_y = \left[\frac{R_b + \Delta R + R_c}{\cos \gamma} \right] \sin(\phi + \gamma) \qquad (9.24)$$

where

$$\gamma = \tan^{-1}\left[\frac{(R_b + \Delta R)\tan(\alpha)}{R_c + R_b + \Delta R} \right] \qquad (9.25)$$

9.9.5 Pivoted Roller Follower

The analytical construction of a pivoted roller follower is similar to the offset translating follower. However, the geometry and definitions are slightly different. Figure 9.34 illustrates the nomenclature used for a cam with a pivoted roller follower.

The following notation is used:

R_L = Length of the follower pivot link

R_P = Distance between the cam center and the pivot location

$\Delta\theta_L$ = Instantaneous angular position of the follower pivot link

ω_L = Instantaneous angular velocity of the follower pivot link

α_L = Instantaneous angular acceleration of the follower pivot link

The main difference with a pivoted follower is that its motion is rotational and the prescribed motion is usually the angular position of the follower versus time, or cam angle. Equation (9.3) gave the relation between the angular displacement of the follower link and the linear displacement of the roller center, point F.

$$\Delta R_F = R_L \sqrt{2(1 - \cos \Delta\theta_L)} \qquad (9.3)$$

The velocity of the follower center can be related to the rotational velocity of the follower link.

$$v_F = R_L \omega_L \qquad (9.26)$$

Again, the angle between the lines connecting the follower center with the cam contact point and the follower center with the cam center varies with the curvature of the cam profile and can be computed as

$$\alpha_L = \tan^{-1}\left[\left(\frac{v}{\omega_{cam}} \right) \left(\frac{1}{(R_f + \Delta R + R_b) - (v/\omega_{cam}) \cos \gamma} \right) \right] \qquad (9.27)$$

As before, the term (v_F/ω_{cam}) is a measure of the rate of change of the follower displacement with respect to the cam angle. In situations where the instantaneous follower velocity is not readily available, the slope of the displacement diagram can be estimated using equation (9.9).

Internal angles are given as

$$\gamma = \cos^{-1}\left[\left(\frac{R_L^2 + (R_b + R_f + \Delta R)^2 - R_p^2}{2(R_L)(R_b + R_b + s)}\right)\right] \quad \textbf{(9.28)}$$

$$\phi = \cos^{-1}\left[\left(\frac{R_p^2 + (R_b + R_f + \Delta R)^2 - R_L^2}{2(R_p)(R_b + R_b + s)}\right)\right] \quad \textbf{(9.29)}$$

$$\beta = \frac{\pi}{2} + \phi + \alpha \quad \textbf{(9.30)}$$

Finally, the x and y coordinates of the cam profile can be given as

$$R_x = -[R_f + R_b + \Delta R]\cos\beta - R_f \sin(\beta - \alpha) \quad \textbf{(9.31)}$$

$$R_y = -[R_f + R_b + \Delta R]\sin\beta - R_f \cos(\beta - \alpha) \quad \textbf{(9.32)}$$

and the pressure angle is given as

$$\delta = \gamma + \alpha - \frac{\pi}{2} \quad \textbf{(9.33)}$$

The x and y coordinates of the milling cutter are given as

$$C_x = [R_f + R_b + \Delta R]\cos\beta - [R_c - R_f]\sin(\beta - \alpha) \quad \textbf{(9.34)}$$

$$C_y = [R_f + R_b + \Delta R]\sin\beta - [R_c - R_f]\cos(\beta - \alpha) \quad \textbf{(9.35)}$$

9.10 CYLINDRICAL CAMS

Although disk cams are the most common type of cam, cylindrical cams are also widely used. As presented in Section 9.2 and illustrated in Figure 9.3b, a cylindrical cam consists of a groove wrapped around a cylinder. A cylindrical cam is a positive motion cam in that the follower is constrained in a groove and an external member is not needed to maintain contact between the follower and the cam. There are many applications in which it is necessary for the cam to exert a positive control of the follower during the rise or fall sequences.

Often, a tapered roller follower is used as shown in Figure 9.3b. It is used because the top edge of the groove travels at a higher speed than the bottom portion. Thus, the taper can compensate for the speed differential and prevent any slipping and skidding action of the roller. When a cylindrical roller is used, it is advisable to use a narrow width to minimize the velocity difference across the face of the roller.

In general, calculation and layout procedures are similar to those for the disk cam. The following sections discuss profile generation techniques for a cylindrical cam with a translating follower. The profile generation for other types of followers are similar.

9.10.1 Graphical Cylindrical Cam Profile Design

The most efficient manner for describing the construction of a cylindrical cam is through an actual construction. Using the displacement diagram from Figure 9.19, a cylindrical cam profile has been constructed and shown in Figure 9.35. The following general procedure is used to construct such a profile:

1. Draw a straight line equal to the circumference of the cylindrical cam.

2. Divide this line into sections corresponding to the reference cam angles on the displacement diagram.

3. Transfer the displacements from the displacement diagram to the lines corresponding to the reference cam angles.

4. Draw the roller follower at the prescribed displacements.

5. Draw a smooth curve tangent to the roller outlines.

6. To accurately construct a profile consistent with the displacement diagram, it may be necessary to transfer additional intermediate points from the rise and fall motions.

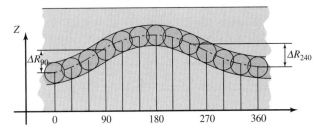

FIGURE 9.35 Cylindrical cam profile design.

9.10.2 Analytical Cylindrical Cam Profile Design

Because a cylindrical cam is wrapped around a cylinder, a cylindrical coordinate system is used to define the groove profile. The angular coordinate, θ, is the angle around the cam, and the z-axis is the axial position on the cam. The angle between the follower centerline and the cam contact point varies with the curvature of the groove profile and can be computed as

$$\alpha_L = \tan^{-1}\left(\frac{v_F}{\omega_{cam}}\right) \quad \textbf{(9.36)}$$

The notation used is the same as in the preceding sections. For a translating follower, this angle is also the pressure angle. Similar to disk cams, the pressure angle should be kept to a minimum and not exceed 30°.

The z-coordinate of the upper groove profile, when the follower center is at ϕ, can be given as

$$R_z = \Delta R + R_f \cos\alpha \quad \textbf{(9.37)}$$

$$\theta = \phi - \tan^{-1}\left(\frac{R_f \cos\alpha}{R_b}\right) \quad \textbf{(9.38)}$$

Here, R_b is the radius of the cylindrical cam.

The z-coordinate of the lower groove profile, when the follower center is at ϕ, can be given as

$$R_z = \Delta R - R_f \cos\alpha \tag{9.39}$$

$$\theta = \phi + \tan^{-1}\left(\frac{R_f \cos\alpha}{R_b}\right) \tag{9.40}$$

The coordinate of the milling cutter is given as

$$C_z = \Delta R \tag{9.41}$$

$$\theta = \phi \tag{9.42}$$

9.11 THE GENEVA MECHANISM

A geneva mechanism is a unique design that produces a repeated indexing motion from constant rotational motion. Because of this motion, the geneva mechanism is commonly classified with cams. A four-station geneva mechanism is illustrated in Figure 9.36.

The geneva mechanism consists of a driving roller and a geneva wheel. The geneva wheel consists of a disk with several radial slots and is fastened to an output shaft. The driving roller is fastened to an arm that, in turn, is fastened to the input shaft. The arm is usually attached to a locking disk that prevents the wheel from rotating when the driving roller is not engaged in a slot. The locking disk fits into a cutout on the wheel.

The motion of the geneva mechanism is characterized by the roller entering a slot in the wheel, indexing the wheel. When the roller exits the slot, the wheel locks into position until the roller enters the next slot. In Figure 9.36a, the roller rotates clockwise and is just about to enter the geneva wheel. In Figure 9.36b, the roller has entered the slot and has turned the wheel counterclockwise. Notice that the locking disk has moved away from the wheel, allowing it to rotate.

When designing a wheel, it is important that the roller enters the slot tangentially. Otherwise, impact loads are created and the mechanism will perform poorly at high speeds or loads. Because of this constraint, the following geometric relationships are derived [Ref. 7]. Refer to Figure 9.36 for definitions of the geometric properties.

$$\beta_o = \frac{360°}{n} \tag{9.43}$$

where
n = Number of stations in the geneva wheel

$$\gamma_o' = 90° - \frac{\beta_o}{2} \tag{9.44}$$

$$a = d\sin\left(\frac{\beta_o}{2}\right) \tag{9.45}$$

$$R = d\cos\left(\frac{\beta_o}{2}\right) \tag{9.46}$$

$$S < d - a \tag{9.47}$$

The kinematics of the geneva wheel can also be analytically determined. The angle of the roller, $\Delta\gamma$, is defined from the start of engagement. The angle of the wheel, measured from the start of engagement, is defined as β and is calculated as

$$\beta = \sin^{-1}\left[\left(\frac{a}{r}\right)\sin(180° - \psi)\right] \tag{9.48}$$

where

$$r = \sqrt{a^2 + d^2 - 2ad\cos(180 - \psi)} \tag{9.49}$$

$$\psi = 180° - \gamma_o + \Delta\gamma \tag{9.50}$$

where
$\Delta\gamma$ = Amount of rotation of driving from the position where the roller has just entered the slot.

The instantaneous velocity and acceleration of the geneva wheel have been found [Ref. 7] by

$$\omega_{\text{wheel}} = \left(\frac{a}{r}\right)(\omega_{\text{input shaft}})\cos(\beta - \psi) \tag{9.51}$$

$$\alpha_{\text{wheel}} = -\left(\frac{a}{r}\right)(\omega_{\text{input shaft}})^2\sin(\beta - \psi)$$
$$-\left(\frac{a}{r}\right)(\alpha_{\text{input shaft}})\cos(\beta - \psi)$$
$$+\left(\frac{a}{r}\right)(\omega_{\text{input shaft}})^2\sin(2\beta - 2\psi) \tag{9.52}$$

These equations are derived using the typical angular sign convention. That is, ω and α are positive when counterclockwise and negative when clockwise.

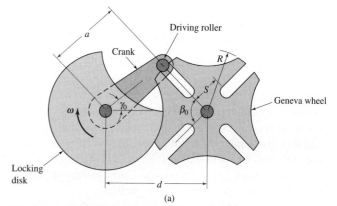

(a)

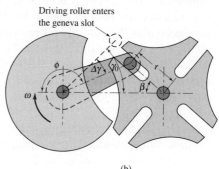

(b)

FIGURE 9.36 Four-station geneva mechanism.

EXAMPLE PROBLEM 9.6

A geneva mechanism has been designed with six stations, as shown in Figure 9.37. The distance between the driving and driven shafts is 80 mm. The driving arm rotates at a constant rate of 80 rpm clockwise. Determine the angular velocity and acceleration of the wheel when the driving arm has rotated 15° from the position where the roller has just entered the slot.

SOLUTION: 1. **Calculate Geneva Geometry**

Equations (9.43) through (9.47) can be used to calculate the geometric properties of this geneva mechanism.

$$\beta_o = \frac{360°}{n} = \frac{360°}{6} = 60°$$

$$\gamma_o = 90° - \frac{\beta_o}{2} = 90° - \frac{60}{2} = 60°$$

$$a = d \sin\left(\frac{\beta_o}{2}\right) = (80\,\text{mm})\sin\left(\frac{60°}{2}\right) = 40\,\text{mm}$$

$$R = d \cos\left(\frac{\beta_o}{2}\right) = (80\,\text{mm})\cos\left(\frac{60°}{2}\right) = 69.3\,\text{mm}$$

$$S < d - a = 80 - 40 = 40\,\text{mm}$$

2. **Calculate Geneva Kinematic Properties**

Equations (9.48) through (9.52) can be used to determine the kinematic relationships when the driving arm has rotated 15° from the position where the roller has just entered the slot.

$$\Delta\gamma = 15°$$

$$\psi = 180° - \gamma_o + \Delta\gamma = 180° - 60° + 15° = 135°$$

$$r = \sqrt{a^2 + d^2 - 2ad\cos(180 - \psi)}$$

$$= \sqrt{(40\,\text{mm})^2 + (80\,\text{mm})^2 - 2(40\,\text{mm})(80\,\text{mm})\cos(45°)} = 58.94\,\text{mm}$$

$$\beta = \sin^{-1}\left[\frac{a}{r}\sin\psi\right] = \sin^{-1}\left[\frac{40\,\text{mm}}{58.94\,\text{mm}}\sin 45°\right] = 28.7°$$

$$\omega_{\text{input shaft}} = 80\,\text{rpm} = -8.4\,\text{rad/s, cw}$$

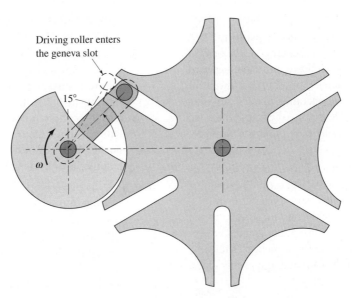

Driving roller enters the geneva slot

15°

ω

FIGURE 9.37 Geneva mechanism for Example Problem 9.6.

$$\omega_{\text{wheel}} = \left(\frac{a}{r}\right)(\omega_{\text{input shaft}})\cos(\beta - \psi) = \left(\frac{40\,\text{mm}}{58.94\,\text{mm}}\right)(-8.4\,\text{rad/s})\cos(28.7° - 135°) = +1.6\,\text{rad/s} = 15.3\,\text{rpm, ccw}$$

$$\alpha_{\text{input shaft}} = 0(\text{constant angular velocity of input shaft})$$

$$a_{\text{wheel}} = -\left(\frac{a}{r}\right)(\omega_{\text{input shaft}})^2\sin(\beta - \alpha) - \left(\frac{a}{r}\right)(\alpha_{\text{input shaft}})\cos(\beta - \alpha)$$

$$+ \left(\frac{a}{r}\right)^2(\omega_{\text{input shaft}})^2\sin(2\beta - 2\alpha)$$

$$= \left(\frac{40\,\text{mm}}{58.94\,\text{mm}}\right)^2(-8.4\,\text{rad/s})^2\sin(28.7° - 135°) - 0$$

$$+ \left(\frac{40\,\text{mm}}{58.94\,\text{mm}}\right)^2(-8.4\,\text{rad/s})^2\sin[2(28.7°) - 2(135°)]$$

$$= +67.1\,\text{rad/s}^2 = +67.1\,\text{rad/s}^2,\text{ccw}$$

PROBLEMS

For graphical solutions, manual techniques can be instructive, but the use of a CAD system is highly recommended.

Graphical Displacement Diagrams

9–1. A cam is required for an automated transfer mechanism. The cam follower must rise outward 1.0 in. with constant velocity in 3.0 s, dwell for 0.5 s, fall with constant velocity in 2.0 s, and then repeat the sequence. Determine the required speed of the cam and graphically plot a follower displacement diagram.

9–2. A cam is required for a reciprocating follower for a pick-and-place robotic arm. The cam follower must rise outward 0.75 in. with constant velocity in 1.4 s, dwell for 2.3 s, fall with constant velocity in 0.8 s, dwell for 1.9 s, and then repeat the sequence. Determine the required speed of the cam and graphically plot a follower displacement diagram.

9–3. A cam drive is required for a shaker platform. This platform is used to test the shipping-worthiness of packaged items. The cam follower must rise outward 1.0 in. with constant acceleration in 0.7 s, dwell for 0.2 s, fall with constant acceleration in 0.5 s, and then repeat the sequence. Determine the required speed of the cam and graphically plot a follower displacement diagram.

9–4. A cam drive is required for a mechanism that feeds papers into a printing press. The cam follower must rise outward 1.0 in. with constant acceleration in 1.7 s, dwell for 0.8 s, fall 0.5 in. with constant acceleration in 0.8 s, dwell for 0.3 s, fall 0.5 in. with constant acceleration in 0.8 s, and then repeat the sequence. Determine the required speed of the cam and graphically plot a follower displacement diagram.

9–5. A cam drive is required for an automated slide on a screw machine that turns intricate parts. The cam follower must rise outward 1.5 in. with constant acceleration in 1.2 s, dwell for 0.7 s, fall 0.5 in. with constant acceleration in 0.9 s, dwell for 0.5 s, fall 1.0 in. with constant acceleration in 1.2 s, and then repeat the sequence. Determine the required speed of the cam and graphically plot a follower displacement diagram.

9–6. A cam drive is used for a mechanism that drives an automated assembly machine. The cam follower must rise outward 13 mm with constant velocity in 3 s, dwell for 3 s, fall 5 mm with constant acceleration in 2 s, dwell for 3 s, fall 8 mm with constant acceleration in 2 s, and then repeat the sequence. Determine the required speed of the cam and graphically plot a follower displacement diagram.

9–7. A cam drive is used for a mechanism that tests the durability of oven doors. The cam follower must rise outward 2 in. with harmonic motion in 1 s, dwell for 0.5 s, fall 2 in. with harmonic motion in 1 s, dwell for 1 s, and then repeat the sequence. Determine the required speed of the cam and graphically plot a follower displacement diagram.

9–8. A cam drive is used for a mechanism that moves a tool in an automated screw machining process. The cam follower must rise outward 24 mm with harmonic motion in 0.2 s, dwell for 0.3 s, fall 10 mm with harmonic motion in 0.3 s, dwell for 0.2 s, fall 14 mm with harmonic motion in 0.2 s, and then repeat the sequence. Determine the required speed of the cam and graphically plot a follower displacement diagram.

9–9. A cam drive is used for a mechanism that packs stuffing into shipping boxes. The cam follower must rise outward 1 in. with cycloidal motion in 1.5 s, fall 1 in. with cycloidal motion in 1 s, dwell for 0.5 s, and then repeat the sequence. Determine the required speed of the cam and graphically plot a follower displacement diagram.

9–10. A cam drive is used for a mechanism incorporated in a shoe-sewing machine. The cam follower must rise outward 0.5 in. with cycloidal motion in 0.7 s, dwell for 0.2 s, fall 0.25 in. with cycloidal motion in 0.5 s, dwell for 0.2 s, fall 0.25 in. with cycloidal motion in 0.5 s and then repeat the sequence. Determine the required speed of the cam and graphically plot a follower displacement diagram.

9–11. A cam drive is required to synchronize two motions on an automated transfer device. The cam follower must rise outward 10 mm with constant acceleration in 90° of cam rotation, dwell for 30°, fall 10 mm with constant acceleration in 180° of cam rotation, and dwell for 60°. Graphically plot a follower displacement diagram.

9–12. A cam is used for an exhaust valve in an engine. The cam follower must rise outward 0.5 in. with harmonic motion in 150° of cam rotation, dwell for 30°, and fall 0.5 in. with harmonic motion in 180° of cam rotation. Graphically plot a follower displacement diagram.

9–13. A cam is used for a newspaper-collating device. The cam follower must rise outward 0.5 in. with cycloidal motion in 120° of cam rotation, dwell for 30°, fall 0.5 in. with cycloidal motion in 120° of cam rotation, dwell for 30°, and fall 0.5 in. with cycloidal motion in 60° of cam rotation. Graphically plot a follower displacement diagram.

Analytical Displacement Diagram

For problems 9–14 through 9–23, determine the speed of the cam and use the motion equations and a spreadsheet to plot a displacement diagram of the follower. Also calculate the maximum velocity and acceleration of the follower.

9–14. Use the required cam follower motion specified in Problem 9–1.

9–15. Use the required cam follower motion specified in Problem 9–2.

9–16. Use the required cam follower motion specified in Problem 9–3.

9–17. Use the required cam follower motion specified in Problem 9–4.

9–18. Use the required cam follower motion specified in Problem 9–5.

9–19. Use the required cam follower motion specified in Problem 9–6.

9–20. Use the required cam follower motion specified in Problem 9–7.

9–21. Use the required cam follower motion specified in Problem 9–8.

9–22. Use the required cam follower motion specified in Problem 9–9.

9–23. Use the required cam follower motion specified in Problem 9–10.

For problems 9–24 through 9–26, use the motion equations and a spreadsheet to plot a displacement diagram of the follower.

9–24. Use the required cam follower motion specified in Problem 9–11.

9–25. Use the required cam follower motion specified in Problem 9–12.

9–26. Use the required cam follower motion specified in Problem 9–13.

Analytical Motion Curves

For problems 9–27 through 9–36, use the motion equations and a spreadsheet to generate plots of the follower displacement, velocity, and acceleration versus time.

9–27. Use the required cam follower motion specified in Problem 9–1.

9–28. Use the required cam follower motion specified in Problem 9–2.

9–29. Use the required cam follower motion specified in Problem 9–3.

9–30. Use the required cam follower motion specified in Problem 9–4.

9–31. Use the required cam follower motion specified in Problem 9–5.

9–32. Use the required cam follower motion specified in Problem 9–6.

9–33. Use the required cam follower motion specified in Problem 9–7.

9–34. Use the required cam follower motion specified in Problem 9–8.

9–35. Use the required cam follower motion specified in Problem 9–9.

9–36. Use the required cam follower motion specified in Problem 9–10.

Graphical Plate Cam Profile Design

9–37. A plate cam must provide the displacement shown in Figure P9.37 to a reciprocating in-line knife-edge follower. The cam must have a base circle of 3.0 in. and rotate clockwise. Graphically construct the profile.

9–38. A plate cam must provide the displacement shown in Figure P9.37 to a reciprocating in-line knife-edge roller follower. The cam must have a base circle of 2.0 in. and rotate counterclockwise. Graphically construct the profile.

9–39. A plate cam must provide the displacement shown in Figure P9.37 to a reciprocating in-line roller follower. The roller diameter is 1.0 in. The cam must have a base circle of 3.0 in. and rotate clockwise. Graphically construct the profile and estimate the largest pressure angle.

9–40. A plate cam must provide the displacement shown in Figure P9.37 to a reciprocating in-line roller follower. The roller diameter is 0.75 in. The cam must have a base circle of 2.0 in. and rotate counterclockwise. Graphically construct the profile and estimate the largest pressure angle.

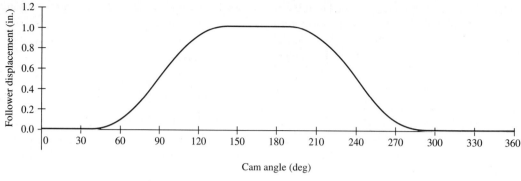

Cam angle (deg)	Follower Displ. (in.)		Cam angle (deg)	Follower Displ. (in.)		Cam angle (deg)	Follower Displ. (in.)
0	0.000		130	0.971		250	0.337
10	0.000		140	0.996		260	0.196
20	0.000		150	1.000		270	0.091
30	0.000		160	1.000		280	0.029
40	0.004		170	1.000		290	0.004
50	0.029		180	1.000		300	0.000
60	0.091		190	0.996		310	0.000
70	0.196		200	0.971		320	0.000
80	0.337		210	0.909		330	0.000
90	0.500		220	0.804		340	0.000
100	0.663		230	0.663		350	0.000
110	0.804		240	0.500		360	0.000
120	0.909						

FIGURE P9.37 Problems 37–44 and 47–54.

9–41. A plate cam must provide the displacement shown in Figure P9.37 to a reciprocating offset roller follower. The follower is positioned in the vertical plane, contacting the top of the cam. The offset distance is 0.75 in. to the left of the cam center. The roller diameter is 1.0 in. The cam must have a base circle of 3.0 in. and rotate clockwise. Graphically construct the profile and estimate the largest pressure angle.

9–42. A plate cam must provide the displacement shown in Figure P9.37 to a reciprocating offset roller follower. The follower is positioned in the vertical plane, contacting the top of the cam. The offset distance is 0.5 in. to the right of the cam center. The roller diameter is 0.75 in. The cam must have a base circle of 2.0 in. and rotate counterclockwise. Graphically construct the profile and estimate the largest pressure angle.

9–43. A plate cam must provide the displacement shown in Figure P9.37 to a reciprocating flat-faced follower. The cam must have a base circle of 5.0 in.

and rotate clockwise. Graphically construct the profile and estimate the largest pressure angle.

9–44. A plate cam must provide the displacement shown in Figure P9.37 to a reciprocating flat-faced follower. The cam must have a base circle of 6.0 in. and rotate counterclockwise. Graphically construct the profile and estimate the largest pressure angle.

9–45. A plate cam must provide the displacement shown in Figure P9.45 to a pivoted roller follower. The length of the follower link is 4 in. and is pivoted 5 in. from the cam rotation axis. The roller diameter is 1 in. The cam must have a base circle of 3.0 in. and rotate clockwise. Graphically construct the profile and estimate the largest pressure angle.

9–46. A plate cam must provide the displacement shown in Figure P9.45 to a pivoted roller follower. The length of the follower link is 3 in. and is pivoted 3.5 in. from the cam rotation axis. The roller

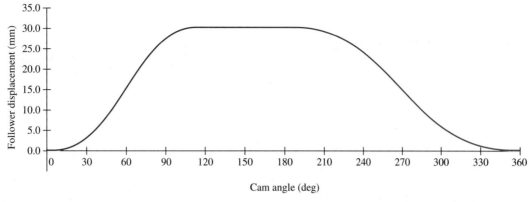

Cam angle (deg)	Follower Displ. (mm)	Cam angle (deg)	Follower Displ. (mm)	Cam angle (deg)	Follower Displ. (mm)
0	0.000	130	30.000	250	21.402
10	0.113	140	30.000	260	18.300
20	0.865	150	30.000	270	15.000
30	2.725	160	30.000	280	11.700
40	5.865	170	30.000	290	8.598
50	10.113	180	30.000	300	5.865
60	15.000	190	29.966	310	3.631
70	19.887	200	29.736	320	1.965
80	24.135	210	29.135	330	0.865
90	27.275	220	28.035	340	0.264
100	29.135	230	26.369	350	0.034
110	29.887	240	24.135	360	0.000
120	30.000				

FIGURE P9.45 Problems 45, 46, 55, and 56.

diameter is 0.75 in. The cam must have a base circle of 2.0 in. and rotate counterclockwise. Graphically construct the profile and estimate the largest pressure angle.

Analytical Plate Cam Profile Design

For problems 9–47 through 9–56, use the specific cam profile equations and a spreadsheet to generate a chart of profile coordinates for every 10° of cam angle.

9–47. Use the cam described in Problem 9–37.

9–48. Use the cam described in Problem 9–38.

9–49. Use the cam described in Problem 9–39.

9–50. Use the cam described in Problem 9–40.

9–51. Use the cam described in Problem 9–41.

9–52. Use the cam described in Problem 9–42.

9–53. Use the cam described in Problem 9–43.

9–54. Use the cam described in Problem 9–44.

9–55. Use the cam described in Problem 9–45.

9–56. Use the cam described in Problem 9–46.

Graphical Cylindrical Cam Design

9–57. A cylindrical cam must provide the displacement shown in Figure P9.37 to a reciprocating roller follower. The roller diameter is 1.0 in. The cylinder diameter is 5 in. and rotates clockwise. Graphically construct the profile and estimate the largest pressure angle.

9–58. A cylindrical cam must provide the displacement shown in Figure P9.37 to a reciprocating roller follower. The roller diameter is 30 mm. The cylinder diameter is 150 mm and rotates clockwise. Graphically construct the profile and estimate the largest pressure angle.

Analytical Cylindrical Cam Design

For problems 9–59 and 9–60, use the cylindrical cam profile equations and a spreadsheet to generate a chart of profile coordinates for every 10° of cam angle.

9–59. Use the cam described in Problem 9–57.

9–60. Use the cam described in Problem 9–58.

Geneva Mechanism Problems

9–61. A geneva mechanism has been designed with four stations. The distance between the driving and driven shafts is 3 in. The driving arm rotates at a constant rate of 60 rpm counterclockwise. Determine the angular velocity and acceleration of the wheel when the driving arm has rotated 25° from the position where the roller has just entered the slot.

9–62. A geneva mechanism has been designed with five stations. The distance between the driving and driven shafts is 60 mm. The driving arm rotates at a constant rate of 70 rpm clockwise. Determine the angular velocity and acceleration of the wheel when the driving arm has rotated 20° from the position where the roller has just entered the slot.

9–63. A geneva mechanism has been designed with six stations. The distance between the driving and driven shafts is 4 in. The driving arm rotates at a constant rate of 90 rpm counterclockwise. Determine the angular velocity and acceleration of the wheel when the driving arm has rotated 25° from the position where the roller has just entered the slot.

CASE STUDIES

9–1. The cam shown in Figure C9.1 is used to feed papers to a printing press. Carefully examine the components of the mechanism, then answer the following leading questions to gain insight into its operation.

1. As shaft *G* is forced to turn clockwise, determine the motion of item *E*.

2. What is the name of the connection between items *E* and *F*?

3. What causes the stack of papers, sitting on item *J*, to remain at a level where a mechanism at *B* can grab them?

4. Why does the radius on item *H* change?

5. What feature allows any rotation of item *H* to be transferred to item *G*?

6. Describe the mechanism that performs the same function as this cam for smaller stacks of papers used in computer printers and copy machines.

9–2. The cam shown in Figure C9.2 drives link *J*, which, in turn, drives another mechanism not shown. Link *A* is pivoted at its bottom to a machine frame. A stud extends from link *A* and through a slot in link *B*.

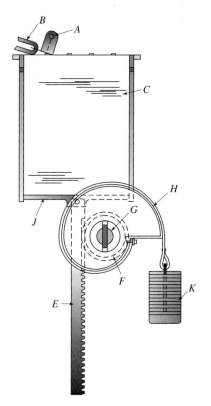

FIGURE C9.1 (Courtesy, Industrial Press.)

Carefully examine the components of the mechanism, then answer the following leading questions to gain insight into its operation.

1. As cam *D* rotates clockwise, describe the motion of link *B*.

2. What type of cam is *D*?

3. What type of follower is *C*?

4. What type of component is item *F*?

5. Describe the action of item *F*.

6. What type of component is item *E*?

7. Describe the function of item *E*.

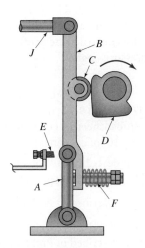

FIGURE C9.2 (Courtesy, Industrial Press.)

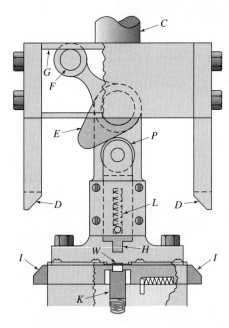

FIGURE C9.3 (Courtesy, Industrial Press.)

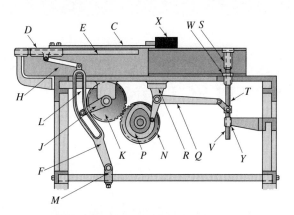

FIGURE C9.4 (Courtesy, Industrial Press.)

8. Describe the cyclical motion of item *B*.
9. What changes would occur to the motion of *B* if item *E* were lengthened?
10. What changes would occur to the motion of *B* if the stud at *E* were shortened?

9–3. The machine shown in Figure C9.3 stamps and forms steel parts. Carefully examine the components of the mechanism, then answer the following leading questions to gain insight into its operation.

1. As rod *C* starts to slide downward, what is the motion of cam *E*?
2. What is the motion of plunger *H*?
3. What happens to a strip of sheet metal clamped at *W*?
4. As rod *C* continues to slide downward, what is the motion of the plunger?
5. What is the motion of slides *I*?
6. What happens to the steel strip at *W*?

7. As rod *C* starts to slide upward, what is the motion of plunger *H*?
8. What is the purpose of this mechanism?
9. Why are springs contacting slides *I*?
10. Why is a spring supporting item *K*?
11. What type of mechanism could drive rod *C*?

9–4. A machine is shown in Figure C9.4. Carefully examine the components of the mechanism, then answer the following leading questions to gain insight into its operation.

1. As gear *K* rotates clockwise, describe the motion of link *F*.
2. Discuss the specifics of the cyclical motion of link *F*.
3. As gear *K* rotates clockwise, describe the motion of slide *D*.
4. As gear *K* rotates clockwise, describe the motion of gear *N*.
5. As gear *K* rotates clockwise, describe the motion of link *Q*.
6. What type of component is item *P* called?
7. Describe the motion to which item *V* is constrained.
8. Discuss exactly the manner in which link *Q* is attached to item *V*.
9. Discuss the cyclical motion of the entire machine.
10. State a need for such a machine.

GEARS: KINEMATIC ANALYSIS AND SELECTION

10.1 INTRODUCTION

Gears are an extremely common component used in many machines. Figure 10.1 illustrates the drive mechanism for the paper feed rollers of a scanner. In this application, an electric motor drives a small gear that drives larger gears to turn the feed rollers. The feed rollers then draw the document into the machine's scanning device.

In general, the function of a gear is to transmit motion from one rotating shaft to another. In the case of the feed drive of Figure 10.1, the motion of the motor must be transmitted to the shafts carrying the rollers. In addition to transmitting the motion, gears are often used to increase or reduce speed, or change the direction of motion from one shaft to the other.

It is extremely common for the output of mechanical power sources, such as electric motors and engines, to be rotating at much greater speeds than the application requires. The fax machine requires that the rollers feed the document through the machine at a rate compatible with the scanning device. However, a typical electric motor rotates at greater speeds than are needed at the rollers. Therefore, the speed of the motor must be reduced as it is transmitted to the feed roller shafts. Also the upper rollers must rotate in the direction opposite to that of the lower rollers. Thus, gears are a natural choice for this application.

Figure 10.2a illustrates two mating spur gears designed to transmit motion between their respective shafts. Figure 10.2b shows two friction rollers or disks that are also designed to transmit motion between the shafts. Such disks are obviously less costly than complex gear configurations. However, the disks rely on friction to transmit forces that may accompany the motion. Because many applications require the transmission of power (both motion and forces), smooth disk surfaces may not be able to generate sufficient frictional forces and thus will slip under larger loads.

To remedy the possibility of slipping, a gear is formed such that the smooth surfaces of the disks are replaced by teeth. The teeth provide a positive engagement and eliminate slipping. From a kinematic viewpoint, the gear pair in

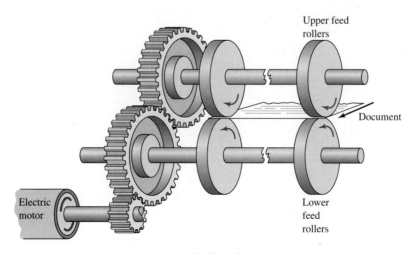

FIGURE 10.1 Feed rollers for a scanner.

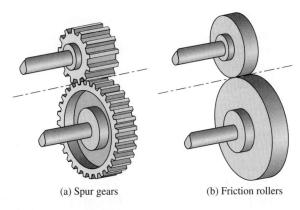

FIGURE 10.2 Gears and rollers.

(a) Spur gears   (b) Friction rollers

Figure 10.2a would replace the disks of Figure 10.2b because the effective diameters are identical.

The principles of general gearing and the associated kinematic relations are presented in this chapter. The focus of this book is on the analysis and design of mechanisms that are necessary to provide the motion required of machinery. Consistent with this mission, the focus of this chapter is on the selection of standard gears to produce the motion required in industrial machinery. Because they are the most widely used and least complicated gear, spur gears are emphasized. The reader is referred to other sources for further detail on gear tooth profiles, manufacture, quality, design for strength, and more complex gears [Refs. 4, 13, 15].

10.2 TYPES OF GEARS

Spur gears are simplest and, hence, the most common type of gear. The teeth of a spur gear are parallel to the axis of rotation. Spur gears are used to transmit motion between parallel shafts, which encompasses the majority of applications. A pair of mating spur gears is illustrated in Figure 10.3a.

A *rack* is a special case of spur gear where the teeth of the rack are not formed around a circle, but laid flat. The rack can be perceived as a spur gear with an infinitely large diameter. When the rack mates with a spur gear, translating motion is produced. A mating rack and gear are illustrated in Figure 10.3b.

Internal or annular gears have the teeth formed on the inner surface of a circle. When mating with a spur gear, the internal gear has the advantage of reducing the distance between the gear centers for a given speed variation. An internal gear mating with a traditional spur gear is illustrated in Figure 10.3c.

Helical gears are similar to, and can be used in the same applications as, spur gears. The difference is that the teeth of a helical gear are inclined to the axis of rotation. The angle of inclination is termed the *helix angle, φ*. This angle provides a more gradual engagement of the teeth during meshing and produces less impact and noise. Because of this smoother action, helical gears are preferred in high-speed applications. However, the helix angle produces thrust forces and bending couples, which are not generated in spur gears. A pair of mating helical gears is illustrated in Figure 10.3d.

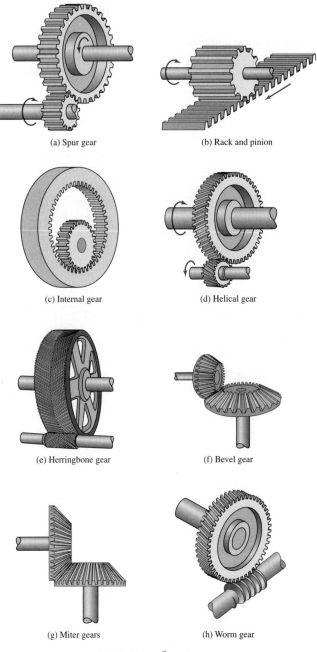

(a) Spur gear   (b) Rack and pinion

(c) Internal gear   (d) Helical gear

(e) Herringbone gear   (f) Bevel gear

(g) Miter gears   (h) Worm gear

FIGURE 10.3 Gear types.

Herringbone gears are used in the same applications as spur gears and helical gears. In fact, they are also referred to as *double helical gears*. The herringbone gear appears as two opposite-hand helical gears butted against each other. This complex configuration counterbalances the thrust force of a helical gear. A herringbone gear is shown in Figure 10.3e.

Bevel gears have teeth formed on a conical surface and are used to transmit motion between nonparallel shafts. Although most of their applications involve connecting perpendicular shafts, bevel gears can also be used in applications that require shaft angles that are both larger and smaller than 90°. As bevel gears mesh, their cones have a common apex. However, the actual cone angle of each gear depends on the gear ratio of the mating gears. Therefore, bevel gears are designed as a set, and replacing one gear to alter the gear ratio is not possible. A pair of mating bevel gears is illustrated in Figure 10.3f.

Miter gears are a special case of bevel gears where the gears are of equal size and the shaft angle is 90°. A pair of mating miter gears is illustrated in Figure 10.3g.

A *worm and worm gear* is used to transmit motion between nonparallel and nonintersecting shafts. The worm has one tooth that is formed in a spiral around a pitch cylinder. This one tooth is also referred to as the *thread* because it resembles a screw thread. Similar to the helical gear, the spiral pitch of the worm generates an axial force that must be supported. In most applications, the worm drives the worm gear to produce great speed reductions. Generally, a worm gear drive is not reversible. That is, the worm gear cannot drive the worm. A mating worm and worm gear are shown in Figure 10.3h.

10.3 SPUR GEAR TERMINOLOGY

As stated, spur gears are the most common type of gear. In addition, the terminology used to describe spur gears also applies to the other types of gears. Therefore, a thorough discussion of spur gear features and terminology is necessary.

The principal gear tooth features for a spur gear are illustrated in Figure 10.4.

The *pitch circle* of a gear is the circle that represents the size of the corresponding friction roller that could replace the gear. These corresponding rollers were illustrated in Figure 10.2b. As two gears mate, their pitch circles are tangent, with a point of contact on the line that connects the center of both circles. The pitch circle is shown in Figure 10.4.

The *pitch point* is the point of contact of the two pitch circles.

The *pitch diameter, d,* of a gear is simply the diameter of the pitch circle. Because the kinematics of a spur gear are identical to an analogous friction roller, the pitch

diameter is a widely referenced gear parameter. However, because the pitch circle is located near the middle of the gear teeth, the pitch diameter cannot be measured directly from a gear.

The *number of teeth, N,* is simply the total number of teeth on the gear. Obviously, this value must be an integer because fractional teeth cannot be used.

The *circular pitch, p,* is the distance measured along the pitch circle from a point on one tooth to the corresponding point on the adjacent tooth of the gear. The circular pitch can be calculated from the number of teeth and the pitch diameter of a gear. The governing equation is

$$p = \frac{\pi d}{N} \tag{10.1}$$

The *base circle* of a gear is the circle from which the curved shape of the gear tooth is constructed. Details on the generation of the curved tooth profile are presented in the following section.

The *base diameter, d_b,* is the diameter of the circle from which the gear tooth profile is derived. The base circle is thoroughly explained in Section 10.4.

The *face width, F,* is the length of the gear tooth parallel with the shaft axis.

The *addendum, a,* is the radial distance from the pitch circle to the top of a gear tooth.

The *dedendum, b,* is the radial distance from the pitch circle to the bottom of a gear tooth.

The *whole depth, h_T,* is the height of a gear tooth and is the sum of the addendum and dedendum.

The *clearance, c,* is the amount that the dedendum exceeds the addendum. This is the room between the top of a gear tooth and the bottom of the mating gear tooth.

The *backlash, B,* is the amount that the width of a tooth space exceeds the thickness of a gear tooth, measured on the pitch circle.

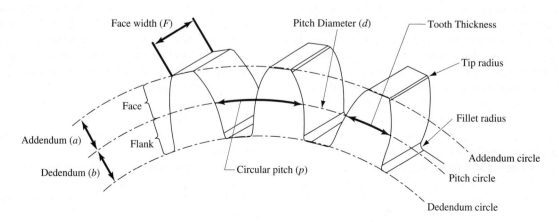

FIGURE 10.4 Spur gear tooth features.

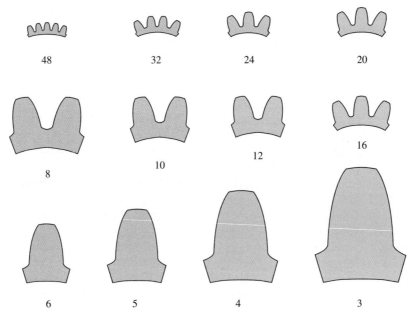

FIGURE 10.5 Standard tooth sizes.

The *diametral pitch, P_d,* or simply *pitch,* actually refers to the tooth size and has become a standard for tooth size specifications. Formally, the diametral pitch is the number of teeth per inch of pitch diameter.

$$P_d = \frac{N}{d} \qquad (10.2)$$

The diametral pitch is a commonly referenced gear parameter in the U.S. Customary Units. Again, it is a relative measure of the size of a gear tooth. The standard tooth sizes and their diametral pitches are shown in Figure 10.5. Although mating gears can have different diameters and number of teeth, *mating gears must have the same diametral pitch.* This should be obvious from the fact that the diametral pitch is a measure of tooth size.

The diametral pitch cannot be measured directly from a gear; yet, it is an extremely common referenced value. Theoretically, it is possible to produce almost any size gear teeth, but in the interest of standardized tooling, the American Gear Manufacturer's Association (AGMA) designated preferred diametral pitches. These standardized pitches are shown in Table 10.1. Although no physical significance exists, preference for standard diametral pitches is given to even integers. Sheet-metal gauges that measure the standard diametral pitches are available. The units of diametral pitch are the reciprocal of inches (in.$^{-1}$); yet it is not customary to specify units when expressing numerical values.

The *module, m,* is a commonly referenced gear parameter in the SI unit system. The module is also a relative measure of tooth size. It is defined as the ratio of pitch diameter to the number of teeth in a gear.

$$m = \frac{d}{N} \qquad (10.3)$$

The module is also a relative measure of tooth size and is theoretically the reciprocal of the diametral pitch. However, because it is referenced in the SI system, the module has units of millimeters. Therefore, the module and diametral pitch are not numerically reciprocal. The relationship between diametral pitch and module accounting for units is

$$m = \frac{25.4}{P_d} \qquad (10.4)$$

As with the diametral pitch, tooling for commercially available metric spur gears is stocked in standardized modules. Common values are shown in Table 10.2.

TABLE 10.1 Standard Diametral Pitches

Coarse Pitch		Fine Pitch	
2	6	20	80
2.25	8	24	96
2.5	10	32	120
3	12	40	150
3.5	16	48	200
4		64	

TABLE 10.2 Standard Metric Modules

1	4	16
1.25	5	20
1.5	6	25
2	8	32
2.5	10	40
3	12	50

Substituting equations (10.2) and (10.3) into (10.1), the circular pitch can also be expressed as

$$p = \frac{\pi d}{N} = \frac{\pi}{P_d} = \pi m \qquad \textbf{(10.5)}$$

The *pressure angle*, ϕ, is the angle between a line tangent to both pitch circles of mating gears and a line perpendicular to the surfaces of the teeth at the contact point. The line tangent to the pitch circles is termed the *pitch line*. The line perpendicular to the surfaces of the teeth at the contact point is termed the *pressure line* or *line of contact*. Therefore, the pressure angle is measured between the pitch line and the pressure line. The pressure angle is illustrated in Figure 10.6.

The pressure angle affects the relative shape of a gear tooth, as shown in Figure 10.7. Although gears can be manufactured in a wide range of pressure angles, most gears are standardized at 20° and 25°. Gears with 14½° pressure angles were widely used but are now considered obsolete. They are still manufactured as replacements for older gear trains still in use. Because the pressure angle affects the shape of a tooth, *two mating gears must also have the same pressure angle.*

Recall that forces are transmitted perpendicular to the surfaces in contact. Therefore, the force acting on a tooth is along the pressure line. As is discussed in the next section, gear teeth are shaped to maintain a constant pressure angle during engagement. Gears with smaller pressure angles efficiently transfer torque and apply lower radial loads onto the shaft and supporting bearings. However, as the pressure angles are reduced, a greater tendency exists for gear teeth to interfere as they engage.

10.4 INVOLUTE TOOTH PROFILES

In order to achieve smooth motion, a gear tooth must have a shape that keeps the driven gear rotating at a constant velocity throughout the engagement and disengagement process. Stated more concisely, gears need to have a constant velocity ratio. This condition requires that the path of gear tooth contact is along a straight line. That line must also intersect the point that is common to both pitch circles. Figure 10.8 illustrates two mating teeth at three intervals of the engagement process. Notice that the contact point traces a straight line, termed the *contact line.* This line also intersects the point that is tangent to both pitch circles, which is necessary for the gears to maintain a constant velocity ratio.

Discovering a tooth shape that fulfils the condition is not a trivial task; however, several forms have been identified as adequate candidates. Of the possible shapes, the involute of a circle has become standard for most gear applications. An involute shape is constructed by unwinding a taut wire from a base circle with diameter d_b. The path traced by the end of the wire is termed the *involute curve of a circle*. An involute profile is illustrated in Figure 10.9a. A segment of this involute curve is then used to form a gear tooth profile.

For an involute profile, the contact line is identical to the pressure line, as described in the previous section. The pressure angle, or inclination of the contact line, is determined from the segment of involute curve used for the gear tooth. The pressure angle increases as the distance between the base circle and the pitch circle increases. This is shown in Figure 10.9b. The relationship between the

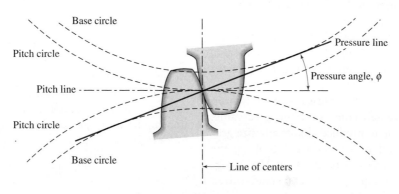

FIGURE 10.6 Pressure angle.

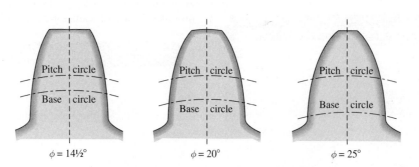

FIGURE 10.7 Pressure angle influence on tooth shapes.

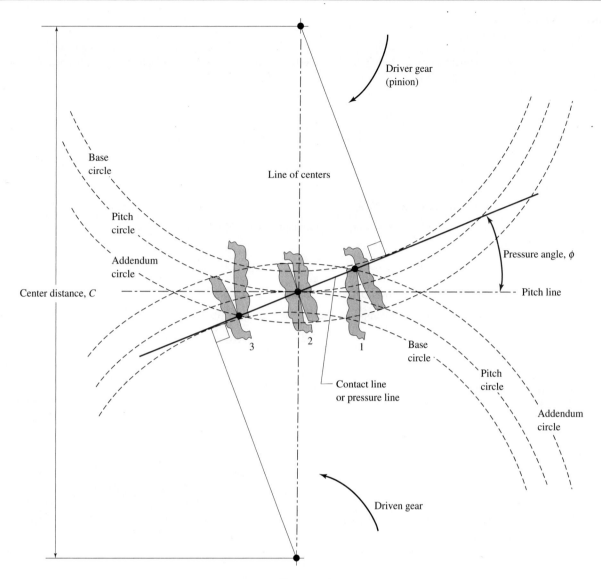

FIGURE 10.8 Gear mating process.

pressure angle, pitch diameter, and base circle diameter is expressed by

$$d_b = d \cos\phi \qquad (10.6)$$

Because the definition of an involute extends only from a base circle, any portion of tooth profile inside the base circle is not an involute. It is common to machine this portion as a radial line and a fillet to the dedendum circle. The portion of the tooth inside the base circle is not designed to be contacted by a mating gear tooth. Such contact would result in interference.

The most serious drawback to using involute gear tooth profiles is the possibility of interference between the tip of the gear tooth and the flank of the mating tooth. This occurs when the smaller gear has a low number of teeth. In certain circumstances, the gear tooth form can be changed from the full-depth shape of Figure 10.9 to avoid interference. Interference and employing alternate profiles are discussed in Section 10.6.3.

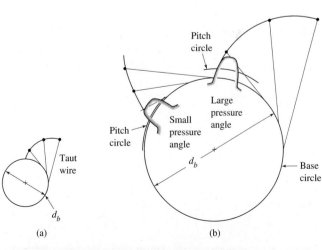

FIGURE 10.9 Involute gear tooth.

EXAMPLE PROBLEM 10.1

A 20° full-depth, involute spur gear with 35 teeth has a diametral pitch of 10. Determine the diameter of the pitch circle, the circular pitch, and the base circle.

SOLUTION:

1. **Calculate the Pitch Diameter**

   The pitch diameter can be computed by rearranging equation (10.2).

   $$d = \frac{N}{P_d} = \frac{35}{10} = 3.5\,\text{in.}$$

2. **Calculate the Circular Pitch**

   The circular pitch can be computed from equation (10.5).

   $$p = \frac{\pi}{P_d} = \frac{\pi}{(10\,\text{in.}^{-1})} = 0.314\,\text{in.}$$

3. **Calculate the Base Circle Diameter**

   The base circle can be computed directly from equation (10.6).

   $$d_b = d\cos\phi = 3.5\,\cos(20°) = 3.289\,\text{in.}$$

This is the diameter of the circle where the involute shape originates. It is not an apparent feature when inspecting an actual gear.

10.5 STANDARD GEARS

Gears can be manufactured using a variety of processes. For metal gears, the most common processes are cutting on shapers or milling machines, casting, and forming through powder-metallurgy processes. Plastic gears are typically injection molded. The reader is referred to sources dedicated to gear manufacture for details on the individual processes [Ref. 13].

Because the majority of processes utilize dedicated tooling, which is unique to each gear size, it is economically desirable to standardize the gear sizes. Standardized gears are readily available in most industrial equipment catalogs. These gears are sold interchangeably and can mesh with other gears having the same diametral pitch and pressure angle. Of course, for this to be accomplished, the manufacturers must follow a standard convention to form the details of the gear tooth profile. The AGMA is the primary organization that oversees this standardization scheme. It is a full-service trade association representing about 400 manufacturers and users of gears and gearing products and suppliers of equipment.

As stated, any two involute gears with the same diametral pitch and pressure angle will mate. Therefore, gear teeth have been standardized based on the diametral pitch and pressure angle. As mentioned in Section 10.3, standard pressure angles are 14½°, 20°, and 25°. The 14½° pressure angle is becoming obsolete and is used mainly for a replacement gear.

The diametral pitch is a measure of tooth size. In applications where the transmitted forces are high, larger teeth, having smaller values of diametral pitch, are required. Gears

TABLE 10.3   AGMA Full-Depth Gear Tooth Specifications

Tooth Feature	Coarse Pitch ($P_d < 20$) 14½° or 20° or 25°	Fine Pitch ($P_d \geq 20$) 20°
Pressure angle, ϕ	14½° or 20° or 25°	20°
Addendum, a	$\dfrac{1.000}{P_d}$	$\dfrac{1.000}{P_d}$
Dedendum, b	$\dfrac{1.250}{P_d}$	$0.002 + \dfrac{1.2}{P_d}$
Working depth, h_k	$\dfrac{2.000}{P_d}$	$\dfrac{2.000}{P_d}$
Whole depth, h_t	$\dfrac{2.250}{P_d}$	$0.002 + \dfrac{2.200}{P_d}$
Circular tooth thickness, t	$\dfrac{1.571}{P_d}$	$\dfrac{1.571}{P_d}$
Fillet radius, r_f	$\dfrac{0.300}{P_d}$	not standardized
Min. clearance, c	$\dfrac{0.250}{P_d}$	$0.002 + \dfrac{0.200}{P_d}$
Clearance (ground tooth), c	$\dfrac{0.350}{P_d}$	$0.002 + \dfrac{0.350}{P_d}$
Min top land width	$\dfrac{0.250}{P_d}$	not standardized
AGMA standard	201.02	207.04
Face width	$\dfrac{12}{P_d}$	$\dfrac{12}{P_d}$

are used in a great range of applications from mechanical watches with low forces to large steel-rolling mills with extremely large forces. Therefore, a wide range of diametral pitches must be available. The standardized values of diametral pitch were given in Table 10.1.

Most gear tooth features, as identified in Section 10.3 and Figure 10.3, are standardized relative to the diametral pitch and pressure angle. The governing relationships are given in Table 10.3. These relationships are maintained by AGMA. AGMA revises and publishes new standards every year, most of which are certified by the American National Standards Institute (ANSI). AGMA standards strongly influence both the American and the world marketplaces.

EXAMPLE PROBLEM 10.2

Consider the 20° full-depth, involute spur gear, with 35 teeth and a diametral pitch of 10, from Example Problem 10.1. Determine the diameter of the addendum circle, dedendum circle, and the clearance.

SOLUTION:

1. **Calculate the Addendum**

   The addendum circle is the outer diameter of the gear. The addendum is the distance from the pitch circle on a gear tooth to the top of the tooth. The standard distance for this gear can be computed from the equations in Table 10.3.

   $$a = \frac{1}{P_d} = \frac{1}{10} = 0.100 \, \text{in.}$$

2. **Calculate the Addendum Diameter**

   Notice that this is the distance between the radii of the pitch circle and the addendum circle. Therefore, the diameter of the addendum circle is offset a distance, a, on both sides of the pitch circle. In Example Problem 10.1, the pitch diameter is 3.5 in. Therefore, the diameter of the addendum circle can be computed as

   $$d_a = d + 2a = 3.5 + 2(0.100) = 3.7 \, \text{in.}$$

3. **Calculate the Dedendum Diameter**

   In a similar fashion, the dedendum is the distance between the radii of the pitch circle and the dedendum circle. Therefore, the dedendum can be computed as

   $$b = \frac{1.25}{P_d} = \frac{1.25}{10} = 0.125 \, \text{in.}$$

   and the dedendum circle diameter is

   $$d_b = d - 2b = 3.5 - 2(0.125) = 3.25 \, \text{in.}$$

4. **Calculate the Amount of Noninvolute Profile on the Tooth**

   Notice that the base circle diameter from this example problem is 3.289 in. Comparing this to the dedendum circle reveals that a short portion of the gear tooth profile is inside the base circle. On a radial basis, the length of this short portion of tooth profile is determined by

   $$\text{Noninvolute radial length} = \frac{(3.289)}{2} - \frac{(3.250)}{2} = 0.019$$

   Recall that the definition of an involute extends only from a base circle. This short portion of tooth profile is not an involute and should not be contacted by the mating gear tooth.

5. **Calculate the Clearance**

   Finally, the clearance is the amount that the dedendum exceeds the addendum. This is the room between the top of a gear tooth and the bottom of the mating gear tooth. The standard distance for this gear can be computed using the equations in Table 10.3.

   $$c = \frac{0.25}{P_d} = \frac{0.25}{10} = 0.025 \, \text{in.}$$

   Notice that the clearance is greater than the distance of the noninvolute portion of the gear tooth. Thus, contact between gear teeth on this portion is not expected.

In practice, the manufactured profile of the gear teeth will deviate from the theoretical form just discussed. Composite error accounts for manufacturing imperfections on the tooth profile and the tooth-to-tooth spacing. AGMA [Standard 2000-A88] defines a spectrum of quality indexes ranging from the lowest (3) to the highest (16) precision. The velocity of the mating gear teeth, which will be discussed in Section 10.7, is one factor that determines the required quality. Obviously, the cost of a gear will be a function of this quality.

10.6 RELATIONSHIPS OF GEARS IN MESH

Two gears in contact were shown in Figure 10.3. As two gears mesh, the smaller gear is commonly termed the *pinion* and the larger is referred to as the *bull gear* or simply the *gear*. Recall that in order for two gears to mate, they must have the same diametral pitch and pressure angle. Relationships of two mating gears are discussed in the following sections.

10.6.1 Center Distance

The *center distance*, *C*, is defined as the center-to-center distance between two mating gears. This is also the distance between the shafts that are carrying the gears. For the common configuration of external gears (Figure 10.3), the distance can be written as

$$C_{\text{external gears}} = r_1 + r_2 = \frac{(d_1 + d_2)}{2} \quad (10.7)$$

because

$$d = \frac{N}{P_d}$$

Equation (10.7) can be rewritten as

$$C_{\text{external gears}} = \frac{(N_1 + N_2)}{2 P_d} \quad (10.8)$$

For internal gears (Figure 10.3c), the center distance is the difference in the pitch radii and can be written as

$$C_{\text{internal gears}} = r_2 - r_1 = \frac{(d_2 - d_1)}{2} = \frac{(N_2 - N_1)}{2 P_d} \quad (10.9)$$

EXAMPLE PROBLEM 10.3

Two 5-pitch, 20° full-depth gears are used on a small construction site concrete mixer. The gears transmit power from a small engine to the mixing drum. This machine is shown in Figure 10.10. The pinion has 15 teeth and the gear has 30 teeth. Determine the center distance.

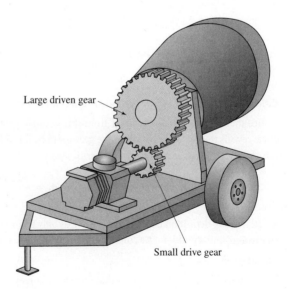

Large driven gear

Small drive gear

FIGURE 10.10 Concrete mixer for Example Problem 10.3.

SOLUTION: 1. *Calculate Pitch Diameters*

The pitch diameters of both gears can be determined from equation (10.2).

$$d_1 = \frac{N_1}{P_d} = \frac{15}{5} = 3.0 \,\text{in.}$$

$$d_2 = \frac{N_2}{P_d} = \frac{30}{5} = 6.0 \, \text{in.}$$

2. **Calculate the Center Distance**

Because these gears are external, the center distance can be found from equation (10.7).

$$C = \frac{(d_1 + d_2)}{2} = \frac{(3.0 \, \text{in.} + 6.0 \, \text{in.})}{2} = 4.5 \, \text{in.}$$

3. **Verify Center Distance**

Of course, equation (10.8) can be used to directly calculate the center distance from the information given.

$$C = \frac{(N_1 + N_2)}{2 P_d} = \frac{(15 + 30)}{2(5)} = 4.5 \, \text{in.}$$

10.6.2 Contact Ratio

The *contact ratio*, m_p, is the average number of teeth that are in contact at any instant. Obviously, the contact ratio must exceed 1 because contact between gears must not be lost. In practice, contact ratios should be greater than 1.2. Robust designs have contact ratios of 1.4 or 1.5. To illustrate the principle, a contact ratio of 1.2 indicates that one pair of teeth is always in contact and a second pair of teeth is in contact 20 percent of the time.

Greater contact ratio values result in smoother action because another gear tooth shares the load for a longer duration during the engaging/disengaging process. In addition, with more teeth sharing the load, greater power may be transmitted. However, the most direct manner in which the contact ratio can be increased is to use larger gears. This is in direct contrast to most design goals of compactness.

Numerically, contact ratio can be expressed as the length of the path of contact divided by the base pitch, p_b. The base pitch, in turn, is defined as the distance between corresponding points of adjacent teeth, measured on the base circle. The base pitch can be computed with the following:

$$p_b = \frac{\pi d_1 \cos \phi}{N_1} = \frac{\pi d_2 \cos \phi}{N_2} \qquad (10.10)$$

Of course, the path of the contact point, of involute gear teeth, is a straight line (Section 10.4). The length of this contact path, Z, is derived by the intersections of the respective addendum circles and the contact line. This length is derived [Ref. 9] as

$$Z = \sqrt{(r_2 + a_2)^2 - (r_2 \cos \phi)^2} - r_2 \sin \phi$$
$$+ \sqrt{(r_1 + a_1)^2 - (r_1 \cos \phi)^2} - r_1 \sin \phi \quad (10.11)$$

Thus, an expression for the contact ratio, in terms of the gear tooth geometry, can be given as

$$m_p = \frac{Z}{p_b} \qquad (10.12)$$

EXAMPLE PROBLEM 10.4

For the concrete mixer gears described in Example Problem 10.3, determine the contact ratio.

SOLUTION: 1. **Calculate the Basic Tooth Properties**

The pitch radii of both gears can be determined from the pitch diameters.

$$r_1 = \frac{d_1}{2} = \frac{3.0 \, \text{in.}}{2} = 1.5 \, \text{in.}$$

$$r_2 = \frac{d_2}{2} = \frac{6.0 \, \text{in.}}{2} = 3.0 \, \text{in.}$$

From Table 10.3, the addendum for 20° full-depth teeth is

$$a = \frac{1}{P_d} = \frac{1}{5} = 0.20 \, \text{in.}$$

2. *Calculate the Base Pitch*

The base pitch is calculated from equation (10.10).

$$p_b = \frac{\pi d_1 \cos\phi}{N_1} = \frac{\pi(3.0 \text{ in.})\cos(20°)}{15} = 0.6890 \text{ in.}$$

3. *Calculate Contact Length*

The length of the contact line is calculated from equation (10.11).

$$\begin{aligned}
Z &= \sqrt{(r_2 + a_2)^2 - (r_2\cos\phi)^2} - r_2\sin\phi + \sqrt{(r_1 + a_1)^2 - (r_1\cos\phi)^2} - r_1\sin\phi \\
&= \sqrt{(3.0 + 0.2)^2 - (3.0\cos 20°)^2} - 3.0\sin 20° \\
&\quad + \sqrt{(1.5 + 0.2)^2 - (1.5\cos 20°)^2} - 1.5\sin 20° \\
&= 0.9255 \text{ in.}
\end{aligned}$$

Then the contact ratio is determined from equation (10.12).

$$m_p = \frac{Z}{pb} = \frac{0.9255 \text{ in.}}{0.6890 \text{ in.}} = 1.3433$$

Although this ratio is acceptable, larger values (1.4–1.5) are desirable.

10.6.3 Interference

As mentioned earlier, the most serious drawback with using involute gear tooth profiles is the possibility of interference. Gear teeth have involute profiles between the base circle and the addendum circle. When a gear with few teeth and small pressure angles is constructed, the dedendum circle is considerably smaller than the base circle of the involute. Therefore, the tooth between the base circle and the dedendum is not an involute. If the mating gear tooth were to contact this portion of the tooth, the fundamental condition for constant velocity ratio would be violated. This condition is termed *interference* and, as it occurs, the teeth can exhibit drastic noise, vibration, and wear.

Interference is induced as designers attempt to make gear assemblies compact by using too few teeth on the gears. Interference commonly occurs when a small gear mates with a much larger one. A relationship that can be used to determine the necessary number of teeth in the gear to avoid interference has been derived [Ref. 9]. Equation (10.13) determines the largest number of teeth in the gear to ensure no interference. The relationship is given as a function of the number of teeth in the mating pinion, along with the pressure angle and addendum size.

$$N_2 < \frac{\{N_1^2 \sin^2\phi - 4k^2\}}{4k - 2N_1 \sin^2\phi} \qquad (10.13)$$

where k is defined from the addendum relation

$$a = \frac{k}{P_d}$$

Equation (10.13) can be used to tabulate suitable combinations of gears that avoid interference. These combinations are given in Table 10.4.

TABLE 10.4 Gear Teeth Combinations to Ensure No Interference

$\phi = 14°$		$\phi = 20°$		$\phi = 25°$	
Number of Pinion Teeth	Maximum Number of Gear Teeth	Number of Pinion Teeth	Maximum Number of Gear Teeth	Number of Pinion Teeth	Maximum Number of Gear Teeth
< 23	Interference	< 13	Interference	< 9	Interference
23	26	13	16	9	13
24	32	14	26	10	32
25	40	15	45	11	249
26	51	16	101	12	∞
27	67	17	1309		
28	92	18	∞		
29	133				
30	219				
31	496				
32	∞				

Note from Table 10.4 that $14\frac{1}{2}°$ pinions with more than 32 teeth can mate with any size gear without interference. Also, any standard $14\frac{1}{2}°$ pinion with fewer than 23 teeth experiences interference, regardless of the size of the mating gear. Such limits can be gathered for other standard pressure angles.

It is apparent from Table 10.4 that an involute pinion with a 25° pressure angle permits usage of gears with fewer teeth without interference. As a result, more compact gear assemblies can be produced. This is the primary reason for the popularity of 25° teeth and the obsolescence of $14\frac{1}{2}°$ teeth.

For the extreme case where a pinion can mate with any other gear, $N_2 = \infty$ can be substituted into equation (10.13). This provides the size of pinion that can mate with any gear. As already mentioned, a $14\frac{1}{2}°$ pinion with 32 teeth exhibits such properties. Once $N_2 = \infty$ is substituted, the following relationship is derived:

$$N_1 > \frac{2k}{\sin^2\phi} \qquad (10.14)$$

It should be noted that a gear with $N_2 = \infty$ would also have an infinite pitch radius. This is the concept behind a rack, as shown in Figure 10.3d. Thus, equation (10.14) must be met to ensure that a gear mates with a rack and avoids interference.

EXAMPLE PROBLEM 10.5

For the gears of the concrete mixer described in Example Problem 10.3, determine whether interference is a concern.

SOLUTION:

1. **Use Interference Table to Check Criteria**

   From Table 10.4, it is observed that a 15 tooth pinion, with 20° full-depth teeth, cannot mate with a gear with more than 45 teeth without interference. With only 30 teeth, interference is not a foreseeable problem.

2. **Use Interference Equation to Check Criteria**

   The same result can be obtained from equation (10.13). From Table 10.3, the addendum is

   $$a = \frac{1}{P}$$

   Therefore,

   $$k = 1$$

   Equation (10.13) can be used to check for interference problems.

   $$N_2 < \frac{\{N_1^2 \sin^2\phi - 4k^2\}}{4k - 2N_1 \sin^2\phi}$$

   $$N_2 < \frac{\{15^2 \sin^2 20° - 4(1)^2\}}{4(1) - 2(15)\sin^2 20°}$$

   $$N_2 < 45.48$$

   The number of teeth on the driven gear, 30, is less than the limiting value of 45.48. Therefore, interference is not a foreseeable problem.

As previously mentioned, the tooth profile can be altered from the full-depth form to allow the mating of small gears with large gears. Of course, this minimizes the total size of a gear system, which is a common design goal. AGMA has included standard provisions for modifying involute profiles.

Stub teeth have been developed with a large pressure angle and short teeth. The stub system has stronger teeth but a working depth usually 20 percent less than full-depth teeth.

Another alternate system is the *long-and-short addendum system*. For these profiles, the addendum on one gear is increased and the addendum on the mating gear is decreased.

The result, of course, is that these nonstandard gears are no longer interchangeable with standard gears. These specialty gears are infrequently used in general machine design and the details are beyond the scope of this text. References should be consulted for discussions of nonstandard profiles [Refs. 4, 9, 13, 15].

10.6.4 Undercutting

Interference can also be avoided by removing the material on the gear tooth between the base circle and dedendum circle. This is the portion of the gear tooth that is not an involute and would interfere with the mating tooth. An undercut gear tooth is shown in Figure 10.11.

Undercutting obviously reduces the strength of the gear, thus reducing the power that can be safely transmitted. In addition, it also reduces the length of contact, which reduces

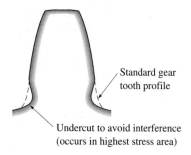

FIGURE 10.11 Undercut gear tooth.

the contact ratio and results in rougher and noisier gear action. Therefore, undercutting should be avoided unless the application absolutely requires a compact gearset. In these cases, advanced kinematic and strength analyses and experiments are necessary to verify proper operation.

10.6.5 Backlash

As stated in Section 10.3, backlash is the amount that the width of a tooth space exceeds the thickness of a gear tooth, measured on the pitch circle. In more practical terms, it is the amount that a gear can turn without its mating gear turning. Although backlash may seem undesirable, some backlash is necessary to provide for lubrication on the gear teeth. Gears that run continuously in one direction can actually have considerable backlash. Gears that frequently start/stop or reverse direction should have closely controlled backlash.

A nominal value of backlash is designed into a gear tooth profile. The amount of backlash determines the thickness of a gear tooth because backlash is a measure of the tooth thickness to the tooth space. Recommended values of backlash are specified by AGMA [Standard 2002-B88]. Although these values are somewhat conservative, general power-transmitting gears have recommended backlash values of

$$\frac{0.05}{P_d} < B_{\text{recommended}} < \frac{0.1}{P_d}$$

For commercially available stock gears, backlash values are considerably higher to allow for greater flexibility in applications. The backlash values of these gears are typically

$$\frac{0.3}{P_d} < B_{\text{stock gears}} < \frac{0.5}{P_d}$$

Therefore, great care must be taken when specifying stock gears for applications with reversing directions or frequent start/stop sequences.

Backlash values are strongly influenced by any variation in the center distance of the gears. Of course, in any production environment, the center distance of two gears varies. However, a deviation in the nominal center distance can be purposely specified by the designer to adjust the backlash to a desired range. The backlash variation ΔB that will be encountered with a variation in the center distance ΔC can be approximated by the following relationship:

$$\Delta B \approx 2(\Delta C)\tan \phi \qquad (10.15)$$

Equation (10.15) can be used with equation (10.7) or (10.8) to specify a center distance that produces backlash values to be maintained in the range given. Reducing the center distance reduces the backlash, and vice versa.

EXAMPLE PROBLEM 10.6

The gears for the concrete mixer described in Example Problem 10.3 are catalog items with a designed backlash of $0.4/P_d$. Specify a center distance that reduces the backlash to an AGMA-recommended value of $0.1/P_d$.

SOLUTION: 1.   ***Calculate Designed Backlash***

$$B_{\text{designed}} = \frac{0.4}{P_d} = \frac{0.4}{5} = 0.08\,\text{in.}$$

2.   ***Calculate Recommended Backlash***

$$B_{\text{recommended}} = \frac{0.1}{P_d} = \frac{0.1}{5} = 0.02\,\text{in.}$$

3.   ***Calculate the Adjusted Center Distance***
Rearranging equation (10.15) gives

$$\Delta C \approx \frac{\Delta B}{(2\tan\phi)} = \frac{(0.02 - 0.08)}{2\tan(20°)} = 0.0824\,\text{in.}$$

From Example Problem 10.3, the nominal center distance was determined as 4.5 in. Therefore, to adjust the backlash value, the center distance should be reduced to

$$C_{\text{adjusted}} = 4.5 - 0.0824 = 4.4176\,\text{in.}$$

It should be emphasized that reducing the center distance will produce an overly tight mesh and may produce excessive noise, overheating, and structural overload. However, some applications require minimal backlash. Of course, testing should be conducted to confirm the design.

10.6.6 Operating Pressure Angle

As mentioned in preceding sections, the pressure angle defines the line of action of the force onto the gear teeth. The designated pressure angle is cut or formed into the gear tooth and affects the actual shape of the tooth (Figure 10.7).

It should be mentioned that as the center distance of the mating gears deviates from the nominal value, the actual pressure angle during operation differs from the designated value. In other words, two 20° gears may actually have a greater pressure angle during operation by increasing the center distance from the nominal value. The relationship that can be used to determine the amount of variance is derived [Ref. 9] as

$$\cos \phi_{\text{operating}} = \left\{ \frac{C_{\text{nominal}}}{C_{\text{operating}}} \right\} \cos \phi_{\text{nominal}} \qquad \textbf{(10.16)}$$

Applications that require precise calculation of the actual force being transmitted should use this operating pressure angle. This reflects the actual performance of the gear forces.

EXAMPLE PROBLEM 10.7

For the gears of the concrete mixer described in Example Problem 10.3, determine the operating pressure angle when the center distance is measured at 4.4176 in. as in Example Problem 10.6.

SOLUTION: From the numbers in Example Problem 10.6 and equation (10.16), the following can be determined:

$$\cos \phi_{\text{operating}} = \left\{ \frac{C_{\text{nominal}}}{C_{\text{operating}}} \right\} \cos \phi_{\text{nominal}} = \left\{ \frac{4.5}{4.4176} \right\} \cos 20° = 0.9572$$

and

$$\phi_{\text{operating}} = 16.82°$$

10.7 SPUR GEAR KINEMATICS

A basic function of gears is to provide a constant velocity ratio between their respective shafts. A pair of gears that has a constant velocity ratio means that the driven gear maintains a uniform speed as long as the driver gear rotates at a constant speed. This condition led to the development of the involute tooth profile. A pair of mating spur gears is shown in Figure 10.12.

Formally, the *velocity ratio, VR,* is defined as the angular speed of the driver gear (gear 1) divided by the angular speed of the driven gear (gear 2).

$$VR = \frac{\omega_{\text{driver}}}{\omega_{\text{driven}}} = \frac{\omega_1}{\omega_2} \qquad \textbf{(10.17)}$$

Because a ratio is valid regardless of units, the velocity ratio can be defined in terms of revolutions per minute, radians per time, or any other convenient set of rotational velocity units. In practice, a velocity ratio of 3 would be indicated as 3:1 and pronounced "three to one." Likewise, a velocity ratio of 1/3 would be indicated as 1:3 and pronounced "one to three."

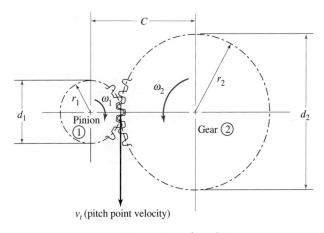

v_t (pitch point velocity)

FIGURE 10.12 Kinematics of meshing gears.

The *pitch line velocity, v_t,* is defined as the magnitude of the velocity of the pitch point of the two mating gears. This velocity is also illustrated in Figure 10.12. It should be apparent that the pitch line velocity of both gears is identical because one gear tooth pushes the mating tooth. Therefore, the pitch line velocity is a linear measure and can be related to the rotational velocities of the gears and their pitch radii using equation (6.5).

$$v_t = r_1\omega_1 = r_2\omega_2 \qquad \textbf{(10.18)}$$

Note that, as in Chapter 6, the angular velocity in this equation must be specified in radians per unit time.

Combining equations (10.17) and (10.18) gives the following relation:

$$\frac{\omega_1}{\omega_2} = \frac{r_1}{r_2} = VR$$

Introducing the pitch diameters,

$$\frac{d_2}{d_1} = \frac{(2r_2)}{(2r_1)} = \frac{r_2}{r_1} = VR$$

and introducing the diametral pitch and number of teeth,

$$\frac{d_2}{d_1} = \frac{\dfrac{N_2}{P_d}}{\dfrac{N_1}{P_d}} = VR$$

Because the diametral pitch of the two gears must be identical for the teeth to mate, P_d can be eliminated from the previous equation, yielding

$$\frac{d_2}{d_1} = \frac{N_2}{N_1} = VR$$

Collecting all the preceding relationships yields a comprehensive expression for the velocity ratio.

$$VR = \frac{\omega_1}{\omega_2} = \frac{r_2}{r_1} = \frac{d_2}{d_1} = \frac{N_2}{N_1} \qquad \textbf{(10.19)}$$

An algebraic sign convention designates the relative direction of gear rotations. In the typical external gearset, the shaft centers are on opposite sides of the common tangent to the pitch circles, which dictates that the gears rotate in opposite directions. To signify this fact, the velocity ratio is given a negative value.

For internal gears, as shown in Figure 10.3c, the shaft centers are on the same side of the common tangent to the pitch circles. This dictates that the gears rotate in the same direction. Thus, the velocity ratio is given a positive value.

As discussed in the introduction, many gears are used in applications where the speed from a power source must be reduced. Therefore, it is typical to have velocity ratios greater than 1. As can be seen from equation (10.17), this indicates that the drive gear rotates faster than the driven gear, which is the case in speed reductions.

EXAMPLE PROBLEM 10.8

A set of gears is used to reduce the speed from an electric motor to a shaft driving a grocery checkout conveyor (Figure 10.13). The gear on the motor shaft is a 10-pitch pinion, has 15 teeth, and drives at 1800 rpm clockwise. Determine the speed of the mating gear, which has 45 teeth. Also calculate the pitch line velocity.

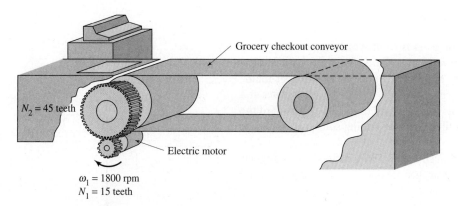

$N_2 = 45$ teeth

Grocery checkout conveyor

Electric motor

$\omega_1 = 1800$ rpm
$N_1 = 15$ teeth

FIGURE 10.13 Checkout conveyor for Example Problem 10.8.

SOLUTION: 1.   *Calculate the Velocity Ratio*

The velocity ratio can be computed from equation (10.19).

$$VR = \frac{N_2}{N_1} = \frac{45}{15} = -3$$

In practice, this value would be commonly expressed as a 3:1 gear ratio. Note that the negative value indicates that the gears rotate in opposite directions. This is consistent with external gears.

2. ***Calculate the Rotational Velocity of the Driven Gear***

The angular velocity of the driven gear can be computed by rearranging equation (10.17).

$$\omega_2 = \frac{\omega_1}{VR} = \frac{1800 \text{ rpm}}{(-3)} = -600 \text{ rpm} = 600 \text{ rpm, counterclockwise}$$

3. ***Calculate the Pitch Line Velocity***

The pitch diameters are computed from equation (10.2).

$$d_1 = \frac{N_1}{P_d} = \frac{15}{10} = 1.5 \text{ in.}$$

$$d_2 = \frac{N_2}{P_d} = \frac{45}{10} = 4.5 \text{ in.}$$

The pitch line velocity can be computed from equation (10.18).

$$v_t = r_1 \omega_1$$

$$r_1 = \frac{1.5}{2} = 0.75 \text{ in.}$$

$$\omega_1 = (1800 \text{ rpm}) \left[\frac{2\pi \text{ rad}}{1 \text{ rev}} \right] = 11309.7 \text{ rad/min}$$

$$v_t = (0.75 \text{ in.}) (11309.7 \text{ rad/min}) = 8482.3 \text{ in./min}$$

Converting units,

$$v_t = 8482.3 \text{ in./min} \left[\frac{1 \text{ ft}}{12 \text{ in.}} \right] = 706.9 \text{ ft/min}$$

As mentioned earlier, AGMA [Standard 2000-A88] has defined a set of quality control numbers, ranging from 3 to 16. These numerical ratings reflect the accuracy of the tooth profile, the tooth-to-tooth spacing, and surface finish. The pitch line velocity, v_t, is one factor that determines the required quality of the gear. Table 10.5 shows the recommended gear quality for drives of precision mechanical systems.

TABLE 10.5 Recommended AGMA Gear Quality

Pitch Line Speed (ft/min, fpm)	Suggested Gear Quality (AGMA Rating)
0–800	6–8
800–2000	8–10
2000–4000	10–12
Over 4000	12–14

10.8 SPUR GEAR SELECTION

In a design situation, gears must be selected to accomplish a desired task. Often this task is to achieve a desired velocity ratio. Because the majority of gears in operation are AGMA standard, the designer only needs to determine the key parameters. These parameters are the diametral pitch, pressure angle, and number of teeth on each gear. Most other gear

TABLE 10.6 Suitable Diametral Pitches for 20° Mild-Steel Gears with Standard Face Width

Power	Pinion rpm								
hp	50	100	300	600	900	1200	1800	2400	3600
0.05	20	20	24	32	32	32	32	32	32
0.10	16	20	20	24	24	24	32	32	32
0.25	12	16	20	20	24	24	24	24	24
0.33	10	12	16	20	20	24	24	24	24
0.50	10	12	16	20	20	20	20	24	24
0.75	8	10	12	16	16	20	20	20	20
1.0	6	10	12	16	16	16	20	20	20
1.5	6	8	12	12	16	16	16	16	20
2.0	6	6	10	12	12	12	16	16	16
3.0	5	6	8	10	12	12	12	12	16
5.0	4	5	6	8	10	10	12	12	12
7.5	4	5	6	8	8	8	10	10	10
10	3	4	6	6	6	8	8	8	10
15	2	4	5	6	6	6	6	6	8
20	2	3	4	5	6	6	6	6	—
25	—	3	4	5	5	5	6	5	—
30	—	2	4	4	5	5	5	—	—
40	—	2	3	4	4	—	—	—	—
50	—	—	3	4	4	—	—	—	—

features can be determined using the AGMA standard relationships presented in previous sections.

10.8.1 Diametral Pitch

In the typical design situation, the first selection parameter is an appropriate diametral pitch. Because the diametral pitch is the relative size of a gear tooth, it stands to reason that the transmitted forces and the gear material properties influence this decision. Precise selection criteria involve calculation of gear tooth stresses and contact pressures. The calculation procedures are outlined in the AGMA specifications. This level of detail is beyond the scope of this text.

Conservative estimates of appropriate diametral pitches can be readily obtained from most commercial gear suppliers. The suppliers use the AGMA standards to determine the power-carrying capabilities of their stock gears. From this data, an estimate of suitable diametral pitch can be made with knowledge of the nominal power that is transmitted by the gear pair, the rotational speed of the pinion, and the gear material. As an example, such data are presented in Table 10.6. This table gives suitable diametral pitches of 20° mild-steel gears with standard face width, based on pinion speed and the power transmitted. Similar tables exist for alternate pressure angles and materials. The use of these tables can be illustrated with an example.

EXAMPLE PROBLEM 10.9

A pair of mild-steel gears is to be selected for the concrete mixer described in Example Problem 10.3. The mixer is driven by a 10-hp engine at a speed of 1800 rpm. Determine an appropriate diametral pitch.

SOLUTION: Mild-steel gears that are capable of handling 10 hp at a pinion speed of 1800 rpm are specified. From interference criteria, Table 10.3 shows that an 18-tooth pinion with a pressure angle of 20° can mate with any other gear. Table 10.6 suggests that a diametral pitch of eight should be used to transfer the power. Therefore, an 18-tooth, mild-steel pinion with a diametral pitch of eight should be suitable. For a more reliable selection, thorough strength analysis should be performed.

10.8.2 Pressure Angle

The second parameter that should be selected is a pressure angle. As mentioned, the standard values of pressure angles are 14½°, 20°, and 25°. Recall that 14½° is recommended only for replacement of other 14½° gears on existing machinery. Gears with pressure angles of 20° are well suited for general applications. Gears with pressure angles of 25° can be smaller without a concern for interference but have less efficient force transmission. Therefore, they are best suited for high-speed and lower-power applications.

10.8.3 Number of Teeth

Finally, the number of gear teeth should be determined. This decision is typically influenced by the desired velocity ratio. In general, smaller gears are preferred because they minimize size, weight, and cost. Of course, the minimum size is determined by interference criteria. The number of teeth on a gear also must be an integer. Although this statement seems obvious, it must be a constant consideration, as obtaining an integer solution can be difficult. In addition, gear manufacturers do not stock gears with tooth increments of one. Table 10.7 lists typical commercially available gears. A specific catalog, such as Boston Gear, Browning Gears, or Martin Sprocket & Gear, should be consulted when deciding on the number of teeth, as more options than listed in Table 10.7 may be available.

TABLE 10.7	Number of Teeth for commercially available stock gears							
32 Diametral Pitch								
12	16	20	28	36	48	64	80	112
14	18	24	32	40	56	72	96	128
24 Diametral Pitch								
12	18	24	30	42	54	72	96	144
15	21	27	36	48	60	84	120	
20 Diametral Pitch								
12	16	24	35	50	80	100	160	
14	18	25	40	60	84	120	180	
15	20	30	45	70	90	140	200	

TABLE 10.7   Continued

16 Diametral Pitch

12	16	24	32	48	64	96	160	
14	18	28	36	56	72	128	192	
15	20	30	40	60	80	144		

12 Diametral Pitch

12	15	20	28	42	60	84	120	168
13	16	21	30	48	66	96	132	192
14	18	24	36	54	72	108	144	216

10 Diametral Pitch

12	16	24	30	45	55	80	120	200
14	18	25	35	48	60	90	140	
15	20	28	40	50	70	100	160	

8 Diametral Pitch

12	16	22	32	44	60	80	112
14	18	24	36	48	64	88	120
15	20	28	40	56	72	96	128

6 Diametral Pitch

12	16	24	33	48	66	96
14	18	27	36	54	72	108
15	21	30	42	60	84	120

5 Diametral Pitch

12	16	24	30	45	70	110	160
14	18	25	35	50	80	120	180
15	20	28	40	60	100	140	

EXAMPLE PROBLEM 10.10

A gear reducer is used on a concept for a small trolling motor for fishing boats. The gears must transmit 5 hp from an electric motor at 900 rpm to the propeller at 320 rpm. Select a set of gears to accomplish this task.

SOLUTION:

1. **Determine a Suitable Diametral Pitch and Pressure Angle**

   Because this application involves general gearing, a pressure angle of 20° is used. Referring to Table 10.6, an estimate of a suitable diametral pitch is

   $$P_d = 10$$

2. **Use the Required Velocity Ratio to Iterate and Determine Appropriate Number of Teeth**

   The required velocity ratio is

   $$VR = \frac{\omega_{\text{driver}}}{\omega_{\text{driven}}} = \frac{900\,\text{rpm}}{320\,\text{rpm}} = 2.8125$$

   Rearranging equation (10.19) yields

   $$N_{\text{driven}} = N_{\text{driver}}\left(\frac{\omega_{\text{driver}}}{\omega_{\text{driven}}}\right) = N_{\text{driver}}\left(\frac{900\,\text{rpm}}{320\,\text{rpm}}\right) = N_{\text{driver}}(2.8125)$$

   Because a smaller assembly is generally preferred, values of pinion (driver) teeth are substituted, beginning with the smallest possible pinion. Note that an iterative procedure must be used because the number of teeth must be an integer. Using

   $$N_{\text{driver}} = 13, \quad N_{\text{driven}} = 13\left(\frac{900}{320}\right) = 36.56$$

   $$N_{\text{driver}} = 14, \quad N_{\text{driven}} = 14\left(\frac{900}{320}\right) = 39.38$$

$$N_{\text{driver}} = 15, \qquad N_{\text{driven}} = 15\left(\frac{900}{320}\right) = 42.14$$

$$N_{\text{driver}} = 16, \qquad N_{\text{driven}} = 16\left(\frac{900}{320}\right) = 45$$

The smallest integer combination is 16 and 45 teeth. Also, from the preceding discussion, a suitable diametral pitch is 10. Table 10.7 confirms that these gears are commercially available.

3. **Calculate the Pitch Diameters and Center Distance**

Finally, the corresponding pitch diameters and center distance are

$$d_1 = \frac{N_1}{P_d} = \frac{16}{10} = 1.6 \text{ in.}$$

$$d_2 = \frac{N_2}{P_d} = \frac{45}{10} = 4.5 \text{ in.}$$

$$C_{\text{external gears}} = \frac{(d_1 + d_2)}{2} = \frac{(1.6 + 4.5)}{2} = 3.05 \text{ in.}$$

Often, gears must be selected to alter the velocity ratio between shafts of an existing machine. A similar situation occurs when the gear shafts must be spaced at a specific distance due to other constraints. Both of these situations place a limit on the center distance of the gears. In these situations, the number of teeth selected for each gear may not be the smallest possible, but those needed to fill the distance between the shafts. Also, a larger tooth than necessary can be used to help fill the distance between shafts. Finally, some deviation from the target ratio may be needed to specify standard gears. In general, the relationships explained throughout this chapter can be used to specify any gearset. The following examples illustrate some possible scenarios.

EXAMPLE PROBLEM 10.11

A pair of gears is powered by an electric motor and used to drive the spindle of a lathe at 200 rpm. This drive system is illustrated in Figure 10.14. The 1-hp motor will be replaced by a more efficient but higher-speed motor, rated at 600 rpm. To accomplish this alteration, a new set of gears must be selected that will maintain the spindle speed at 200 rpm. However, the gears are mounted in an elaborate housing that cannot be modified. Therefore, the center distance between the gears must remain at 7.5 in. Specify a set of gears that can be used.

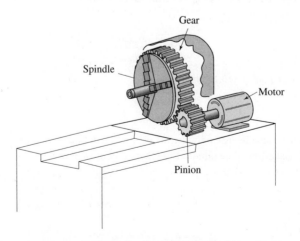

FIGURE 10.14 Lathe drive for Example Problem 10.11.

SOLUTION:

1. **Specify the Gear Ratio and Center Distance**

   The main parameters in this problem are the velocity ratio and the center distance. The required velocity ratio is

   $$VR = \frac{\omega_{driver}}{\omega_{driven}} = \frac{600\,\text{rpm}}{200\,\text{rpm}} = 3.0$$

   Therefore,

   $$VR = \frac{d_2}{d_1} = 3.0$$

   which can be rewritten as

   $$d_2 = 3d_1$$

   In addition, the center distance is

   $$C_{external\ gears} = \frac{(d_1 + d_2)}{2} = 7.5$$

2. **Determine the Required Diameter of the Gears**

   Using these relationships, appropriate pitch diameters can be algebraically determined by

   $$C_{external\ gears} = \frac{d_1 + d_2}{2} = \frac{d_1 + 3d_1}{2} = \frac{4d_1}{2} = 7.5$$

   Solving,

   $$d_1 = 3.75\text{ in.}$$

   and

   $$d_2 = 3(3.75) = 11.25\text{ in.}$$

3. **Determine an Appropriate Diametral Pitch**

   The problem now reduces to finding a suitable diametral pitch and number of teeth that result in the required pitch diameters. Because this application involves general gearing, a pressure angle of 20° is used. Referring to Table 10.6, an estimate of a suitable diametral pitch is 14. Therefore, only values of $P_d \leq 14$ are considered. By relating the pitch diameter, diametral pitch, and number of teeth, the following can be calculated:

   $$N_{driver} = (d_1)(P_d) = 3.75\,P_d$$

   $$N_{driven} = (VR)N_{driver} = 3N_{driver} = 3(3.75\,P_d) = 11.25\,P_d$$

   Diametral pitches of 14 and lower are substituted into these two equations. Recall that only the solutions having an integer number of teeth are valid. Iterating through all combinations, only three are feasible.

   The best alternative would depend on availability of standard gears, cost, and weight of the gearset. Notice that the output speed will be exactly 200 rpm. In many situations, the driven speed cannot be exactly obtained. The next problem illustrates such a case.

EXAMPLE PROBLEM 10.12

A gear-driven exhaust fan and housing is shown in Figure 10.15. To improve the air flow, the speed of the fan needs to be increased to 460 rpm, but it must be as close to this speed as possible. The existing 3-hp motor will be used, which operates at 1750 rpm. The housing should not be altered, which has a bearing system with a center distance of 9.5 in. Select a set of gears for this application.

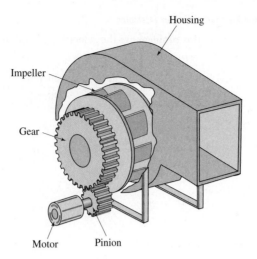

FIGURE 10.15 Exhaust fan for Example Problem 10.12.

SOLUTION: 1.   *Specify the Gear Ratio and Center Distance*

As in Example Problem 10.11, the main parameters in this problem are the velocity ratio and the center distance. The required velocity ratio is

$$VR = \frac{\omega_{\text{driver}}}{\omega_{\text{driven}}} = \frac{1750\,\text{rpm}}{460\,\text{rpm}} = 3.80$$

This design scenario is complicated by a nonfractional velocity ratio. It will be impossible to obtain a driven speed at exactly 460 rpm using an integer number of gear teeth. This is solved by rounding the velocity ratio to a fractional value.

$$VR = \left(\frac{d_2}{d_1}\right) \approx 3.75$$

This rounding will yield a driven speed of

$$\omega_{\text{driven}} = \left(\frac{\omega_{\text{driver}}}{VR}\right) = \left(\frac{1750\,\text{rpm}}{3.75}\right) = 466\text{ rpm}$$

Assuming that the fan operates properly at this speed,

$$d_2 = 3.75\,d_1$$

As before, the center distance is

$$C_{\text{external gears}} = \frac{(d_1 + d_2)}{2} = 5.5\text{ in.}$$

2.   *Determine the Required Diameters for the Gears*

Using these relationships, appropriate pitch diameters can be algebraically determined.

$$C_{\text{external gears}} = \frac{(d_1 + 3.75d_1)}{2} = \frac{4.75d_1}{2} = 9.5\text{ in.}$$

Solving,

$$d_1 = 4\text{ in.}$$

and

$$d_2 = 3.75(4) = 15\text{ in.}$$

3. **Determine an Appropriate Diametral Pitch**

As before, the problem now reduces to finding a suitable diametral pitch and number of teeth that result in the required pitch diameters. As mentioned, finding integer teeth is improbable because of the decimal velocity ratio. An interactive solution is required.

Because this application involves general gearing, a pressure angle of 20° is used. Referring to Table 10.6, an estimate of a suitable diametral pitch is 12. Therefore, only values of $P_d \leq 12$ are considered. By relating the velocity, pitch diameter, diametral pitch, and number of teeth, the following can be calculated:

$$N_{driver} = (d_1)(P_d) = 2P_d$$

$$N_{driven} = (VR)\,N_{driver} = 3.75\,N_{driver} = 3.75(2P_d) = 7.5\,P_d$$

Diametral pitches of 12 and lower are substituted into these two equations. Recall that only solutions having an integer number of teeth are valid. Calculating all combinations, three are feasible:

Other feasible combinations exist with gears having a diametral pitch lower than 8. Again, the best alternative would depend on the availability of standard gears, cost, and weight of the gearset.

10.9 RACK AND PINION KINEMATICS

A gear rack was briefly discussed in Section 10.2 and illustrated in Figure 10.3b. It is used to convert rotational motion of a pinion to translating motion of the rack. The most noteworthy application is the rack-and-pinion steering in automobiles. In this application, the rotational motion from the steering wheel pushes the rear of the front wheels, steering the car in a new direction; thus, the motion transfers from rotational to linear. A rack and pinion can also be operated such that the linear motion of the rack rotates the pinion.

As briefly mentioned in Section 10.5.3, a rack is a special case of a spur gear. As the diameter of a gear becomes very large, the local profile of the teeth resembles a rack. In fact, mathematically, a rack can be treated as a spur gear with an infinite pitch diameter. Therefore, all geometric properties that were introduced for spur gears also apply to a rack. The only difference is that instead of referring to a pitch diameter, a rack has a pitch line.

From a kinematic standpoint, the rotational motion of the pinion and the linear motion of the rack can be related through concepts presented in Chapter 6, equation (6.5). The rack displacement equation can be given as

$$\Delta R_{rack} = r(\Delta\theta) = \frac{(d_{pinion})(\Delta\theta_{pinion})}{2} \quad \textbf{(10.20)}$$

where $\Delta\theta_{pinion}$ must be specified in radians. The magnitude of the linear velocity of the rack is given as

$$v_{rack} = \omega_{pinion}r_{pinion} = \frac{(d_{pinion})(\omega_{pinion})}{2} \quad \textbf{(10.21)}$$

EXAMPLE PROBLEM 10.13

A rack and pinion is used on a drill press as shown in Figure 10.16. The 16-pitch pinion has 16 teeth. Determine the distance that the handle (and pinion) must be rotated in order to advance the drill 0.75 in.

SOLUTION: From equation (10.20), the rotation of the pinion is desired.

$$\Delta\theta_{pinion} = \frac{2\Delta R_{rack}}{d_{pinion}}$$

And

$$d_{pinion} = \frac{N_{pinion}}{P_d} = \frac{16}{16} = 1.0 \text{ in.}$$

$$\Delta\theta_{pinion} = \frac{2(0.75\,\text{in.})}{(1.0\,\text{in.})} = 1.5 \text{ rad}$$

Converting to degrees,

$$\Delta\theta_{pinion} = 1.5\,\text{rad}\left(\frac{180°}{\pi\,\text{rad}}\right) = 85.94°$$

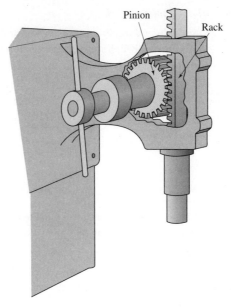

FIGURE 10.16 Rack and pinion drill press.

EXAMPLE PROBLEM 10.14

For the drill press described in Example Problem 10.13, determine the speed that the pinion must be rotated in order to advance the drill at a rate of 12 in./min.

SOLUTION: From equation (10.21), the rotational speed of the pinion is determined by

$$\omega_{\text{pinion}} = \frac{2v_{\text{rack}}}{d_{\text{pinion}}} = \frac{2(12 \text{ in./min})}{(10 \text{ in.})} = 24 \text{ rad/min}$$

Converting to revolutions per minute,

$$\omega_{\text{pinion}} = 24 \text{ rad/min}\left(\frac{1 \text{ rev}}{2\pi \text{ rad}}\right) = 3.82 \text{ rpm}$$

10.10 HELICAL GEAR KINEMATICS

Helical gears were introduced in Section 10.2 and illustrated in Figure 10.3d. The development of helical gears actually resulted from machinists discovering that stepped gears ran smoother and quieter than spur gears. A stepped gear consisted of a number of thin spur gears placed side by side, with each gear rotated a small angle relative to the adjacent gear. The resulting stacked gear did not exhibit the same large impact that two teeth usually have when they come into contact (e.g., ordinary spur gears).

Helical gears are the extreme case of stepped gears, where the teeth are not stepped but inclined to the axis of the gear. When used on parallel shafts, helical gears provide overlapping tooth contact. That is, when the front edge of a tooth comes into contact and begins to take the transmitted load, the back edge of the previous tooth is also in contact. This results in smoother and quieter operation, as a tooth loads gradually. For these reasons, helical gears are often preferred, even though they are more difficult to manufacture and, consequently, more expensive.

Helical gears are designated as either right-hand or left-hand, depending on the slope of the inclined teeth. A helical gear with teeth that slope down toward the left is designated as a left-hand helix. Conversely, a helical gear with teeth that slope down toward the right is designated as a right-hand helix. The upper helical gear illustrated in Figure 10.3d is a left-hand gear.

Helical gears can also be used on nonparallel shafts without altering the inherent geometry. Such a configuration is termed *crossed helical gears*. However, with crossed configurations, the forces required to drive the gearset increase dramatically with the shaft angle. Therefore, such configurations are recommended for lower-power transmitting applications.

The geometric and kinematic relationships for helical gears are very similar to spur gears. The major difference is the definition of a *helix angle*, ϕ, which is the angle of inclination of the teeth. This angle is illustrated with the left-hand helical gear shown in Figure 10.17.

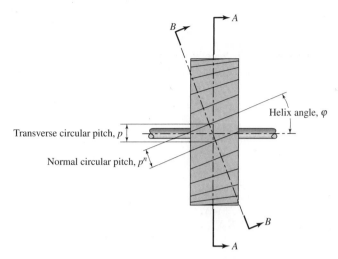

FIGURE 10.17 Helical gear geometry.

A cross-sectional view through a helical gear, perpendicular to the gear axis, appears identical to a spur gear. This view is generated by section A–A in Figure 10.17, termed the *transverse section*. The tooth geometric properties defined for spur gears can be used for helical gears. To avoid confusion, these properties are designated as transverse properties. The transverse circular pitch, transverse pressure angle, and transverse diametral pitch are identical to the corresponding spur gear definitions. The transverse circular pitch is shown in Figure 10.17.

Some additional geometric properties are defined by viewing a cross section, normal to the gear teeth. This view would be generated by section B–B in Figure 10.17, termed the *normal section*.

The *normal circular pitch*, p^n, is defined as the distance between corresponding points on a gear, measured on the pitch circle and normal to the gear tooth. The normal circular pitch is also shown in Figure 10.17. The normal circular pitch can be related to the transverse circular pitch through trigonometry.

$$p^n = p \cos \phi \qquad (10.22)$$

The *normal diametral pitch*, P_d^n, is defined using the normal circular pitch in a similar fashion as equation (10.5).

$$P_d^n = \frac{\pi}{p^n} \qquad (10.23)$$

A *normal module, m^n*, is similarly defined as

$$m^n = \pi p^n \qquad (10.24)$$

Also from trigonometry,

$$P_d = P_d^n \cos \phi \qquad (10.25)$$

$$m = \frac{m^n}{\cos \phi} \qquad (10.26)$$

A *normal pressure angle, θ^n*, is also defined from the tooth form in this normal view. The normal pressure angle

can also be related to the transverse pressure angle by the following:

$$\tan \phi^n = \tan \phi \cos \phi \qquad (10.27)$$

Helical gears are rarely used interchangeably and, therefore, no standard tooth systems exist like those described for spur gears. The preferred dimensions are usually dependent on the manner in which the helical gear is formed. When the gear is cut through a hobbing operation, the normal diametral pitch should conform to the standards listed in Table 10.1. Conversely, when a gear is cut on a shaper, the transverse diametral pitch should conform to values listed in Table 10.1.

The helix angle for most gears varies between 15° and 45°. Because the teeth are at an angle to the shaft, a thrust load is produced with mating helical gears. The thrust force varies directly with the tangent of the helix angle, and therefore, larger helix angles require sufficient axial gear and shaft support.

For parallel shaft applications, the velocity ratio presented in equation (10.19) is also applicable to helical gears. Two additional requirements, beyond those for spur gears, for proper meshing of helical gears include:

1. The gears must have equal helix angles.

2. The helix on the two mating gears must be opposite hand. That is, one gear must have a left-hand helix and the other a right-hand one.

The presence of the helix angle also aids in the avoidance of interference. An equation similar to equation (10.13) has been derived for helical gears. Thus, the minimum number of pinion teeth that can be used, mating with any size gear, without interference concerns is written as follows:

$$N_1 > \frac{2k \cos \varphi}{\sin^2 \phi} \qquad (10.28)$$

Values generated from this equation are condensed into Table 10.8.

TABLE 10.8   Minimum Helical Gear Teeth to Avoid Interference

Helix Angle	Normal Pressure Angle, ϕ^n		
	$14\frac{1}{2}°$	20°	25°
0 (spur gear)	32	17	12
5°	32	17	12
10°	31	17	12
15°	29	16	11
20°	27	15	10
22.5°	25	14	10
25°	24	13	9
30°	21	12	8
35°	18	10	7
40°	15	8	6
45°	12	7	5

EXAMPLE PROBLEM 10.15

In order to reduce the noise in a gear drive, two 12-pitch gears with 20 and 65 teeth are to be replaced with helical gears. The new set of gears must have the same velocity ratio. Because the same housing will be used, the center distance must also remain the same. Assume that the helical gears will be formed with a hob.

SOLUTION:

1. ***Calculate Desired Velocity Ratio and Center Distance***

   The original velocity ratio and center distance must be computed as follows:

   $$VR = \frac{N_{driven}}{N_{driver}} = \frac{65}{20} = 3.25$$

   $$C_{external\ gears} = \frac{(N_1 + N_2)}{2P_d} = \frac{(20 + 65)}{2(12)} = 3.4\,\text{in.}$$

2. ***Determine an Appropriate Diametral Pitch***

   Because the gears will be cut with a hob, the normal diametral pitch should conform to the standards listed in Table 10.1. The original gears had a diametral pitch of 12; thus, it is assumed that the teeth have sufficient strength. The helical gears then are selected with a normal diametral pitch of 12.

3. ***Determine Appropriate Number of Teeth***

   By substituting equation (10.22) into equation (10.7), the following calculations can be made:

   $$C_{external\ gears} = \frac{(N_1 + N_2)}{2P_d^n \cos\varphi} = \frac{(N_1 + N_2)}{2(12\cos\varphi)} = 3.4\,\text{in.}$$

   Also,

   $$\frac{N_2}{N_1} = 3.25$$

   Therefore,

   $$\frac{(N_1 + 3.25\,N_1)}{24\cos\varphi} = 3.4$$

   which reduces to

   $$\cos\varphi = \frac{N_1}{19.2}$$

   This equation reveals that N_1 must be less than 19.2 for this application. By trial, the following combinations are considered in Table 10.9.

   The first solution to generate integer numbers for both teeth will be used. A 16-tooth pinion and a 52-tooth gear having a helical angle of 33.55° are selected. Notice that from the interference criteria in Table 10.8, a normal pressure angle of either 20° or 25° can be used.

TABLE 10.9	Iterations for Example Problem 10.15			
Pinion Teeth N_1	Gear Teeth N_2	Normal Diametral Pitch P_d^n	Helix Angle φ	Diametral Pitch P_d
19	61.75	12	8.27	11.88
18	58.50	12	20.36	11.25
17	55.25	12	27.70	9.62
16	52	12	33.55	9.00

10.11 BEVEL GEAR KINEMATICS

Bevel gears were introduced in Section 10.2 and illustrated in Figure 10.3f. Bevel gears are used for transmitting motion between two shafts that intersect. One of the most important properties of a bevel gear arrangement is the *shaft angle*, Σ. The shaft angle is defined as the angle between the centerlines of the supporting shafts. Common bevel gear applications consist of shafts that intersect at right angles or have a shaft angle of 90°.

As discussed in Section 10.1 and illustrated in Figure 10.2, spur gears exhibit the same kinematics as two friction rollers. In a similar fashion, bevel gears can be replaced by two friction cones. With this conical geometry, the depth of the gear teeth tapers from the outside toward the middle. Most geometric tooth features used with spur gears, such as the pitch diameter and addendum, apply to bevel gears. This can be seen from the axial section of the two mating bevel gears shown in Figure 10.18. Because the tooth tapers, tooth features are measured at the outside edge of the tooth.

The angular velocity ratio, as presented for spur gears in equation (10.19), is also applicable to bevel gears. The diametral pitch and pressure angle also have the same definition as spur gears and must be identical for bevel gears to mate. The diametral pitch for bevel gears typically follows the standard values as presented in Table 10.1. Most bevel gears are made with a pressure angle of 20°; however, the tooth form is usually not an involute due to the difficulty in manufacturing. Alternate profiles have been developed, have been trademarked by vendors, and serve as competitive features.

In addition to diametral pitch and pressure angle, bevel gears are classified by their *pitch angle*, γ. The pitch angle is the generating angle of the cone upon which the gear is constructed. The pitch angles are labeled for the two mating gears shown in Figure 10.18. The pitch angle of each gear is a function of the velocity ratio and can be given as

$$\tan \gamma_{\text{pinion}} = \frac{\sin \Sigma}{\left\{ \cos \Sigma + \left(\dfrac{N_{\text{gear}}}{N_{\text{pinion}}} \right) \right\}} \quad (10.29)$$

$$\tan \gamma_{\text{gear}} = \frac{\sin \Sigma}{\left\{ \cos \Sigma + \left(\dfrac{N_{\text{pinion}}}{N_{\text{gear}}} \right) \right\}} \quad (10.30)$$

Because the pitch cone is a function of the velocity ratio, a single bevel gear cannot be replaced to alter the ratio, as was the case for spur gears. Thus, bevel gears are sold as a set.

In Figure 10.18, it is apparent that the sum of the pitch angles for the two mating gears must equal the shaft angle. Thus

$$\Sigma = \gamma_{\text{pinion}} + \gamma_{\text{gear}} \quad (10.31)$$

A miter gear, as shown in Figure 10.3g, is a special case of a bevel gear with a shaft angle of 90° and a velocity ratio of 1. Using equations (10.29) and (10.30), the pitch angle for both miter gears equals 45°.

The mounting of bevel gears is critical. For ideal mating, the apex of the cones for both gears must be at the same location. Any deviation could cause excessive backlash or interference. Due to the inherent geometry of bevel gears, at least one gear must be attached to the end of a cantilevered shaft. This configuration lends itself to excessive deflections, which can also result in problems with backlash.

Axial thrust loads developed by mating bevel gears always tend to separate the gears. This can contribute to shaft deflection and must also be considered. Of course, the shaft support bearings must also be configured to withstand this thrust force.

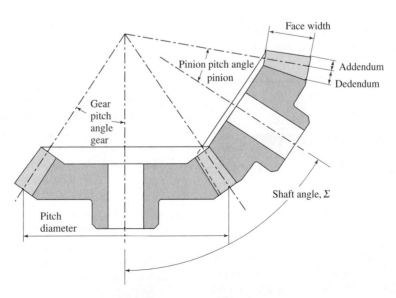

FIGURE 10.18 Mating bevel gears.

EXAMPLE PROBLEM 10.16

A pair of bevel gears have 18 and 27 teeth and are used on shafts that intersect each other at an angle of 70°. Determine the velocity ratio and the pitch angles of both gears.

SOLUTION: 1.   *Calculate Velocity Ratio*

The velocity ratio can be computed from equation (10.17).

$$VR = \frac{N_{\text{gear}}}{N_{\text{pinion}}} = \frac{27\,\text{teeth}}{18\,\text{teeth}} = 1.5$$

2.   *Calculate Pitch Angles*

The pitch angles can be computed from equations (10.29) and (10.30).

$$\tan\gamma_{\text{pinion}} = \frac{\sin\Sigma}{\left[\cos\Sigma + \left(\dfrac{N_{\text{gear}}}{N_{\text{pinion}}}\right)\right]}$$

$$\gamma_{\text{pinion}} = \tan^{-1}\left[\frac{\sin(70°)}{(\cos 70°) + (27/18)}\right] = 27.02°$$

$$\tan\gamma_{\text{gear}} = \frac{\sin\Sigma}{\left[\cos\Sigma + \left(\dfrac{N_{\text{pinion}}}{N_{\text{gear}}}\right)\right]}$$

$$\gamma_{\text{gear}} = \tan^{-1}\left[\frac{\sin(70°)}{(\cos 70°) + (18/27)}\right] = 42.98°$$

10.12 WORM GEAR KINEMATICS

A worm and worm gear is described in Section 10.2 and illustrated in Figure 10.3h. A worm and worm gear is used to transfer motion between nonparallel and nonintersecting shafts. With a worm gearset, large velocity ratios can be obtained in a rather limited space. The small gear is termed the *worm*, and the larger is termed the *worm gear*, *worm wheel*, or simply the *gear*.

The worm resembles a screw, and often the teeth on the worm are referred to as threads (Figure 10.3h). Worms are commonly available with single, double, and quadruple threads. Thus, the *number of worm teeth (threads)*, N_w, is an important property. The concept of multiple threads superimposed on a single worm is illustrated in Figure 10.19.

FIGURE 10.19 Multiple thread concept.

The tooth form of the worm gear is typically an involute. It is also common to cut the teeth concave across the face so they better conform to the cylindrical worm. This technique is termed *enveloping worm gear teeth*. It is an attempt to provide a larger contact patch on which the forces are transferred. The worm may also be cut with a concave length so it better conforms to the round worm gear. When both options are incorporated, the worm gearset is known as *double-enveloping* and provides a larger contact patch and greater power transmission. For such configurations, the worm and worm gear are not interchangeable and thus are sold as a set.

The worm gear is actually an extreme case of a helical gear with a large helix angle, which wraps the tooth around the gear. Therefore, the worm is described by all the geometric properties of a helical gear given in Section 10.6. The values of diametral pitch typically conform to the standards in Table 10.1. The pressure angles also conform to the 14½°, 20°, and 25° standards used with helical gears. In practice, the pressure angle is also selected based on the lead angle of the worm, as will be discussed later.

The worm is described by the number of threads, the *worm pitch diameter*, d_w, the *pitch*, p_w, and the *lead angle*, λ. The worm pitch diameter is determined similar to that in spur gears, as the diameter of the circle that remains tangent to the pitch diameter of the worm gear. The worm pitch is also similar to the definition for spur gears and is the distance between corresponding points on adjacent teeth (threads). These worm geometric properties are illustrated in Figure 10.20.

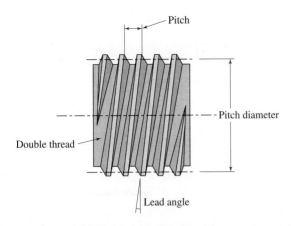

Pitch

Pitch diameter

Double thread

Lead angle

FIGURE 10.20 Worm geometry.

Also shown in Figure 10.20 is the lead angle, which is the angle of inclination of the teeth (threads). It can be computed from a trigonometric relationship to the other worm features.

$$\tan\lambda = \frac{N_w p_w}{\pi d_w} \tag{10.32}$$

For a mating worm gearset, the pitch of the worm must be the same as the pitch of the worm gear. Thus, from equation (10.1),

$$p_w = p_{\text{gear}} = \frac{\pi}{P_d} \tag{10.33}$$

For shafts that are at 90°, which is the usual case, the lead angle of the worm must equal the helix angle of the worm gear.

The velocity ratio of a worm gearset is computed as the number of teeth on the worm gear divided by the number of threads of the worm.

$$VR = \frac{N_{\text{gear}}}{N_w} \tag{10.34}$$

This is also identical to the spur gear application.

In most gearsets, the worm is the driver, thereby making the set a speed reducer. Most sets are irreversible in that the worm cannot turn the gear because a substantial friction force develops between the teeth. Irreversible drives are also referred to as *self-locking*. Worms must have a lead angle greater than approximately 10° to be able to drive the mating worm gear. This would result in a *reversible gearset*, but it is highly uncommon.

Although irreversibility may sound like a pitfall, distinct advantages exist. For example, lifting equipment typically requires that the load be held in an upward position, even as the power source is removed, such as a motor being turned off. Because the worm cannot rotate the worm gear, the load is locked in an upright position. This braking action is used in several mechanical devices, such as hoists, jacks, and lifting platforms. For these cases, the strength of the teeth and the predictability of friction must be analyzed to ensure safety.

EXAMPLE PROBLEM 10.17

A worm gearset is needed to reduce the speed of an electric motor from 1800 rpm to 50 rpm. Strength considerations require that 12-pitch gears be used, and it is desired that the set be self-locking. Select a set that accomplishes this task.

SOLUTION: 1.   ***Identify Appropriate Number of Teeth***

The velocity ratio can be computed from equation (10.17).

$$VR = \frac{\omega_{\text{worm}}}{\omega_{\text{gear}}} = \frac{1800\,\text{rpm}}{50\,\text{rpm}} = 36$$

When a single-thread worm is selected, the worm gear must have

$$N_{\text{gear}} = \frac{VR}{N_w} = \frac{(36)}{(1)} = 36 \text{ teeth}$$

From equation (10.33), and using a diametral pitch of 12, the pitch of the worm is determined by

$$p_w = \frac{\pi}{P_d} = \frac{\pi}{12} = 0.2618 \text{ in.}$$

2. ***Calculate the Size of the Gearset***

Because self-locking is desired, a conservative lead angle of 5° is used. Equation (10.32) is used to determine the following:

$$\tan\lambda = \frac{N_w p_w}{\pi d_w}$$

$$\tan 5° = \frac{(1)(0.2618)}{\pi d_w}$$

Solving,

$$d_w = 1.0499 \text{ in.}$$

The pitch diameter of the worm gear is

$$d_{\text{gear}} = \frac{N_{\text{gear}}}{P_d} = \frac{36 \text{ teeth}}{12} = 3.0 \text{ in.}$$

Finally, the center distance is

$$C = \frac{(d_{\text{worm}} + d_{\text{gear}})}{2} = \frac{(1.0499 + 3.0)}{2} = 2.0250 \text{ in.}$$

10.13 GEAR TRAINS

A gear train is a series of mating gearsets. Gear trains are commonly used to achieve large speed reductions. Many mechanical power sources, such as engines, turbines, and electric motors, operate efficiently at high speeds (1800–10,000 rpm). Many uses for this power, such as garage door openers, automotive drive wheels, and ceiling fans, require low speeds (10–100 rpm) for operation. Therefore, a desire to achieve large-velocity reductions is common, and the use of gear trains is very common.

For example, it may be desired to reduce the speed of a shaft from 1800 rpm to 10 rpm. Thus, a velocity reduction of 180:1 is required. If this reduction were attempted with one gearset, equation (10.19) would reveal that the driven gear would be 180 times larger than the drive gear. Obviously, the driven gear would be tremendously large, heavy, and expensive.

A second, more logical option is to reduce the speed in steps, through a series of gear pairs. This strategy cascades the rotational velocities toward the desired output velocity. This is exactly the logic behind gear trains.

When multiple gear pairs are used in a series, the overall velocity ratio is termed a *train value—TV*. The train value is defined as the input velocity to the gear train divided by the output velocity from the train. This is consistent with the definition of a velocity ratio. A train value is the product of the velocity ratio of the individual mating gear pairs that comprise the train. In equation form, it is stated as

$$TV = \frac{\omega_{\text{in}}}{\omega_{\text{out}}} = (VR_1)(VR_2)(VR_3) \dots \quad \textbf{(10.35)}$$

The algebraic sign resulting from the multiplication of individual velocity ratios determines the relative rotational direction of input and output shafts. Positive values reveal that the input and output shafts rotate in the same direction, and negative values indicate opposite rotation.

EXAMPLE PROBLEM 10.18

A gear train is shown in Figure 10.21. The gears have the following properties:

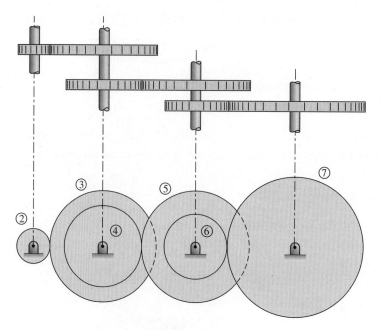

FIGURE 10.21 Gear train for Example Problem 10.18.

Gear 2: $N_2 = 12$ teeth and $P_d = 12$

Gear 3: $d_3 = 2.5$ in.

Gear 4: $N_4 = 15$ teeth

Gear 5: $d_5 = 3.0$ in. and $P_d = 10$

Gear 6: $d_6 = 1.5$ in. and $P_d = 8$

Gear 7: $N_7 = 32$ teeth

Determine the rotational velocity of gear 7 as gear 2 drives at 1800 rpm counterclockwise. Also determine the distance between the shafts that carry gears 2 and 7.

SOLUTION: 1.   *Calculate Consistent Gear Dimensions*

In order to calculate the train value, consistent properties of the gears must be determined. For this problem, gear pitch diameters are used and must be computed.

$$d_2 = \frac{N_2}{P_d} = \frac{12}{12} = 1 \text{ in.}$$

Gear 4 mates with gear 5 and must have an identical diametral pitch.

$$d_4 = \frac{N_4}{P_d} = \frac{15}{10} = 1.5 \text{ in.}$$

Likewise, gear 7 mates with gear 6 and must have an identical diametral pitch.

$$d_7 = \frac{N_7}{P_d} = \frac{32}{8} = 4 \text{ in.}$$

2.   *Calculate Velocities and Ratios*

The train value can then be computed as

$$TV = (VR_{2-3})(VR_{4-5})(VR_{6-7}) = \left(-\frac{d_3}{d_2}\right)\left(-\frac{d_5}{d_4}\right)\left(-\frac{d_7}{d_6}\right)$$

$$= \left(-\frac{2.5 \text{ in.}}{1 \text{ in.}}\right)\left(-\frac{3 \text{ in.}}{1.5 \text{ in.}}\right)\left(-\frac{4 \text{ in.}}{1.5 \text{ in.}}\right) = -13.33$$

The speed of gear 7 can be determined through this train value.

$$\frac{\omega_2}{\omega_7} = TV$$

$$\omega_7 = \frac{\omega_2}{TV} = \frac{1800 \text{ rpm}}{(-13.33)} = -135 \text{ rpm} = 135 \text{ rpm, cw}$$

The center distance between gears 2 and 7 can be determined by stacking the pitch radii from all gears between 2 and 7. This can be seen in Figure 10.21.

$$C = r_2 + r_3 + r_4 + r_5 + r_6 + r_7$$

$$= \left(\frac{1 \text{ in.}}{2}\right) + \left(\frac{2.5 \text{ in.}}{2}\right) + \left(\frac{1.5 \text{ in.}}{2}\right) + \left(\frac{3 \text{ in.}}{2}\right) + \left(\frac{1.5 \text{ in.}}{2}\right) + \left(\frac{4 \text{ in.}}{2}\right) = 6.75 \text{ in.}$$

EXAMPLE PROBLEM 10.19

Design a gear train that yields a train value of +300:1. From interference criteria, no gear should have fewer than 15 teeth and, due to size restrictions, no gear can have more than 75 teeth.

SOLUTION: 1.   *Break Train Value into Individual Velocity Ratios*

With the restrictions placed on gear size used in this train, the maximum individual velocity ratio is determined by

$$VR_{max} = \frac{N_2}{N_1} = \frac{75}{15} = 5$$

As with all design problems, more than one possible solution exists. Because a train value is the product of individual velocity ratios, one solution can be obtained by factoring the train value into values no greater than the maximum individual velocity ratios. For this problem, no factor can be greater than 5.

$$TV = 300 = (-5)(-60) = (-5)(-5)(12) = (-5)(-5)(-4)(-3)$$

Therefore, a gear train with gear pairs that have individual velocity ratios of –5, –5, –4, and –3 nets a train value of 300. A negative value is used for the individual velocity ratios because it is desirable to use the more common external gears.

2. **Identify the Number of Teeth for Each Gear**

$$VR_{1-2} = -5, \text{use external gears with } N_1 = 15 \text{ and } N_2 = 75$$

$$VR_{3-4} = -5, \text{use external gears with } N_3 = 15 \text{ and } N_4 = 75$$

$$VR_{5-6} = -4, \text{use external gears with } N_5 = 15 \text{ and } N_6 = 60$$

$$VR_{7-8} = -3, \text{use external gears with } N_7 = 15 \text{ and } N_8 = 45$$

In general, when using external gears that produce opposite rotations, an even number of gear pairs must be used to produce a positive train value. Because the solution for this example has four gear pairs, the output rotation occurs in the same direction as the input.

10.14 IDLER GEARS

Consider the gear train shown in Figure 10.22. Notice that the middle gear mates with the small gear to form the first ratio. The middle gear also mates with the large gear to form a second ratio. As always, the train value can be computed as the product of the velocity ratios.

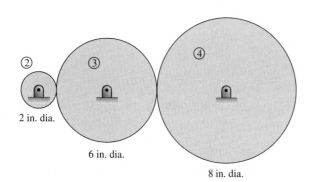

2 in. dia.

6 in. dia.

8 in. dia.

FIGURE 10.22 Gear train with an idler gear.

$$TV = (VR_{2-3})(VR_{3-4}) = \left(-\frac{d_3}{d_2}\right)\left(-\frac{d_4}{d_3}\right)$$

$$TV = \left(-\frac{6}{2}\right)\left(-\frac{8}{6}\right) = +\frac{8}{2} = +4$$

Notice that d_3 appears in both the numerator and the denominator. In this situation, the influence of the middle gear is negated. This gear arrangement creates a train value of

$$TV = \left(-\frac{d_3}{d_2}\right)\left(-\frac{d_4}{d_3}\right) = +\frac{d_4}{d_2}$$

Therefore, the train value is dependent only on the size of the first and last gears. The diameter, or the number of teeth, of the center gear does not influence the train value. The center gear is termed an *idler gear*. Its function is to

alter the direction of the output motion, yet not affect the magnitude of that motion. To illustrate this function, consider an arrangement where gear 2 mates directly with gear 4. The resulting train value would be

$$TV = (-VR_{2-4}) = -\frac{d_4}{d_2}$$

Thus, the idler gear serves to reverse the direction of the output. As mentioned, the size of the idler gear does not influence the kinematics of the train. In practice, this idler gear can be sized to conveniently locate the centers of the input and output gears. Of course, because all three gears mesh, they must have identical diametral pitches and pressure angles.

10.15 PLANETARY GEAR TRAINS

The gear trains presented in preceding sections all had gear centers attached to fixed bodies. With planetary gear trains, this restriction is removed. In these trains, a link that holds the center of the gears is allowed to move. A planetary gear train, which is also called an *epicyclic train*, is shown in Figure 10.23.

Planetary trains can be used to achieve large speed reductions in a more compact space than a conventional gear train. However, a greater benefit is the ability to readily alter the train value. Because all links are capable of moving, one can alter the train value by holding different gears or carriers. In practice, switching the fixed link is accomplished with brake or clutch mechanisms, thus releasing one link and fixing another. For this reason, planetary gear trains are very common in automotive transmissions.

Because the motion can resemble the planets rotating about the sun in our solar system, the term *planetary gear train* was applied to this system. Expanding on this comparison, the center gear is called the *sun*. Gears that revolve around the sun are called *planets*. A carrier holds the planet gears in orbit around the sun. Finally, the train is commonly encased in the internal gear termed the *ring gear*. These gears are labeled in Figure 10.23.

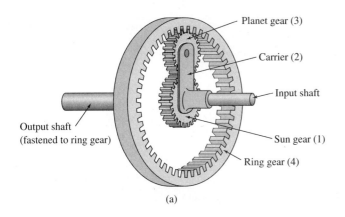

(a)

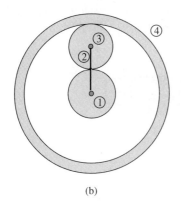

(b)

FIGURE 10.23 A planetary gear train.

10.15.1 Planetary Gear Analysis by Superposition

The motion of a planetary gear train is not always as intuitive as fixed-center trains. As gears and carriers rotate, the motion can appear rather complex. To analyze the motion of a planetary gear train, the method of superposition can be used to "step through" the gear movements.

The method of superposition consists of the following:

Step One

The first step is to relax the constraint on the fixed link and temporarily assume that the carrier is locked. Turn the previously fixed gear one revolution and calculate the effect on the entire train.

Step Two

The second step is to free all constraints and record the movement of rotating each link one revolution in the opposite direction of the rotation in step one. As this motion is combined with the motion in the first step, the superimposed motion of the fixed gear equals zero.

Step Three

The motion of all links is determined by combining the rotations from the first two steps. Finally, velocities are proportional to the rotational movements.

Stated in general terms, this method seems complex. However, it is rather straightforward. The method is best illustrated with an example problem.

EXAMPLE PROBLEM 10.20

A planetary gear train is illustrated in Figure 10.24. The carrier (link 2) serves as the input to the train. The sun (gear 1) is the fixed gear and has 30 teeth. The planet gear (gear 3) has 35 teeth. The ring gear serves as the output from the train and has 100 teeth. Determine the rotational velocity of all members of this gear train when the input shaft rotates at 1200 rpm clockwise.

SOLUTION: 1. *Complete Step 1*

The first step is to temporarily fix the carrier then compute the motions of all gears as the previously fixed gear rotates one revolution. Thus, the following can be determined:

Gear 1 rotates one revolution.

$$\Delta\theta_1 = +1 \text{ rev}$$

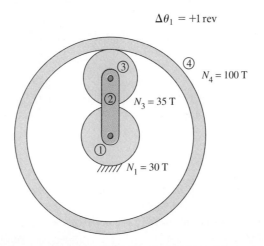

FIGURE 10.24 Planetary train for Example Problem 10.20.

Gear 3 rotates (VR_{1-3}) as much as gear 1.

$$\Delta\theta_3 = (VR_{1-3})(\Delta\theta_1) = \left(-\frac{30}{35}\right)(+1\,\mathrm{rev}) = -0.857\ \mathrm{rev}$$

Gear 4 rotates (VR_{3-4}) as much as gear 3.

$$\Delta\theta_4 = (VR_{3-4})(\Delta\theta_3) = (VR_{3-4})(V_{1-3})(\Delta\theta_1) = \left(\frac{+35}{100}\right)\left(-\frac{30}{35}\right)(+1\,\mathrm{rev}) = -0.3\ \mathrm{rev}$$

2. **Complete Step 2**

   The second step rotates all links –1 revolution. This returns the sun gear to its original position, yielding a net movement of zero.

3. **Complete Step 3**

   The method of superposition involves combining these two motions, resulting in the actual planetary gear train motion. Thus the rotations from both steps are algebraically added together. The two steps are summarized in Table 10.10.

TABLE 10.10	Tabulating Planetary Gear Analysis for Example Problem 10.20			
Link	**Sun**	**Planet**	**Ring**	**Carrier**
Step 1: Rotate with fixed carrier	+1	−0.857	−0.3	0
Step 2: Rotate all links	−1	−1	−1	−1
Step 3: Total rotations	0	−1.857	−1.3	−1

4. **Determine Velocities of All Links**

   The velocities can be determined by using the ratios of angular displacements.

   $$\omega_{\mathrm{sun}} = \left(\frac{\Delta\theta_{\mathrm{sun}}}{\Delta\theta_{\mathrm{carrier}}}\right)\omega_{\mathrm{carrier}} = \left(\frac{0}{-1}\right)(1200\,\mathrm{rpm}) = 0\,\mathrm{rpm}$$

   $$\omega_{\mathrm{planet}} = \left(\frac{-1.857}{-1}\right)\omega_{\mathrm{carrier}} = (+1.857)(1200\,\mathrm{rpm}) = +2228\,\mathrm{rpm} = 2228\,\mathrm{rpm, cw}$$

   $$\omega_{\mathrm{ring}} = \left(\frac{-1.3}{-1}\right)\omega_{\mathrm{carrier}} = (+1.3)(1200\,\mathrm{rpm}) = +1560\,\mathrm{rpm} = 1560\,\mathrm{rpm, cw}$$

EXAMPLE PROBLEM 10.21

A planetary gear train is illustrated in Figure 10.25. The carrier (link 2) serves as the input to the train. The ring gear (gear 1) is the fixed gear and has 120 teeth. The planet gear (gear 4) has 40 teeth. The sun gear (gear 3) serves as the output from the train and has 30 teeth. Determine the rotational velocity of all members of this gear train when the input shaft rotates at 1200 rpm clockwise.

SOLUTION: 1. **Complete Steps 1–3**

   The first step is to temporarily fix the carrier. Then compute the motions of all gears as the previously fixed gear rotates one revolution.

   The second step rotates all links –1 revolution. This returns the ring gear to its original position, yielding a net movement of zero.

   The two steps are summarized in Table 10.11.

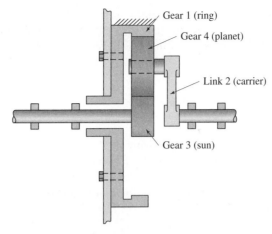

FIGURE 10.25 Planetary train for Example Problem 10.21.

TABLE 10.11	Solution Steps for Example Problem 10.21			
Link	**Sun (Gear 3)**	**Planet (Gear 4)**	**Ring (Gear 1)**	**Carrier (Gear 2)**
Step 1: Rotate with fixed carrier	$\left(\dfrac{120}{40}\right)\left(\dfrac{-40}{30}\right) = -4.0$	$\left(\dfrac{120}{40}\right) = 3.0$	+1	0
Step 2: Rotate all links	−1	−1	−1	−1
Step 3: Total rotations	−5.0	+2.0	0	−1

2. **Compute Velocities of All Links**

   The velocities can be determined by using the ratios of angular displacements.

   $$\omega_{\text{ring}} = \left(\frac{\Delta\theta_{\text{ring}}}{\Delta\theta_{\text{carrier}}}\right)\omega_{\text{carrier}} = \left(\frac{0}{-1}\right)(0\,\text{rpm}) = 0\ \text{rpm}$$

   $$\omega_{\text{carrier}} = 1200\,\text{rpm, clockwise}$$

   $$\omega_{\text{planet}} = \left(\frac{+2.0}{-1}\right)\omega_{\text{carrier}} = (-2.0)(1200\,\text{rpm}) = -2400\,\text{rpm} = 2400\,\text{rpm, counterclockwise}$$

   $$\omega_{\text{sun}} = \left(\frac{-5.0}{-1}\right)\omega_{\text{carrier}} = (+5.0)(1200\,\text{rpm}) = +6000\,\text{rpm} = 6000\ \text{rpm, cw}$$

10.15.2 Planetary Gear Analysis by Equation

Along with the tabular method, the motion of a planetary gear train can also be analyzed through an equation that is derived from the relative angular velocities. To develop the formula method, the motion of the mating gears is examined relative to the carrier. Thus, kinematic inversion is used to view the train is as if the carrier were fixed. One gear on the end of the train is designated the first gear. The gear on the opposite end of the train is designated the last gear.

The train is comprised of meshing gear pairs consisting of driver and driven gears. The first gear is designated as a driver gear and the last gear is a driven gear. The intermediate gears are appropriately identified depending on whether they drive or are driven. In computing the ratio for each pair, the ratio is negative for mating external gears and positive for gears having an internal mesh.

Shifting focus to absolute velocities, the first gear has an angular velocity designated ω_F and the last gear has an angular velocity designated ω_L. The carrier has an angular velocity ω_{carrier}. The relationship between the angular velocities and number of teeth in the train is given as follows.

$$\frac{\omega_{L/\text{carrier}}}{\omega_{F/\text{carrier}}} = \tag{10.36}$$

$$\pm \frac{\text{product of number of teeth on driver gears}}{\text{product of number of teeth on driven gears}}$$

$$= \frac{\omega_L - \omega_{\text{carrier}}}{\omega_F - \omega_{\text{carrier}}}$$

Equation (10.36) can be solved for any of the angular velocity terms, knowing the other two. Often, either the first gear, last gear, or the carrier is fixed and a zero is substituted for that term. While less complicated than the tabular method, this formula method is limited to cases where a path of meshes links the first and last gears. The method is illustrated in the following example problems.

EXAMPLE PROBLEM 10.22

A planetary gear train was illustrated in Figure 10.24. The carrier (link 2) serves as the input to the train. The sun (gear 1) is the fixed gear and has 30 teeth. The planet gear (gear 3) has 35 teeth. The ring gear serves as the output from the train and has 100 teeth. In Example Problem 10.20 the rotational velocity of the ring gear was determined to be 1560 rpm clockwise, as the input shaft rotates at 1200 rpm clockwise. Use the formula method to verify this result.

SOLUTION:

1. **Specify the first and last gear.**

   The sun (gear 1) will be designated the first gear. Being on the other end of the train, the ring gear (gear 4) will be designated as the last gear.

2. **Substitute Gear Ratios into the Planetary Train Formula.**

   Gear 1 (first) mates with gear 3, which in turn mates with gear 4 (last). Substituting into Equation (10.36) gives:

   $$\left(-\frac{N_1}{N_3}\right)\left(+\frac{N_3}{N_4}\right) = \frac{\omega_L - \omega_{carrier}}{\omega_F - \omega_{carrier}}$$

3. **Identify the Angular Velocity Terms.**

   The sun is fixed, giving $\omega_F = 0$. The carrier rotates at 1200 rpm clockwise. Taking clockwise to be a negative direction, $\omega_{carrier} = -1200$. The ring gear must be determined, thus $\omega_L = ?$

4. **Substitute Values into the Planetary Train Formula and Solve.**

   Substituting values into Equation (10.36) gives:

   $$\left(-\frac{30}{35}\right)\left(+\frac{35}{100}\right) = \frac{\omega_L - (-1200)}{0 - (-1200)}$$

   Solving,

   $$\omega_L = 1200\left(-\frac{30}{35}\right)\left(+\frac{35}{100}\right) - 1200 = -1560 = 1560 \text{ rpm, cw}$$

EXAMPLE PROBLEM 10.23

A planetary gear train was illustrated in Figure 10.25. The carrier (link 2) serves as the input to the train. The ring (gear 1) is the fixed gear and has 120 teeth. The planet gear (gear 4) has 40 teeth. The sun gear (gear 3) serves as the output from the train and has 30 teeth. In Example Problem 10.21 the rotational velocity of the sun gear was determined to be 6000 rpm clockwise, as the input shaft rotates at 1200 rpm clockwise. Use the formula method to verify this result.

SOLUTION:

1. **Specify the first and last gear.**

   The sun (gear 3) will be designated the first gear. Being on the other end of the train, the ring gear (gear 1) will be designated as the last gear.

2. **Substitute Gear Ratios into the Planetary Train Formula.**

   Gear 2 (first) mates with gear 4, which in turn mates with gear 1 (last). Substituting into Equation (10.36) gives:

   $$\left(-\frac{N_3}{N_4}\right)\left(+\frac{N_4}{N_1}\right) = \frac{\omega_L - \omega_{carrier}}{\omega_F - \omega_{carrier}}$$

3. **_Identify the Angular Velocity Terms._**

The ring is fixed, giving $\omega_L = 0$. The carrier rotates at 1200 rpm clockwise. Taking clockwise to be a negative direction, $\omega_{carrier} = -1200$. The sun gear must be determined, thus $\omega_F = ?$

4. **_Substitute Values into the Planetary Train Formula and Solve._**

Substituting values into Equation (10.36) gives:

$$\left(-\frac{30}{40}\right)\left(+\frac{40}{120}\right) = \frac{0 - (-1200)}{\omega_F - (-1200)}$$

Solving,

$$\omega_F = 1200\left(-\frac{40}{30}\right)\left(+\frac{120}{40}\right) - 1200 = -6000 = 6000 \text{ rpm, cw}$$

PROBLEMS

Spur Gear Geometry

For Problems 10–1 through 10–4, determine the following:

a. The pitch circle diameter
b. The diameter of the base circle
c. The diameter of the addendum circle
d. The diameter of the dedendum circle
e. The circular pitch

10–1. A 20°, full-depth, involute spur gear with 18 teeth has a diametral pitch of 12.

10–2. A 20°, full-depth, involute spur gear with 48 teeth has a diametral pitch of 8.

10–3. A 14½° full-depth, involute spur gear with 40 teeth has a diametral pitch of 16.

10–4. A 25° spur gear with 21 teeth has a metric module of 4. Determine the pitch circle diameter.

For Problems 10–5 through 10–8, determine the following:

a. The center distance
b. The contact ratio
c. Whether interference will occur
d. A center distance that reduces backlash from a vendor value of $0.4/P_d$ to an AGMA-recommended value of $0.1/P_d$

10–5. Two 12-pitch, 20°, full-depth, involute spur gears are used on an industrial circular saw for cutting timber. The pinion has 18 teeth and the gear has 42.

10–6. Two 4-pitch, 20°, full-depth, involute spur gears are used on a tumbling machine for deburring steel stamped parts. The pinion has 12 teeth and the gear has 28.

10–7. Two plastic, 48-pitch, 25°, full-depth, involute spur gears are used on an electric shaver. The pinion has 18 teeth and the gear has 42.

10–8. Two 16-pitch, 14½° full-depth, involute spur gears are used on a machine shop lathe. The pinion has 16 teeth and the gear has 72.

For Problems 10–9 through 10–14, determine the following:

a. Their pitch diameters
b. The center distance

10–9. Two mating 12-pitch gears have 18 external and 48 internal teeth, respectively.

10–10. Two mating 20-pitch gears have 15 external and 60 internal teeth, respectively.

10–11. Two mating gears have 18 and 48 teeth, respectively, and a center distance of 4.125.

10–12. Two mating gears have 20 and 45 teeth, respectively, and a center distance of 3.25.

10–13. An 8-pitch, 18-tooth pinion mates with an internal gear of 64 teeth.

10–14. A 12-pitch, 24-tooth pinion mates with an internal gear of 108 teeth.

Gear Kinematics

For Problems 10–15 through 10–18, determine the following:

a. The speed of the gear
b. The pitch line velocity

10–15. An 8-pitch, 18-tooth pinion rotates clockwise at 1150 rpm and mates with a 64-tooth gear.

10–16. A 20-pitch, 15-tooth pinion rotates clockwise at 1725 rpm and mates with a 60-tooth gear.

10–17. A 6-pitch, 21-tooth pinion rotates clockwise at 850 rpm and mates with a 42-tooth gear.

10–18. A 24-pitch, 24-tooth pinion rotates clockwise at 1725 rpm and mates with a 144-tooth gear.

Gear Selection with an Established Center Distance

10–19. Two 10-pitch gears are to be mounted 12 in. apart and have a velocity ratio of 5:1. Find the pitch diameters and the number of teeth on both gears.

10–20. Two 16-pitch gears are to be mounted 3.75 in. apart and have a velocity ratio of 4:1. Find the pitch diameters and the number of teeth on both gears.

Gear Selection with a Set Center Distance

10–21. Two 32-pitch gears are to be mounted 2.25 in. apart and have a velocity ratio of 8:1. Find the pitch diameters and the number of teeth on both gears.

10–22. Two gears are to be mounted 5 in. apart and have a velocity ratio of 4:1. Find appropriate pitch diameters, diametral pitches, and the number of teeth on both gears that will be suitable.

10–23. Two gears are to be mounted 3.5 in. apart and have a velocity ratio of 6:1. Find appropriate pitch diameters, diametral pitches, and the number of teeth on both gears that will be suitable.

10–24. Two gears are to be mounted 10 in. apart and have a velocity ratio of 3:1. Find appropriate pitch diameters, diametral pitches, and the number of teeth on both gears that will be suitable.

Catalog Gear Selection

10–25. A pair of 20°, mild-steel gears are to be selected for an application where they need to transfer 5 hp. The pinion drives at 1800 rpm and the gear must rotate as close to 480 rpm as possible. Determine an appropriate set of catalog gears, from Table 10.7, for this application.

10–26. A pair of 20°, mild-steel gears are to be selected for an application where they need to transfer 2.5 hp. The pinion drives at 1500 rpm and the gear must rotate as close to 500 rpm as possible. Determine an appropriate set of catalog gears, from Table 10.7, for this application.

10–27. A pair of 20°, mild-steel gears are to be selected for an application where they need to transfer 8 hp. The pinion drives at 1500 rpm and the gear must rotate as close to 200 rpm as possible. Determine an appropriate set of catalog gears, from Table 10.7, for this application.

10–28. A pair of 20°, mild-steel gears are to be selected for an application where they need to transfer 10 hp. The pinion drives at 800 rpm and the gear must rotate as close to 180 rpm as possible. Determine an appropriate set of catalog gears, from Table 10.7, for this application.

10–29. A pair of 20°, mild-steel gears are to be selected for an application where they need to transfer 1 hp. The pinion drives at 1725 rpm and the gear must rotate as close to 560 rpm as possible. Determine an appropriate set of catalog gears, from Table 10.7, for this application.

10–30. A pair of 20°, mild-steel gears are to be selected for an application where they need to transfer 10 hp. The pinion drives at 1175 rpm and the gear must rotate as close to 230 rpm as possible. Determine an appropriate set of catalog gears, from Table 10.7, for this application.

10–31. A pair of 20°, mild-steel gears are to be selected for an application where they need to transfer 10 hp. The pinion drives at 1175 rpm and the gear must rotate as close to 170 rpm as possible. Determine an appropriate set of catalog gears, from Table 10.7, for this application.

10–32. A pair of 20°, mild-steel gears are to be selected for an application where they need to transfer 3 hp. The pinion will be driven at 1750 rpm and the gear must rotate as close to 290 rpm as possible. Determine an appropriate set of catalog gears, from Table 10.7, for this application.

10–33. A pair of 20°, mild-steel gears are to be selected for an application where they need to transfer 20 hp. The pinion drives at 825 rpm and the gear must rotate as close to 205 rpm as possible. Determine an appropriate set of catalog gears, from Table 10.7, for this application.

Rack and Pinion

10–34. A rack and pinion will be used for a height adjustment on a camera stand. The 24-pitch pinion has 18 teeth. Determine the angle that the handle (and pinion) must rotate to raise the camera 5 in.

10–35. A rack and pinion will be used to lower a drill on a drill press. The 16-pitch pinion has 20 teeth. Determine the angle that the handle (and pinion) must rotate to raise the drill 3 in.

10–36. An 8-pitch, 18-tooth pinion is used to drive a rack. Determine the distance that the rack travels when the pinion rotates 3 revolutions.

10–37. A 12-pitch, 24-tooth pinion is used to drive a rack. Determine the distance that the rack travels when the pinion rotates 5 revolutions.

10–38. A rack and pinion will be used for a steering mechanism. The 12-pitch pinion has 18 teeth. Determine the required speed of the pinion if the rack must be driven at a rate of 50 in./min.

10–39. A rack and pinion will be used for a steering mechanism. The 10-pitch pinion has 20 teeth. Determine the required speed of the rack if the pinion rotates at a rate of 80 rpm.

Helical Gears

For Problems 10–40 and 10–41, determine the following:

    a. The pitch diameters
    b. The normal diametral pitch
    c. The normal circular pitch
    d. Whether interference is a problem

10–40. A pair of helical gears has a 20° pressure angle, a 45° helix angle, and an 8 diametral pitch. The pinion has 16 teeth and the gear has 32 teeth.

10–41. A pair of helical gears has a 14½° pressure angle, a 30° helix angle, and a 12 diametral pitch. The pinion has 16 teeth and the gear has 48 teeth.

10–42. In order to reduce the noise of a gear drive, two 8-pitch spur gears with 20 and 40 teeth are to be replaced with helical gears. The new set must have the same velocity ratio and center distance. Specify two helical gears, which will be formed on a hob, to accomplish the task.

10–43. In order to reduce the noise of a gear drive, two 12-pitch spur gears with 18 and 54 teeth are to be replaced with helical gears. The new set must have the same velocity ratio and center distance. Specify two helical gears, which will be formed on a hob, to accomplish the task.

Bevel Gears

10–44. A pair of bevel gears has 20 and 75 teeth and is used on shafts that intersect each other at an angle of 90°. Determine the velocity ratio and the pitch angles of both gears.

10–45. A pair of bevel gears has 20 and 75 teeth and is used on shafts that intersect each other at an angle of 60°. Determine the velocity ratio and the pitch angles of both gears.

10–46. A pair of bevel gears has 18 and 90 teeth and is used on shafts that intersect each other at an angle of 75°. Determine the velocity ratio and the pitch angles of both gears.

Worm Gears

10–47. A worm gearset is needed to reduce the speed of an electric motor from 3600 to 60 rpm. Strength considerations require that 16-pitch gears be used, and it is desired that the set be self-locking. Specify a set that accomplishes the task.

10–48. A worm gearset is needed to reduce the speed of an electric motor from 1800 to 18 rpm. Strength considerations require that 12-pitch gears be used, and it is desired that the set be self-locking. Specify a set that accomplishes the task.

10–49. A worm gearset is needed to reduce the speed of an electric motor from 3600 to 40 rpm. Strength considerations require that 20-pitch gears be used, and it is desired that the set be self-locking. Specify a set that accomplishes the task.

Gear Trains

10–50. A gear train is shown in Figure P10.50. The gears have the following properties: $N_2 = 18$ teeth; $N_3 = 72$ teeth and $P_d = 10$; $N_4 = 16$ teeth and $P_d = 8$; and $N_5 = 48$

FIGURE P10.50 Problems 50 and 51.

teeth. Determine the velocity of gear 5 as gear 2 drives at 1200 rpm clockwise. Also determine the center distance between gears 2 and 5.

10–51. A gear train is shown in Figure P10.50. The gears have the following properties: $N_2 = 20$ teeth and $P_d = 10$; $d_3 = 6$ in.; $d_4 = 2$ in., and $P_d = 8$; and $N_5 = 48$ teeth. Determine the velocity of gear 5 as gear 2 drives at 1800 rpm counterclockwise. Also determine the center distance between gears 2 and 5.

10–52. A gear train is shown in Figure P10.52. The gears have the following properties: $N_2 = 15$ teeth; $N_3 = 90$ teeth and $P_d = 16$; $N_4 = 15$ teeth; $N_5 = 75$ teeth; $N_6 = 75$ teeth and $P_d = 12$; $N_7 = 15$ teeth; and $N_8 = 60$ teeth and $P_d = 8$. Determine the velocity of gear 8 as gear 2 drives at 3600 rpm clockwise. Also determine the center distance between gears 2 and 8.

FIGURE P10.52 Problems 52 and 53.

10–53. A gear train is shown in Figure P10.52. The gears have the following properties: $N_2 = 16$ teeth and $P_d = 16$; $d_3 = 8$ in.; $d_4 = 1.5$ in.; $N_5 = 50$ teeth and $P_d = 10$; $d_6 = 5.5$ in.; $N_7 = 1.5$ in. and $P_d = 8$; and $N_8 = 56$ teeth. Determine the velocity of gear 8 as gear 2 drives at 1200 rpm counterclockwise. Also determine the center distance between gears 2 and 8.

10–54. A gear train is shown in Figure P10.54. The gears have the following properties: $N_1 = 20$ teeth and $P_d = 16$; $d_2 = 8$ in.; and $d_3 = 1.5$ in. and $P_d = 10$. Determine the distance that the rack moves for each revolution of gear. Also determine the center distance between gears 1 and 3.

10–55. A gear train is shown in Figure P10.54. The gears have the following properties: $N_1 = 18$ teeth and

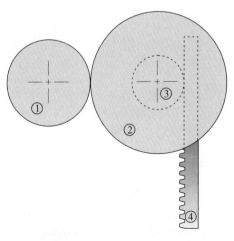

FIGURE P10.54 Problems 54–56.

$P_d = 20$; $d_2 = 5.5$ in.; and $d_3 = 2.5$ in. and $P_d = 8$. Determine the required speed of gear 1 for the rack to move at a rate of 50 in./min.

10–56. A gear train is shown in Figure P10.54. The gears have the following properties: $d_1 = 2.5$ in.; $N_2 = 75$ teeth and $P_d = 10$; and $N_3 = 24$ teeth. Determine the required diametral pitch of the rack for the rack to move 0.5 in. for each revolution of gear 1.

10–57. A gear train is shown in Figure P10.57. The gears have the following properties: $N_1 = 16$ teeth and $P_d = 16$; $d_2 = 8$ in.; $N_3 = 20$ teeth; and $N_4 = 50$ teeth. Determine the velocity of gear 4 as gear 1 drives at 1800 rpm.

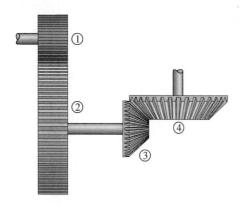

FIGURE P10.57 Problems 57 and 58.

10–58. For the gear train shown in Figure P10.57, the gears have the following properties: $N_1 = 17$ teeth and $P_d = 20$; $d_2 = 4$ in.; $N_3 = 18$ teeth; and $N_4 = 36$ teeth. Determine the required velocity of gear 1 for gear 4 to drive at 380 rpm.

10–59. A gear train is shown in Figure P10.59. The gears have the following properties: $N_{\text{worm}} = 1$ thread; $N_2 = 45$ teeth; $N_3 = 18$ teeth and $P_d = 16$; $d_4 = 6$ in.; and $N_5 = 80$ teeth. Determine the velocity of gear 5 as gear 1 drives at 1800 rpm. Also determine the center distance between gears 2 and 5.

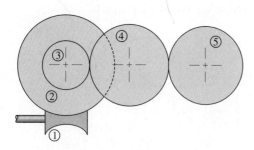

FIGURE P10.59 Problems 59 and 60.

10–60. For the gear train shown in Figure P10.59, the gears have the following properties: $N_{\text{worm}} = 2$ threads; $N_2 = 60$ teeth; $N_3 = 18$ teeth and $P_d = 12$; $d_4 = 6$ in.; and $N_5 = 54$ teeth. Determine the required velocity of gear 1 (the worm) to enable gear 5 to drive at 28 rpm. Also determine the center distance between gears 2 and 5.

Gear Train Design

10–61. Devise a gear train with a train value of 400:1. Specify the number of teeth in each gear. From interference criteria, no gear should have fewer than 17 teeth. Due to size restrictions, no gear should be larger than 75 teeth. Also sketch the concept of the train.

10–62. Devise a gear train with a train value of –200:1. Specify the number of teeth in each gear. From interference criteria, no gear should have fewer than 17 teeth. Due to size restrictions, no gear should be larger than 75 teeth. Also sketch the concept of the train.

10–63. Devise a gear train with a train value of –900:1. Specify the number of teeth in each gear. From interference criteria, no gear should have fewer than 17 teeth. Due to size restrictions, no gear should be larger than 75 teeth. Also sketch the concept of the train.

Gear-Driven Mechanisms

10–64. A casement window opening mechanism is shown in Figure P10.64. The gears have the following properties: $d_1 = 1$ in.; $N_2 = 30$ teeth and $P_d = 20$; $N_3 = 18$ teeth and $P_d = 18$; and $d_4 = 4$ in. Starting at the configuration shown, with $\beta = 20°$, graphically determine (using either manual drawing techniques or CAD) the angular rotation of the window when the crank rotates one revolution.

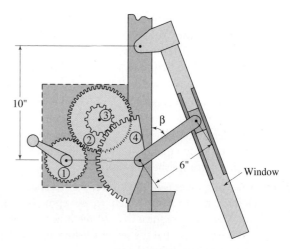

FIGURE P10.64 Problems 64–67.

10–65. For the casement window shown in Figure P10.64, analytically determine the angular rotation of the window when the crank rotates one revolution using the configuration shown ($\beta = 20°$).

10–66. For the casement window mechanism shown in Figure P10.64, the gears have the following properties: $d_1 = 0.75$ in.; $N_2 = 48$ teeth and $P_d = 32$; $N_3 = 16$ teeth and $P_d = 32$; $d_4 = 4$ in. Starting at the configuration shown ($\beta = 20°$), graphically determine (using either manual drawing techniques or CAD) the rotational speed with which the window opens when the crank rotates at a constant rate of 20 rpm.

10–67. For Problem 10–66, analytically determine the rotational speed with which the window opens from the configuration shown ($\beta = 20°$) when the crank rotates at a constant rate of 20 rpm.

Planetary Gear Trains

10–68. A planetary gear train is shown in Figure P10.68. The carrier (link 2) serves as the input to the train. The sun (gear 1) is fixed and has 16 teeth with a diametral pitch of 16. The planet gear (gear 3) has a 2-in. pitch diameter. The ring serves as the output from the train and has a 5-in. pitch diameter. Determine the rotational velocity of all members of this gear train when the input shaft rotates at 1800 rpm clockwise.

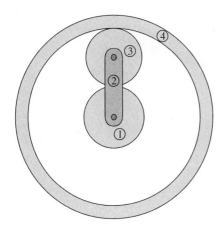

FIGURE P10.68 Problems 68 and 69.

10–69. For the planetary gear train shown in Figure P10.68, the carrier (link 2) serves as the input to the train. The sun (gear 1) serves as the output gear and has 18 teeth with a diametral pitch of 12. The planet gear (gear 3) has a 2.5-in. pitch diameter. The ring gear is fixed and has a 6.5-in. pitch diameter. Determine the rotational velocity of all members of this gear train when the input shaft rotates at 800 rpm counterclockwise.

10–70. A planetary gear train is shown in Figure P10.70. The carrier (link 2) serves as the input to the train. The sun (gear 1) is fixed and has a 1.25-in. pitch diameter with a diametral pitch of 16. Gear 3 has 42 teeth and gear 4 has 21 teeth. Gear 5 has 32 teeth and is keyed to the same shaft as gear 4. Gear 5 mates with the ring gear (gear 6), which serves as the output from the train and has 144 teeth. Determine the rotational velocity of all members of this gear train when the input shaft rotates at 680 rpm clockwise.

10–71. A planetary gear train is shown in Figure P10.70. The carrier (link 2) serves as the input to the train. The sun (gear 1) serves as the output from the train and has a 1.0-in. pitch diameter with a diametral pitch of 20. Gear 3 has 45 teeth and gear 4 has 20 teeth. Gear 5 has 30 teeth and is keyed to the same shaft as gear 4. Gear 5 mates with the ring

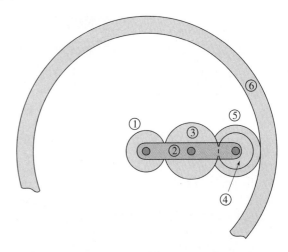

FIGURE P10.70 Problems 70 and 71.

gear (gear 6), which is fixed and has 150 teeth. Determine the rotational velocity of all members of this gear train when the input shaft rotates at 1125 rpm counterclockwise.

10–72. A planetary gear train is shown in Figure P10.72. The carrier (link 2) serves as the input to the train. Gear 2 is fixed and has 48 teeth with a diametral pitch of 12. Gear 1 has 24 teeth, gear 3 has a pitch diameter of 2.5 in., and gear 4 has 35 teeth and a diametral pitch of 10. Determine the rotational velocity of all members of this gear train when the input shaft rotates at 900 rpm clockwise.

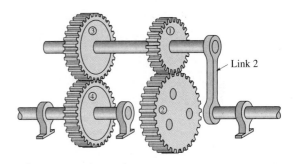

FIGURE P10.72 Problems 72 and 73.

10–73. A planetary gear train is shown in Figure P10.72. The carrier (link 2) serves as the input to the train. Gear 2 is fixed and has a 4.0-in. pitch diameter with a diametral pitch of 10. Gear 1 has 25 teeth, gear 3 has a pitch diameter of 2.5 in., and gear 4 has 32 teeth and a diametral pitch of 8. Determine the rotational velocity of all members of this gear train when the output shaft rotates at 210 rpm clockwise.

CASE STUDIES

10–1. A mechanism utilizing two spur gears and a rack is shown in Figure C10.1. Carefully examine the components of the mechanism, then answer the following leading questions to gain insight into its operation.

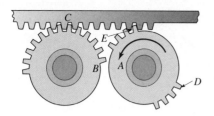

FIGURE C10.1 (Courtesy, Industrial Press.)

1. As segment gear *A* rotates counterclockwise from the position shown, what is the motion of rack *C*?
2. As segment gear *A* rotates counterclockwise from the position shown, what is the motion of the gear *B*?
3. As gear *A* rotates until tooth *E* disengages with rack *C*, what motion is exhibited in gear *B*?
4. What is the entire range of motion for gear *B*?
5. What is the entire range of motion for rack *C*?
6. What is the purpose of this mechanism?
7. What are possible operating problems with this mechanism?

10–2. A device from a wire-forming machine is shown in Figure C10.2. Link *B* and spur gear *C* are keyed to the same shaft. Likewise, link *E* and spur gear *D* are keyed to the same shaft. Carefully examine the components of the mechanism, then answer the following leading questions to gain insight into its operation.

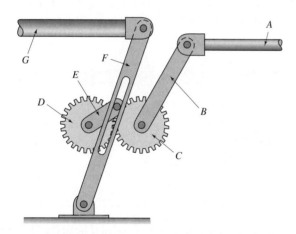

FIGURE C10.2 (Courtesy, Industrial Press.)

1. As link *A* moves to the left, what is the motion of link *B*?
2. As link *A* moves to the left, what is the motion of gear *C*?
3. As link *A* moves to the left, what is the motion of gear *D*?
4. As link *A* moves to the left, what is the motion of link *E*?
5. As link *A* moves to the left, what is the motion of link *F*?
6. As link *A* moves to the left, what is the motion of link *G*?
7. Describe specifically the motion given to *G* as link *A* reciprocates back and forth.

8. How would the motion of link *G* be altered if the mechanism were assembled such that everything appeared identical except link *E* rotated clockwise 90°?

10–3. A device that controls the motion of a gear attached to gear *D* is shown in Figure C10.3. Carefully examine the components of the mechanism, then answer the following leading questions to gain insight into its operation.

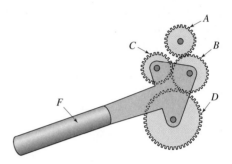

FIGURE C10.3 (Courtesy, Industrial Press.)

1. As gear *A* rotates clockwise, what is the motion of gear *B*?
2. As gear *A* rotates clockwise, what is the motion of gear *C*?
3. As gear *A* rotates clockwise, what is the motion of gear *D*?
4. As the handle *F* is forced upward, what happens to the mating gears?
5. As gear *A* rotates clockwise, what are the motions of gears *B*, *C*, and *D*?
6. What is the purpose of this mechanism?
7. What problems may occur when operating this mechanism?

10–4. A device that drives a piston (*G*) is shown in Figure C10.4. Carefully examine the components of the mechanism, then answer the following leading questions to gain insight into its operation.

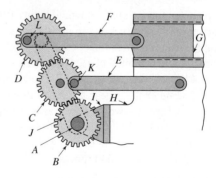

FIGURE C10.4 (Courtesy, Industrial Press.)

1. As gear *B* rotates clockwise, what is the motion of gear *C*?
2. As gear *B* rotates clockwise, what is the motion of gear *D*?

3. If link *J* were hinged at *A*, but were not attached to gear *B*, what motion would link *J* exhibit and what would cause this motion?
4. What is the motion of the center of gear *D*?
5. What is the motion of piston *G*?
6. What is the purpose of this mechanism?
10–5. A device is shown in Figure C10.5. Shaft *C* is a free running fit through gears *H* and *J*, but item *K* is attached with a pin to the shaft. Carefully examine the components of the mechanism, then answer the following leading questions to gain insight into its operation.

1. As shaft *G* rotates as shown, which direction does gear *H* rotate?
2. What type of gears are *F*, *J*, and *H*?
3. As shaft *G* rotates as shown, what is the motion of item *A*?

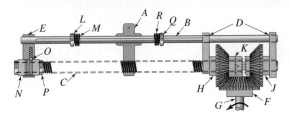

FIGURE C10.5 (Courtesy, Industrial Press.)

4. As item *A* contacts collar *L*, what changes occur to the motion of the mechanism?
5. What is the purpose of item *O*?
6. Why are there set screws on collars *L* and *Q*?
7. What is the purpose of this mechanism?

BELT AND CHAIN DRIVES

11.1 INTRODUCTION

The primary function of a belt or chain drive is identical to that of a gear drive. All three of these mechanisms are used to transfer power between rotating shafts. However, the use of gears becomes impractical when the distance between the shafts is large. Both belt and chain drives offer the flexibility of efficient operation at large and small center distances.

Consider the chain on a bicycle. This mechanism is used to transfer the motion and forces of the rotating pedal assembly to the rear wheel. The distance between these two rotating components is considerable, and a gear drive would be unreasonable. Additionally, the velocity ratio of the chain drive can be readily altered by relocating the chain to an alternate set of sprockets. Thus, a slower pedal rotation but greater forces are needed to maintain the identical rotation of the rear wheel. The velocity ratio of a belt drive can be similarly altered. Changing a velocity ratio on a gear drive is a much more complex process, as in an automotive transmission.

Belt and chain drives are commonly referred to as flexible connectors. These two types of mechanisms can be "lumped together" because the kinematics are identical. The determination of the kinematics and forces in belt and chain drives is the purpose of this chapter. Because the primary motion of the shafts is pure rotation, graphical solutions do not provide any insight. Therefore, only analytical techniques are practical and are introduced in this chapter.

11.2 BELTS

The function of a belt drive is to transmit rotational motion and torque from one shaft to another, smoothly, quietly, and inexpensively. Belt drives provide the best overall combination of design flexibility, low cost, low maintenance, ease of assembly, and space savings.

Compared to other forms of power transmission, belt drives have these advantages:

- They are less expensive than gear or chain drives.
- They have flexible shaft center distances, where gear drives are restricted.
- They operate smoothly and with less noise at high speeds.
- They can be designed to slip when an overload occurs in the machine.
- They require no lubrication, as do chains and gears.
- They can be used in more than one plane.
- They are easy to assemble and install and have flexible tolerances.
- They require little maintenance.
- They do well in absorbing shock loading.

Belts are typically made of continuous construction of materials, such as rubberized fabric, rubberized cord, reinforced plastic, leather, and fabric (i.e., cotton or synthetic fabric). Many belt shapes are commercially available and are listed here.

1. A *flat belt* is shown in Figure 11.1a. This belt is the simplest type but is typically limited to low-torque applications because the driving force is restricted to pure friction between the belt and the pulley.

2. A *V-belt* is shown in Figure 11.1b. This is the most widely used type of belt, particularly in automotive and industrial machines. The V shape causes the belt to wedge tightly into the pulley, increasing friction and allowing higher operating torque.

3. A *multi-V-belt* is shown in Figure 11.1c. This belt design is identical to several V-belts placed side by side but is integrally connected. It is used to increase the amount of power transferred.

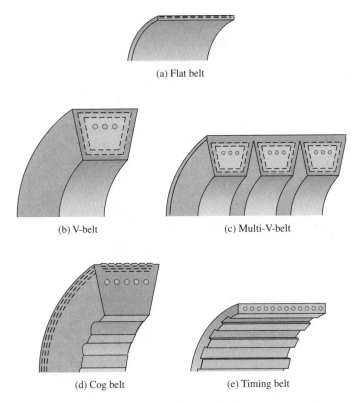

FIGURE 11.1 Types of belts.

4. A *cog belt* is shown in Figure 11.1d. This belt design is similar to a V-belt but has grooves formed on the inner surface. This feature increases belt flexibility, allowing the belt to turn smaller radii. Thus, it can be used on smaller pulleys, reducing the size of the drive.

5. A *timing belt* is shown in Figure 11.1e. This belt design has gear-like teeth that engage with mating teeth on the pulleys. This arrangement combines the flexibility of a belt with the positive grip of a gear drive. This belt is widely used in applications where relative positioning of the respective shafts is desired.

Pulleys, more appropriately referred to as *sheaves*, are the wheels that are connected to the shafts and carry the belt. The pulleys have a groove around the outside, with a shape to match that of the belt. A V-belt sheave is shown in Figure 11.2. Industrial sheaves are machined from steel or

cast iron, depending on diameter. For lighter service, sheaves may be made from aluminum, plastic, or die-cast zinc. The construction is either solid or spoked, also depending on size. Large sheaves are typically spoked and constructed from cast iron.

Sheaves are classified with a pitch diameter, which is the diameter slightly smaller than the outside of the groove, corresponding to the location of the center of the belt. Commercially stocked sheaves are commonly sold in fractional-inch inside groove diameters. Table 11.1 illustrates the available sheave diameters.

When belts are in operation, they stretch over time. Machines that utilize a belt drive need some feature that can compensate for the belt stretch, such as an adjustable motor base, or an idler pulley. An *idler pulley* is used to maintain constant tension on the belt. It is usually placed on the slack side of the belt and is preloaded, usually with springs, to keep the belt tight.

As stated, V-belts are the most widely used type of belt. Commercially available industrial V-belts are made to one of the standard sizes shown in Figure 11.3. Of course, the larger cross sections are able to transmit greater power. Often, several belts are used on multiple-groove pulleys to increase

TABLE 11.1 Commercially Available Sheaves

Sheave Pitch Diameters (in.)

3V Belt		5V Belt		8V Belt
2.2	5.3	4.3	8.4	12.3
2.3	5.6	4.5	8.9	13.0
2.5	6.0	4.8	9.2	13.8
2.6	6.5	4.9	9.7	14.8
2.8	6.9	5.1	10.2	15.8
3.0	8.0	5.4	11.1	16.8
3.1	10.6	5.5	12.5	17.8
3.3	14.0	5.8	13.9	18.8
3.6	19.0	5.9	15.5	19.8
4.1	25.0	6.2	16.1	21.0
4.5	33.5	6.3	18.5	22.2
4.7		6.6	20.1	29.8
5.0		6.7	23.5	39.8
		7.0	25.1	47.8
		7.1	27.9	52.8
		7.5		57.8
		8.1		63.8

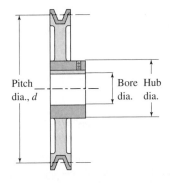

FIGURE 11.2 Single-groove V-belt sheave.

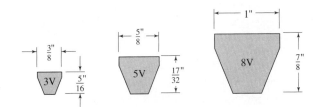

FIGURE 11.3 Industrial narrow-section V-belts.

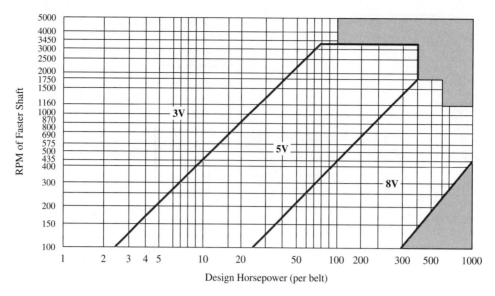

FIGURE 11.4 Industrial V-Belt selection chart.

the amount of power transmitted by the belt drive. A guide to V-belt selection is given in Figure 11.4. The power values are listed "per belt." When the belt drive must transfer 6 hp using a three-groove belt, each of the three belts must be capable of carrying 2 hp.

It must be noted that Figure 11.4 gives only a rough guide to selecting an appropriate belt size. It is important to select the most suitable belt drive based on a detailed study of the application and the power transmission requirements. These detailed selection procedures are given in the manufacturers' catalogs.

11.3 BELT DRIVE GEOMETRY

A belt drive is intended to provide a constant velocity ratio between the respective shafts. A sketch of the basic geometry in a belt drive is shown in Figure 11.5.

As stated, the *pitch diameter, d,* of the sheave is measured to the point in the groove where the center of the belt sits. This is slightly smaller than the outside diameter of the sheave. Note that the diameters shown for the sheaves in Figures 11.2 and 11.5 are the pitch diameters.

The *center distance, C,* is the distance between the center of the driver and driven sheaves. Of course, this is also the distance between the two shafts coupled by the belt drive.

Small center distances can cause fatigue, with frequent maximum loading on the belt sections as it enters the small sheave. Large center distances, with the long unsupported span, can cause belt whip and vibrations. Normal center distances for V-belts should be in the following range:

$$d_2 < C < 3(d_1 + d_2)$$

The *belt length, L,* is the total length of the belt. Specifically, the outside length is usually specified. This is the dimension obtained by wrapping a tape measure around the outside of the belt in the installed position. Belts are commercially available at specified lengths. Table 11.2 illustrates the available lengths for industrial V-belts. The center distance and pitch diameters can be mathematically related [Ref. 2].

$$L = 2C + \frac{\pi}{2}(d_2 + d_1) + \frac{(d_2 - d_1)^2}{4C} \quad \textbf{(11.1)}$$

and

$$C = \frac{B + \sqrt{B^2 - 32(d_2 - d_1)^2}}{16} \quad \textbf{(11.2)}$$

where

$$B = 4L - 2\pi(d_2 + d_1) \quad \textbf{(11.3)}$$

The *angle of contact, θ,* is a measure of the angular engagement of the belt on each sheave. It can be computed for each sheave as follows:

$$\theta_1 = 180° - 2\sin^{-1}\left\{ \frac{d_2 - d_1}{2C} \right\} \quad \textbf{(11.4)}$$

$$\theta_2 = 180° + 2\sin^{-1}\left\{ \frac{d_2 - d_1}{2C} \right\} \quad \textbf{(11.5)}$$

The power ratings for commercially available belts, as shown in Figure 11.4, are for drives with sheaves of the same size. Thus, the "rated" angle of contact is 180°. For smaller

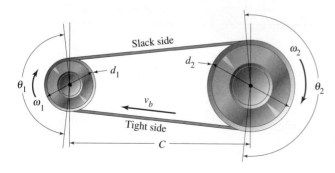

FIGURE 11.5 Belt drive geometry.

TABLE 11.2	Commercially Available V-Belt Lengths (in.)		
3V Belt Lengths			
25.0	40.0	63.0	100.0
26.5	42.5	67.0	106.0
28.0	45.0	71.0	112.0
30.0	47.5	75.0	118.0
31.5	50.0	80.0	125.0
33.5	53.0	85.0	132.0
35.5	56.0	90.0	140.0
37.5	60.0	95.0	
5V Belt Lengths			
50.0	90.0	160.0	280.0
53.0	95.0	170.0	300.0
56.0	100.0	180.0	315.0
60.0	106.0	190.0	335.0
63.0	112.0	200.0	355.0
67.0	118.0	212.0	
71.0	125.0	224.0	
75.0	132.0	236.0	
80.0	140.0	250.0	
85.0	150.0	265.0	
8V Belt Lengths			
100.0	160.0	236.0	355.0
112.0	170.0	250.0	400.0
118.0	180.0	265.0	450.0
125.0	190.0	280.0	
132.0	200.0	300.0	
140.0	212.0	315.0	
150.0	224.0	335.0	

per time, or any other convenient set of rotational velocity units. Using the same logic as the derivation of equation (10.19) yields the following equation:

$$\frac{\omega_1}{\omega_2} = \frac{r_2}{r_1} = VR$$

Introducing the pitch diameters gives

$$\frac{d_2}{d_1} = \frac{2r_2}{2r_1} = \frac{r_2}{r_1} = VR$$

Thus, a comprehensive definition of a velocity ratio is given as

$$VR = \frac{\omega_1}{\omega_2} = \frac{r_2}{r_1} = \frac{d_2}{d_1} \qquad (11.7)$$

Notice that for the typical arrangement, as shown in Figure 11.3, the sheaves rotate in the same direction. Crossed drives or serpentine drives, as shown in Figure 11.6, can be used to reverse the direction of sheave rotation.

Many industrial applications require belts to reduce the speed of a power source. Therefore, it is typical to have velocity ratios greater than 1. As can be seen from equation (11.6), this indicates that the drive sheave rotates faster than the driven sheave, which is the case in speed reductions.

The *belt speed*, v_b, is defined as the linear velocity of the belt. The magnitude of this velocity corresponds to the magnitude of the linear velocity of a point on the pitch diameter of each sheave. Therefore, the belt speed can be related to the

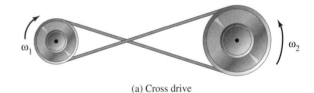

(a) Cross drive

(b) Serpentine drive

FIGURE 11.6 Alternate forms of belt drives.

angles, the amount of friction that can be developed around the sheave is reduced, and therefore, the amount of power that a belt can transfer is reduced. Table 11.3 shows the percent of actual rated power that can be transferred by a belt riding over a sheave with a contact angle smaller than 180°. Belt manufacturers suggest keeping the contact angle greater than 120° when possible.

11.4 BELT DRIVE KINEMATICS

In a manner identical to gear drives, the *velocity ratio*, *VR*, is defined as the angular speed of the driver sheave (sheave 1) divided by the angular speed of the driven sheave (sheave 2).

$$VR = \frac{\omega_{\text{driver}}}{\omega_{\text{driven}}} = \frac{\omega_1}{\omega_2} \qquad (11.6)$$

Because a ratio is valid regardless of units, the velocity ratio can be defined in terms of revolutions per minute, radians

TABLE 11.3	Reduced Power Capability with Contact Angle					
Angle of Contact, θ	180°	160°	140°	120°	100°	80°
Actual Capability (% of rated power)	100	95	89	82	74	63

rotational velocities of the sheaves and their pitch radii using equation (6.5).

$$v_b = r_1\omega_2 = \frac{d_1}{2}\omega_1 = r_2\omega = \frac{d_2}{2}\omega_2 \qquad \textbf{(11.8)}$$

Note that, as in Chapter 6, the angular velocity in this equation must be specified in radians per unit time.

A belt transfers maximum power at speeds of 4000 to 5000 fpm (ft/min). Therefore, it is best to design a belt drive to operate in this range. Large sheaves for industrial use are cast iron and typically are limited to a maximum belt speed of 6500 fpm. This is because the inertial forces created by the normal acceleration become excessive. Special balance may be needed for speeds exceeding 5000 fpm, as vibration can be caused by the centrifugal acceleration. Finally, another type of drive, specifically chains, is typically more desirable for speeds under 1000 fpm.

EXAMPLE PROBLEM 11.1

A belt drive is used to transmit power from an electric motor to a compressor for a refrigerated truck. The compressor must still operate when the truck is parked and the engine is not running. The 10-hp electric motor is rated at 3550 rpm, and the motor sheave diameter is 5 in. The compressor sheave has 7.5 in. diameter. Determine the appropriate industrial belt size, the operating speed of the compressor, and the belt speed.

SOLUTION:

1. *Select an Appropriate Belt Size*

   With a 10-hp motor driving at 3550 rpm, Figure 11.4 suggests using a 3V belt.

2. *Calculate Driver Sheave Speed*

   From equation (11.8), the velocity ratio is determined by

   $$VR = \frac{d_2}{d_1} = \frac{7.5\,\text{in.}}{5\,\text{in.}} = 1.5$$

   The speed of the compressor can be determined from rewriting equation (11.8).

   $$\omega_2 = \frac{d_1\omega_1}{d_2} = \frac{(5\,\text{in.})(3550\,\text{rpm})}{(7.5\,\text{in.})} = 2367\,\text{rpm}$$

   Units of the motor shaft speed are converted to radians per unit time.

   $$\omega_1 = 3550\,\text{rev/min}\left(\frac{2\pi\,\text{rad}}{1\,\text{rev}}\right) = 22{,}305\,\text{rad/min}$$

3. *Calculate Belt Speed*

   The belt speed can be calculated from equation (11.7).

   $$v_b = \left(\frac{d_1}{2}\right)\omega_1 = \frac{5\,\text{in.}}{2}(22{,}305\,\text{rad/min}) = 55{,}762\,\text{in./min} = 4647\,\text{fpm}$$

EXAMPLE PROBLEM 11.2

A belt drive is required to reduce the speed of an electric motor for a grinding wheel, as shown in Figure 11.7. The 50-hp electric motor is rated at 1725 rpm, and a grinding wheel speed of approximately 600 rpm is desired. Determine an appropriate belt size and find suitable sheave diameters of stock pulleys listed in Table 11.1. Also select a suitable belt length from Table 11.2 and calculate the corresponding center distance.

SOLUTION:

1. *Determine Appropriate Belt Size*

   With a 50-hp motor driving at 1725 rpm, Figure 11.4 suggests using a 5V belt.

2. *Determine Ideal Diameter for Driver Sheave*

   The respective shaft speeds are as follows:

   $$\omega_1 = 1725\,\text{rev/min}\left(\frac{2\pi\,\text{rad}}{1\,\text{rev}}\right) = 10{,}838\,\text{rev/min}$$

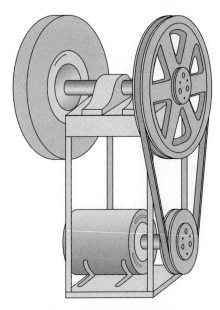

FIGURE 11.7 Grinding wheel for Example Problem 11.2.

$$\omega_2 = 600 \text{ rev/min} \left(\frac{2\pi \, \text{rad}}{1 \, \text{rev}} \right) = 3770 \text{ rev/min}$$

A belt speed of 4000 to 5000 fpm is optimal. Rewriting equation (11.7) yields the following:

$$d_1 = 2\left(\frac{v_b}{\omega_1} \right) = 2\left(\frac{4500 \text{ ft/min}}{10{,}838 \text{ rad/min}} \right)$$

$$= 0.83 \, \text{ft} = 9.96 \, \text{in.}$$

3. *Select Available Sheave*

   Selecting a driver sheave of 10.20 in. from Table 11.1 yields a belt speed of

   $$v_b = \frac{d_1}{2} \omega_1 = \frac{10.20 \, \text{in.}}{2} (10{,}838 \, \text{rad/min}) = 55{,}274 \, \text{in./min} = 4606 \text{ fpm}$$

4. *Select Available Driven Sheave*

   From equation (11.8), the desired velocity ratio is determined by

   $$VR = \frac{\omega_1}{\omega_2} = \frac{10{,}838 \text{ rad/min}}{3770 \text{ rad/min}} = 2.87$$

   And the resulting driven sheave diameter is calculated as follows:

   $$d_2 = (VR)(d_1) = 2.87(10.2 \, \text{in.}) = 29.3 \, \text{in.}$$

   The closest stock sheave of 27.9 in. is selected. Rewriting equation (11.8), the actual grinding wheel speed is

   $$\omega_2 = \frac{\omega_1 d_1}{d_2} = \frac{(1725 \text{ rpm})(10.2 \, \text{in.})}{27.9 \, \text{in.}} = 630 \text{ rpm}$$

5. *Select an Available Belt*

   The suggested center distance for belt drives is within the following range

   $$d_2 < C < 3(d_1 + d_2)$$

   $$27.9 \, \text{in.} < C < 114.3 \, \text{in.}$$

A mid-value of 72 in. is tentatively selected. Substituting into equation (11.1) gives

$$L = 2C + \frac{\pi}{2}(d_2 + d_1) + \frac{(d_2 - d_1)^2}{4C}$$

$$= 2(72 \text{ in.}) + \frac{\pi}{2}(27.9 + 10.2) + \frac{(27.9 - 10.2)^2}{4(72)} = 204.9 \text{ in.}$$

Because a standard length of belt is desired, a 212-in. belt will be selected from Table 11.2. Equations (11.3) and (11.4) are used to calculate the required, actual center distance.

$$C = \frac{B + \sqrt{B^2 - 32(d_2 - d_1)^2}}{16}$$

$$= \frac{580.2 + \sqrt{(580.2)^2 - 32(27.9 - 10.2)^2}}{16} = 71.98 \text{ in.}$$

where

$$B = 4L - 2\pi(d_2 + d_1)$$

$$= 4(204.9) - 2\pi(27.9 + 10.2) = 580.2 \text{ in.}$$

11.5 CHAINS

As with belts, chain drives are used to transmit rotational motion and torque from one shaft to another, smoothly, quietly, and inexpensively. Chain drives provide the flexibility of a belt drive with the positive engagement feature of a gear drive. Therefore, chain drives are well suited for applications with large distances between the respective shafts, slow speed, and high torque.

Compared to other forms of power transmission, chain drives have the following advantages:

- They are less expensive than gear drives.
- They have no slippage, as with belts, and provide a more efficient power transmission.
- They have flexible shaft center distances, whereas gear drives are restricted.
- They are more effective at lower speeds than belts.
- They have lower loads on the shaft bearings because initial tension is not required as with belts.
- They have a longer service life and do not deteriorate with factors such as heat, oil, or age, as do belts.
- They require little adjustment, whereas belts require frequent adjustment.

11.5.1 Types of Chains

Chains are made from a series of interconnected links. Many types of chain designs are commercially available and are listed here.

1. A *roller chain* is shown in Figure 11.8a. This is the most common type of chain used for power transmission. Large roller chains are rated to over 600 hp. The roller chain design provides quiet and efficient operation but must be lubricated.

2. A *multiple-strand roller chain* is shown in Figure 11.8b. This design uses multiple standard roller chains built into parallel strands. This increases the power capacity of the chain drive.

(a) Roller chain

(b) Multiple-strand roller chain

(c) Offset sidebar roller chain

(d) Silent chain

FIGURE 11.8 Types of chains.

3. An *offset sidebar roller chain* is shown in Figure 11.8c. This is less expensive than a roller chain but has slightly less power capability. It also has an open construction that allows it to withstand dirt and contaminants, which can wear out other chains. These chains are often used on construction equipment.

4. An *inverted tooth, silent chain* is shown in Figure 11.8d. This is the most expensive chain to manufacture. It can be efficiently used in applications that require high-speed, smooth, and quiet power transmission. Lubrication is required to keep these chains in reliable operation. They are common in machine tools, pumps, and power drive units.

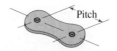

FIGURE 11.9 Chain pitch.

or $1\frac{1}{2}$ inches. The larger-pitch chains have greater power capacity. Roller chain pitch selection is dependant on the power transmitted and speed of the system. A general guide to selecting an appropriate chain pitch is given in Figure 11.10. Manufacturers' catalogs provide detailed procedures to select the most suitable chain drive based on a detailed study of the application and the power transmission requirements.

11.5.2 Chain Pitch

Technical organizations maintain standards (e.g., ANSI standard B29-1) for the design and dimensions of power transmission chains to allow interchangeability. Roller chains are classified by a *pitch, p,* which is the distance between the pins that connect the adjacent links. The pitch is illustrated in Figure 11.9. Roller chains have a size designation ranging from 25 to 240. This size designation refers to the pitch of the chain, in eightieths of an inch. Thus, a 120 chain has a pitch of 120/80

11.5.3 Multistrand Chains

In a similar fashion to belts, multiple-strand chains can be used to increase the amount of power transmitted by the chain drive. However, a multiple-strand chain does not provide a direct multiple of the single-strand capacity. When the chain drive requires multiple strands, equation (11.9) is used to calculate the power transmitted through each chain. A multistrand factor has been experimentally determined and is tabulated in Table 11.4.

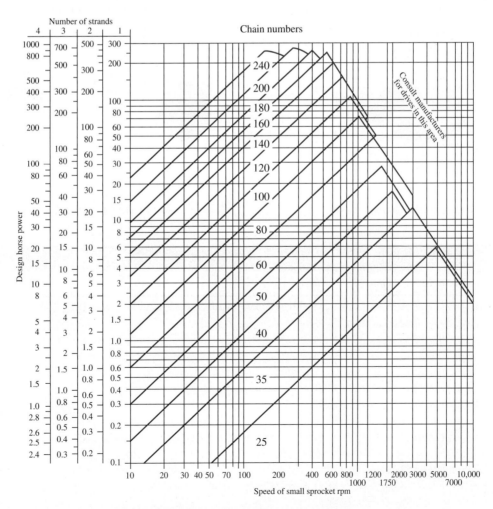

FIGURE 11.10 Chain pitch selection guide.

TABLE 11.4 **Multistrand Factor**							
Number of Roller Chain Strands	2	3	4	5	6	8	10
Multistrand Factor	1.7	2.5	3.3	3.9	4.6	6.2	7.5

$$\text{Power per chain} = \frac{\text{total power transmitted}}{\text{multistrand factor}} \quad (11.9)$$

The vertical axis of Figure 11.10 displays the power capacity based on different numbers of strands. Equation (11.9) has already been implemented in generating Figure 11.10.

11.5.4 Sprockets

Sprockets are the toothed wheels that connect to the shaft and mate with the chain. The teeth on the sprocket are designed with geometry to conform to the chain pin and link. The shape of the teeth varies with the size of the chain and the number of teeth. A sprocket designed to mate with a roller chain is shown in Figure 11.11.

Sprockets are commonly referenced by the corresponding chain size and the number of teeth. Commercially available sprockets are given in Table 11.5. As with gears and

sheaves, the pitch diameter is an important kinematic property. The pitch diameter is the diameter across the middle of the sprocket teeth, which corresponds to the centerline of the chain. It can be determined from the chain size and number of teeth, as will be presented in the next section.

11.6 CHAIN DRIVE GEOMETRY

The basic geometry in a chain drive is virtually identical to that of a belt drive, as shown in Figure 11.12.

The *number of teeth*, N, in the sprocket is a commonly referenced property. It is generally recommended that sprockets have at least 17 teeth, unless they operate at very low speeds—under 100 rpm. Of course, a higher number of teeth will result in a bigger sprocket. The larger sprocket should normally have no more than 120 teeth.

As stated, the *pitch diameter*, d, of the sprocket is measured to the point on the teeth where the center of the chain rides. This is slightly smaller than the outside diameter of the

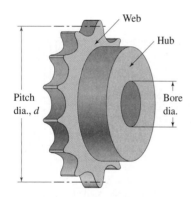

FIGURE 11.11 Roller chain sprocket.

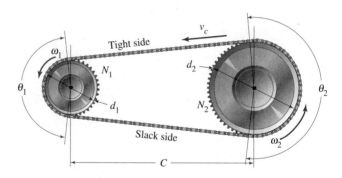

FIGURE 11.12 Chain drive geometry.

TABLE 11.5 **Commercially Available Single-Strand Sprockets**	
Chain Size	**Number of Teeth on the Sprocket**
25	8 through 30, 32, 34, 35, 36, 40, 42, 45, 48, 54, 60, 64, 65, 70, 72, 76, 80, 84, 90, 95, 96, 102, 112, 120
35	4 through 45, 48, 52, 54, 60, 64, 65, 68, 70, 72, 76, 80, 84, 90, 95, 96, 102, 112, 120
40	8 through 60, 64, 65, 68, 70, 72, 76, 80, 84, 90, 95, 96, 102, 112, 120
50	8 through 60, 64, 65, 68, 70, 72, 76, 80, 84, 90, 95, 96, 102, 112, 120
60	8 through 60, 62, 63, 64, 65, 66, 67, 68, 70, 72, 76, 80, 84, 90, 95, 96, 102, 112, 120
80	8 through 60, 64, 65, 68, 70, 72, 76, 78, 80, 84, 90, 95, 96, 102, 112, 120
100	8 through 60, 64, 65, 67, 68, 70, 72, 74, 76, 80, 84, 90, 95, 96, 102, 112, 120
120	9 through 45, 46, 48, 50, 52, 54, 55, 57, 60, 64, 65, 67, 68, 70, 72, 76, 80, 84, 90, 96, 102, 112, 120
140	9 through 28, 30, 31, 32, 33, 34, 35, 36, 37, 39, 40, 42, 43, 45, 48, 54, 60, 64, 65, 68, 70, 72, 76, 80, 84, 96
160	8 through 30, 32 through 36, 38, 40, 45, 46, 50, 52, 53, 54, 56, 57, 60, 62, 63, 64, 65, 66, 68, 70, 72, 73, 80, 84, 96
180	13 through 25, 28, 35, 39, 40, 45, 54, 60
200	9 through 30, 32, 33, 35, 36, 39, 40, 42, 44, 45, 48, 50, 51, 54, 56, 58, 59, 60, 63, 64, 65, 68, 70, 72
240	9 through 30, 32, 35, 36, 40, 44, 45, 48, 52, 54, 60

sprocket. Note that the diameters shown for the sprockets in Figure 11.11 are the pitch diameters. The pitch diameter of a sprocket with N teeth for a chain with a pitch of p is determined by

$$d = \frac{p}{\sin(180°/N)} \quad (11.10)$$

The *center distance*, C, is the distance between the center of the driver and driven sprockets. Of course, this is also the distance between the two shafts coupled by the chain drive. In typical applications, the center distance should be in the following range:

$$30p < C < 50p$$

The *chain length* is the total length of the chain. Because the chain is comprised of interconnected links, the chain length must be an integral multiple of the pitch. It is preferable to have an odd number of teeth on the driving sprocket (17, 19, . . .) and an even number of pitches (links) in the chain to avoid a special link. The chain length, L, expressed in number of links, or pitches, can be computed as

$$L = \frac{2C}{p} + \frac{(N_2 + N_1)}{2} + \left\{ \frac{p(N_2 - N_1)^2}{4\pi^2 C} \right\} \quad (11.11)$$

The center distance for a given chain length can be computed as

$$C = \frac{p}{4}\left[L - \frac{(N_2 + N_1)}{2} + \sqrt{\left\{ L - \frac{(N_2 + N_1)}{2} \right\}^2 - \frac{8(N_2 - N_1)^2}{4\pi^2}} \right] \quad (11.12)$$

It should be restated that the chain length, L, in Equation (11.12) must be stated in the number of links.

The *angle of contact*, θ, is a measure of the angular engagement of the chain on each sprocket. It can be computed as

$$\theta_1 = 180° - 2\sin^{-1}\left\{ \frac{p(N_2 - N_1)}{2C} \right\} \quad (11.13)$$

$$\theta_2 = 180° + 2\sin^{-1}\left\{ \frac{p(N_2 - N_1)}{2C} \right\} \quad (11.14)$$

Chain manufacturers suggest keeping the angle of contact greater than 120° when possible.

Finally, when in operation, chains have a tight side and a slack side. In most applications, chain drives should be designed so that the slack side is on the bottom or lower side. Due to the direction of shaft rotation and the relative positions of the drive and driven shafts, the arrangement shown in Figure 11.12 has the slack side on the bottom.

11.7 CHAIN DRIVE KINEMATICS

Once again, the *velocity ratio*, VR, is defined as the angular speed of the driver sprocket (sprocket 1) divided by the angular speed of the driven sprocket (sprocket 2). Using the same derivations as for gear and belt drives, the velocity ratio consists of

$$VR = \frac{\omega_{\text{driver}}}{\omega_{\text{driven}}} = \frac{\omega_1}{\omega_2} = \frac{d_2}{d_1} = \frac{N_2}{N_1} \quad (11.15)$$

Because a ratio is valid regardless of units, the velocity ratio can be defined in terms of revolutions per minute, radians per time, or any other convenient set of rotational velocity units. Many industrial applications require chains to reduce the speed of a power source. Therefore, it is typical to have velocity ratios greater than 1. As can be seen from equation (11.15), this indicates that the drive sprocket rotates faster than the driven sprocket, which is the case in speed reductions.

Similar to belts, the linear velocity of the chain, or *chain speed*, is defined as v_c. The magnitude of this velocity corresponds to the magnitude of the linear velocity of a point on the pitch diameter of each sprocket. As with belt speed, chain speed can be computed by

$$v_c = \frac{d_1}{2}\omega_1 = \frac{d_2}{2}\omega_2 \quad (11.16)$$

In equation (11.16), the rotation velocities must be stated in radians per unit time.

Lubrication for the chain is important in maintaining long life for the drive. Recommended lubrication methods are primarily dictated by the speed of the chain. The recommended lubrication is as follows:

- Low speed ($v_c < 650$ fpm): manual lubrication, where the oil is periodically applied to the links of the chain.

- Moderate speed ($650 < v_c < 1500$ fpm): bath lubrication, where the lowest part of the chain dips into a bath of oil.

- High speed (1500 fpm $< v_c$): oil stream lubrication, where a pump delivers a continuous stream onto the chain.

EXAMPLE PROBLEM 11.3

A single-strand roller chain drive connects a 10-hp engine to a lawn waste chipper/shredder, as shown in Figure 11.13. As the engine operates at 1200 rpm, the shredding teeth should rotate at 240 rpm. The drive sprocket has 18 teeth. Determine an appropriate pitch for the chain, the number of teeth on the driven sprocket, the pitch diameters of both sprockets, and the chain speed. Also indicate the number of links in a suitable chain and specify the required center distance.

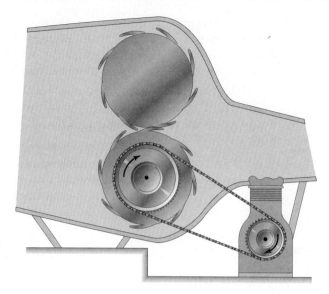

FIGURE 11.13 Chipper/shredder for Example Problem 11.3.

SOLUTION: 1. *Select an Appropriate Chain Pitch*

With a 10-hp engine driving a sprocket at 1200 rpm, Figure 11.10 specifies that a 40-pitch, single-strand chain is appropriate.

2. *Determine the Number of Teeth on the Driven Sprocket*

By rewriting equation (11.15), the number of teeth needed on the driven sprocket can be determined.

$$N_2 = N_1\left(\frac{\omega_1}{\omega_2}\right) = 18\left\{\frac{1200 \text{ rpm}}{240 \text{ rpm}}\right\} = 90 \text{ teeth}$$

Notice from Table 11.5 that a 90-tooth sprocket is commercially available for a No. 40 chain.

3. *Determine the Pitch Diameter of the Sprockets*

A No. 40 roller chain has a pitch of

$$p = \frac{40}{80} = 0.5 \text{ in.}$$

From equation (11.10), the pitch diameters of the sprockets are

$$d_1 = \frac{p}{\sin\left(\dfrac{180°}{N_1}\right)} = \frac{0.5 \text{ in.}}{\sin\left(\dfrac{180°}{18 \text{ teeth}}\right)} = 2.88 \text{ in.}$$

$$d_2 = \frac{p}{\sin\left(\dfrac{180°}{N_2}\right)} = \frac{0.5 \text{ in.}}{\sin\left(\dfrac{180°}{90 \text{ teeth}}\right)} = 14.33 \text{ in.}$$

4. *Calculate the Chain Speed*

The chain speed can be calculated from equation (11.16).

$$\omega_1 = 1200 \text{ rev/min}\left(\frac{2\pi \text{ rad}}{1 \text{ rev}}\right) = 10{,}838 \text{ rad/min}$$

$$v_c = \left(\frac{d_1}{2}\right)\omega_1 = \left(\frac{2.88 \text{ in.}}{2}\right)10{,}833 \text{ rad/min} = 15{,}603 \text{ in./min} = 1300 \text{ fpm}$$

With a chain speed of 1300 fpm, a bath lubrication system for the chain is desired.

5. *Determine an Appropriate Center Distance*

The suggested center distance for a chain drive is

$$30p < C < 50p$$

$$15 \text{ in.} < C < 25 \text{ in.}$$

A mid-value of 20 in. is tentatively selected. Substituting into equation (11.11) gives

$$L = \frac{2C}{P} + \frac{(N_2 + N_1)}{2} + \left\{ \frac{p(N_2 - N_1)^2}{4\pi^2 C} \right\}$$

$$= \frac{2(20 \text{ in.})}{(0.5 \text{ in.})} + \frac{(90 + 18)}{2} + \left\{ \frac{0.5 \text{ in.}(90 - 18)^2}{4\pi^2(20 \text{ in.})} \right\} = 137.28 \text{ links}$$

An even 138 links will be specified. The corresponding, actual center distance is computed from equation (11.12).

$$C = \frac{p}{4} \left[L - \frac{(N_2 + N_1)}{2} + \sqrt{\left\{ L - \frac{(N_2 + N_1)}{2} \right\}^2 - \frac{8(N_2 - N_1)^2}{4\pi^2}} \right]$$

$$= \frac{(0.5 \text{ in.})}{4} \left[(138) - \frac{(90 + 18)}{2} + \sqrt{\left\{ (138) - \frac{(90 + 18)}{2} \right\}^2 - \frac{8(90 - 18)^2}{4\pi^2}} \right] = 20.187 \text{ in.}$$

PROBLEMS

Belt Kinematics

11–1. A motor, operating clockwise at 1725 rpm, is coupled through a belt drive to rotate the crank of an industrial sewing machine. The diameter of the motor sheave is 3.5 in. and the sheave on the sewing machine crank is 8 in. Determine the speed of the driven sheave and the belt speed.

11–2. A motor, operating clockwise at 1150 rpm, is coupled through a belt drive to rotate an industrial exhaust fan. The diameter of the motor sheave is 5 in. and the sheave on the fan is 12 in. Determine the speed of the driven sheave and the belt speed.

11–3. An engine is coupled through a belt drive to rotate an air compressor, which must operate at 1200 rpm clockwise. The diameter of the motor sheave is 4 in. and the sheave on the compressor is 10 in. Determine the required speed of the engine and the belt speed.

11–4. A motor, operating counterclockwise at 1125 rpm, is coupled through a belt drive to rotate a drill press. The diameter of the motor sheave is 2.5 in. and the sheave on the drill spindle is 9 in. Determine the speed of the driven sheave and the belt speed.

11–5. A motor, operating counterclockwise at 1750 rpm, is coupled through a belt drive to furnace blower. The diameter of the motor sheave is 6.5 in. and the sheave on the drill spindle is 10.5 in. Determine the speed of the driven sheave and the belt speed.

11–6. An engine is coupled through a belt drive to rotate a generator, which must rotate at 1800 rpm counterclockwise. The diameter of the engine sheave is 6 in. and the sheave on the generator is 9 in. Determine the required speed of the engine sheave and the belt speed.

Belt Drive Geometry

11–7. Two sheaves have diameters of 3.5 in. and 8 in. and their center distance is 23 in. Compare the center distance to the ideal range and determine the associated belt length. Also determine the angle of contact over the smaller sheave.

11–8. Two sheaves have diameters of 5 in. and 12 in. Determine the center distance of a drive utilizing a 72-in. belt and compare that center distance to the ideal range. Also determine the angle of contact over the smaller sheave.

11–9. Two sheaves have diameters of 8 in. and 12 in. Determine the center distance of a drive utilizing an 88-in. belt and compare that center distance to the ideal range. Also determine the angle of contact over the smaller sheave.

11–10. Two sheaves have diameters of 8 in. and 24 in. Determine the center distance of a drive utilizing a 104-in. belt and compare that center distance to the ideal range. Also determine the angle of contact over the smaller sheave.

Belt Drive Selection

11–11. A belt drive is desired to couple the motor with a mixer for processing corn syrup. The 10-hp electric motor is rated at 3550 rpm and the mixer must operate as close to 900 rpm as possible. Select an appropriate belt size, commercially available sheaves, and a belt for this application. Also calculate the actual belt speed and the center distance.

11–12. A belt drive is desired to couple the motor with a mixer for processing corn syrup. The 25-hp electric motor is rated at 950 rpm and the mixer must operate as close to 250 rpm as possible. Select an appropriate belt size, commercially available sheaves, and a belt for this application. Also calculate the actual belt speed and the center distance.

11–13. A belt drive is desired to couple the motor with a conveyor. The 5-hp electric motor is rated at 1150 rpm and the conveyor driveshaft must operate as close to 250 rpm as possible. Select an appropriate belt size, commercially available sheaves, and a belt for this application. Also calculate the actual belt speed and the center distance.

11–14. A belt drive is desired to couple an engine with the blade shaft of a commercial lawn mower. The 10-hp engine is to operate at 2000 rpm and the blade shaft must operate as close to 540 rpm as possible. Select an appropriate belt size, commercially available sheaves, and a belt for this application. Also calculate the actual belt speed and the center distance.

11–15. A belt drive is desired to couple an engine with the drive system of a snowmobile. The 70-hp engine is to operate at 3000 rpm and the driveshaft must operate as close to 750 rpm as possible. Select an appropriate belt size, commercially available sheaves, and a belt for this application. Also calculate the actual belt speed and the center distance.

11–16. A belt drive is desired to couple an electric motor with the spool of a winch. The 2-hp electric motor operates at 200 rpm and the spool must operate as close to 60 rpm as possible. Select an appropriate belt size, commercially available sheaves, and a belt for this application. Also calculate the actual belt speed and the center distance.

Chain Kinematics

11–17. The shaft of a gearbox is coupled through a No. 60 chain drive, rotating a concrete mixer at 180 rpm clockwise. The drive sprocket has 19 teeth and the sprocket on the mixer has 84. Determine the speed of the drive sprocket, the chain speed, and the recommended lubrication method.

11–18. A gearmotor, operating clockwise at 220 rpm, is coupled through a No. 40 chain drive to a liquid agitator. The drive sprocket has 19 teeth and the sprocket on the mixer has 50. Determine the speed of the driven sprocket, the chain speed, and the recommended lubrication method.

11–19. A gearmotor, operating clockwise at 180 rpm, is coupled through a No. 80 chain drive to an escalator. The drive sprocket has 27 teeth and the sprocket on the escalator has 68. Determine the speed of the driven sprocket, the chain speed, and the recommended lubrication method.

11–20. The shaft of a gearbox is coupled through a No. 100 chain drive, rotating a driveshaft for a pulp screen at a paper-producing plant. The screen driveshaft rotates at 250 rpm clockwise. The drive sprocket has 25 teeth and the sprocket on the mixer has 76. Determine the speed of the drive sprocket, the chain speed, and the recommended lubrication method.

Chain Drive Geometry

11–21. Two sprockets, for a No. 60 chain, have 17 and 56 teeth. The chain has 120 links. Determine the pitch diameter of each sprocket, their center distance, and the angle of contact over the smaller sprocket. Also compare the center distance to the ideal range.

11–22. Two sprockets, for a No. 80 chain, have 19 and 64 teeth. The chain has 140 links. Determine the pitch diameter of each sprocket, their center distance, and the angle of contact over the smaller sprocket. Also compare the center distance to the ideal range.

11–23. Two sprockets, for a No. 40 chain, have 21 and 84 teeth. The chain has 200 links. Determine the pitch diameter of each sprocket, their center distance, and the angle of contact over the smaller sprocket. Also compare the center distance to the ideal range.

11–24. Two sprockets, for a No. 120 chain, have 25 and 72 teeth. The chain has 150 links. Determine the pitch diameter of each sprocket, their center distance, and the angle of contact over the smaller sprocket. Also compare the center distance to the ideal range.

Chain Drive Selection

11–25. A chain drive is desired to couple a gearmotor with a bucket elevator. The 40-hp gearmotor will operate at 350 rpm and the elevator drive shaft must operate as close to 60 rpm as possible. Select an appropriate chain size, commercially available sprockets, and number of links for a chain. Also calculate the actual chain speed and the center distance.

11–26. A chain drive is desired on a corn picker to couple a hydraulic motor with a driveshaft. The 30-hp motor will operate at 550 rpm and the driveshaft must operate as close to 100 rpm as possible. Select an appropriate chain size, commercially available sprockets, and number of links for a chain. Also calculate the actual chain speed and the center distance.

11–27. A chain drive is desired to couple an engine and gearbox with the drive wheels on an all-terrain vehicle. When the 130-hp engine/gearbox will output 600 rpm, the drive axle must operate as close to 140 rpm as possible. Select an appropriate chain size, commercially available sprockets, and number of links for a chain. Also calculate the actual chain speed and the center distance.

11–28. A chain drive is desired to couple a gearmotor with a screw drive on a press. The 50-hp gearmotor will operate at 600 rpm and the screw drive must operate as close to 100 rpm as possible. Select an appropriate chain size, commercially available sprockets, and number of links for a chain. Also calculate the actual chain speed and the center distance.

11–29. A chain drive is desired to couple an engine and gearbox with the auger on a snowblower. When the 8-hp engine/gearbox will output 300 rpm, the auger must operate as close to 40 rpm as possible. Select an appropriate chain size, commercially available sprockets, and number of links for a chain. Also calculate the actual chain speed and the center distance.

11–30. A chain drive is desired to couple a gearmotor with a parts storage carousel. The 10-hp gearmotor will operate at 425 rpm and the carousel shaft must operate as close to 75 rpm as possible. Select an appropriate chain size, commercially available sprockets, and number of links for a chain. Also calculate the actual chain speed and the center distance.

CASE STUDIES

11–1. The device shown in Figure C11.1 drives a chute that funnels individual beverage bottles into 12-pack containers. Pin *C* is rigidly attached to one link of the chain. Yoke *D* is rigidly welded to rod *E*, which extends to the chute (not shown). Carefully examine the components of the mechanism, then answer the following leading questions to gain insight into its operation.

1. As sprocket *A* drives clockwise, determine the motion of sprocket *B*.
2. As sprocket *A* drives clockwise, specify the instantaneous motion of pin *C*.
3. As sprocket *A* drives clockwise, specify the instantaneous motion of yoke *D*.
4. How far must sprocket *A* rotate to move pin *C* onto a sprocket?
5. What happens to the motion of rod *E* when the pin rides onto the sprocket?
6. What happens to rod *E* when pin *C* is on the top portion of the chain drive?
7. Discuss the overall motion characteristics of rod *E*.

11–2. The sheave shown in Figure C11.2 drives shaft *A*, which drives a log splitter (not shown). Notice that the sheave is split into two halves, labeled *B* and *C*. These two halves are threaded together at *D*. Carefully examine the components of the mechanism, then answer the following leading questions to gain insight into its operation.

1. As handle *H* rotates, what is the motion of shaft *I*?
2. As handle *H* rotates, what is the motion of the right half, *C*, of the sheave?
3. What is the resulting effect on the sheave by rotating handle *H*?
4. What is item *J* and what is its function?

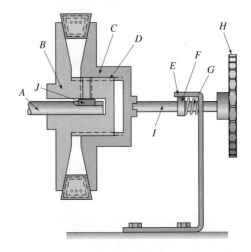

FIGURE C11.2 (Courtesy, Industrial Press.)

5. What is the purpose of item *F* and must it stay in contact with item *E*?
6. What is item *G* and what is its function?
7. What would you call such a device?

11–3. The sheave shown in Figure C11.3 drives a mechanism (not shown) in a machine that shakes paint cans. These machines are used to thoroughly mix the paint at the time of purchase and are common at most paint retail locations. Carefully examine the components of the mechanism, then answer the following leading questions to gain insight into its operation.

1. As tab *B* is forced upward into lever *C*, determine the motion of item *D*.
2. As tab *B* is forced upward into lever *C*, determine the resulting action of sheave *G*.
3. As tab *B* is forced upward into lever *C*, determine the resulting action of the paint-shaking mechanism.
4. Item *A* is the door to the paint-shaking compartment; discuss the purpose of the mechanism.
5. Discuss the reason for so many notches, *E*, in sheave *G*.
6. Discuss the purpose of spring *H*.
7. What would you call such a device?

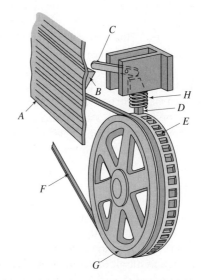

FIGURE C11.3 Mechanism for Case Study 11.3

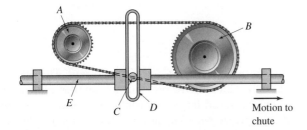

FIGURE C11.1 (Courtesy, Industrial Press.)

Motion to chute

SCREW MECHANISMS

12.1 INTRODUCTION

In general, screw mechanisms are designed to convert rotary motion to linear motion. Consider a package for a stick deodorant. As the knob turns, the deodorant stick either extends or retracts into the package. Inside the package, a screw turns, which pushes a nut and the deodorant stick along the thread. Thus, a "disposable" screw mechanism is used in the deodorant package. This same concept is commonly used in automotive jacks, some garage door openers, automotive seat adjustment mechanisms, and milling machine tables.

The determination of the kinematics and forces in a screw mechanism is the purpose of this chapter. Because the motion of a nut on a thread is strictly linear, graphical solutions do not provide any insight. Therefore, only analytical techniques are practical and are introduced in this chapter.

12.2 THREAD FEATURES

For a screw to function, there must be two mating parts, one with an internal thread and the other with an external thread. The external threads are turned on the surface of a shaft or stud, such as a bolt or screw. The internal threads can be tapped into a part, such as a cast housing or, more

commonly, inside a nut. Whenever possible, the selection of a thread should be standard to improve interchangeability for maintenance or replacement. Threads, whether internal or external, are classified with the following features.

The two most common features of a thread are the pitch and pitch diameter. The *pitch, p,* is the distance measured parallel to the screw axis from a point on one thread to the corresponding point on the adjacent thread. The *pitch diameter, d,* is the diameter measured from a point halfway between the tip and root of the thread profile through the axis and to the corresponding point on the opposite side. Figure 12.1 illustrates these properties.

Other pertinent features of a screw thread (Figure 12.1) include the *major diameter,* the *minor diameter,* the *lead angle,* and the *included angle.* In the U.S. Customary Unit System, the number of *threads per inch, n,* along the length of the screw is more commonly used than the pitch. The threads-per-inch value is related to the pitch through the following equation:

$$n = \frac{1}{p} \qquad (12.1)$$

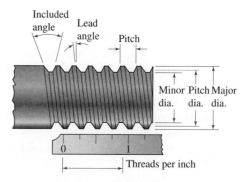

FIGURE 12.1 Thread profile.

12.3 THREAD FORMS

Thread form defines the shape of the thread. The thread features introduced in the previous section were illustrated on a unified thread form. Regardless, these definitions are applicable to all thread forms. The most popular thread forms include unified, metric, square, and ACME threads.

12.3.1 Unified Threads

Unified threads are the most common threads used on fasteners and small mechanisms. They are also commonly used for positioning mechanisms. Figure 12.2a illustrates the profile of a unified thread. It is described as a sharp, triangular tooth. The dimensions of a unified thread have been standardized and are given in Table 12.1. Unified threads are designated as either coarse pitch (UNC) or fine pitch (UNF).

A standard unified thread is specified by the size, threads per inch, and coarse or fine pitch. Standard thread designations would appear as

10–32 UNF
1/2–13 UNC

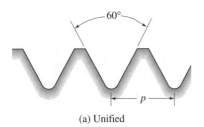

(a) Unified

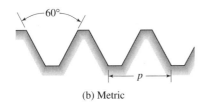

(b) Metric

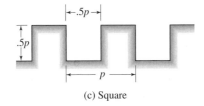

(c) Square

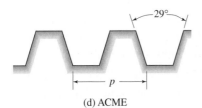

(d) ACME

FIGURE 12.2 Thread forms.

12.3.2 Metric Threads

Metric thread forms are also described as sharp, triangular shapes, but with a flat root. However, the standard dimensions are convenient metric values and coordinated through the International Organization for Standardization (ISO). The thread shape is shown in Figure 12.2b. Standard metric thread dimensions are given in Table 12.2.

A standard metric thread is specified by the metric designation "M," the nominal major diameter, and pitch. A standard thread designation would appear as

M10 × 1.5

12.3.3 Square Threads

Square threads, as the name implies, are a square, flat-top thread. They are strong and were originally designed to transfer power. A square thread form is shown in Figure 12.2c. Although they efficiently transfer large loads, these threads are difficult to machine with perpendicular sides. The square threads have been generally replaced by ACME threads.

12.3.4 ACME Threads

ACME threads are similar to square threads, but with sloped sides. They are commonly used when rapid movement is required or large forces are transmitted. An ACME thread form is shown in Figure 12.2d. The standard ACME screw thread dimensions are given in Table 12.3. This thread is the most common form used in screw mechanisms for industrial machines. Its advantages are low cost and ease of manufacture. Its disadvantages include low efficiencies, as will be discussed later, and difficulty in predicting service life.

12.4 BALL SCREWS

Ball screws have also been designed to convert rotary motion of either the screw or nut into relatively slow linear motion of the mating member along the screw axis. However, a ball screw has drastically less friction than a traditional screw configuration. The sliding contact between the screw and nut has been replaced with rolling contact of balls in grooves along the screw. Thus, a ball screw requires less power to drive a load. A ball screw is shown in Figure 12.3.

The operation of a ball screw is smooth because the rolling balls eliminate the "slip-stick" motion caused by the friction of a traditional screw and nut. However, because of the low friction of a ball screw, a brake must usually be used to hold the load in place.

The kinematics of a ball screw are identical to those of a traditional screw. Therefore, a distinction is not required when performing a kinematic analysis. The following concepts apply to both traditional and ball screws.

12.5 LEAD

In determining the motion of a screw mechanism, the lead of the screw is a critical parameter and must be understood. The *lead, L,* is the distance along the screw axis that a nut travels with one revolution of the screw. For most screws, the lead is identical to the pitch. However, screws are available with single or double threads. Thus, the *number of threads, N_t,* superimposed on a screw is an important property. The concept of multiple threads superimposed on a single screw is illustrated in Figure 12.4.

TABLE 12.1   Standard Unified Thread Dimensions

		Coarse Threads			Fine Threads		
Size	Nominal Major Diameter (in.)	Threads per inch, n	Pitch (in.) $p = \dfrac{1}{n}$	Nominal Pitch Diameter (in.)	Threads per inch, n	Pitch (in.) $p = \dfrac{1}{n}$	Nominal Pitch Diameter (in.)
0	0.0600	—	—	—	80	0.0125	0.0519
1	0.0730	64	0.0156	0.0629	72	0.0139	0.0640
2	0.0860	56	0.0179	0.0744	64	0.0156	0.0759
3	0.0990	48	0.0208	0.0855	56	0.0179	0.0874
4	0.1120	40	0.0250	0.0958	48	0.0208	0.0985
5	0.1250	40	0.0250	0.1088	44	0.0227	0.1102
6	0.1380	32	0.0313	0.1177	40	0.0250	0.1218
8	0.1640	32	0.0313	0.1437	36	0.0278	0.1460
10	0.1900	24	0.0417	0.1629	32	0.0313	0.1697
12	0.2160	24	0.0417	0.1889	28	0.0357	0.1928
¼	0.2500	20	0.0500	0.2175	28	0.0357	0.2268
⁵⁄₁₆	0.3125	18	0.0556	0.2764	24	0.0417	0.2854
⅜	0.3750	16	0.0625	0.3344	24	0.0417	0.3479
⁷⁄₁₆	0.4375	14	0.0714	0.3911	20	0.0500	0.4050
½	0.5000	13	0.0769	0.4500	20	0.0500	0.4675
⁹⁄₁₆	0.5625	12	0.0833	0.5084	18	0.0556	0.5264
⅝	0.6250	11	0.0909	0.5660	18	0.0556	0.5889
¾	0.7500	10	0.1000	0.6850	16	0.0625	0.7094
⅞	0.8750	9	0.1111	0.8028	14	0.0714	0.8286
1	1.0000	8	0.1250	0.9188	12	0.0833	0.9459
1¼	1.2500	7	0.1429	1.1572	12	0.0833	1.1959
1½	1.5000	6	0.1667	1.3917	12	0.0833	1.4459
1¾	1.7500	5	0.2000	1.6201	—	—	—
2	2.0000	4½	0.2222	1.8557	—	—	—

The lead can be computed as

$$L = N_t p \qquad (12.2)$$

A *lead angle*, λ, is shown in Figure 12.1 and is defined as the angle of inclination of the threads. It can be computed from a trigonometric relationship to the other screw features.

$$\tan\lambda = \frac{N_t p}{\pi d} = \frac{L}{\pi d} \qquad (12.3)$$

When a screw thread is very steep and has large lead angles, the torque required to push a load along a screw can become large. Typical screws have lead angles that range from approximately 2° to 6°. Additionally, small lead angles prohibit a load to "slide down a screw" due to gravity. The friction force and shallow thread slope combine to lock the load in place. This is known as self-locking and is desirable for lifting devices. For example, a car jack requires that the load be held in an upward position, even as the power source is removed. When the thread is self-locking, the load is locked in an upright position. This braking action is used in several mechanical devices, but the strength of the thread

and the predictability of friction must be analyzed to ensure safety. Mathematically, the condition that must be met for self-locking is as follows:

$$\mu > \tan\lambda \qquad (12.4)$$

In equation (12.4), μ is the coefficient of friction of the nut–thread interface. For traditional threads, common values for the coefficient of friction are

$\mu = 0.10$ for very smooth, well-lubricated surfaces

$\mu = 0.15$ for general machined screws with well-lubricated surfaces

$\mu = 0.20$ for general machined screws with ordinary surfaces

Special surface treatments and coatings can reduce these values by at least half. Ball screws, with inherent low friction, are virtually never self-locking.

12.6 SCREW KINEMATICS

From a kinematic viewpoint, the screw joint connects two bodies and couples two degrees of freedom. Typically, the joint

TABLE 12.2 Standard Metric Thread Dimensions

Nominal Major Diameter (mm)	Coarse Threads		Fine Threads	
	Pitch (mm) $p = \dfrac{1}{n}$	Nominal Pitch Diameter (mm)	Pitch (mm) $p = \dfrac{1}{n}$	Nominal Pitch Diameter (mm)
1	0.25	0.84	—	—
1.6	0.35	1.37	0.20	1.47
2	0.40	1.74	0.25	1.84
2.5	0.45	2.20	0.35	2.27
3	0.50	2.67	0.35	2.77
4	0.70	3.54	0.50	3.67
5	0.80	4.47	0.50	4.67
6	1.00	5.34	0.75	5.51
8	1.25	7.18	1.00	7.34
10	1.50	9.01	1.25	9.18
12	1.75	10.85	1.25	11.18
16	2.00	14.68	1.50	15.01
20	2.50	18.35	1.50	19.01
24	3.00	22.02	2.00	22.68
30	3.50	27.69	2.00	28.68
36	4.00	33.36	3.00	34.02
42	4.50	39.03	—	—
48	5.00	44.70	—	—

TABLE 12.3 Standard ACME Thread Dimensions

Nominal Major Diameter (in.)	Threads per inch, n	Pitch (in.) $p = \dfrac{1}{n}$	Nominal Pitch Diameter (in.)
$\frac{1}{4}$	16	0.0625	0.2043
$\frac{5}{16}$	14	0.0714	0.2614
$\frac{3}{8}$	12	0.0833	0.3161
$\frac{7}{16}$	12	0.0833	0.3783
$\frac{1}{2}$	10	0.1000	0.4306
$\frac{5}{8}$	8	0.125	0.5408
$\frac{3}{4}$	6	0.1667	0.6424
$\frac{7}{8}$	6	0.1667	0.7663
1	5	0.2000	0.8726
$1\frac{1}{8}$	5	0.2000	0.9967
$1\frac{1}{4}$	5	0.2000	1.1210
$1\frac{3}{8}$	4	0.2500	1.2188
$1\frac{1}{2}$	4	0.2500	1.3429
$1\frac{3}{4}$	4	0.2500	1.5916
2	4	0.2500	1.8402
$2\frac{1}{4}$	3	0.3333	2.0450
$2\frac{1}{2}$	3	0.3333	2.2939
$2\frac{3}{4}$	3	0.3333	2.5427
3	2	0.5000	2.7044
$3\frac{1}{2}$	2	0.5000	3.2026
4	2	0.5000	3.7008
$4\frac{1}{2}$	2	0.5000	4.1991
5	2	0.5000	4.6973

is configured such that one body will translate with a rotational input from the other body. Depending on the constraints of the two bodies, the following relative motions are possible:

I. *Translation of the nut as the screw rotates:* Occurs when the screw is unable to translate and nut is unable to rotate.

II. *Translation of the screw as the nut rotates:* Occurs when the nut is unable to translate and screw is unable to rotate.

III. *Translation of the screw as it rotates:* Occurs when the nut is fully constrained against any motion.

IV. *Translation of the nut as it rotates:* Occurs when the screw is fully constrained against any motion.

Regardless of the actual system configuration, the relative motion is the same. A given rotation produces a resulting translation. Therefore, equations are developed to describe the relative motion, and the absolute motion can be determined when examining the actual system configuration. A notation is made where

A is the part that is allowed to rotate.

B is the other part joined by the screw joint.

As previously defined, the lead of a screw is the distance along the screw axis that a nut travels with one revolution of the screw. Therefore, the magnitude of the relative displacement of *B* relative to *A* is calculated as follows:

$$\Delta R_{B/A} = L\,\Delta\theta_A \qquad (12.5)$$

FIGURE 12.3 Ball screw. (Courtesy, Warner Electric.)

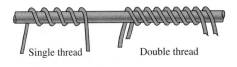

Single thread   Double thread

FIGURE 12.4 Multiple thread concept.

Differentiating yields equations for the magnitude of the velocity and acceleration.

$$v_{B/A} = L\omega_A \qquad (12.6)$$

$$a_{B/A} = L\alpha_A \qquad (12.7)$$

Note that the lead, L, is specified as the relative displacement per revolution. Therefore, in this instance, the angular motion must be specified in revolutions. Thus, ω_A should be specified in revolutions per minute (or second), and α_A should be specified in revolutions per squared minute (or squared second).

The direction of the relative motion depends on the *hand designation* of the thread. Screws and the mating nuts are classified as either right-hand or left-hand. A right-hand thread is most common. For this threaded joint, the screw advances into the nut when the screw rotates clockwise. A right-hand thread slopes downward to the left on an external thread when the axis is horizontal. The slope is opposite on an internal thread. The opposite, left-hand configuration produces the opposite motion.

The following examples illustrate the determination of screw kinematics.

EXAMPLE PROBLEM 12.1

A screw-driven slide, shown in Figure 12.5, is used on a production machine that moves a saw blade to cut the sprue off raw castings. A single thread, ¾ −6 ACME screw shaft moves the slide. The screw is rotated at 80 rpm, moving the slide to the right. Determine the speed of the slide. Also determine the number of revolutions to move the slide 3.5 in.

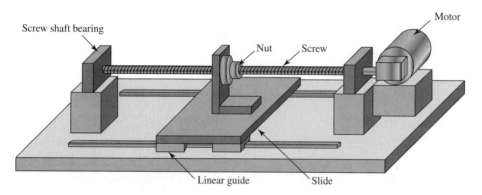

FIGURE 12.5 Slide for Example Problem 12.1.

SOLUTION:

1. **Determine Relative Motion**

    In this configuration, the motor rotates the screw in the bearings, but shoulders on the shaft prevent the screw translating. The nut is held against rotation, but is allowed to translate along the linear guides. This is case I as previously described. The following notation will be used:

    Part A is the screw.
    Part B is the nut.

2. **Calculate Screw Geometry**

    A single-thread, ¾ −6 ACME screw has the following properties:

    Number of threads: $N_t = 1$ thread/rev
    Number of threads per inch: $n = 6$

    Pitch: $p = \dfrac{1}{n} = \dfrac{1}{6} = 0.167$ in./thread
    Lead: $L = N_t\, p = 0.167$ in./rev

3. **Determine Screw Displacement**

    The angular displacement of the screw to produce a 3.5 in. linear displacement of the nut and slide is determined by rearranging equation (12.5).

    $$\Delta\theta_A = \frac{\Delta R_{B/A}}{L} = \frac{3.5\,\text{in.}}{0.167\,\text{in./rev}} = 20.96\,\text{rev}$$

    In the absence of further information, it is assumed that this is a standard, right-hand thread. Therefore, the screw must turn counterclockwise, as viewed from the right end, to move the nut to the right.

4. **Determine Nut Velocity**

The linear velocity of the nut can be calculated from equation (12.6).

$$\mathbf{V}_{B/A} = L\omega_A = (0.167\,\text{in./rev})\,(80\,\text{rev/min}) = 13.36\,\text{in./min} \rightarrow$$

Because the screw is constrained against translation, this computed velocity is the absolute velocity of the nut.

EXAMPLE PROBLEM 12.2

A screw-operated press is shown in Figure 12.6. The screw has a single ½ ×10 ACME thread, both in a right-hand and left-hand orientation, as shown. The handle rotates counterclockwise at 45 rpm to drive the pressure plate downward. In the position shown, with $\beta = 25°$, determine the velocity of the pressure plate.

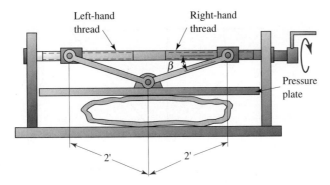

FIGURE 12.6 Press for Example Problem 12.2.

SOLUTION: 1. **List Screw Properties**

A single-thread, ½ ×10 ACME screw has the following properties:

Number of teeth per inch: $n = 10$

Pitch: $p = \dfrac{1}{n} = \dfrac{1}{10} = 0.10\,\text{in.}$

Number of threads: $N_t = 1$

Lead: $L = N_t p = 0.10\,\text{in./rev}$

2. **Sketch a Kinematic Diagram and Identify Degrees of Freedom**

A kinematic diagram of this mechanism is shown in Figure 12.7. By calculating the mobility of the mechanism, five links are identified. There are also four pin joints. Therefore,

$$n = 5 \quad j_p = 5\,(3\,\text{pins and 2 sliding joints}) \quad j_b = 0$$

and

$$M = 3(n - 1) - 2j_p - j_b = 3(5 - 1) - 2(5) - 0 = 12 - 10 = 2$$

With two degrees of freedom, both nuts must be driven. The screw configuration shown in Figure 12.6 does drive both nuts.

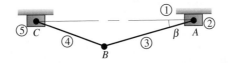

FIGURE 12.7 Kinematic diagram for Example Problem 12.2

3. **Determine Velocity of Nuts**

From Figure 12.6, the screw is free to rotate but is fixed against axial displacement. With the opposite-hand threads, the two nuts also move in opposite directions. Therefore, the relative velocity of the nut with respect to the screw equation (12.6) is the absolute velocity of each advancing nut. As the screw rotates with a velocity of 45 rpm, the nut advances at a rate of

$$v_{\text{nut/screw}} = L\omega_{\text{screw}} = (0.10 \text{ in./rev})(45 \text{ rev/min}) = 4.5 \text{ in./min}$$

Thus,

$$\mathbf{V}_A = 4.5 \text{ in./min} \leftarrow \quad \text{and} \quad \mathbf{V}_C = 4.5 \text{ in./min} \rightarrow$$

4. **Determine Velocity of the Plate**

A velocity equation can be written as

$$\mathbf{V}_B = \mathbf{V}_A +> \mathbf{V}_{B/A} = \mathbf{V}_C +> \mathbf{V}_{B/C}$$

A velocity diagram is formed from both velocity equations. Notice that because of symmetry, the displacement and velocity of B is vertical (Figure 12.8).

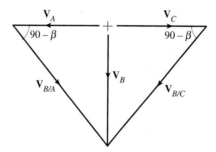

FIGURE 12.8 Velocity diagram for Example Problem 12.2.

Using trigonometry, the following relationship can be written:

$$\tan(90 - \beta) = \frac{v_B}{v_A}$$

For the case shown,

$$\beta = 25°$$

$$\mathbf{V}_B = v_A \tan(90 - \beta) = (4.5 \text{ in./min}) \tan(90° - 20°) = 12.4 \text{ in./min} \downarrow$$

12.7 SCREW FORCES AND TORQUES

The torque and force acting on a screw and nut assembly are shown in Figure 12.9.

The relationships between the force and torque have been derived [Ref. 2] and are a strong function of the coefficient of friction, μ, between the thread and nut. Friction was discussed in Section 12.5. A substantial amount of energy can be lost to friction when using a threaded mechanism.

The first case to study is one in which the motion of the nut occurs in the opposite direction from the applied force acting on a nut. This is commonly referred to as a case of lifting or pushing a load. The required torque to accomplish this motion is calculated as follows:

$$T = \left(\frac{Fd}{2}\right)\left[\frac{(L + \pi\mu d)}{(\pi d - \mu L)}\right] \qquad (12.8)$$

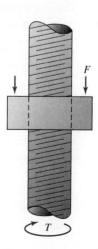

FIGURE 12.9 Force and torque on a screw.

where:
  F = magnitude of applied force on nut
  d = pitch diameter of threads
  L = lead of threads
  μ = coefficient of friction between the nut and threads

The second case to be studied is one in which the motion of the nut is in the same direction as the force acting on the nut. In essence, the load assists the motion of the nut. This is commonly referred to as a case of *lowering a load*. The required torque to accomplish this motion is as follows:

$$T = \left(\frac{Fd}{2}\right)\left[\frac{(\pi\mu d - L)}{(\pi d + \mu L)}\right] \quad (12.9)$$

An *efficiency*, e, can be defined as the percent of power that is transferred through the threads to the nut. It is the ratio of torque required to raise the load in the absence of friction to the torque required to raise a load with friction. Again, a closed-form equation has been derived for efficiency.

$$e = \left(\frac{L}{\pi d}\right)\left[\frac{(\pi d\cos\alpha - \mu L)}{(\pi\mu d + L\cos\alpha)}\right] \quad (12.10)$$

In addition to the previously defined quantities, the *included thread angle*, α, is used. This angle was illustrated in Figure 12.2. Standard values include

  Unified thread: $\alpha = 30°$
  Metric thread: $\alpha = 30°$
  Square thread: $\alpha = 0°$
  ACME thread: $\alpha = 14.5°$

Threaded screws typically have efficiencies between 20 and 50 percent. Thus, a substantial amount of energy is lost to friction. As opposed to threaded screws, ball screws have efficiencies in excess of 90 percent. For ball screws, the operational torque equations can be estimated as

$$T = 0.177\,FL \text{(To raise a load)} \quad (12.11)$$

$$T = 0.143\,FL \text{(To lower a load)} \quad (12.12)$$

EXAMPLE PROBLEM 12.3

A screw jack mechanism is shown in Figure 12.10. A belt/sheave is used to rotate a nut, mating with a single-thread, 1–5 ACME screw, to raise the jack. Notice that a pin is used in a groove on the screw to prevent the screw from rotating. The nut rotates at 300 rpm. Determine the lifting speed of the jack, the torque required, and the efficiency of the jack.

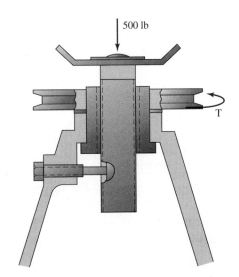

FIGURE 12.10 Jack for Example Problem 12.3.

SOLUTION: 1. *List Screw Properties*

A single-thread, 1×5 ACME screw has the following properties:

Number of teeth per inch: $n = 5$

Pitch: $p = \dfrac{1}{n} = \dfrac{1}{5} = 0.20$ in.

Number of threads: $N_t = 1$

Lead: $L = N_t\,p = 0.20$ in./rev

2. **Calculate Velocity of Screw**

In Figure 12.10, the nut is fixed from translation. Therefore, the velocity computed with equation (12.6) is that of the advancing screw. As the nut rotates with a velocity of 300 rpm, the thread advances through the nut at a rate of

$$\mathbf{V}_{screw} = L\omega_{nut} = (0.20 \text{ in./rev})(300 \text{ rev/min}) = 60 \text{ in./min} \uparrow$$

3. **Calculate Required Torque**

The torque required to raise the load is dependent on the estimated coefficient of friction between the threads and nut. Because this jack configuration is used in standard industrial settings, a coefficient of friction of 0.2 is assumed. In Table 12.3, the nominal pitch diameter for 1×5 ACME threads is 0.8726 in. Also, for all ACME threads, the included angle is 29°. Therefore, the torque can be computed from equation (12.8).

$$T = \left(\frac{Fd}{2}\right)\frac{(L + \pi\mu d)}{(\pi d - \mu L)}$$

$$= \frac{(500 \text{ lb})(0.8726 \text{ in.})}{2}\left\{\frac{[(0.20 + \pi(0.2)(0.8726)]}{[\pi(0.8726) - (0.2)(0.2)]}\right\} = 60.4 \text{ in.} - \text{lb}$$

4. **Calculate Efficiency**

Finally, efficiency can be computed from equation (12.10).

$$e = \left(\frac{L}{\pi d}\right)\left[\frac{(\pi d \cos\alpha - \mu L)}{(\pi\mu d + L\cos\alpha)}\right]$$

$$= \frac{(0.2)}{\pi(0.8726)}\left\{\frac{[\pi(0.8726)\cos(29) - (0.2)(0.2)]}{[\pi(0.2)(0.8726) + (0.2)(\cos 29°)]}\right\} = 0.24$$

An efficiency of 0.24 reveals that only 24 percent of the power transferred to the nut is delivered into lifting the weight. The remaining 76 percent is lost in friction. If these values are not acceptable, a ball screw could be substituted for the ACME thread. A ball screw has not only an efficiency of approximately 90 percent but also a significantly higher cost. However, recall that a ball screw is not self-locking and does not maintain the load at an elevated level.

12.8 DIFFERENTIAL SCREWS

A differential screw is a mechanism designed to provide very fine motions. Although they can be made in several forms, one common form is shown in Figure 12.11. This particular differential screw configuration consists of two different threads on the same axis and one sliding joint.

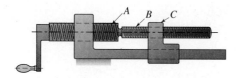

FIGURE 12.11 A differential screw.

In Figure 12.11, as the handle turns one revolution, thread A rotates one revolution and advances a distance equal to the lead of thread A. Of course, the motion of thread B is identical to thread A because it is machined onto the same shaft as A. Thus, thread B also rotates one revolution and advances a distance equal to the lead of thread A. As thread B rotates one revolution, nut C is retracted because it is unable

to rotate. Specifically, as thread B rotates one revolution, nut C is retracted a distance equal to the lead of thread B. However, because thread B already advances one revolution, the net motion of nut C is the difference between the lead of threads A and B. Thus, this screw arrangement with different leads is called the *differential screw*.

For differential screws, the kinematic relationships among the magnitude of the angular and the linear motion can be modified as follows:

$$\Delta R_{nut} = (L_A - L_B)\Delta\theta_{screw} \qquad (12.13)$$

$$v_{nut} = (L_A - L_B)\omega_{screw} \qquad (12.14)$$

$$a_{nut} = (L_A - L_B)\alpha_{screw} \qquad (12.15)$$

Again note that the lead, L, is specified as the nut displacement per revolution. Therefore, in this rare instance, the angular motion should be specified in revolutions.

When the leads of the two threads are close, small motions of the nut can be produced. This configuration is popular for fine adjustments of precision equipment at a relatively low cost.

EXAMPLE PROBLEM 12.4

A device that is intended to gauge the length of parts is shown in Figure 12.12. The concept utilizes a differential screw, such that the rotation of knob A slides nut D until it is firmly pressed against part E. Nut D also has a pointer that can be used to determine the length of part E. The objective is to configure the system such that one rotation of knob A causes a 0.1-mm traverse of nut D. Select threads B and C to accomplish this requirement.

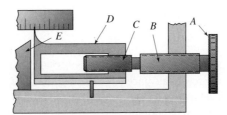

FIGURE 12.12 Measuring device for Example Problem 12.4.

SOLUTION: Use the following differential screw kinematics equation:

$$\Delta R_{\text{nut}} = (L_B - L_C)\,\Delta\theta_{\text{screw}}$$

$$0.1\text{ mm} = (L_B - L_C)(1\text{ rev})$$

$$(L_B - L_C) = 0.1\text{ mm/rev}$$

Several arrangements are possible. The standard threads listed in Table 12.2 are utilized. These have a single thread, so the lead and pitch are identical. Then two threads need to be selected that have a difference in pitch of 0.1 mm. Although a few options are feasible, arbitrarily select coarse pitch threads.

For thread B: M5 × 0.8
For thread C: M4 × 0.7

12.9 AUGER SCREWS

Many centuries ago, Archimedes ingeniously applied a screw mechanism to lifting water, which is now known as the "Archimedes Screw" (Figure 12.13). As the screw rotates, each thread of the screw transports a certain amount of water. With this screw mechanism, the mating nut is actually the fluid being transported. This form is still utilized today to transport many different types of material. Common applications include transporting molten plastic into molds, moving salt from dump trucks through spreaders for icy winter roads, digging fence post holes in soil, and moving cattle feed through long troughs. This screw mechanism is more commonly referred to as an *auger*.

FIGURE 12.13 Archimedes screw.

The kinematic equations presented in equations (12.16), (12.17), and (12.18) can be used to determine the motion of the material being transported, given the motion of the auger. Consistent with standard screws, a pitch or lead of an auger blade is defined. The volumetric transport rate is then a function of the clearance between auger blades, which traps the material being transported. This can be mathematically written as

Volume through auger
= (volume trapped between auger blades) $(L_{\text{auger}})\,\Delta\theta_{\text{screw}}$
(12.16)

Volumetric flow through auger =
(volume trapped between auger blades) $(L_{\text{auger}})\omega_{\text{screw}}$
(12.17)

Volumetric acceleration through auger =
(volume trapped between auger blades) $(L_{\text{auger}})\alpha_{\text{screw}}$
(12.18)

PROBLEMS

Screw Thread Geometry

12–1. Compute the lead, and lead angle, of a ¼ –20 UNC thread. Also determine whether it is self-locking when the thread is of general machined quality.

12–2. Compute the lead, and lead angle, of a ¼ –28 UNC thread. Also determine whether it is self-locking when the thread is of general machined quality.

12–3. Compute the lead, and lead angle, of an M16 × 2.0 thread. Also determine whether it is self-locking when the thread is of general machined quality.

12–4. Compute the lead, and lead angle, of a 1– ⅛ –5 ACME thread. Also determine whether it is self-locking when the thread is of general machined quality.

Screw-Driven Displacement

12–5. A stick deodorant package utilizes a screw to advance and retract the stick. The thumb wheel rotates a standard ¼ –20 thread that moves the nut and deodorant stick. Determine the distance that the stick advances when the thumb wheel is rotated 3 revolutions.

12–6. A tension-testing machine is shown in Figure P12.6. A single-thread, 2–4 ACME screw moves the nut. Determine the displacement of the raising ram when the screw is rotated 10 revolutions.

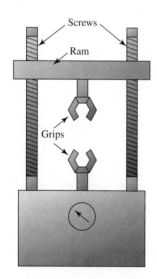

FIGURE P12.6 Problems 6, 18, and 25.

12–7. A garage door opener is shown in Figure P12.7. A single-thread, 1–5 ACME screw moves the nut. Determine the displacement of the bottom of the lowering door when the screw is rotated 10 revolutions.

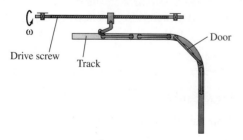

FIGURE P12.7 Problems 7, 19, and 26.

12–8. An adjustable work table is shown in Figure P12.8. The input shaft is coupled, through a set of bevel gears, to a nut. The nut rotates, pushing a screw up and down. The bevel gears have a ratio of 5:1. The screw has a ½ –13 UNC thread. Determine the displacement of the raising table when the input shaft is rotated 10 revolutions.

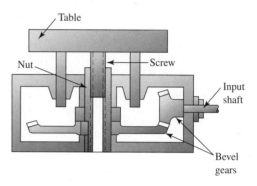

FIGURE P12.8 Problems 8, 20, and 27.

12–9. A screw-operated press is shown in Figure P12.9. The screw has a single ½ –10 ACME thread, both in right-hand and left-hand orientation, as shown. The press is initially configured with $\beta = 25°$. Graphically determine the displacement of the lowering pressure plate when the crank is rotated 20 revolutions.

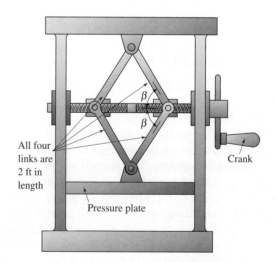

FIGURE P12.9 Problems 9–11 and 21.

12–10. The press described in Problem 12–9 is initially configured with $\beta = 45°$. Graphically determine the displacement of the lowering pressure plate when the crank is rotated 15 revolutions.

12–11. The press described in Problem 12–9 is initially configured with $\beta = 65°$. Graphically determine the displacement of the lowering pressure plate when the crank is rotated 30 revolutions.

12–12. A ¾ –10 UNC threaded rod drives a platform as shown in Figure P12.12. Graphically determine the displacement of the lowering platform when the crank is rotated 12 revolutions.

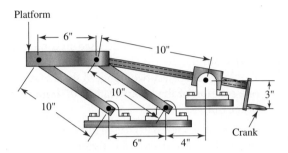

FIGURE P12.12 Problems 12, 13, and 22.

12–13. For the platform described in Problem 12–12, graphically determine the displacement of the raising platform when the crank is rotated 8 revolutions.

12–14. The motor shown in Figure P12.14 rotates a ¾ –10 UNC threaded rod to a tilt platform used for flipping crates. Graphically determine the angular displacement of the raising table and the linear displacement of the top edge when the motor rotates 25 revolutions.

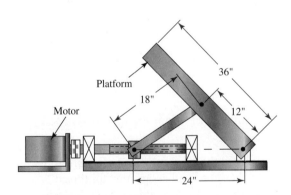

FIGURE P12.14 Problems 14, 15, 23, and 28.

12–15. For the platform described in Problem 12–14, graphically determine the angular displacement of the lowering table and the linear displacement of the top edge when the motor rotates 15 revolutions.

12–16. The height and angle of the drawing table, shown in Figure P12.16, is adjusted by rotating the crank. The crank rotates a screw, moving the nut and altering the L distance. The screw has a ⅞ –14 UNF thread. The table is initially configured with $L = 9$ in.

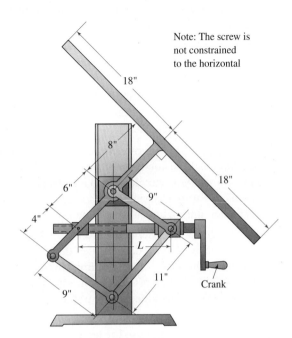

FIGURE P12.16 Problems 16, 17, 24.

Graphically determine the displacement of the top and bottom edges of the raising table when the crank is rotated 5 revolutions.

12–17. The drawing table, shown in Figure P12.16, is initially configured with $L = 8$ in. The screw has a ⅞ –14 UNF thread. Graphically determine the displacement of the top and bottom edges of the lowering table when the crank is rotated 30 revolutions.

Screw-Driven Velocity

12–18. The screw in the tension-testing machine described in Problem 12–6 rotates at 40 rpm, lowering the ram. Determine the linear velocity of the ram.

12–19. The screw in the garage door opener described in Problem 12–7 rotates at 1200 rpm, opening the door. Determine the linear velocity of the bottom of the door.

12–20. The input shaft of the work table described in Problem 12–8 rotates at 600 rpm, raising the table. Determine the linear velocity of the table.

12–21. The screw in the press described in Problem 12–9 rotates at 45 rpm, lowering the pressure plate. The press is configured with $\beta = 25°$. Determine the linear velocity of the pressure plate.

12–22. The crank in the platform described in Problem 12–12 rotates at 30 rpm, raising the platform. Determine the linear velocity of the platform.

12–23. The motor in the tilt platform mechanism described in Problem 12–14 rotates at 1800 rpm, raising the platform. Determine the angular velocity of the platform.

12–24. The crank of the drawing table described in Problem 12–16 rotates at 20 rpm, lowering the table. The table is initially configured with $L = 9$ in. Determine the velocity of the top and bottom edges of the table.

Screw-Driven Acceleration

12–25. The screw in the tension-testing machine described in Problem 12–6 rotates at 40 rpm, lowering the ram. The motor is shut down and it will take 1.7 s to completely stop. Determine the linear acceleration of the ram during the shut-down period.

12–26. The garage door opener described in Problem 12–7 is activated to open the door. It takes the motor 0.7 s to achieve the steady-state running speed of 1200 rpm. At the instant of activation, determine the linear acceleration of the bottom of the door.

12–27. The motor for the work table described in Problem 12–8 is turned on to raise the table. It takes the 20 revolutions for the input shaft to achieve its steady-state running speed of 1200 rpm. Determine the linear acceleration of the table.

12–28. The motor in the tilt platform mechanism described in Problem 12–14 rotates at 900 rpm, raising the platform, then is shut down. It will take 4 revolutions to completely stop. Determine the angular velocity and acceleration of the platform.

Screw Force and Torque

12–29. A ½ – in. standard ACME thread is used on a C-clamp. This thread is of general machine quality with minimal lubricant. For the clamp to exert a 500-lb force on the materials being clamped together, determine the torque required.

12–30. Estimate the efficiency of the C-clamp described in Problem 12–29.

12–31. A jack uses a double-thread ACME thread with a major diameter of 25 mm and a pitch of 5 mm. The jack is intended to lift 4000 N. Determine
    the lead angle,
    whether the jack is self-locking,
    the torque to raise the load,
    the torque to lower the load, and
    the efficiency of the jack.

12–32. A jack uses a double-thread 1–5 ACME thread. The jack is intended to lift 2000 lb. Determine
    the lead angle,
    whether the jack is self-locking,
    the torque to raise the load,
    the torque to lower the load, and
    the efficiency of the jack.

12–33. For the table in Problem 12–8, the thread is of general machine quality with minimal lubricant. The table supports 75 lb; determine the torque that must be transferred to the nut.

12–34. Estimate the efficiency of the screw used in the table described in Problem 12–16.

Differential Screws

12–35. A differential screw is to be used in a measuring device similar to the one described in Example Problem 12.4. Select two standard threads such that one rotation of the knob creates a 0.5-mm traverse of the nut.

12–36. A differential screw is to be used in a measuring device similar to the one described in Example Problem 12.4. Select two standard threads such that one rotation of the knob creates a 0.25-in. traverse of the nut.

12–37. A differential screw is to be used in a measuring device similar to the one described in Example Problem 12.4. Select two standard threads such that one rotation of the knob creates a 0.05-in. traverse of the nut.

CASE STUDIES

12–1. The device shown in Figure C12.1 utilizes a screw mechanism. Carefully examine the components of the mechanism, then answer the following leading questions to gain insight into the operation.

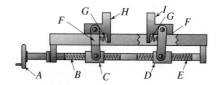

FIGURE C12.1 (Courtesy, Industrial Press.)

1. What is the hand designation for thread *B*?
2. What is the hand designation for thread *E*?
3. When handle *A* rotates counterclockwise, what is the motion of nut *C*?
4. When handle *A* rotates counterclockwise, what is the motion of slide *H*?
5. When handle *A* rotates counterclockwise, what is the motion of nut *D*?
6. When handle *A* rotates counterclockwise, what is the motion of slide *I*?
7. What is the function of both links labeled *F*?
8. What is component *G* and what is its function?
9. What is the function of this device, and what would you call it?

12–2. The device shown in Figure C12.2 utilizes a screw mechanism. Carefully examine the components of

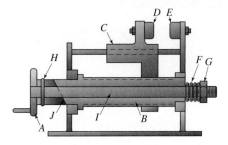

FIGURE C12.2 (Courtesy, Industrial Press.)

the mechanism, then answer the following leading questions to gain insight into the operation.

1. As handwheel *A* rotates counterclockwise, what is the motion of shaft *I*?

2. What is the function of pin *H*?
3. As handwheel *A* rotates counterclockwise, what is the motion of threaded sleeve *B*?
4. What actually couples the motion of shaft *I* and threaded sleeve *B*?
5. Threaded sleeve *B* has right-hand threads; as handwheel *A* rotates counterclockwise, what is the motion of nut *C*?
6. What happens to this device when pads *D* and *E* make contact?
7. What component is *F* and what is its function?
8. What is the function of this device?
9. What would happen if nut *G* were tightened?
10. What would happen to this device if the interface *J* were designed with a more vertical slope?

STATIC FORCE ANALYSIS

13.1 INTRODUCTION

The general function of any machine is to transmit motion and forces from an actuator to the components that perform the desired task. Consider an escalator used in many commercial buildings: Electrical power is fed into motors, which drive mechanisms that move and fold the stairs. The task is to safely and efficiently move people up and down multilevel buildings.

Up to this point in the book, the sole focus was on the motion of a machine. This chapter, and the next, is dedicated to an introduction to machine forces. A critical task in the design of machines is to ensure that the strength of the links and joints is sufficient to withstand the forces imposed on them. Therefore, a full understanding of the forces in the various components of a machine is vital.

As mentioned in Chapter 7, an inertial force results from any accelerations present in a linkage. This chapter deals with force analysis in mechanisms without accelerations, or where the accelerations can be neglected. This condition is termed *static equilibrium*. Static equilibrium is applicable in many machines where the changes in

movement are gradual or the mass of the components is negligible. These include clamps, latches, support linkages, and many hand-operated tools, such as pliers and cutters.

The following chapter deals with force analysis in mechanisms with significant accelerations. In many high-speed machines, the inertial forces created by the motion of a machine exceed the forces required to perform the intended task. This condition is termed *dynamic equilibrium*. The analysis of dynamic equilibrium uses many concepts of static equilibrium. Thus, static equilibrium (Chapter 13) is presented before proceeding to dynamic equilibrium (Chapter 14).

13.2 FORCES

A *force,* **F**, is a vector quantity that represents a pushing or pulling action on a part. Pulling a child in a wagon implies that a force (pulling action) is applied to the handle of the wagon. Being a vector, this force is defined by a magnitude F, and a direction of the pulling action. In the U.S. Customary System, the common unit for the magnitude of a force is the avoirdupois pound or simply pound (lb). In the International System, the primary unit used is the Newton (N).

One of the most common operations is the determination of the net effect of several forces. Two or more forces that are applied to a part can be combined to determine the resulting effect of the forces. Combining forces to find a resultant is a procedure identical to adding displacement, velocity, or acceleration vectors. This was presented in Sections 3.9 and 3.11. Conversely, one force can be broken into two components along the orthogonal axis. This was presented in Section 3.10. Often, utilizing the components of a force, along a set of convenient axes, facilitates analysis. Being vector quantities, forces can be manipulated through all the methods illustrated in Chapter 3.

13.3 MOMENTS AND TORQUES

A moment, or torque, is the twisting action produced by a force. Pushing on the handle of a wrench produces an action that tends to rotate a nut on a bolt. Thus, the force causes a twisting action around the center of a bolt. This resulting action is termed a *moment* or *torque*. Figure 13.1 illustrates such a force, causing a twisting action.

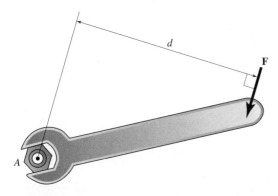

FIGURE 13.1 The definition of a moment or torque.

Where F is the magnitude of the force, d is the perpendicular distance between the force and a reference point A. Moments are expressed in the units of force multiplied by distance. In the U.S. Customary System, the common units for moments are inch-pound (in. lb) or foot-pound (lb-ft). In the International System, the common units used are Newton-millimeters (Nmm) or Newton-meters (Nm).

The moment of a force not only has a magnitude but also a direction, depending on the relative position of the force and the reference point. The direction of a moment, or twisting action, relative to the reference point is simply designated as clockwise or counterclockwise. This direction is consistent with the twisting direction of the force around the reference point. The twisting action of the force illustrated in Figure 13.1, relative to the nut, is a clockwise moment. Moments are conventionally considered positive when acting counterclockwise and negative when acting clockwise.

The difference between a moment and a torque is subtle. A moment is any twisting action of a force. A torque is a specific type of a moment. In machine applications, a torque is any moment where the point of reference is at the center of a shaft or other pin-type connection.

In planar mechanics, a moment is a property that is stated relative to a reference point. For the wrench in Figure 13.1, the objective of the force is to deliver a twisting action to the nut. Therefore, a reference point is appropriately placed at the center of the nut, point A. The magnitude of a moment relative to point A, M_A, created by a force can be calculated as

$$M_A = (F)(d) \qquad \textbf{(13.1)}$$

EXAMPLE PROBLEM 13.1

A mechanism to automatically open a door exerts a 37-lb force on the door, applied in a direction as shown in Figure 13.2. Determine the moment, relative to the pivot of the door, created by the force.

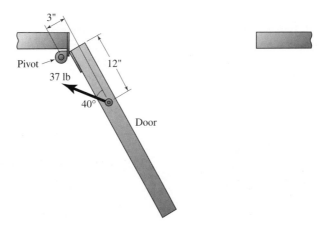

FIGURE 13.2 Door for Example Problem 13.1.

SOLUTION: 1. *Calculate Perpendicular Distance*

The moment can be computed from equation (13.1). Although the force is given, the geometry of the door must be examined to determine the perpendicular distance, d. The geometry has been isolated and broken into two triangles in Figure 13.3. Notice that both triangles were constructed to be right triangles. The common side to the two triangles, labeled as side c, can be determined with the known data for the upper triangle. From the Pythagorean theorem,

$$c = \sqrt{(12 \text{ in.})^2 + (3 \text{ in.})^2} = 12.37 \text{ in.}$$

The included angle, β, can also be found from the trigonometric relations.

$$\beta = \tan^{-1}\left(\frac{3 \text{ in.}}{12 \text{ in.}}\right) = 14.0°$$

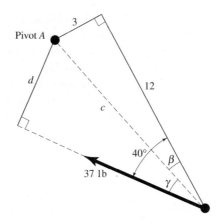

FIGURE 13.3 Door geometry for Example Problem 13.1.

Focusing on the lower triangle, the included angle, γ, can be found because the total angle was given as 40°; thus,

$$\gamma = 40° - \beta = 26°$$

Finally, the perpendicular distance can be determined from the trigonometric relations of the lower triangle.

$$d = c\sin(\gamma) = (12.37\,\text{in.})\sin(26°) = 5.42\,\text{in.}$$

2. **Calculate Moment**

    The moment, relative to the pivot A, is calculated from equation (13.1). The direction is consistent with the twisting action of the force relative to the pivot A, which in this case is clockwise.

    $$M_A = F(d) = 37\,\text{lb}\,(5.42\,\text{in.}) = 200.5\,\text{in.\,lb, cw}$$

ALTERNATIVE SOLUTION:

1. **Resolve Force into Rectangular Components**

    In the previous solution, the calculation of a perpendicular distance was rather complex. An alternative solution can be used that involves defining a convenient coordinate system that is aligned with the given dimensions. The components of the original force $\mathbf{F}$ are identified as $\mathbf{F}^1$ and $\mathbf{F}^2$ and shown in Figure 13.4.

    The magnitude of $\mathbf{F}^1$ and $\mathbf{F}^2$ can be computed as

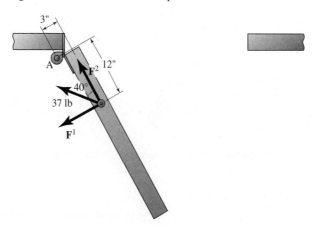

FIGURE 13.4 Force components for Example Problem 13.1.

$$\mathbf{F}^1 = (37\,\text{lb})\sin 40° = 23.8\,\text{lb}$$

$$\mathbf{F}^2 = (37\,\text{lb})\cos 40° = 28.3\,\text{lb}$$

2. **Identify Perpendicular Distance for Each Component**

    Notice that both of these components cause a moment relative to point A. However, the perpendicular distance for each moment is apparent. From Figure 13.4, it is seen that the perpendicular distance for $\mathbf{F}^1$ and $\mathbf{F}^2$ is 12 in. and 3 in., respectively. Also notice that $\mathbf{F}^1$ causes a clockwise turning action around point A. The moment caused by $\mathbf{F}^2$ is counterclockwise. The traditional sign convention is to assign counterclockwise moments a positive value.

3. **Calculate Moment**

Calculating the moment relative to point A,

$$\mathbf{M_A} = -\mathbf{F^1}(12\,\text{in.}) + \mathbf{F^2}(3\,\text{in.})$$

$$= -[(23.8\,\text{lb})(12\,\text{in.})] + [(28.3\,\text{lb})(3\,\text{in.})]$$

$$= -200.5\,\text{in.\,lb}$$

$$= 200.5\,\text{in.\,lb, cw}$$

13.4 LAWS OF MOTION

Sir Isaac Newton developed three laws of motion that serve as the basis of all analysis of forces acting on machines and components. These laws are stated as follows:

FIRST LAW: Every object remains at rest, or moves with constant velocity, unless an unbalanced force acts upon it.

SECOND LAW: A body that has an unbalanced force has

a. Acceleration that is proportional to the force,

b. Acceleration that is in the direction of the force, and

c. Acceleration that is inversely proportional to the mass of the object.

THIRD LAW: For every action, there is an equal and opposite reaction.

All of these laws are utilized in the study of mechanisms. However, in this chapter dealing with static force analysis, only the first and third laws are applicable. The following chapter incorporates the second law into the analysis.

13.5 FREE-BODY DIAGRAMS

To fully understand the safety of a machine, all forces that act on the links should be examined. It is widely accepted that the best way to track these forces is to construct a free-body diagram. A *free-body diagram* is a picture of the isolated part, as if it were floating freely. The part appears to be floating because all the supports and contacts with other parts have been removed. All these supports and contacts are then replaced with forces that represent the action of the support. Thus, a free-body diagram of a part shows *all* the forces acting on the part.

13.5.1 Drawing a Free-Body Diagram

Figure 13.5 illustrates a free-body diagram of an isolated link. Notice that this part is designated as link 3. It is essential that all forces are shown on the free-body diagram. A convenient notation is to label the forces consistent with the link number that is being acted upon and the link number that is

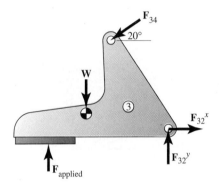

FIGURE 13.5 Free-body diagram.

driving the action. Thus, a force designated as $\mathbf{F}_{34}$ is a force on link 3 from the contact of link 4.

Because forces are vectors, determination of a force requires knowledge of the magnitude and direction of that force. If the direction of a force is known, it should be indicated on the free-body diagram. This is the case for $\mathbf{F}_{34}$ in Figure 13.5. When the direction of a force is not known, it is common to draw two orthogonal components of the unknown force. These two components represent the two items that need to be determined for full understanding of the force. Notice that this is the case for $\mathbf{F}_{32}$ in Figure 13.5.

The following steps can assist in systematically drawing a free-body diagram:

I. Isolate the component(s) that must be studied.

II. Draw the component as if it were floating freely in space by removing all visible supports and physical contact that it has with other objects.

III. Replace the supports, or physical contacts, with the appropriate force and/or moments, which have the same effect as the supports.

13.5.2 Characterizing Contact Forces

Establishing the supporting forces takes some care. As a general rule, if the nature of the contact prevents motion in a certain direction, there must be a supporting force in that

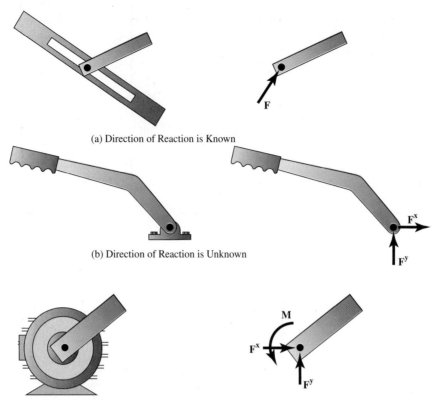

(a) Direction of Reaction is Known

(b) Direction of Reaction is Unknown

(c) Reaction Prohibits Translation and Rotation

FIGURE 13.6 Reaction forces.

direction. The types of reactions can be divided into three groups corresponding to the type of physical contacts.

a. *Direction of Reaction is Known:* Components in this group include rollers, sliders, pins in slots, and cables. Each of these supports can prevent motion in only one direction. Reactions in this group involve only one unknown, namely the magnitude of the reaction force. Figure 13.6a illustrates this type of contact.

b. *Direction of Reaction is Unknown:* Components in this group include frictionless pins, hinges, and sliders on rough surfaces. Each of these supports can prevent

translation in both planar directions. Reactions in this group involve two unknowns, usually shown as the x- and y-components of the reaction force. Figure 13.6b illustrates this type of contact.

c. *Reaction Prohibits Rotation:* Components in this group include fixed supports and pin joints at an actuator (motor or engine). Each of these supports can prevent translation in both planar directions and free rotation. Reactions in this group involve three unknowns, usually shown as the x- and y-components of the reaction force and a reaction moment. Figure 13.6c illustrates this type of contact.

EXAMPLE PROBLEM 13.2

An engine hoist is shown in Figure 13.7. The engine being raised weighs 250 lb. Draw a free-body diagram of the entire hoist.

SOLUTION: In order to construct a free-body diagram of the entire hoist, it first must be isolated and drawn as if it were floating freely in space. This is done by removing the floor, as it is the only body that supports the hoist. The engine is also removed, as it is not an integral part of the hoist.

Once the hoist is redrawn without the engine and floor, reaction forces must be placed at the contact points of the removed items. First, because the engine weighs 250 lb, a force with a known magnitude and direction replaces the effect of the engine.

Second, the action of the floor must be replicated. The front roller falls into case (a), where the direction of the reaction force is known. Any roller on a smooth surface prevents translation in a direction perpendicular to the smooth surface. The reaction at the front roller is labeled $\mathbf{F}_{21A}$.

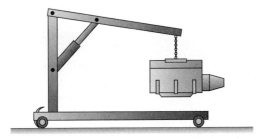

FIGURE 13.7 Engine hoist for Example Problem 13.2.

The rear roller differs in that a braking device is implemented. In addition to exhibiting a vertical reaction force, the wheel-and-brake configuration also prevents translation along the floor. Therefore, the reaction at the rear roller has both x- and y-components. The reactions at the rear roller are labeled $\mathbf{F}^x_{21B}$ and $\mathbf{F}^y_{21B}$. A completed free-body diagram is shown in Figure 13.8.

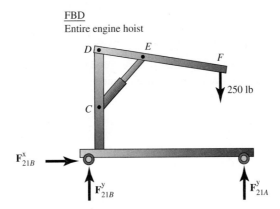

FBD
Entire engine hoist

FIGURE 13.8 Free-body diagram for Example Problem 13.2.

13.6 STATIC EQUILIBRIUM

Newton's first law applies to all links that are at rest or moving at constant velocity; thus, the condition is referred to as *static equilibrium*. For an object to be in static equilibrium, the following two necessary and sufficient conditions must be met:

Condition I:
The combination, or resultant, of all external forces acting on the object is equivalent to zero and does not cause it to translate. Mathematically, the first condition of equilibrium can be summarized as

$$\Sigma \mathbf{F} = 0 \qquad (13.2)$$

This condition indicates that all the external forces acting on the component are balanced. The symbol Σ implies the summation of all forces acting on a free-body diagram. As introduced in Chapter 3, forces are vectors and equation (13.2) can be written as

$$\mathbf{F}_1 +> \mathbf{F}_2 +> \mathbf{F}_3 +> \ldots +> \mathbf{F}_N = 0$$

All of the methods for vector manipulation that were introduced in Chapter 3 can be used with this vector equation to solve for unknown forces. Either graphical or analytical methods can be used, but force analysis is typically better suited for analytical methods. Therefore, the first

condition of static equilibrium can be resolved into components, yielding two algebraic equations.

$$\Sigma \mathbf{F}^x = 0 \qquad (13.3)$$

$$\Sigma \mathbf{F}^y = 0 \qquad (13.4)$$

Condition II:
The moment due to any external force is canceled by the moments of the other forces acting on the object and do not cause it to rotate about any point. The second condition of equilibrium can be mathematically summarized as

$$\Sigma M_A = 0 \qquad (13.5)$$

This condition indicates that all the moments acting on the component are balanced. The location of point A is arbitrary.

13.7 ANALYSIS OF A TWO-FORCE MEMBER

A special case of equilibrium, which is of considerable interest, is that of a member that is subjected to only two forces. This type of machine component is termed a *two-force member*. Many mechanism links, particularly couplers and connecting rods, are two-force members. A two-force member is shown in Figure 13.9.

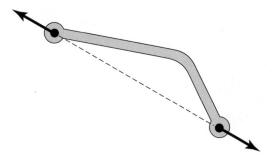

FIGURE 13.9 Two-force member.

In order for a two-force member to be in equilibrium the two forces must:

1. Have the same magnitude,

2. Act along the same line, and

3. Be opposite in sense.

Because the two forces must act along the same line, the only line that can satisfy this constraint is the line that extends between the points where the forces are applied. Thus, a link with only two forces simply exhibits either tension or compression.

This fact can be extremely useful in force analysis. When the locations of the forces are known, the direction of the forces are defined. When the magnitude and sense of a single force are known, the other force's magnitude and sense can be immediately determined. Thus, the analysis of a two-force member is simple.

EXAMPLE PROBLEM 13.3

A novelty nutcracker is shown in Figure 13.10. A force of 5 lb is applied to the top handle, as shown, and the mechanism does not move (static). Draw a free-body diagram and determine the forces on each link. For this analysis, the weight of each link can be considered negligible.

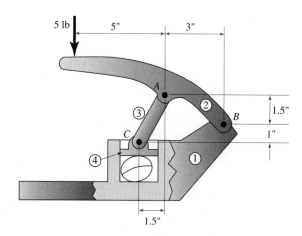

FIGURE 13.10 Nutcracker for Example Problem 13.3.

SOLUTION: 1. **Sketch the Free-Body Diagrams for the Mechanism Links**

Notice that link 3 *(AC)* is a simple link, containing only two pin joints. In addition, no other force is acting on this link. Thus, it is a two-force member and the forces acting on the link must be equal and along the line that connects the two pins. The free-body diagram for link 3 is shown as Figure 13.11a. As stated previously, the notation used is that F_{32} is a force applied to link 3 as a result of contact from link 2.

Being a two-force member, the direction of the two forces, F_{34} and F_{32}, is along the line that connects the two pins. The angle of inclination, θ, of this line can be determined.

$$\theta = \tan^{-1}\left(\frac{2.5}{1.5}\right) = 59.0°$$

Link 2 is also a simple link that contains only two pin joints; however, an additional force is applied to the handle. Thus, this link is not a two-force member. Newton's third law stipulates that a force that is acting at *A* will be equal and opposite to F_{32}. Thus, the direction of F_{23} is known as a result of Figure 13.11a. The general pin joint at point *B* dictates that two reaction forces will be used. The free-body diagram for link 2 is shown as Figure 13.11b.

Link 4 has sliding contact with link 1. Neglecting any friction force, this contact force will act perpendicular to the contact surface. The contact force from the nut itself will similarly act perpendicular to the mating

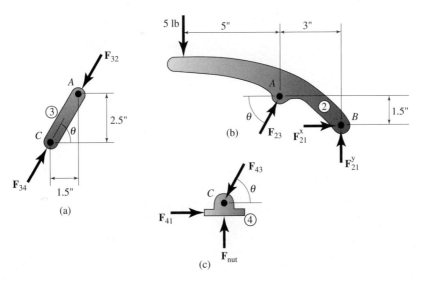

FIGURE 13.11 Free-body diagrams for Example Problem 13.3.

surface. Also, Newton's third law stipulates that a force acting at B will be equal and opposite to $\mathbf{F}_{34}$. Thus, the direction of $\mathbf{F}_{43}$ is known as a result of Figure 13.11a. The free-body diagram for link 4 is shown as Figure 13.11c.

2. **Solve the Equilibrium Equations for Link 2**

Link 2 is examined first because it contains the applied force. The three unknown forces on this link (Figure 13.11b) are solved by using the three equilibrium equations.

$\xrightarrow{+}\ \Sigma \mathbf{F}^x = 0$:

$$F_{23}\cos 59.0° + \mathbf{F}_{21}^x = 0$$

$+\uparrow\ \Sigma \mathbf{F}^y = 0$:

$$F_{23}\sin 59.0° + \mathbf{F}_{21}^y - 5\,\text{lb} = 0$$

$+\!\!\curvearrowleft\ \Sigma M_B = 0$:

$$(5\,\text{lb})(8\,\text{in.}) - (\mathbf{F}_{23}\cos 59.0°)(1.5\,\text{in.}) - (\mathbf{F}_{23}\sin 59.0°)(3\,\text{in.}) = 0$$

Solving the three equations yields

$$F_{23} = +11.96\,\text{lb} = 11.96\ \text{lb}\ \nearrow 59°$$

$$\mathbf{F}_{21}^x = -6.16\,\text{lb} = 6.16\ \text{lb}\leftarrow$$

$$\mathbf{F}_{21}^y = -5.25\,\text{lb} = 5.25\ \text{lb}\downarrow$$

3. **Solve the Equilibrium Equations for Link 3**

Because link 3 is a two-force member (Figure 13.11a), the equilibrium equations dictate that the forces have the same magnitude, act along the same line, and are opposite in sense. Of course, Newton's third law dictates that $\mathbf{F}_{32} = \mathbf{F}_{23}$. Thus, the forces acting on link 3 are

$$F_{32} = 11.96\,\text{lb}\ \overline{59°}\!\swarrow$$

$$F_{34} = 11.96\,\text{lb}\ \nearrow\overline{59°}$$

4. **Solve the Equilibrium Equations for Link 4**

The free-body diagram of link 4 (Figure 13.11c) will reveal the force exerted on the nut. Of course, Newton's third law dictates that $\mathbf{F}_{34} = \mathbf{F}_{43}$. Because the forces on link 4 all converge at a point, the moment equation of equilibrium does not apply. The two unknown forces on this link are solved by using two component equilibrium equations.

$\xrightarrow{+} \Sigma \mathbf{F}^x = 0:$

$$\mathbf{F}_{41} - \mathbf{F}_{43} \cos 59.0° = 0$$

$+\uparrow \Sigma \mathbf{F}^y = 0:$

$$\mathbf{F}_{\text{nut}} - \mathbf{F}_{43} \sin 59.0° = 0$$

Solving the two equations yields

$$\mathbf{F}_{41} = +6.16 \, \text{lb} = 6.16 \, \text{lb} \rightarrow$$

$$\mathbf{F}_{\text{nut}} = +10.25 \, \text{lb} = 10.25 \, \text{lb} \uparrow$$

EXAMPLE PROBLEM 13.4

Figure 13.12 shows a mechanism used to crush rocks. The 60-mm mechanism crank is moving slowly, and inertial forces can be neglected. In the position shown, determine the torque required to drive the 60-mm crank and crush the rocks.

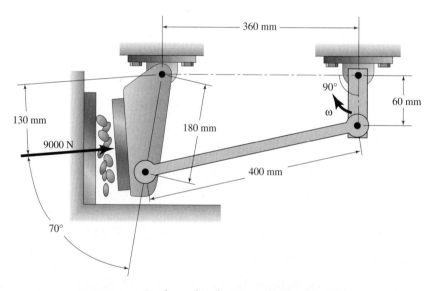

FIGURE 13.12 Rock crusher for Example Problem 13.4.

SOLUTION: 1. *Sketch Free-Body Diagrams for the Mechanism Links*

Using trigonometry to determine the internal angles of this four-bar mechanism can be completed as discussed in Chapter 4. An alternative approach is to construct the kinematic diagram using CAD. The internal angles were measured and the results are shown in Figure 13.13. Notice that link 3 (*BC*) is a simple link, containing only two pin joints. In addition, no other force is acting on this link. Thus, it is a two-force member and the forces acting on the link must be equal and along the line that connects the two pins. The free-body diagram for link 3 is shown as Figure 13.14a. As stated previously, the notation used is that $\mathbf{F}_{32}$ is a force applied to link 3 as a result of contact from link 2.

Link 2 is also a simple link that contains only two pin joints; however, a drive torque is applied to the link at the shaft (point *A*). Thus, this link is not a two-force member. Newton's third law stipulates that a force that is acting on link 2 at point *B* will be equal and opposite to $\mathbf{F}_{32}$. Thus, the direction of $\mathbf{F}_{23}$ is known as a result of Figure 13.14c. The general pin joint at point *A* dictates that two reaction forces will be used. The free-body diagram for link 2 is shown as Figure 13.14b.

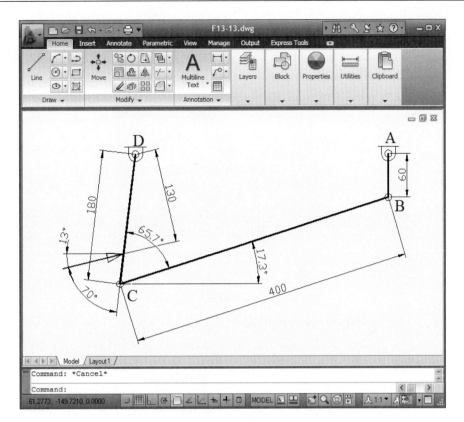

FIGURE 13.13 CAD layout for Example Problem 13.4.

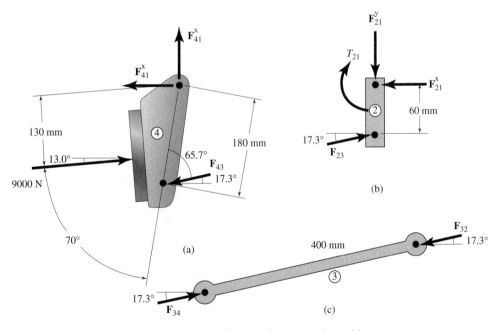

FIGURE 13.14 Free-body diagrams for Example Problem 13.4.

Link 4 also contains two pin joints but is not a two-force member. The rock-crushing force is applied at a third point on the link. The angle of this crushing force, from the horizontal, can be calculated from the angles shown in Figure 13.13. Aligning angles along the horizontal gives

$$180° - 97.0° - 70° = 13.0°$$

Also, Newton's third law stipulates that a force that is acting onto link 4 at point C will be equal and opposite to F_{34}. The general pin joint at point D dictates that two reaction forces must be used. The free-body diagram for link 4 is shown as Figure 13.14a.

2. ***Solve the Equilibrium Equations for Link 4***

Link 4 is examined first because it contains the applied force. The three unknown forces on this link (Figure 13.11a) are solved by using the three equilibrium equations.

$\overset{+}{\rightarrow}$ $\Sigma F^x = 0$:

$$(9000 \text{ N}) \cos 13.0° - F_{43} \cos 17.3° - F_{41}^x = 0$$

$+\uparrow$ $\Sigma F^y = 0$:

$$(9000 \text{ N}) \sin 13.0° - F_{43} \sin 17.3° + F_{41}^y = 0$$

$+\circlearrowleft$ $\Sigma M_D = 0$:

$$(9000 \text{ N})(130 \text{ mm}) - (F_{43} \sin 65.7°)(180 \text{ mm}) = 0$$

Solving the three equations yields

$$F_{43} = +7132 \text{ N} = 7132 \text{ N} \quad \overline{17.3°}\!\!\diagdown$$

$$F_{41}^x = +1960 \text{ N} = 1960 \text{ N} \leftarrow$$

$$F_{41}^y = +96.3 \text{ N} = 96.3 \text{ N} \uparrow$$

3. ***Solve the Equilibrium Equations for Link 3***

Because link 3 is a two-force member (Figure 13.14c), the equilibrium equations dictate that the forces have the same magnitude, act along the same line, and are opposite in sense. Of course, Newton's third law dictates that $F_{34} = F_{43}$. Thus, the forces acting on link 3 are

$$F_{34} = 7132 \text{ N} \quad \diagup\!\!\overline{17.3°}$$

$$F_{32} = 7132 \text{ N} \quad \overline{17.3°}\!\!\diagdown$$

4. ***Solve the Equilibrium Equations for Link 4***

The free-body diagram of link 2 (Figure 13.14b) will reveal the instantaneous torque required to operate the device. Of course, Newton's third law dictates that $F_{23} = F_{32}$.

$\overset{+}{\rightarrow}$ $\Sigma F^x = 0$:

$$-F_{21}^x + F_{23} \cos 17.3° = 0$$

$+\uparrow$ $\Sigma F^y = 0$:

$$-F_{21}^y + F_{23} \sin 17.3° = 0$$

$+\circlearrowleft$ $\Sigma M^A = 0$:

$$-T_{21} + (F_{23} \cos 17.3°)(60 \text{ mm}) = 0$$

Solving the three equations yields

$$F_{21}^x = +6809 \text{ N} = 6809 \text{ N} \leftarrow$$

$$F_{21}^y = +2121 \text{ N} = 212 \text{ N} \downarrow$$

$$T_{21} = +408{,}561 \text{ Nmm} = 409 \text{ Nm, cw}$$

Because the torque is the desired value, solving only the moment equation was necessary.

13.8 SLIDING FRICTION FORCE

As stated in Section 13.5, a contact force, as a result of a sliding joint, always acts perpendicular to the surface in contact. This contact force is commonly referred to as a *normal force* because it acts perpendicular to the surfaces in contact.

When friction cannot be neglected in machine analysis, an additional force, *friction force*, F_f, is observed. Friction always acts to impede motion. Therefore, a friction force acts on a sliding link, perpendicular to the normal force, and in a direction opposite to the motion (velocity).

For a stationary object, friction works to prevent motion until the maximum attainable friction has been reached. This maximum value is a function of a *coefficient of friction*, μ. The coefficient of friction is a property that is determined experimentally and is dependent on the materials and surface conditions of the contacting links. Average values of friction coefficients for common materials are given in Table 13.1. The magnitude of the friction force that acts on sliding components is calculated as

$$F_f = \mu N \qquad (13.6)$$

As mentioned, for moving objects, the friction force acts opposite to the direction of the relative sliding motion.

TABLE 13.1 Approximate Coefficients of Sliding Friction

		Dry	Lubricated
Hard steel	On hard steel	0.45	0.08
	On Babbitt	0.35	0.15
Mild steel	On mild steel	0.60	0.12
	On bronze	0.34	0.17
	On brass	0.44	—
	On copper lead	0.36	0.15
	On cast iron	0.23	0.13
	On lead	0.95	0.30
	On aluminum	0.50	—
	On laminated plastic	0.35	0.05
	On Teflon	—	0.04
Cast iron	On cast iron	0.15	0.07
	On bronze	0.22	0.07
	On brass	0.30	—
	On copper	0.29	—
	On zinc	0.21	—
Aluminum	On aluminum	1.40	—

EXAMPLE PROBLEM 13.5

The scotch-yoke mechanism shown in Figure 13.15 is used in a valve actuator. As fluid is pumped into the cylinder, the increased pressure drives the mechanism and applies a torque to the output shaft. This torque can be used to actuate (open and close) valves. At the instant shown, the pressure load on the piston is 25 lb. Determine the torque generated on the output shaft. The coefficient of friction between the follower pin and crosshead slot is 0.15.

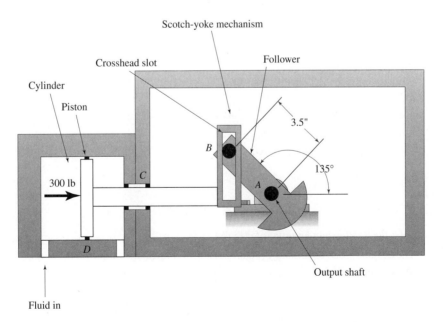

FIGURE 13.15 Valve actuator for Example Problem 13.5.

SOLUTION: 1. *Sketch a Kinematic Diagram of the Mechanism*

The kinematic diagram of this scotch-yoke mechanism is shown in Figure 13.16.

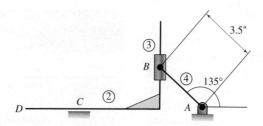

FIGURE 13.16 Kinematic diagram for Example Problem 13.5.

2. **Sketch Free-Body Diagrams of the Mechanism Links**

Link 2 is the piston/rod assembly/crosshead slot. Link 4 is the follower. Notice that link 3 is not a tangible link. It is used in a kinematic simulation to separate the revolute joint on the follower and the sliding joint in the crosshead slot. Thus, the mechanism is modeled with all lower-order joints. The kinematic diagram has four links, two pin joints, and two sliding joints, and consequently one degree of freedom. The driver for this mechanism is the movement of the fluid into the cylinder.

The free-body diagrams for links 2 and 4 are shown in Figure 13.17. Link 3 is not required for force analysis. Notice that a friction force is shown opposing relative motion. The directions may seem confusing and warrant further explanation.

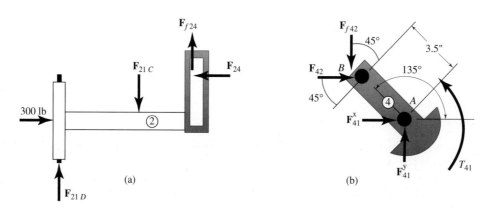

FIGURE 13.17 Free-body diagrams for Example Problem 13.5.

Consider link 4 (Figure 13.17b). The pin moves upward relative to the crosshead slot. Therefore, friction will act to prevent this motion of the pin by acting downward. Similarly, consider link 2 (Figure 13.17a). The slot moves downward relative to the pin (recall the definition of relative motion). Therefore, friction will act to prevent this motion of the slot by acting upward.

3. **Solve the Equilibrium Equations for Link 2**

Link 2 (Figure 13.17a) is examined first because it contains the applied force. For this analysis, only the x-equilibrium equation is required.

$$\overset{+}{\rightarrow} \ \Sigma F^x = 0:$$

$$F_{24} = 300 \text{ lb} \leftarrow$$

4. **Solve the Equilibrium Equations for Link 4**

The free-body diagram of link 4 (Figure 13.17b) will reveal the torque on the output shaft. Of course, Newton's first law dictates that $\mathbf{F}_{42} = \mathbf{F}_{24}$.

$$\mathbf{F}_{f42} = \mu \mathbf{F}_{42} = (0.15)(300 \text{ lb}) = 45 \text{ lb}$$

The torque can be determined by using the moment equilibrium equation.

$$+\circlearrowleft \ \Sigma M_A = 0:$$

$$-(\mathbf{F}_{42}\cos 45°)(3.5 \text{ in.}) + (\mu \mathbf{F}_{42}\cos 45°)(3.5 \text{ in.}) + T_{21} = 0$$

$$- [(300 \text{ lb}) \cos 45°] (3.5 \text{ in.}) + [(45 \text{ lb}) \cos 45° (3.5 \text{ in.})] + T_{21} = 0$$

Finally, the torque exerted on the output shaft is

$$T_{21} = +631 \text{ lb-in.} = 631 \text{ lb-in., cw}$$

PROBLEMS

Resultant Force

13–1. Determine the resultant for the forces shown in Figure P13.1 when $\beta = 25°$.

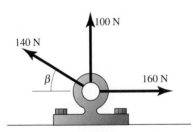

FIGURE P13.1 Problems 1–3.

13–2. Determine the resultant for the forces shown in Figure P13.1 when $\beta = 65°$.

13–3. Determine the resultant for the forces shown in Figure P13.1 when $\beta = 105°$.

Moment of a Force

13–4. A force is applied to a box wrench as shown in Figure P13.4. Determine the moment, relative to the center of the nut, when $\beta = 90°$.

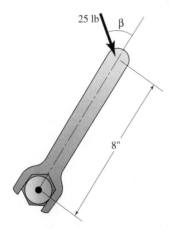

FIGURE P13.4 Problems 4–6.

13–5. A force is applied to a box wrench as shown in Figure P13.4. Determine the moment, relative to the center of the nut, when $\beta = 75°$.

13–6. A force is applied to a box wrench as shown in Figure P13.4. Determine the moment, relative to the center of the nut, when $\beta = 110°$.

13–7. A force that is applied to a control lever is shown in Figure P13.7. Determine the moment, relative to the pivot block, when $\beta = 0°$.

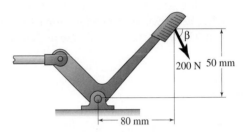

FIGURE P13.7 Problems 7–9.

13–8. A force that is applied to a control lever is shown in Figure P13.7. Determine the moment, relative to the pivot block, when $\beta = 60°$.

13–9. A force that is applied to a control lever is shown in Figure P13.7. Determine the moment, relative to the pivot block, when $\beta = 130°$.

Static Machine Forces

13–10. Figure P13.10 shows an overhead lift device. If a 600-lb force is suspended from the top boom while the mechanism is stationary, determine the force required in the cylinder. The top boom weighs 80 lb and the weight of the cylinder is negligible.

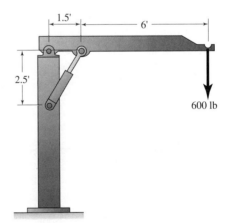

FIGURE P13.10 Problem 10.

13–11. Figure P13.11 shows a mechanism that raises packages in a transfer mechanism. If a 100-N package sits on the horizontal link while the mechanism is stationary, determine the torque required from the motor. The weights of the links are negligible.

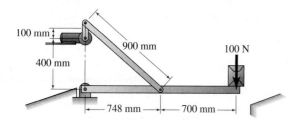

FIGURE P13.11 Problem 11.

13–12. Figure P13.12 shows a mechanism that is used to shear thin-gauge sheet metal. If a 200-N force is applied as shown, determine the force on the sheet metal. The weights of the links are negligible.

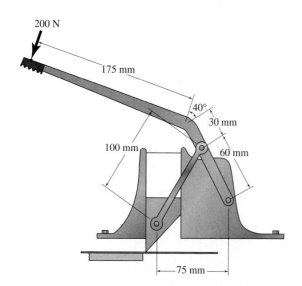

FIGURE P13.12 Problem 12.

13–13. Figure P13.13 shows an adjustable platform used to load and unload freight trucks. Currently, a 1200-lb crate is located as shown. Draw a free-body diagram for each link. The platform weighs 400 lb, and the weight of all other links is considered insignificant.

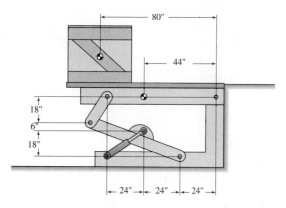

FIGURE P13.13 Problem 13.

13–14. The clamp shown in Figure P13.14 has a rated load of 1500 lb. Determine the compressive force this creates in the threaded rod, *AB*.

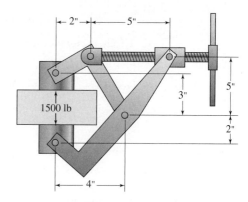

FIGURE P13.14 Problem 14.

13–15. A utility lift vehicle is shown in Figure P13.15. Determine the force required by the hydraulic cylinder to maintain the position of the bucket.

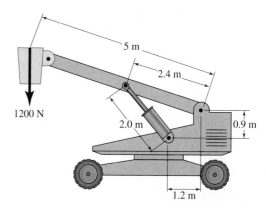

FIGURE P13.15 Problem 15.

13–16. A front loader is shown in Figure P13.16. Determine the force required from both hydraulic cylinders to maintain the shovel position.

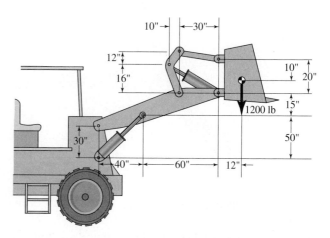

FIGURE P13.16 Problem 16.

13–17. A 500-lb crate is supported on a lift table, as shown in Figure P13.17. Determine the force required in the hydraulic cylinder to keep the platform in the position shown.

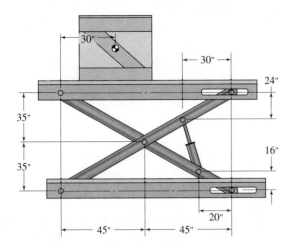

FIGURE P13.17 Problem 17.

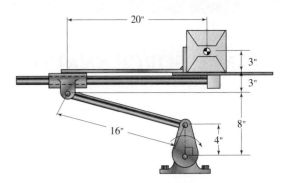

FIGURE P13.19 Problem 19.

Determine the instantaneous torque required to operate this mechanism. It operates at a low speed, so inertial forces are negligible.

13–18. Figure P13.18 illustrates a refuse truck capable of moving a dumpster from a lowered position, as shown, to a raised and rotated position. Gravity removes the contents into the truck box. The dumpster weighs 2400 lb and is shared equally by the two front forks. Determine the force in the two hydraulic cylinders.

CASE STUDY

13–1. Figure C13.1 shows a mechanism that gives motion to plunger J. Carefully examine the components of the mechanism, then answer the following leading

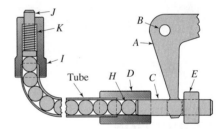

FIGURE C13.1 (Courtesy, Industrial Press.)

questions to gain insight into its operation.

1. As lever A is rotated, what type of motion does item C exhibit?
2. What type of connection do items A and C have?
3. What type of motion does ball H have?
4. What type of motion does plunger J have?
5. What is the purpose of spring K?
6. What is the purpose of item E?
7. What is the purpose of this mechanism?
8. Compare this mechanism to another mechanical concept that serves the same purpose.

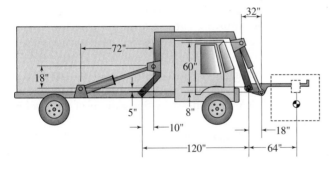

FIGURE P13.18 Problem 18.

13–19. Figure P13.19 shows a materials handling mechanism that slides 8-lb packages along a counter. The coefficient of kinetic friction between the package and counter is 0.25. The coefficient of kinetic friction between the collar and rod is 0.10.

DYNAMIC FORCE ANALYSIS

14.1 INTRODUCTION

During the design of a machine, determining the operating forces is critical. Consider the development of an automotive windshield wiper system. A key task is the selection of an electric motor that will drive the wipers. The torque required to operate the system is the main attribute in this selection process. Different scenarios must be considered, such as the fact that the car might be parked under a maple tree. Increased wiper friction due to the tree sap will demand greater motor torque. A common scenario occurs during periods of heavy rain. The wipers will be operated on a high-speed setting. As the wipers oscillate at increased speeds, large accelerations will result. Inertial forces will be created from the high accelerations. These forces may be large enough to damage the components of the wiper system. In fact, the inertial forces created by the motion of many high-speed machines exceed the forces required to perform the intended task. In a reciprocating engine, such as an automobile engine, the inertial forces can be greater than the force produced by the gas pressure. In a gas turbine, the inertial force on the bearings due to an unbalanced rotor can be magnitudes greater than the weight of the rotor.

Thus, for machines with significant accelerations, dynamic force analysis is necessary. The previous chapter dealt with force analysis in mechanisms without accelerations. This

chapter deals with force analysis in machines with significant accelerations. This condition is termed *dynamic equilibrium.* The analysis of dynamic equilibrium uses many concepts of static equilibrium. Therefore, an understanding of the topics presented in the previous chapter is necessary prior to studying this chapter.

14.2 MASS AND WEIGHT

Mass and weight are not identical. *Mass, m,* is a measure of the amount of material in an object. Mass can also be described as the resistance of an object to acceleration. It is more difficult to "speed up" an object with a large mass.

The *weight,* **W**, of an object is a measure of the pull of gravity on it. Thus, weight is a force directed toward the center of the earth. The *acceleration of gravity, g,* varies depending on the location relative to a gravitational pull. Thus, the weight of an object will vary. Mass, however, is a quantity that does not change with gravitational pull. As stated, it is used to describe the amount of material in a part.

The magnitude of weight and mass can be related through Newton's gravitational law.

$$\mathbf{W} = mg \tag{14.1}$$

In most analyses on earth, the acceleration of gravity is assumed to be

$$g = 32.2\,\text{ft/s}^2 = 386.4\,\text{in./s}^2 = 9.81\,\text{m/s}^2 = 9810\,\text{mm/s}^2$$

This assumption is applicable to all machines and mechanisms discussed in this book. Of course, in the case of designing machines for use in outer space, a different gravitational pull would exist.

Mass and weight are often confused in the U.S. Customary System; it is most convenient to use a derived unit for mass, which is the slug. This unit directly results from the use of equation (14.1)

$$\text{slug} = [\text{lb/ft/s}^2] = \text{lb s}^2/\text{ft}$$

Occasionally, the pound-mass (lb_m) is also used as a measure of mass. It is the mass that weighs 1 pound on the surface of the earth. Assuming that the standard value of gravity applies, the pound-mass can be converted to slugs by

$$1\,\text{slug} = 32.2\,\text{lb}_m$$

Generally stated, any calculation in the U.S. Customary System should use the unit of slug for mass. In the International System, the common unit used for mass is the kilogram ($\text{kg} = \text{N}\,\text{s}^2/\text{m}$).

14.3 CENTER OF GRAVITY

The *center of gravity, cg,* of an object is the balance point of that object. That is, it is the single point at which the object's weight could be held and be in balance in all directions. For parts made of homogeneous material, the *cg* is the three-dimensional, geometric center of the object. For many simple parts, such as a cylinder, the geometric center is apparent. Locating the center of gravity becomes important in force analysis because this is the location of the

force of gravity, or weight. In dynamic force analysis, any inertia effects due to the acceleration of the part will also act at this point.

For complex parts, the location of the center of gravity is not obvious. A common method of determining the center of gravity is to divide the complex part into simple shapes, where the center of gravity of each is apparent. The composite center of gravity can be determined from a weighted average of the coordinates of the individual *cg*s. For example, the *x*-component of the center of gravity of a composite shape can be found from the following equation:

$$x_{cg\,\text{total}} = \frac{m_1 x_{cg\,1} + m_2 x_{cg\,2} + m_3 x_{cg\,3} + \dots}{m_1 + m_2 + m_3 + \dots} \quad (14.2)$$

Because the acceleration due to gravity will be the same for the entire body, weight can be substituted for mass in equation (14.2). Of course, similar equations can be written for the *y*- and *z*-coordinates of the center of gravity.

EXAMPLE PROBLEM 14.1

The part shown in Figure 14.1 is made from steel (0.283 lb/in.^{3}). Determine the coordinates of the center of gravity.

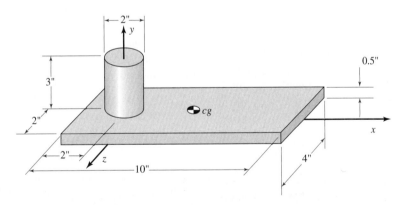

FIGURE 14.1 Part for Example Problem 14.1.

SOLUTION: 1. **Divide the Link into Basic Shapes**

This part can be readily divided into two components. The lower plate will be designated as component 1, and the upper shaft will be designated as component 2.

2. **Calculate the Weight of the Basic Shapes**

The weight of parts is determined by calculating the volume of the parts and multiplying by the density of steel.

$$W_1 = 10.283\,\text{lb/in.}^3\,[(10\,\text{in.})(4\,\text{in.})(0.5\,\text{in.})] = 5.66\,\text{lb}$$

$$W_2 = (0.283\,\text{lb/in.}^3)\left[\frac{\pi}{4}(2\,\text{in.})^2(3\,\text{in.})\right] = 5.33\,\text{lb}$$

These weights and *cg* coordinates are organized in Table 14.1.

TABLE 14.1	Basic Shapes Data for Example Problem 14.1			
Component	Weight (lb)	x_{cg} (in.)	y_{cg} (in.)	z_{cg} (in.)
1	5.66	$(10/2 - 2) = 3$	$(0.5/2) = 0.25$	0
2	5.33	0	$(0.5/2) = 0.25$	0

3. *Use Equation (14.2) to Calculate the Center of Gravity*

The coordinates of the center of gravity are found.

$$x_{cg\ total} = \frac{W_{part\,1}x_{cg\ part\,1} + W_{part\,2}x_{cg\ part\,2}}{W_{part\,1} + W_{part\,2}}$$

$$= \frac{(5.66\ \text{lb})(3\ \text{in.}) + (5.33\ \text{lb})(0\ \text{in.})}{(5.66 + 5.33)\ \text{lb}} = 1.545\ \text{in.}$$

$$y_{cg\ total} = \frac{W_{part\,1}y_{cg\ part\,1} + W_{part\,2}y_{cg\ part\,2}}{W_{part\,1} + W_{part\,2}}$$

$$= \frac{(5.66\ \text{lb})(0.25\ \text{in.}) + (5.33\ \text{lb})(2\ \text{in.})}{(5.66 + 5.33)\ \text{lb}} = 1.099\ \text{in.}$$

The center of gravity of both parts lies on the z-axis. Therefore, the center of gravity of the composite (total) part will also lie on the z-axis. Therefore,

$$z_{cg\ total} = 0$$

14.4 MASS MOMENT OF INERTIA

The *mass moment of inertia, I,* of a part is a measure of the resistance of that part to rotational acceleration. It is more difficult to "speed up" a spinning object with a large mass moment of inertia. Mass moment of inertia, or simply moment of inertia, is dependent on the mass of the object along with the shape and size of that object. In addition, inertia is a property that is stated relative to a reference point (or axis when three dimensions are considered). This reference point is commonly the center of gravity of the part.

Figure 14.2 illustrates a general solid object. Notice that a small element of the object has been highlighted. The mass moment of inertia of this small element is determined by multiplying its mass, *dm*, by the square of the distance, *r*, to a reference axis, *z*. This distance is the perpendicular distance from the axis to the arbitrary element *dm*.

The mass moment of inertia of the entire object is the sum of all particles that comprise the object. Mathematically, the moment of inertia is expressed as

$$I_z = \int r^2\, dm \tag{14.3}$$

Because the definition involves *r*, the value of the mass moment of inertia is different for each axis. For example, consider a slender rod. The mass moment of inertia relative to its longitudinal axis will be small because *r* is small because *r* is small for each element of the rod. For an axis that is perpendicular to the rod, the moment of inertia will be large because *r* is large for the outermost elements.

Mass moment of inertia is expressed in the units of mass times squared length. In the U.S. Customary System, the common units are slug-squared feet (slug ft^2), which convert to pound-feet-squared seconds (lb ft s^2). In the International System, the common units used are kilogram-squared meters (kg m^2).

14.4.1 Mass Moment of Inertia of Basic Shapes

Equation (14.2) has been used to derive equations for primary shapes. Table 14.2 gives these equations, which can be used to compute the mass moment of inertia for common solid shapes of uniform density.

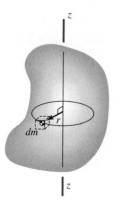

FIGURE 14.2 A general solid object.

TABLE 14.2   Mass Moments of Inertia

Shape Name	Rendering	Mass Moment of Inertia
Cylinder		$I_x = \dfrac{1}{2}[mr^2]$ $I_y = \dfrac{1}{12}[m(3r^2 + l^2)]$ $I_z = \dfrac{1}{12}[m(3r^2 + l^2)]$
Slender rod		$I_x = 0$ $I_y = \dfrac{1}{12}[ml^2]$ $I_z = \dfrac{1}{12}[ml^2]$
Thin disk		$I_x = \dfrac{1}{2}[mr^2]$ $I_y = \dfrac{1}{4}[mr^2]$ $I_z = \dfrac{1}{4}[mr^2]$
Rectangular block		$I_x = \dfrac{1}{12}[m(w^2 + h^2)]$ $I_y = \dfrac{1}{12}[m(w^2 + l^2)]$ $I_z = \dfrac{1}{12}[m(h^2 + l^2)]$

EXAMPLE PROBLEM 14.2

The part in Figure 14.3 weighs 3 lb. Determine the mass moment of inertia of the part, relative to an *x*-axis at the center of the part.

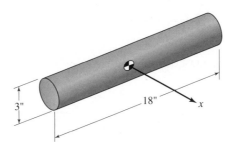

FIGURE 14.3 Part for Example Problem 14.2.

SOLUTION: 1. ***Determine the Mass of the Part***

The part weighs 3 lb and it is assumed to be used on the earth's surface. The mass can be calculated from equation (14.1).

$$m = \frac{\mathbf{W}}{g} = \frac{3\,\text{lb}}{32.2\,\text{ft/s}^2} = 0.093\,\text{slug}$$

2. *Calculate the Mass Moment of Inertia (Solid Cylinder)*

    In a true sense, this part is a solid cylinder with

    $$r = 1.5 \text{ in.} = 0.125 \text{ ft}$$

    $$l = 18 \text{ in.} = 1.5 \text{ ft}$$

    The z-axis in Table 14.2 is equivalent to the x-axis in this analysis. The mass moment of inertia relative to this axis at the center of the part is

    $$l_x = \frac{1}{12}[m(3r^2 + l^2)] = \frac{1}{12}[0.093 \text{ slug} (3(0.125 \text{ ft})^2 + (1.5 \text{ ft})^2)]$$

    $$= 0.0178 \text{ slug ft}^2 = 0.0178 \text{ lb ft s}^2$$

3. *Calculate the Mass Moment of Inertia (Slender Rod)*

    This part may be approximated as a slender rod. Using this assumption, the mass moment of inertia is calculated from Table 14.2 as

    $$I_x = \frac{1}{12}[m(l)^2] = \frac{1}{12}[0.093 \text{ slug} (1.5 \text{ ft})^2]$$

    $$= 0.0174 \text{ slug ft}^2 = 0.0174 \text{ lb ft s}^2$$

    The slender rod assumption underestimates the actual mass moment of inertia by only 1.15 percent. Apparently this part could be approximated as a slender rod.

14.4.2 Radius of Gyration

Occasionally, the moment of inertia of a part about a specified axis is reported in handbooks using the *radius of gyration, k.* Conceptually, the radius of gyration is the distance from the center of gravity to a point where the entire mass could be concentrated and have the same moment of inertia.

The radius of gyration can be used to compute the mass moment of inertia by

$$I = mk^2 \qquad (14.4)$$

The radius of gyration is expressed in units of length. In the U.S. Customary System, the common units are feet (ft) or inches (in.). In the International System, the common units used are meters (m) or millimeters (mm).

14.4.3 Parallel Axis Theorem

Mass moment of inertia is stated relative to an axis. Occasionally, the mass moment of inertia is desired relative to an alternate, parallel axis. A parallel axis transfer equation has been derived [Ref. 11] to accomplish this task. To transfer the mass moment of inertia from the x-axis to a parallel x'-axis, the transfer equation is

$$I_{x'} = I_x \pm md^2 \qquad (14.5)$$

The value d in equation (14.5) is the perpendicular distance between the two axes. Notice that the second term in equation (14.5) can be either added or subtracted. The term is added when the reference axis is moved away from the center of gravity of the basic shape. Conversely, the term is subtracted when the transfer is toward the center of gravity.

EXAMPLE PROBLEM 14.3

For the part shown in Figure 14.4, determine the mass moment of inertia of the part relative to an x-axis at the end of the part.

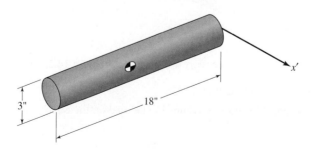

FIGURE 14.4 Part for Example Problem 14.3.

SOLUTION: The mass moment of inertia through the center of the part was determined in Example Problem 14.2 as

$$I_x = 0.0178 \text{ slug ft}^2$$

The distance of the transfer from the center to the end of the part is

$$d = 9 \text{ in.} = 0.75 \text{ ft}$$

Equation (14.5) can be used to transfer the reference axis to the end of the part. Notice that the second term is added because the transfer is away from the center of gravity.

$$I_{x'} = I_x + md^2 = 0.0178 \text{ slug ft}^2 + (0.093 \text{ slug})(0.75 \text{ ft})^2$$

$$= 0.0701 \text{ slug ft}^2 = 0.0701 \text{ lb ft s}^2$$

14.4.4 Composite Bodies

In practice, parts cannot always be simply approximated by the basic shapes shown in Table 14.2. However, for more complex parts, the determination of the moment of inertia can be done by dividing the complex parts into several basic shapes from Table 14.2. The mass moment of inertia of each basic shape is computed relative to an axis through the center of the entire part. Finally, the total mass moment of inertia is determined by combining the values from the individual shapes.

EXAMPLE PROBLEM 14.4

The part in Figure 14.5 is made from steel. Determine the mass moment of inertia of the part, relative to a y-axis at the center of the part.

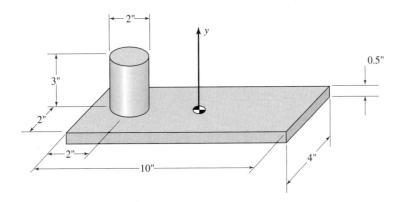

FIGURE 14.5 Part for Example Problem 14.4.

SOLUTION: 1. ***Identify the Basic Shapes and Determine Their Mass***

The part can be divided into two component shapes, as in Example Problem 14.1. Using the weights determined in that problem, the mass of the two parts is

$$m_1 = \frac{W_1}{g} = \frac{5.66 \text{ lb}}{32.2 \text{ ft/s}^2} = 0.176 \text{ slug}$$

$$m_2 = \frac{W_2}{g} = \frac{5.33 \text{ lb}}{32.2 \text{ ft/s}^2} = 0.165 \text{ slug}$$

2. ***Determine the Centroidal Mass Moment of Inertia of the Basic Shapes***

Component 1 is a rectangular block and component 2 is a cylinder. Using Table 14.2, the mass moment of inertia of each part is determined relative to their individual centers of gravity.
Component 1:

$$I_y = \frac{1}{12}[m(w^2 + l^2)] = \frac{1}{12}[0.176 \text{ slug }[(4 \text{ in.})^2 + (10 \text{ in.})^2]$$

$$= 0.701 \text{ slug in.}^2 = 0.0118 \text{ slug ft}^2 = 0.0118 \text{ lb ft s}^2$$

Component 2:

$$I_y = \frac{1}{12}[m(r^2)] \quad \text{(longitudinal axis)} = \frac{1}{12}[0.165 \text{ slug}(1 \text{ in.})^2]$$

$$= 0.0138 \text{ slug in.}^2 = 0.0001 \text{ slug ft}^2 = 0.0001 \text{ lb ft s}^2$$

3. **Utilize the Parallel Axis Theorem**

The center of gravity information, determined in Example Problem 14.1, will be used to determine the moment of inertia for each component relative to the composite center of gravity. The parallel axis theorem is used to accomplish this. Notice that the perpendicular distance between y-axes is along the x-direction. Component 1:

$$d_1 = (3.0 - 1.099) \text{ in.} = 1.901 \text{ in.} = 0.158 \text{ ft}$$

$$I_{y'(\text{component 1})} = I_{y(\text{component 1})} + m_1 d_1^2 = 0.0118 \text{ slug ft}^2 + (0.176 \text{ slug})(0.158 \text{ ft})^2$$

$$= 0.0162 \text{ slug ft}^2 = 0.0162 \text{ lb ft s}^2$$

Component 2:

$$d_2 = (1.099 - 0) \text{ in.} = 1.099 \text{ in.} = 0.092 \text{ ft}$$

$$I_{y'(\text{component 2})} = I_{y(\text{component 2})} + m_2 d_2^2 = 0.0001 \text{ slug ft}^2 + (0.165 \text{ slug})(0.0923 \text{ ft})^2$$

$$= 0.0015 \text{ slug ft}^2 = 0.0015 \text{ lb ft s}^2$$

4. **Calculate the Composite Mass Moment of Inertia**

$$I_{y'} = I_{y'(\text{component 1})} + I_{y'(\text{component 2})}$$

$$= (0.0162 + 0.0015) \text{ slug ft}^2 = 0.0177 \text{ slug ft}^2 = 0.0177 \text{ lb ft s}^2$$

14.4.5 Mass Moment of Inertia—Experimental Determination

One popular experimental method of determining the mass moment of inertia of a part is to swing the part as a pendulum. This method is illustrated in Figure 14.6.

FIGURE 14.6 Mass moment of inertia experiment.

If the part is displaced a small angle and released, it will oscillate. The moment of inertia can be determined by measuring the time to complete one oscillation, Δt. The mass moment of inertia of the part relative to an axis through the center of gravity, has been derived [Ref. 11] as

$$I_{cg} = mr_{cg}\left[\left(\frac{\Delta t}{2\pi}\right)^2 g - r_{cg}\right] \quad (14.6)$$

14.5 INERTIAL FORCE

Section 13.4 listed Newton's three principal laws of mechanics. The second law is critical for all parts that experience acceleration. It is stated as

SECOND LAW: A body that has an unbalanced force has

a. An acceleration that is proportional to the force,

b. An acceleration that is in the direction of the force, and

c. An acceleration that is inversely proportional to the mass of the object.

For linear motion, this law can be stated in terms of the acceleration of the link's center of gravity, $\mathbf{A}_g$; thus,

$$\Sigma \mathbf{F} = m\mathbf{A}_g \quad (14.7)$$

Equation (14.7) can be rewritten as

$$\Sigma \mathbf{F} -> m\mathbf{A}_g = 0 \quad (14.8)$$

Notice that the subtraction symbol ($->$) is used because both force and acceleration are vectors.

The second term in equation (14.8) is referred to as the inertia of a body. This term is defined as an *inertial force,* $\mathbf{F}_g^i$

$$\mathbf{F}_g^i = -> m\mathbf{A}_g \qquad (14.9)$$

The negative sign indicates that the inertial force opposes acceleration (it acts in the opposite direction of the acceleration). Inertia is a passive property and does not enable a body to do anything except oppose acceleration.

This notion is commonly observed. Imagine pounding on the gas pedal in an automobile, violently accelerating the vehicle. Envision the tendency for your head to lurch backward during the acceleration. This is the inertial force, acting in an opposite direction to the acceleration of the automobile. Further, the extent of the lurch is

proportional to the magnitude of acceleration. Similarly, as the brakes in an automobile are slammed, decelerating the vehicle, your head lurches forward, again opposing the acceleration of the automobile. This is Newton's second law in practice.

Equation (14.8) can be rewritten as

$$\Sigma \mathbf{F} +> \mathbf{F}_g^i = 0 \qquad (14.10)$$

This concept of rewriting equation (14.7) in the form of equation (14.8) is known as *d'Alembert's principle.* Using d'Alembert's principle in force analysis is referred to as *the inertia–force method of dynamic equilibrium.* It allows for analysis of accelerating links, using the same methods that are used in a static analysis.

EXAMPLE PROBLEM 14.5

The compressor mechanism shown in Figure 14.7 is driven clockwise by a DC electric motor at a constant rate of 600 rpm. In the position shown, the cylinder pressure is 45 psi. The piston weighs 0.5 lb, and the coefficient of friction between the piston and the compressor cylinder is 0.1. The weight of all other links is negligible. At the instant shown, determine the torque required from the motor to operate the compressor.

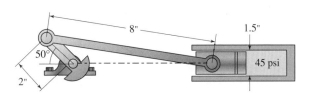

FIGURE 14.7 Mechanism for Example Problem 14.5.

SOLUTION: 1. ***Draw a Kinematic Diagram and Identify the Degrees of Freedom***

This is a common in-line, slider-crank mechanism, having a single degree of freedom. A scaled kinematic diagram is shown in Figure 14.8a.

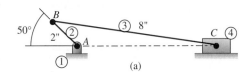

FIGURE 14.8 Diagrams for Example Problem 14.5.

2. ***Decide on a Method to Achieve the Required Motor Torque***

Because the piston is the only component without negligible weight, the inertial force, and the acceleration, of this component must be determined. The acceleration of the piston (link 4) is strictly translational and is identical to the motion of point *C*. Such acceleration analyses have been extensively presented in Chapter 7.

Once the acceleration of the piston has been obtained, the subsequent inertial forces can be calculated. Finally, free-body diagrams and the corresponding equations can be used to determine the required torque.

3. **Determine the Velocity of Points B and C**

This type of analysis was extensively discussed in the earlier chapters of the book. The 2-in. crank is rotating at 600 rpm. The velocity of point B is

$$\omega_2 = \frac{\pi}{30}\,(600\ \text{rev/min}) = 62.8\ \text{rad/s, cw}$$

$$\mathbf{V}_B = \omega_2\,r_{AB} = (62.8\ \text{rad/s})\,(2\ \text{in.}) = 125.6\ \text{in./s}\quad \angle 40°$$

The direction of $\mathbf{V}_B$ is perpendicular to link 2 and consistent with the direction of ω_2, up and to the left. Using CAD, a vector can be drawn to scale, from the velocity diagram origin, to represent this velocity. The relative velocity equation for points B and C can be written as

$$\mathbf{V}_C = \mathbf{V}_B + {>}\ \mathbf{V}_{C/B}$$

A completed velocity diagram is shown in Figure 14.8b. Scaling the vector magnitudes from the diagram,

$$\mathbf{V}_C = 80.5\ \text{in./s} \rightarrow$$

$$\mathbf{V}_{C/B} = 82.2\ \text{in./s}\quad \overline{79°}$$

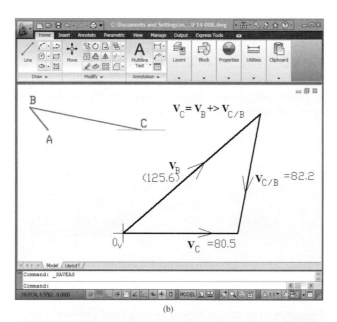

(b)

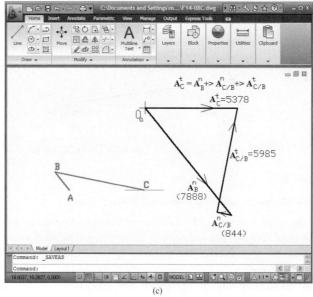

(c)

FIGURE 14.8 Continued

4. *Calculate Acceleration Components*

The next step is to construct an acceleration diagram, which includes points B and C. Calculating the magnitudes of the known accelerations,

$$A_B^n = \frac{(V_B)^2}{r_{AB}} = \frac{(125.6\,\text{in./s})^2}{2.0\,\text{in.}} = 7888\,\text{in./s}^2 \ \underline{\diagdown 50°} \quad \text{(directed toward the center of rotation, point } A\text{)}$$

$$a_B^t = 0 \quad \text{(no angular acceleration of the 2-in. crank)}$$

$$A_{C/B}^n = \frac{(V_{C/B})^2}{r_{BC}} = \frac{(82.2\,\text{in./s})^2}{8.0\,\text{in.}} = 844\,\text{in./s}^2 \ \underline{\diagup 11°} \quad \text{(directed from } C \text{ toward } B, \text{ measured from CAD)}$$

Note that point A does not have a normal acceleration because the motion is strictly translational.

5. *Construct an Acceleration Diagram*

The relative acceleration equation for points B and C can be written as

$$A_C^n +> A_C^t = A_B^n +> A_B^t +> A_{C/B}^n +> A_{C/B}^t$$

The completed acceleration diagram is shown in Figure 14.8c.

6. *Measure the Piston Acceleration*

Scaling the vector magnitudes from the diagram,

$$A_{C/B}^t = 5985\,\text{in./s}^2 \ \nearrow 79°$$

$$A_C^t = 5378\,\text{in./s}^2 \ \rightarrow$$

Because the tangential acceleration of point B is in the same direction as the velocity, the piston is accelerating (speeding up), not decelerating.

7. *Calculate the Inertial Force*

Because the piston is the only link of considerable weight, its inertial force is computed by combining equations (14.9) and (14.1).

$$F_{g4}^i = -> m_4 A_{g4} = \frac{W_4}{g}(-> A_{g4})$$

$$= \frac{(0.5\,\text{lb})}{386\,\text{in./s}^2}(5378\,\text{in./s}^2) = 6.96\,\text{lb} \leftarrow$$

Because the piston does not encounter rotational acceleration, rotational inertia is not observed.

8. *Sketch Free-Body Diagrams of the Mechanism Links*

Notice that link 3 *(BC)* is a simple link, containing only two pin joints. In addition, no other force is acting on this link. Thus, it is a two-force member, and the forces acting on the link must be equal and along the line that connects the two pins. The free-body diagram for link 3 is shown as Figure 14.8d. As before, the notation used is that F_{32} is a force that is applied to link 3 as a result of contact from link 2.

   Link 2 is also a simple link; it contains only two pin joints. However, a moment (torque) is also applied to this crank. Thus, this link is not a simple, two-force member. Newton's third law stipulates that a force that is acting at B will be equal and opposite to F_{32}. Thus, the direction of F_{23} is known as a result of Figure 14.8d. The angle between links 2 and 3 was measured from the CAD model. A general pin joint at point A dictates that two reaction forces will be present. The free-body diagram for link 2 is shown as Figure 14.8e.

   Link 4 has sliding contact with link 1. This contact force will act perpendicular to the contact surface. The force from the compressed gas will, similarly, act perpendicular to the piston surface. A friction force will oppose the motion (velocity) of link 4. Also, Newton's third law stipulates that a force that is acting at C will be equal and opposite to F_{34}. Thus, the direction of F_{43} is known as a result of Figure 14.8d. The free-body diagram for link 4 is shown as Figure 14.8f.

9. *Solve the Dynamic Equilibrium Equations for Link 4*

Link 4 is examined first because it contains the applied force. The gas force is calculated as

$$F_{gas} = p_{gas} A_{piston} = p_{gas}\left[\frac{\pi(d_{piston})^2}{4}\right] = 45\,\text{lb/in.}^2\left[\frac{\pi(1.5\,\text{in.})^2}{4}\right] = 79.5\,\text{lb} \leftarrow$$

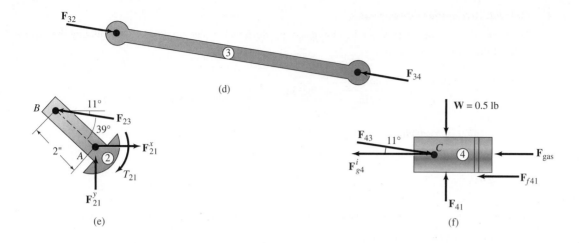

(d)

(e)   (f)

The friction force is

$$\mathbf{F}_f = \mu \mathbf{F}_{41} = 0.1\,\mathbf{F}_{41}$$

The two unknown forces on this link (Figure 14.8f) are solved by using the following equilibrium equations:

$\xrightarrow{+}$ $\Sigma \mathbf{F}^x + \ > \mathbf{F}^i = 0$:

$$\mathbf{F}_{43} \cos 11.0^\circ - \mathbf{F}_{gas} - \mathbf{F}_{g4}^i - \mathbf{F}_f = 0$$

$+\uparrow$ $\Sigma \mathbf{F}^y = 0$:

$$- \mathbf{F}_{43} \cos 11.0^\circ + \mathbf{F}_{41} - 0.5\,\text{lb} = 0$$

Solving these equations yields

$$\mathbf{F}_{43} = +89.8\,\text{lb} = 89.8\,\text{lb} \quad \diagdown 11^\circ$$

$$\mathbf{F}_{41} = +16.6\,\text{lb} = 16.6\,\text{lb}\uparrow$$

10. **Solve for Equilibrium of Link 3**

Because link 3 is a two-force member (Figure 14.8d.), the equilibrium equations dictate that the forces have the same magnitude, act along the same line, and are opposite in sense. Of course, Newton's third law dictates that $\mathbf{F}_{32} = \mathbf{F}_{23}$. Thus, the forces acting on link 3 are

$$\mathbf{F}_{34} = 89.8\,\text{lb} \quad \underline{11^\circ}\diagdown$$

$$\mathbf{F}_{32} = 89.8\,\text{lb} \quad \diagdown\underline{11^\circ}$$

11. **Solve for Equilibrium of Link 2**

The free-body diagram of link 2 (Figure 14.8e) will reveal the required motor torque. Of course, Newton's third law dictates that $\mathbf{F}_{32} = \mathbf{F}_{23}$. The unknown forces and moment on this link are solved using the following equilibrium equations:

$\xrightarrow{+}$ $\Sigma \mathbf{F}^x = 0$:

$$\mathbf{F}_{21}^x - \mathbf{F}_{23} \cos 11^\circ = 0$$

$+\uparrow$ $\Sigma \mathbf{F}^y = 0$:

$$\mathbf{F}_{21}^y + \mathbf{F}_{23} \sin 11^\circ = 0$$

$+\curvearrowleft$ $\Sigma M_A = 0$:

$$- T_{21} + (\mathbf{F}_{23} \sin 39^\circ)(2\,\text{in.}) = 0$$

Solving the three equations yields

$$\mathbf{F}_{21}^x = +88.1 \text{ lb} = 88.1 \text{ lb} \rightarrow$$

$$\mathbf{F}_{21}^y = -17.1 \text{ lb} = 17.1 \text{ lb} \downarrow$$

$$T_{21} = +113.0 \text{ lb in.} = 113.0 \text{ lb in., cw}$$

Because the torque is the desired value, solving only the moment equation was necessary.

14.6 INERTIAL TORQUE

The concept of an inertial force, as described in equation (14.7), is an extension of Newton's second law for linear motion. For rotational motion, the second law can be summarized in terms of rotational acceleration and moment of inertia, relative to an axis through the center of gravity.

$$\Sigma M_g = I_g \alpha \qquad (14.11)$$

Again, the subscript "g" refers to the reference point at the link's center of gravity.

Similarly to linear motion, equation (14.11) can be rewritten as

$$\Sigma M_g -> T_g^{\,i} = 0 \qquad (14.12)$$

Notice that the subtraction symbol $(->)$ is used because the directions of the moment and angular acceleration must be accounted for. The second term in equation (14.12) is

termed the *angular inertia of a body.* This term is used to define an *inertial torque, T_g^i*:

$$T_g^i = -> I_g \alpha \qquad (14.13)$$

Again, the negative sign indicates that the inertial torque is directed opposite to the angular acceleration.

Equation (14.12) can be rewritten as

$$\Sigma M +> T_g^{\,i} = 0 \qquad (14.14)$$

Equation (14.14) is termed the *moment equation of dynamic equilibrium.* It is the rotational equivalent of d'Alembert's principle described in Section 14.5. It allows for analysis of accelerating links, using the same methods as are used in a static analysis.

The following example problem will combine several of the dynamic force analysis concepts presented in this chapter.

EXAMPLE PROBLEM 14.6

The mechanism shown in Figure 14.9 is used to lower and retract the landing gear on small airplanes. The wheel assembly link weighs 100 lb, with a center of gravity as shown. The radius of gyration of the assembly, relative to the center of gravity, has been experimentally determined as 1.2 ft. The motor link is rotating counterclockwise at 3 rad/s and accelerating at 10 rad/s². For mass property estimation, the motor crank will weigh approximately 15 lb and will be 2 ft long, 1 ft wide, and 0.25 ft thick. The connecting link is estimated to weigh 20 lb and can be modeled as a 3.5-ft slender rod. Determine all forces acting on the joints of all links and the torque required to drive the motor link.

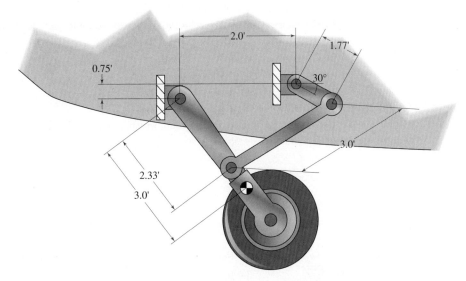

FIGURE 14.9 Landing gear for Example Problem 14.6.

SOLUTION: 1. **Draw a Kinematic Diagram and Identify the Degrees of Freedom**

This mechanism is the common four-bar linkage, having a single degree of freedom. A kinematic diagram is given in Figure 14.10a.

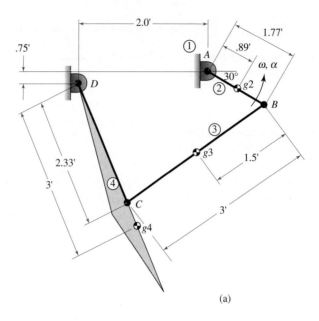

(a)

2. **Decide on a Method to Achieve the Required Motor Torque**

Because all links have significant weight, the acceleration of the center of gravity of all links must be determined. Such acceleration analysis has been extensively presented in Chapter 7. Once the accelerations have been established, the subsequent inertial forces and torques can be calculated. Finally, free-body diagrams and the corresponding equations can be used to determine the required torque.

3. **Determine the Velocity of Points B and C**

This type of analysis was extensively discussed in the earlier chapters of the book. The 1.77-ft crank is rotating at 3 rad/s. The velocity of point B is

$$\mathbf{V}_B = \omega_2 \, r_{AB} = (3 \, \text{rad/s})(1.77 \, \text{ft}) = 5.31 \, \text{ft/s} \; \angle 60°$$

The direction of $\mathbf{V}_B$ is perpendicular to link 2 and consistent with the direction of ω_2, up and to the left. Using CAD, a vector can be drawn to scale, from the velocity diagram origin, to represent this velocity.

The relative velocity equation for points B and C can be written as

$$\mathbf{V}_C = \mathbf{V}_B + > \mathbf{V}_{C/B}$$

The vector diagram is constructed in Figure 14.10b. Scaling the vector magnitudes from the diagram,

$$\mathbf{V}_C = 5.00 \, \text{ft/s} \; \angle 30.7°$$

$$\mathbf{V}_{C/B} = 2.63 \, \text{ft/s} \; \angle 51.4°$$

4. **Calculate Acceleration Components**

The next step is to construct an acceleration diagram, which includes points B and C. Calculating the magnitudes of the known accelerations,

$$\mathbf{A}_B^n = \frac{(\mathbf{V}_B)^2}{r_{AB}} = \frac{(5.31 \, \text{ft/s})^2}{1.77 \, \text{ft}} = 15.93 \, \text{ft/s}^2 \; \angle 30° \qquad \text{(directed toward the center of rotation, point } A\text{)}$$

$$\mathbf{A}_B^t = \alpha_2 \, r_{AB} = (10 \, \text{rad/s}^2)(1.77 \, \text{ft}) = 17.70 \, \text{ft/s}^2 \; \angle 60° \qquad \text{(perpendicular to link 2, consistent with } \alpha_2\text{)}$$

$$\mathbf{A}_{C/B}^n = \frac{(\mathbf{V}_{C/B})^2}{r_{BC}} = \frac{(2.63 \, \text{ft/s})^2}{3.0 \, \text{ft}} = 2.30 \, \text{ft/s}^2 \; \angle 38.6° \qquad \text{(directed from } C \text{ toward } B, \text{ measured from CAD)}$$

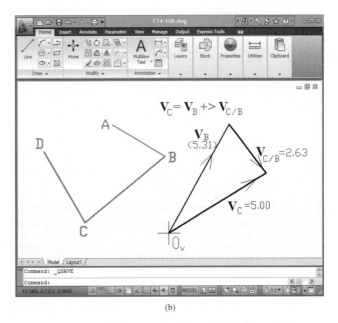

(b)

FIGURE 14.10 Diagrams for Example Problem 14.6.

$$\mathbf{A}_C^n = \frac{(\mathbf{V}_C)^2}{r_{CD}} = \frac{(500\,\text{ft/s}^2)}{2.33\,\text{ft}} = 10.72\,\text{ft/s}^2 \ \underset{59.3°}{\nearrow} \quad \begin{array}{l}\text{(directed from } C \text{ toward } D,\\ \text{measured from } CAD)\end{array}$$

5. **Construct an Acceleration Diagram**

   The relative acceleration equation for points B and C can be written as

   $$\mathbf{A}_C^n + > \mathbf{A}_C^t = \mathbf{A}_B^n + > \mathbf{A}_B^t + > \mathbf{A}_{C/B}^n + > \mathbf{A}_{C/B}^t$$

   The acceleration polygon is constructed and shown in Figure 14.10c. Notice that the concept of the acceleration image, as presented in Section 7.10, was used to determine the acceleration of the center of gravity of the three moving links.

6. **Measure the Acceleration of the Center of Gravity of All Links**

   Scaling the vector magnitudes from the diagram,

   $$\mathbf{A}_{C/B}^t = 12.28\,\text{ft/s}^2 \ \underset{51.4°}{\searrow} \qquad \mathbf{A}_C^t = 11.60\,\text{ft/s}^2 \ \underset{30.6°}{\nearrow}$$

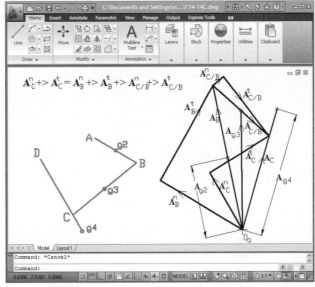

(c)

$$\mathbf{A}_{g2} = 11.91 \, \text{ft/s}^2 \; \underleftarrow{78.0°} \qquad \mathbf{A}_{g3} = 19.21 \, \text{ft/s}^2 \; \underleftarrow{89.4°}$$

$$\mathbf{A}_{g4} = 20.32 \, \text{ft/s}^2 \; \nearrow 73.4°$$

The angular accelerations of the links can then be determined.

$$\alpha_3 = \frac{a_{C/B}^t}{r_{BC}} = \frac{12.28 \, \text{ft/s}^2}{3.0 \, \text{ft}} = 4.1 \, \text{rad/s}^2, \text{counterclockwise}$$

$$\alpha_4 = \frac{a_C^t}{r_{CD}} = \frac{11.60 \, \text{ft/s}^2}{2.33 \, \text{ft}} = 5.0 \, \text{rad/s}^2, \text{counterclockwise}$$

7. **Calculate Mass Properties**

    The motor crank can be considered a rectangular block. From Table 14.2, the mass moment of inertia, at the center of mass, relative to an axis normal to the broad side of the link is

$$I_{g2} = \frac{1}{12} \, [m(\mathbf{W}^2 + l^2)] = \frac{1}{12} \left(\frac{15 \, \text{lb}}{32.2 \, \text{ft/s}^2} \right) [(2 \, \text{ft})^2 + (1 \, \text{ft})^2] = 0.194 \, \text{lb ft s}^2$$

    The connecting arm can be considered a slender rod. From Table 14.2, the mass moment of inertia at the center of mass relative to an axis normal to the length of the link is

$$I_{g3} = \frac{1}{12} \, [ml^2] = \frac{1}{12} \left(\frac{20 \, \text{lb}}{32.2 \, \text{ft/s}^2} \right) (3.5 \, \text{ft})^2 = 0.634 \, \text{lb ft s}^2$$

    The radius of gyration of the wheel assembly has been experimentally determined. From equation (14.4), the mass moment of inertia at the center of mass relative to an axis normal to the length of the assembly is

$$I_{g4} = ml^2 = \left(\frac{100 \, \text{lb}}{32.2 \, \text{ft/s}^2} \right) (1.2 \, \text{ft})^2 = 4.472 \, \text{lb ft s}^2$$

8. **Calculate the Inertial Force**

    For the three moving links, the inertial force is computed by combining equations (14.9) and (14.1).

$$\mathbf{F}_{g2}^i \, -> \, m_2 \mathbf{A}_{g2} = \frac{\mathbf{W}_2}{g} \, (-> \, \mathbf{A}_{g2})$$

$$= \frac{(15 \, \text{lb})}{32.2 \, \text{ft/s}^2} \, (11.91 \, \text{ft/s}^2) = 5.55 \, \text{lb} \; \underleftarrow{78.0°}$$

$$\mathbf{F}_{g3}^i = \, -> \, m_3 \mathbf{A}_{g3} = \frac{\mathbf{W}_3}{g} \, (-> \, \mathbf{A}_{g3})$$

$$= \frac{(20 \, \text{lb})}{32.2 \, \text{ft/s}^2} \, (19.21 \, \text{ft/s}^2) = 11.93 \, \text{lb} \; \underleftarrow{89.4°}$$

$$\mathbf{F}_{g4}^i = \, -> \, m_4 \mathbf{A}_{g4} = \frac{\mathbf{W}_4}{g} \, (-> \, \mathbf{A}_{g4})$$

$$= \frac{(100 \, \text{lb})}{32.2 \, \text{ft/s}^2} \, (20.32 \, \text{ft/s}^2) = 63.11 \, \text{lb} \; \overline{73.4°}$$

9. **Calculate the Inertial Torque**

    For the three moving links, the inertial torque is computed with equation (14.13).

$$T_{g2}^i = \, -> \, I_{g2}\alpha_2 = (0.194 \, \text{lb ft s}^2)(10 \, \text{rad/s}^2)$$

$$= 1.94 \, \text{ft lb, cw}$$

$$T_{g3}^i = \, -> \, I_{g3}\alpha_3 = (0.634 \, \text{lb ft s}^2)(4.1 \, \text{rad/s}^2)$$

$$= 2.60 \, \text{ft lb, cw}$$

$$T_{g4}^i = - > I_{g4}\alpha_4 = (4.472\,\text{lb ft s}^2)(5\,\text{rad/s}^2)$$

$$= 22.36\,\text{ft lb, cw}$$

10. **Sketch Free-Body Diagrams of the Mechanism Links**

Because the weight of all links is to be included in the analysis, there are no two-force members. Thus, all contact forces at the joints are general and are represented by their orthogonal components. The free-body diagram of link 4 is shown in Figure 14.10d. The free-body diagram of link 3 is shown in Figure 14.10e. Of course, Newton's third law, declaring that $\mathbf{F}_{34}$ and $\mathbf{F}_{43}$ have the same magnitude and opposing directions, still applies. Finally, the free-body diagram of link 2 is shown in Figure 14.10f. Because each link has more than three unknown forces, the equilibrium equations from all links will need to be solved simultaneously.

11. **Generate Equilibrium Equations for Link 4**

The following dynamic equilibrium equations are generated from the free-body diagram of link 4 (Figure 14.10d).

$$\xrightarrow{+} \quad \Sigma \mathbf{F}^x + > \mathbf{F}_g^i = 0:$$

$$\mathbf{F}_{41}^x - \mathbf{F}_{43}^x - \mathbf{F}_{g4}^i \cos 73.4° = 0$$

$$\mathbf{F}_{41}^x - \mathbf{F}_{43}^x - 18.03\,\text{lb} = 0$$

$$+\uparrow \quad \Sigma \mathbf{F}^y + > \mathbf{F}_g^i = 0:$$

$$\mathbf{F}_{41}^y - \mathbf{F}_{43}^y - \mathbf{W}_4 - \mathbf{F}_{g4}^i \sin 73.4° = 0$$

$$\mathbf{F}_{41}^y - \mathbf{F}_{43}^y - 160.48\,lb = 0$$

$$+\circlearrowleft \quad \Sigma M_D + > T_g^i = 0:$$

$$- \mathbf{F}_{43}^x [2.33\,\text{ft}\,(\sin 59.4°)] - \mathbf{F}_{43}^y [2.33\,\text{ft}\,(\cos 59.4°)] - \mathbf{W}_4 [3.0\,\text{ft}(\cos 59.4°)]$$

$$- \mathbf{F}_{g4}^i\,[\cos(73.4° - 30.6°)]\,[3.0\,\text{ft}] - T_{g4}^i = 0$$

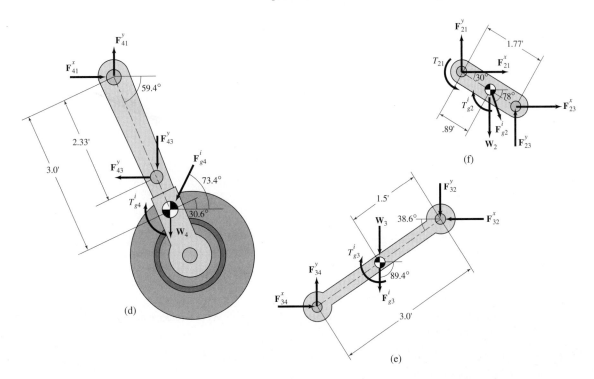

FIGURE 14.10 Continued

Substituting values gives

$$-2.000\,\mathbf{F}_{43}^{x} - 1.186\,\mathbf{F}_{43}^{y} - 313.98 \text{ ft lb} = 0$$

12. **Generate Equilibrium Equations for Link 3**

The following dynamic equilibrium equations are generated from the free-body diagram of link 3 (Figure 14.10e).

$\overset{+}{\longrightarrow}\ \Sigma\mathbf{F}^{x} + > \mathbf{F}_{g}^{i} = 0$:

$$\mathbf{F}_{34}^{x} - \mathbf{F}_{32}^{x} - \mathbf{F}_{g3}^{i}\cos 89.4° = 0$$

$$\mathbf{F}_{34}^{x} - \mathbf{F}_{32}^{x} - 0.13 \text{ lb} = 0$$

$+\uparrow\ \Sigma\mathbf{F}^{y} + > \mathbf{F}_{g}^{i} = 0$:

$$\mathbf{F}_{34}^{y} - \mathbf{F}_{32}^{y} - \mathbf{W}_{3} - \mathbf{F}_{g3}^{i}\sin 89.4° = 0$$

$$\mathbf{F}_{34}^{y} - \mathbf{F}_{32}^{y} - 31.93 \text{ lb} = 0$$

$+\!\!\smallfrown\ \Sigma M_{B} + > T_{g}^{i} = 0$:

$$\mathbf{F}_{34}^{x}\,[3.0\,\text{ft}\,(\cos 38.6°)] - \mathbf{F}_{34}^{y}\,[3.0\,\text{ft}\,(\sin 38.6°)]\,(2.33 \text{ ft}) + \mathbf{W}_{3}\,[1.5\,\text{ft}\,(\cos 38.6°)]$$

$$+ \mathbf{F}_{g3}^{i}\,[\cos(38.6° - 0.6°)]\,[1.5 \text{ ft}] - T_{g3}^{i} = 0$$

Substituting values gives

$$2.344\,\mathbf{F}_{34}^{x} - 1.872\,\mathbf{F}_{34}^{y} + 34.95 \text{ ft lb} = 0$$

13. **Generate Equilibrium Equations for Link 2**

The following dynamic equilibrium equations are generated from the free-body diagram of link 2 (Figure 14.10f).

$\overset{+}{\longrightarrow}\ \Sigma\mathbf{F}^{x} + > \mathbf{F}_{g}^{i} = 0$:

$$\mathbf{F}_{23}^{x} + \mathbf{F}_{21}^{x} + \mathbf{F}_{g2}^{i}\cos 78° = 0$$

$$\mathbf{F}_{23}^{x} + \mathbf{F}_{21}^{y} + 1.15 \text{ lb} = 0$$

$+\uparrow\ \Sigma\mathbf{F}^{y} + > \mathbf{F}_{g}^{i} = 0$:

$$\mathbf{F}_{23}^{y} + \mathbf{F}_{21}^{y} - \mathbf{W}_{2} - \mathbf{F}_{g2}^{i}\sin 78° = 0$$

$$\mathbf{F}_{23}^{y} + \mathbf{F}_{21}^{y} - 20.43 \text{ lb} = 0$$

$+\!\!\smallfrown\ \Sigma M_{A} + > T_{g}^{i} = 0$:

$$T_{21} + \mathbf{F}_{23}^{x}\,[1.77\,\text{ft}\,(\sin 30°)] + \mathbf{F}_{23}^{y}\,[1.77\,\text{ft}\,(\cos 30°)] - \mathbf{W}_{2}\,[0.89\,\text{ft}\,(\cos 30°)]$$

$$\mathbf{F}_{g2}^{i}\,[\sin(78° - 30°)]\,[0.89 \text{ ft}] - T_{g2}^{i} = 0$$

Substituting values gives

$$T_{21} + 0.885\,\mathbf{F}_{23}^{x} + 1.533\,\mathbf{F}_{23}^{y} - 17.17 \text{ ft lb} = 0$$

14. **Solve the Equilibrium Equations**

A total of nine equilibrium equations have been generated. As previously stated, Newton's third law stipulates that the following magnitudes are equal.

$$\mathbf{F}_{43}^{x} = \mathbf{F}_{34}^{x} \qquad \mathbf{F}_{43}^{y} = \mathbf{F}_{34}^{y}$$

$$\mathbf{F}_{23}^{x} = \mathbf{F}_{32}^{x} \qquad \mathbf{F}_{23}^{y} = \mathbf{F}_{23}^{y}$$

Therefore, nine unknown quantities remain. Solving the nine equilibrium equations, simultaneously, gives the following results:

$$\mathbf{F}_{41}^{x} = -78.41 \text{ lb} = 78.41 \text{ lb} \leftarrow$$

$$\mathbf{F}_{41}^{y} = +58.38 \text{ lb} = 58.38 \text{ lb} \uparrow$$

$$\mathbf{F}_{43}^{x} = -96.44 \text{ lb} = 96.44 \text{ lb} \rightarrow \quad \text{and} \quad \mathbf{F}_{34}^{x} = 96.44 \text{ lb} \leftarrow$$

$$\mathbf{F}_{43}^{y} = -102.09 \text{ lb} = 102.09 \text{ lb} \uparrow \quad \text{and} \quad \mathbf{F}_{34}^{y} = 102.09 \text{ lb} \downarrow$$

$$\mathbf{F}_{32}^{x} = -96.32 \text{ lb} = 96.32 \text{ lb} \rightarrow \quad \text{and} \quad \mathbf{F}_{23}^{x} = 96.32 \text{ lb} \leftarrow$$

$$\mathbf{F}_{32}^{y} = -134.03 \text{ lb} = 134.03 \text{ lb} \uparrow \quad \text{and} \quad \mathbf{F}_{23}^{y} = 134.03 \text{ lb} \downarrow$$

$$\mathbf{F}_{21}^{x} = +95.17 \text{ lb} = 95.17 \text{ lb} \rightarrow$$

$$\mathbf{F}_{21}^{y} = +154.46 \text{ lb} = 154.46 \text{ lb} \uparrow$$

$$T_{21} = +307.88 \text{ ft lb} = 307.88 \text{ ft lb, cw}$$

PROBLEMS

Mass and Mass Moment of Inertia

14–1. The mass of a connecting rod from an internal combustion engine has been determined to be 2.3 kg. Compute the weight of the rod.

14–2. A robotic gripper was weighed at 4.5 lb. Determine the mass of the gripper.

14–3. A robotic gripper was weighed at 4.5 lb and has a radius of gyration relative to a certain axis at the center of gravity of 5 in. Determine the mass moment of inertia of the part relative to this axis.

14–4. A 6-kg mechanism link has a radius of gyration relative to a certain axis at the center of gravity of 150 mm. Determine the mass moment of inertia of the part relative to this axis.

14–5. For the part shown in Figure P14.5, calculate the mass moment of inertia and the radius of gyration about a centroidal longitudinal axis of a 14-in.-long shaft that weighs 5 lb and has a diameter of 0.625 in.

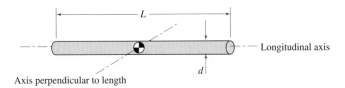

FIGURE P14.5 Problems 5–8.

14–6. For the part shown in Figure P14.5, calculate the mass moment of inertia and the radius of gyration about a centroidal longitudinal axis of a 1200-mm-long shaft that has a mass of 100 kg and a diameter of 50 mm.

14–7. The part shown in Figure P14.5 is a solid cylinder 2 ft in diameter, 3 ft long, and weighing 48 lb. Determine the mass moment of inertia about its centroidal axial axis.

14–8. The part shown in Figure P14.5 is a solid cylinder 2 ft in diameter, 3 ft long, and weighing 48 lb. Determine the mass moment of inertia about a centroidal axis, perpendicular to its length.

14–9. The part shown in Figure P14.9 is a slender rod, 14 in. long, rotating about an axis perpendicular to its length and 3 in. from its center of gravity. Knowing that the rod weighs 2 lb and has a diameter of 1.25 in., determine its mass moment of inertia about that axis.

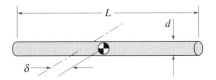

FIGURE P14.9 Problems 9 and 10.

14–10. The part in Figure P14.9 is a slender rod, 0.4 m long, rotating about an axis perpendicular to its length and 0.12 m from its center of gravity. Knowing that the rod has a mass of 6 kg, determine its mass moment of inertia about that axis.

14–11. Determine the moment of inertia of the steel link ($\rho = 0.183 \text{ lb/in.}^{3}$) shown in Figure P14.11 with respect to the y-axis.

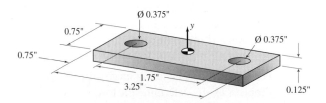

FIGURE P14.11 Problem 11.

14–12. Determine the moment of inertia of the steel link ($\rho = 0.183 \, \text{lb/in.}^3$) shown in Figure P14.12 with respect to the *y*-axis.

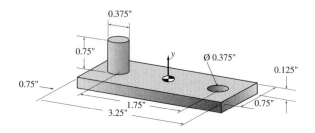

FIGURE P14.12 Problem 12.

Inertial Forces

14–13. The compressor mechanism shown in Figure P14.13 is driven clockwise by a DC electric motor at a constant rate of 800 rpm. In the position shown, the cylinder pressure is 70 psi, and the piston weighs 0.75 lb. The coefficient of friction between the piston and the compressor cylinder is 0.1. The weight of all other links is negligible. At the instant shown, determine the torque required from the motor to operate the compressor.

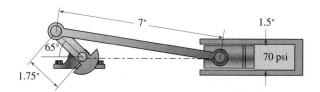

FIGURE P14.13 Problems 13–15.

14–14. For the compressor mechanism described in Problem 14–13, determine the torque required from the motor if the motor is rotating at 800 rpm and accelerating at a rate of 5000 rad/s².

14–15. For the compressor mechanism described in Problem 14–13, determine the torque required from the motor if the motor is rotating at 800 rpm and decelerating at a rate of 5000 rad/s².

14–16. The materials handling mechanism, shown in Figure P14.16, slides 4-kg packages along a counter. The machine operates with the crank rotating counterclockwise at a constant rate of 120 rpm. The coefficient of kinetic friction between the package and counter is 0.15. The weight of all the mechanism links is negligible. Determine the instantaneous torque required from the motor to operate this mechanism.

14–17. For the materials handling mechanism described in Problem 14–16, determine the torque required from the motor if the motor is rotating at 120 rpm and accelerating at a rate of 100 rad/s².

14–18. For the materials handling mechanism described in Problem 14–16, determine the torque required from the motor if the motor is rotating at 120 rpm and decelerating at a rate of 100 rad/s².

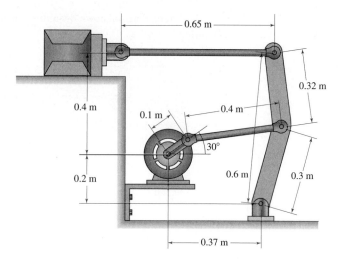

FIGURE P14.16 Problem 16.18.

Inertial Torques

Figure P14.19 shows a link that weighs 4 lb and is rotating clockwise at 20 rad/s. For Problems 14–19 and 14–20, determine the magnitude of the inertial force and the inertial torque at the center of gravity if:

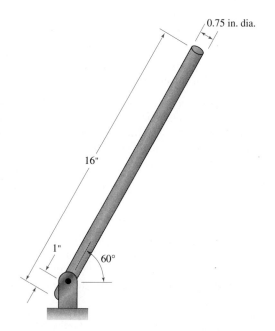

FIGURE P14.19 Problems 19 and 20.

14–19. The link accelerates at 600 rad/s².

14–20. The link decelerates at 600 rad/s².

Figure P14.21 shows a 10-kg link that rotates counterclockwise at 15 rad/s. Determine the magnitude of the inertial force and the inertial torque at the center of gravity if:

14–21. The link accelerates at 400 rad/s².

14–22. The link decelerates at 400 rad/s².

14–23. Figure P14.23 shows a slider-crank mechanism. Link 2 rotates clockwise at a constant 200 rad/s. The weight of link 2 is negligible, link 3 is 3 lb, and link 4 is 2 lb. The radius of gyration of link 3 relative to the

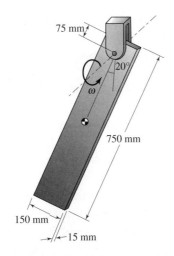

FIGURE P14.21 Problems 21 and 22.

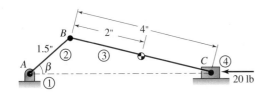

FIGURE P14.23 Problems 23 and 24.

center of gravity is 3 in. For $\beta = 45°$, determine the following:

1. The linear acceleration of link 4 and the center of gravity of link 3,

2. The angular acceleration of link 3,

3. The inertial force and torque of the coupler link,

4. The pin forces at B and C, and

5. The torque to drive the mechanism in this position.

14–24. Repeat Problem 14–23 with $\beta = 120°$.

14–25. Figure P14.25 shows a four-bar mechanism. Link 2 rotates counterclockwise at a constant 10 rad/s. The weight of links 2 and 3 is negligible, and link 4 is 17 kg. The radius of gyration of link 4 relative to the center of gravity is 45 mm. For $\beta = 45°$, determine the following:

1. The linear acceleration of the center of gravity of link 4,

2. The angular acceleration of link 4,

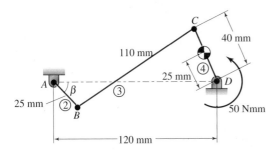

FIGURE P14.25 Problems 25 and 26.

3. The inertial force and torque of link 4,

4. The pin forces at B and C, and

5. The torque to drive the mechanism in this position.

14–26. Repeat Problem 14–25 with $\beta = 90°$.

14–27. Figure P14.27 shows a small hydraulic jack. At this instant, a 10-lb force is applied to the handle. This causes the 3.5-in. link to rotate clockwise at a constant rate of 5 rad/s. The weight of links 2 and 3 is negligible, and link 4 is 1.5 lb. Determine the following:

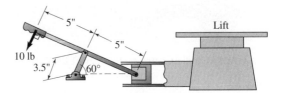

FIGURE P14.27 Problem 27.

1. The linear acceleration of the piston,

2. The inertial force of link 4,

3. The pin forces, and

4. The force developed on the piston due to the hydraulic fluid.

14–28. Figure P14.28 shows a mechanism for a transfer conveyor. The driving link rotates counterclockwise at a constant rate of 25 rpm. The box weighs 50 lb as shown. The weight of the driving link and the coupler are negligible. The weight of the conveyor link is 28 lb and the center of gravity is at its midspan. The radius of gyration of the conveyor link relative to the center of gravity is 26 in. For $\beta = 30°$, graphically determine the following:

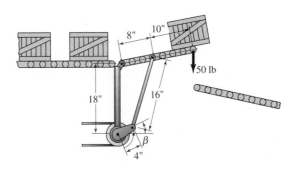

FIGURE P14.28 Problems 28 and 29.

1. The linear acceleration of the center of gravity of the conveyor link,

2. The rotational acceleration of the conveyor link,

3. The inertial force and torque of the conveyor link,

4. The pin forces, and

5. The torque required to drive the mechanism.

14–29. Repeat Problem 14–28 with $\beta = 100°$.

CASE STUDY

C14–1 Figure C14.1 shows a mechanism that gives motion to slides *C* and *D* and is used in a wire-stripping machine. Carefully examine the components of the

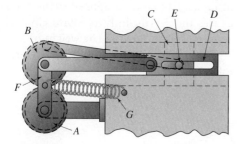

FIGURE C14.1 (Courtesy, Industrial Press.)

mechanism, then answer the following leading questions to gain insight into its operation.

1. As gear *A* rotates clockwise, describe the motion of gear *B*.

2. As gear *A* rotates clockwise, what is the immediate motion of slide *C*?

3. Discuss the action that takes place as pin *E* reaches the end of the slot.

4. Discuss precisely the continual motion of slides *C* and *E*.

5. Discuss how this motion could possibly be used in a wire-stripping machine.

6. What is the purpose of spring *G*?

7. How would the mechanism change if a "stiffer" spring were installed?

ANSWERS TO SELECTED EVEN-NUMBERED PROBLEMS

Chapter 1

1–26. $n = 4, j_p = 4, j_h = 0, M = 1$

1–28. $n = 4, j_p = 4, j_h = 0, M = 1$

1–30. $n = 4, j_p = 4, j_h = 0, M = 1$

1–32. $n = 6, j_p = 7, j_h = 0, M = 1$

1–34. $n = 4, j_p = 4, j_h = 0, M = 1$

1–36. $n = 4, j_p = 4, j_h = 0, M = 1$

1–38. $n = 6, j_p = 7, j_h = 0, M = 1$

1–40. $n = 6, j_p = 7, j_h = 0, M = 1$

1–42. $n = 6, j_p = 7, j_h = 0, M = 1$

1–44. $n = 9, j_p = 11, j_h = 0, M = 2$

1–46. $n = 4, j_p = 4, j_h = 0, M = 1$

1–48. $n = 8, j_p = 10, j_h = 0, M = 1$

1–50. $n = 6, j_p = 7, j_h = 0, M = 1$

1–52. Crank-rocker

1–54. Crank-rocker

Chapter 3

3–2. $A = 17.3$ in.

3–4. $R = 12$ in.

3–6. $s = 156.6$ mm

3–8. $x = 11.5$ in. $y = 16.4$ in.

3–10. $s = 175$ mm

3–12. $L = 8$ ft, 8 in.

3–14. $h = 11.3$ ft

3–16. $y = 11.7$ ft

3–18. $h = 83.1$ in.

3–20. $\mathbf{R} = 24.18 \quad \angle 18.1°$

3–22. $\mathbf{R} = 212.13 \quad 25.0°$

3–24. $\mathbf{R} = 221.20 \quad 13.5°$

3–26. $\mathbf{R} = 24.18 \quad 18.1°$

3–28. $\mathbf{R} = 212.13 \quad 25.0°$

3–30. $\mathbf{R} = 221.2 \quad 13.4°$

3–32. $\mathbf{J} = 8.074 \quad 68.3°$

3–34. $\mathbf{J} = 5.587 \quad \angle 13.76°$

3–36. $\mathbf{J} = 212.13 \quad 65.0°$

3–38. $\mathbf{J} = 8.074 \quad 68.3°$

3–40. $\mathbf{J} = 5.587 \quad \angle 13.8°$

3–42. $\mathbf{J} = 212.13 \quad 65.0°$

3–44. $\mathbf{J} = 26.094 \quad 11.4°$

3–46. $\mathbf{J} = 109.76 \quad 24.1°$

3–48. $\mathbf{J} = 101.68 \quad 18.3°$

3–52. $\mathbf{J} = 26.10 \quad 11.4°$

3–54. $\mathbf{J} = 109.8 \quad 24.1°$

3–56. $\mathbf{J} = 101.68 \quad 18.3°$

3–58. $\mathbf{C} = 19.22 \quad \mathbf{E} = 17.52$

3–60. $\mathbf{B} = 8.81 \quad \mathbf{C} = 117.7$

3–62. $\mathbf{D} = 38.12 \quad \mathbf{F} = 238.9$

Chapter 4

4–2. $\Delta x = 2.189$ in. $\rightarrow$

4–4. $\Delta R_P = 8.420$ in. $\quad 27.5°$

4–6. $\Delta R_{piston} = 47.10$ mm $\leftarrow$

4–8. $\Delta \theta_{crank} = 23.0°$, ccw

4–10. $\Delta R_{end} = 2.029$ in. $\quad 55.1°$

4–12. $\Delta \theta_{handle} = 22.2°$, ccw

4–14. $\Delta \theta_{ram} = 17.6°$, ccw

4–16. $\Delta R_{end} = 22.644$ in. $\quad 44.9°$

4–18. $\Delta \theta_{handle} = 34.4°$, ccw

4–20. $\Delta \theta_{wheel} = 16.3°$, cw

4–22. $\Delta R_{end} = 203.4 \quad 73.9°$

4–24. $\Delta R_{carrier} = 249.7$ mm $\quad \angle 45.5°$

4–26. $\Delta R_{box} = 0.579$ m $\downarrow$

4–28. $\Delta L_{cylinder} = 1.566$ in.

4–30. $\Delta R_{claw} = 29.62$ mm $\quad 85.2°$

4–32. $\Delta L_{spring} = 1.118$ in., shorter

4–34. $\Delta \theta_{ram} = 3.03°$, cw

4–36. $\Delta \theta_{bed} = 14.0°$, ccw

4–38. $\Delta R_P = 7.247 \quad 10.0°$

4–40. $\Delta R_{piston} = 66.82$ mm $\leftarrow$

4–42. $\Delta R_{stamp} = 1.570$ in. $\downarrow$

4–44. $\Delta \theta_{ram} = 14.4°$, cw

4–46. $\Delta \theta_{top\ handle} = 16.8°$, cw

4–48. $\Delta L_{cylinder} = 68.1$ mm, shorter

4–50. $\Delta R_{box} = 0.362$ m $\downarrow$

4–52. $\Delta R_{claw} = 30.87$ mm $\quad \angle 86.52°$

4–54. $\Delta \theta_{ram} = 5.5°$, cw

4–56. $(\Delta R_{piston})_{max} = 90.0$ mm

4–58. $(\Delta \theta_{ram})_{max} = 46.3°$

4–60. $(\Delta \theta_{wheel\ assy})_{max} = 57.6°$

4–62. $(\Delta \theta_{ram})_{max} = 29.5°$

4–64. $(\Delta R_{blade})_{max} = 1.513$ in.

4–66. $(\Delta \theta_{wiper\ arm})_{max} = 72.8°$

4–68. $(\Delta R_{slide\ pin})_{max} = 44.50$ mm

Chapter 5

5–2. $\beta = 49.1°, \omega = 109$ rpm

5–4. $t_1 = 0.188s, t_2 = 0.142$ s

5–6. $t_1 = 0.067$ s, $t_2 = 0.53$ s

5–8. $Q = 1.714, \omega = 63.2$ rpm

5–10. $Q = 2.083, \omega = 162.2$ rpm

5–12. $L_2 = 4$ mm, $\omega = 750$ rpm

5–14. $\beta = 20°$, $\omega = 100$ rpm

5–16. $\beta = 12.6°$, $\omega = 4286$ rpm

5–18. $\beta = 8.6°$, $\omega = 1818$ rpm

5–20. $\beta = 0°$, $\omega = 17.14$ rpm

5–22. $\beta = 19.3°$, $\omega = 33.3$ rpm

5–24. $\beta = 16.36°$, $\omega = 200$ rpm

5–26. $\beta = 8.57°$, $\omega = 300$ rpm

5–28. $\beta = 6.92°$, $\omega = 40$ rpm

5–30. $\beta = 49.09°$, $\omega = 17.6$ rpm

5–32. $\beta = 51.43°$, $\omega = 240$ rpm

Chapter 6

6–2. $\Delta t = 37.5$ s

6–4. $v = 22.37$ min/hr

6–6. $\omega_{min} = .0167$ rpm

6–8. $\Delta R_{total} = 72$ in.

6–10. $\Delta R_{total} = 7.5$ in.

6–12. $V = 125.66$ ft/min

6–14. $V_B = 90$ ft/s $\searrow 20°$

6–16. $V_{B/A} = 5.94$ ft/s $\nearrow 59.4°$

6–18. $V_{A/B} = 15.72$ ft/s $25.9° \searrow$

6–20. $V_{piston} = 272.55$ in./s $\leftarrow$

6–22. $V_{blade} = 59.63$ in./s $\leftarrow$

6–24. $V_{blade} = 5.94$ in./s $\downarrow$

6–26. $\omega_{wiper\ blade} = 2.50$ rad/s, ccw

6–28. $\omega_{bath} = 0.85$ rad/s, cw

6–30. $\omega_{segment\ gear} = 2.232$ rad/s, ccw

6–32. $V_{blade} = 112.91$ mm/s $\uparrow$

6–34. $V_{cylinder} = 4.39$ in./s, compressing

6–36. $V_{right\ piston} = 150.68$ in./s $\nearrow 45°$

6–38. $V_{package} = 775$ mm/s $\rightarrow$

6–40. $V_{platform} = 15.99$ ft/min $\nearrow 12.9°$

6–42. $V_{blade} = 112.64$ in./s $\rightarrow$

6–44. $\omega_{wiper\ arm} = 1.88$ rad/s, cw

6–46. $\omega_{segment\ gear} = 3.827$ rad/s, cw

6–48. $V_{cylinder} = 4.39$ in./s, compressing

6–50. $V_{package} = 953$ mm/s $\rightarrow$

6–74. $V_{piston} = 230.3$ in./s $\leftarrow$

6–76. $V_{blade} = 4.82$ in./s $\downarrow$

6–78. $\omega_{bath} = 1.245$ rad/s, cw

6–80. $\omega_{blade} = 0.071$ rad/s, ccw

6–82. $V_{right\ piston} = 288.3$ in./s $45° \searrow$

6–84. $V_{platform} = 12.79$ ft/min $\nearrow 12.9°$

6–86. $V_{blade} = 91.11$ in./s $\rightarrow$

6–88. $\omega_{wiper\ arm} = 31.46$ rad/s, cw

6–90. $\omega_{segment\ gear} = 5.421$ rad/s, cw

6–92. $V_{Cylinder} = 2.64$ in./s, compressing

6–94. $V-2$. $A_{vehicle} = 10.60$ ft/s^2

Chapter 7

7–4. $\Delta R_A = 40$ mm $\uparrow$

7–6. $\alpha_{cam} = 9.82$ rad/s^2, cw

7–8. $\alpha_{shaft} = 7.85$ rad/s^2, cw

7–10. $\Delta R_{actuator} = 85$ in.

7–12. $\Delta R_{linear\ motor} = 15$ in. $\downarrow$

7–14. $\Delta R_{actuator} = 10$ in.

7–16. $A_B^n = 17{,}770$ in/s^2 $\searrow 70°$

7–18. $A_B = 22{,}872$ in./s^2 $\searrow 71.0°$

7–20. $A_A = 5158$ in./s^2 $6.9° \searrow$

7–22. $A_{A/B} = 25.46$ mm/s^2 $11.3° \nearrow$

7–24. $A_{C/B} = 1.35$ ft/s^2 $42.3° \searrow$

7–26. $A_{piston} = 31{,}341$ in./s^2 $\rightarrow$

7–28. $A_{piston} = 37{,}194$ in./s^2 $\rightarrow$

7–30. $A_{needle} = 29{,}271$ mm/s^2 $\uparrow$

7–32. $A_{blade} = 58.97$ in./s^2 $\leftarrow$

7–34. $A_{blade} = 103.73$ in./s^2 $\leftarrow$

7–36. $\alpha_{horse} = 5.22$ rad/s^2, cw

7–38. $\alpha_{nozzle} = 9.80$ rad/s^2, ccw

7–40. $\alpha_{nozzle} = 78.55$ rad/s^2, cw

7–42. $\alpha_{wheel\ assy} = 2.08$ rad/s^2, cw

7–44. $A_{piston} = 93{,}195$ in./s^2 $\rightarrow$

7–60. $A_{B3/B2}^c = 900$ mm/s^2 $\nearrow 30°$

7–62. $A_{B3/B2}^c = 900$ mm/s^2 $\searrow 60°$

Chapter 8

(spreadsheet/program results at shown at 120° crank angle)

8–2. $\Delta R_4 = 123.9$ mm at $\theta_2 = 120°$

8–4. $\theta_4 = 16.6°$ at $\theta_2 = 120°$

8–6. $V_4 = -2577$ mm/s at $\theta_2 = 120°$

8–8. $\omega_4 = 3.27$ rad/s at $\theta_2 = 120°$

8–10. $A_4 = 2349$ mm/s^2 at $\theta_2 = 120°$

8–12. $\omega_4 = 204.4$ rad/s at $\theta_2 = 120°$

Chapter 9

9–14. $\omega_{cam} = 10.9$ rpm, $V_{max} = 0.5$ in./s

9–16. $\omega_{cam} = 42.9$ rpm, $V_{max} = 4.0$ in./s

9–18. $\omega_{cam} = 13.3$ rpm, $V_{max} = 2.5$ in./s

9–20. $\omega_{cam} = 17.1$ rpm, $V_{max} = 3.1$ in./s

9–22. $\omega_{cam} = 20$ rpm, $V_{max} = 2.0$ in./s

9–28. $\omega_{cam} = 9.4$ rpm, $V_{max} = 0.94$ in./s

9–30. $V_{max} = 1.25$ in./s, $A_{max} = 3.13$ in./s^2

9–32. $V_{max} = 8.0$ mm/s, $A_{max} = 8.0$ mm/s^2

9–34. $V_{max} = 189$ mm/s, $A_{max} = 2961$ mm/s^2

9–36. $V_{max} = 1.43$ in./s, $A_{max} = 6.41$ in./s^2

9–62. $\omega_{out} = 7.46$ rad/s, $\alpha_{out} = 123$ rad/s^2

Chapter 10

10–2. $p = 0.393$ in.

10–4. $D = 84$ mm

10–6. $m_p = 1.53$

10–8. $m_p = 1.47$

10–10. $C = 1.125$ in.

10–12. $D_1 = 2$ in., $D_2 = 4.5$ in.

10–14. $C = 3.5$ in.

10–16. $\mathbf{V}_t = 67.7$ in./s

10–18. $\mathbf{V}_t = 90.3$ in./s

10–20. $N_1 = 24$ teeth, $N_2 = 96$ teeth

10–22. $P_d = 8$, $N_1 = 16$, $N_2 = 64$

10–24. $P_d = 4$, $N_1 = 20$, $N_2 = 60$

10–26. $P_d = 12$

10–28. $D_1 = 3.0$ in., $D_2 = 13.33$ in.

10–30. $D_1 = 2.0$ in., $D_2 = 10.25$ in.

10–32. $P_d = 12$, $D_1 = 4.0$ in., $D_2 = 24.0$ in.

10–34. $\Delta\theta = 2.12$ rev

10–36. $\Delta s = 42.4$ in.

10–38. $\omega = 10.6$ rpm

10–40. $p^n = 0.28$ in.

10–42. $P_d = 7.6$, $N_1 = 19$, $N_2 = 38$

10–44. $\gamma_p = 14.9°$, $\gamma_g = 75.1°$

10–46. $\gamma_p = 10.4°$, $\gamma_g = 64.6°$

10–48. $P_d = 12$, $N_w = 2$, $N_g = 50$, $\lambda = 5°$

10–50. $\omega_5 = 100$ rpm, cw; $c = 8.5$ in.

10–52. $\omega_8 = 30$ rpm, cw; $C = 17.97$ in.

10–54. $s_4 = 0.74$ in., $c = 4.625$ in.

10–56. $P_d = 8$

10–58. $\omega_1 = 3576$ rpm

10–60. $\omega_1 = 2520$ rpm, $C = 9$ in.

10–62. $N = 17{-}68,\ 17{-}68,\ 18{-}45,\ 18{-}45,\ 17{-}34$

10–64. $\Delta\theta_{\text{window}} = 8°$ccw

10–66. $\Delta\theta_{\text{window}} = 10.8°$ccw

10–68. $\omega_4 = 2160$ rpm, cw

10–70. $\omega_6 = 536$ rpm, cw

10–72. $\omega_8 = 378$ rpm, ccw

Chapter 11

11–2. $\omega_{\text{out}} = 479$ rpm, cw

11–4. $\omega_{\text{out}} = 313$ rpm, ccw

11–6. $\omega_{\text{in}} = 2700$ rpm, ccw

11–8. $c = 22.375$ in., $T = 162°$

11–10. $c = 25.618$ in., $\theta = 144°$

11–12. 5V belt

11–14. 3V belt

11–16. 3V belt

11–18. $\omega_{\text{out}} = 84$ rpm, cw

11–20. $\omega_{\text{in.}} = 760$ rpm, cw

11–22. $c = 48.724$ in., $\theta = 125°$

11–24. $c = 52.424$ in., $\theta = 96°$

11–26. No. 80 Chain

11–28. No. 100 Chain

Chapter 12

12–2. $p = .0357$ in., $\lambda = 2.87°$

12–4. $p = .020$ in., $\lambda = 3.65°$

12–6. $\Delta\mathbf{R}_{\text{ram}} = 2.5$ in. $\uparrow$

12–8. $\Delta\mathbf{R}_{\text{table}} = 0.154$ in. $\uparrow$

12–10. $\Delta\mathbf{R}_{\text{plate}} = 2.756$ in. $\downarrow$

12–12. $\Delta\mathbf{R}_{\text{platform}} = 2.564$ in. $\nearrow 29.7°$

12–14. $\Delta\mathbf{R}_{\text{end}} = 6.445$ in. $\nearrow 28.3°$

12–16. $\Delta\mathbf{R}_{\text{front end}} = 0.921$ in. $\nearrow 86.7°$

12–18. $\mathbf{V}_{\text{nut}} = 0.167$ in./s $\downarrow$

12–20. $\mathbf{V}_{\text{table}} = 0.154$ in./s $\uparrow$

12–22. $\mathbf{V}_{\text{platform}} = 7.236$ in./s $\nearrow 32.8°$

12–34. $e = 26\%$

12–36. $e = 47.2\%$

12–38. $e = 24.5\%$

12–40. M10 $\times$ 1.50 and M8 $\times$ 1.25

Chapter 13

13–2. $R = 248 \nearrow 66°$

13–4. $M = 200$ in. lbs, cw

13–6. $M = 188$ in. lbs, cw

13–8. $M = 18.9$ Nm, cw

13–10. $\mathbf{F}_{\text{cyl}} = 3733$ lbs(C)

13–12. $\mathbf{F}_{\text{metal}} = 868$ N $\downarrow$

13–14. $\mathbf{F}_{\text{screw}} = 1200$ lb(C)

13–16. Front cyl = 2137 lbs(C)

Rear cyl = 7182 lbs(C)

13–18. Front cyl = 5000 lbs(T)

Rear cyl = 11,110 lbs(C)

Chapter 14

14–2. $m = 0.14$ slugs

14–4. $I = 0.135$ kg m^2

14–6. $I_x = 31.25$ kg m^2

14–8. $I_z = 1.49$lb ft s^2

14–10. $I_z = 0.166$ kg m^2

14–12. $I_{y\,cg} = 0.00626$ lb in. s^2

14–14. $T_{\text{motor}} = 199.49$ in. lbs, cw

14–16. $T_{\text{motor}} = 14.04$ Nm, ccw

14–18. $T_{\text{motor}} = 10.22$ Nm, ccw

14–20. $\mathbf{F}^i_{cg} = 52.25$ lbs $\nearrow 3.7°$

$T^i_{cg} = 132.5$ in. lbs, cw

14–22. $\mathbf{F}^i_{cg} = 1377$ N $\searrow 49.4°$

$T^i_{cg} = 19.5$ Nm, ccw

REFERENCES

1. Barton, Lyndon, *Mechanism Analysis: Simplified Graphical and Analytical Techniques*, 2nd ed., Marcel Dekker Inc., New York, 1993.

2. Baumeister, Theodore III, Avallone, Eugene, and Sadegh, Ali, *Mark's Standard Handbook for Mechanical Engineers*, 11th ed., McGraw-Hill Book Company, New York, 2006.

3. Chironis, Nicholas and Sclater, Neil, *Mechanisms and Mechanical Drives Sourcebook*, 4th ed., McGraw-Hill Book Company, New York, 2007.

4. Erdman, Aurthur, Sandor, George, and Kota, Sridhar, *Mechanism Design, Vol 1: Analysis and Synthesis*, 4th ed., Prentice Hall, Upper Saddle River, NJ, 2001.

5. Kepler, Harold, *Basic Graphical Kinematics*, 2nd ed., McGraw-Hill Book Company, New York, 1973.

6. Jensen, Preben, *Cam Design and Manufacture*, 2nd ed., Marcel Dekker, New York, 1987.

7. Jensen, Preben, *Classical Modern Mechanisms for Engineers and Inventors*, Marcel Dekker, Inc., New York, 1991.

8. Jones, Franklin, Holbrook, Horton, and Newell, John, *Ingenious Mechanisms for Designers and Inventors, Vols. I–IV*, Industrial Press Inc, New York, 1930.

9. Mabie, Hamilton and Reinholtz, Charles, *Mechanisms and Dynamics of Machinery*, 4th ed., John Wiley and Sons Inc., New York, 1987.

10. Martin, George, *Kinematics and Dynamics of Machines*, 2nd ed., Waveland Press Inc., Long Groove, IL, 2002.

11. Norton, Robert, *Design of Machinery*, 4th ed., McGraw-Hill Book Company, New York, 2008.

12. Uicker, John, Pennock, Gordon, and Shigley, Joseph, *Theory of Machines and Mechanisms*, 4th ed., Oxford University Press, New York, 2010.

13. Townsend, Dennis and Dudley, Darle, *Dudley's Gear Handbook*, 2nd ed., McGraw-Hill Book Company, New York, 1991.

14. Waldron, Kenneth and Kinzel, Gary, *Kinematics, Dynamics, and Design of Machinery*, 2nd ed., John Wiley and Sons Inc., Hoboken, NJ, 2004.

15. Wilson, Charles and Sadler, Peter, *Kinematics and Dynamics of Machinery*, 3rd ed., Pearson Education, Upper Saddle River, NJ, 2003.

16. *Working Model Demonstration Guide*, Knowledge Revolution Inc., San Mateo, CA, 1995.

INDEX

Note: The letter 'i' and 't' followed by locators refers to illustrations and tables cited in the text